Hadrons and Hadronic Matter

NATO ASI Series

Advanced Science Institutes Series

A series presenting the results of activities sponsored by the NATO Science Committee, which aims at the dissemination of advanced scientific and technological knowledge, with a view to strengthening links between scientific communities.

The series is published by an international board of publishers in conjunction with the NATO Scientific Affairs Division

A	**Life Sciences**	Plenum Publishing Corporation
B	**Physics**	New York and London
C	**Mathematical and Physical Sciences**	Kluwer Academic Publishers Dordrecht, Boston, and London
D	**Behavioral and Social Sciences**	
E	**Applied Sciences**	
F	**Computer and Systems Sciences**	Springer-Verlag
G	**Ecological Sciences**	Berlin, Heidelberg, New York, London,
H	**Cell Biology**	Paris, and Tokyo

Recent Volumes in this Series

Volume 227—Dynamics of Polyatomic Van der Waals Complexes
edited by Nadine Halberstadt and Kenneth C. Janda

Volume 228—Hadrons and Hadronic Matter
edited by D. Vautherin, F. Lenz, and J. W. Negele

Volume 229—Geometry and Thermodynamics: Common Problems of Quasi-Crystals, Liquid Crystals, and Incommensurate Systems
edited by J.-C. Tolédano

Volume 230—Quantum Mechanics in Curved Space-Time
edited by Jürgen Audretsch and Venzo de Sabbata

Volume 231—Electronic Properties of Multilayers and Low-Dimensional Semiconductor Structures
edited by J. M. Chamberlain, L. Eaves, and J.-C. Portal

Volume 232—New Frontiers in Quantum Electrodynamics and Quantum Optics
edited by A. O. Barut

Volume 233—Radiative Corrections: Results and Perspectives
edited by N. Dombey and F. Boudjema

Volume 234—Constructive Quantum Field Theory II
edited by G. Velo and A. S. Wightman

Series B: Physics

Hadrons and Hadronic Matter

Edited by

D. Vautherin

Université Paris-Sud
Orsay, France

F. Lenz

Universität Erlangen
Erlangen, Federal Republic of Germany

and

J. W. Negele

Massachusetts Institute of Technology
Cambridge, Massachusetts

Plenum Press
New York and London
Published in cooperation with NATO Scientific Affairs Division

Proceedings of a NATO Advanced Study Institute
on Hadrons and Hadronic Matter,
held August 8-18, 1989,
in Cargèse, France

Library of Congress Cataloging-in-Publication Data

NATO Advanced Study Institute on Hadrons and Hadronic Matter (1989 : Cargèse, France)
Hadrons and hadronic matter / edited by D. Vautherin, F. Lenz, and J.W. Negele.
p. cm. -- (NATO ASI series. Series B, Physics ; vol. 228)
"Proceedings of a NATO Advanced Study Institute on Hadrons and Hadronic Matter, held August 8-18, 1989, in Cargèse, France"--Verso CIP t.p.
"Published in cooperation with NATO Scientific Affairs Division."
Includes bibliographical references and index.
ISBN-13: 978-1-4684-1338-0 e-ISBN-13: 978-1-4684-1336-6
DOI: 10.1007/978-1-4684-1336-6
1. Hadrons--Congresses. 2. Nuclear structure--Congresses. I. Vautherin, Dominique. II. Lenz, F. (Frieder) III. Negele, John W. IV. North Atlantic Treaty Organization. Scientific Affairs Division. V. Title. VI. Series: NATO ASI series. Series B, Physics ; v. 228.
QC793.5.H32N37 1989
539.7'216--dc20 90-43049
CIP

Softcover reprint of the hardcover 1st edition 1990
A Division of Plenum Publishing Corporation
233 Spring Street, New York, N.Y. 10013

PREFACE

The present NATO Advanced Study Institute held in CARGESE (Corsica) from August 8th to August 18th, 1989 was devoted to Hadronic Physics. We tried to give this school a key educational role in this new and rapidly developing interdisciplinary field. We hope that the combination of the lectures and the open atmosphere of scientific exchange and inquiry afforded by the Cargèse format has provided a unique educational and scientific opportunity for students and has brought together all the relevant concepts and issues for frontier research in this field.

We would like to express our gratitude to NATO for its generous financial support which made this Institute possible. We also wish to thank Dr. Luis V. Da Cunha, Director of the Scientific Affairs Division, for his valuable comments and advice.

We acknowledge the support of the Institut de Physique Nucléaire et de Physique des Particules (France), the Commissariat à l'Energie Atomique (France), and the U.S. National Science Fundation, for the attribution of travel grants.

Our special appreciation is due to Frédérique Dykstra for her oustanding organizational work throughout the preparation and duration of this conference

It is also a pleasure to thank the Université de Nice for making available the facilities of the Cargèse Scientific Institute.

The pictures of the lecturers included in the present volume were kindly provided by one of the participants, Dr.R.Janner.

Last but not least we would like to express our gratitude to Marie-France Hanseler, Director of the Cargèse Institute and to the staff of the Institute, in particular Chantal Ariano and Joseph Antoine Ariano for making our stay in Cargèse extremely enjoyable.

Frieder LENZ
John NEGELE
Dominique VAUTHERIN

CONTENTS

QUARKS AND GLUONS IN HADRONS AND NUCLEI

F. E. Close

Oak Ridge National Laboratory*, Oak Ridge, TN 37831-6373
and
University of Tennessee, Knoxville, TN 37996-1200

ABSTRACT

These lectures discuss (1) the particle-nuclear interface — a general introduction to the ideas and application of colored quarks in nuclear physics, (2) color, the Pauli principle, and spin flavor correlations — this lecture shows how the magnetic moments of hadrons relate to the underlying color degree of freedom, and (3) the proton's spin — a quark model perspective. This lecture reviews recent excitement which has led some to claim that in deep inelastic polarized lepton scattering very little of the spin of a polarized proton is due to its quarks.

THE PARTICLE-NUCLEAR INTERFACE

Once upon a time nuclear physics was the study of nucleons and pions vibrating and oscillating at the center of the atom; particle physics was the study of nucleons and pions interacting and producing resonances. From the latter, people gradually realized that hadrons are built from quarks; the fundamental rules governing their interactions were deduced ("QCD" — quantum chromodynamics) and the similarities with QED suggested the possibility that all of the natural forces can be described in a grand unified theory. Today, particle physics deals with questions ranging from the origins of matter in the first microseconds of the universe, whose experimental investigation requires the energies of the SSC, down to the quark structure of protons and nuclei that can be studied at relatively low energies.

Nuclear physics theory has taken QCD on board. There are still detailed studies going on in nuclear excitations and there are important overlaps with nuclear astrophysics. The field is very rich. The attempt to understand nuclear phenomena at the quark level causes many nuclear physicists to be concerned with the same sort of problems as their colleagues in particle physics.

*Operated by Martin Marietta Energy Systems, Inc. under Contract DE-AC05-84OR21400 with the U.S.D.O.E.

Hadrons and Hadronic Matter
Edited by D. Vautherin *et al.*
Plenum Press, New York, 1990

During the last three years, there has been a blossoming in this overlap area which I would like to call "hadron physics". I would like to make the following definitions.

Nuclear Physics deals with the collective properties of the many-body nucleus consisting of nucleons. Hadron physics deals with the structure and interaction dynamics of those hadrons. Particle physics uses these particles as tools with which to elucidate deeper truths, seeking the origins of matter, the nature of mass, and the several other parameters which presently have to be invoked *ad hoc* (the weak mixing angles, fermion masses, etc.).

With tongue in cheek, one might contrast the two extremes. In extreme high-energy particle physics, there is infinite theorizing but almost no data; in the nuclear structure field, we have copious data but no truly fundamentally useful theory. In hadron physics we have much data and the hope of confronting them with the fundamental QCD theory of interacting quarks and gluons. This is stimulating but also difficult. A major problem is confinement of colored quarks and gluons within the nucleons — this is simulated on computers where space-time is described as a discrete lattice, but its origins analytically are still rather poorly understood.

This confinement phenomenon is also the catch-22 of "deriving" nuclear physics from QCD. As Bob Jaffe once remarked, "Looking for evidence of quarks in nuclei is like looking for the mafia in Sicily: everyone knows they are there, but it's hard to find the evidence." The quarks are confined in nucleons, and so any successful description of the nuclear structure must reduce to that of quarks clustered inside individual nucleons. Thus we have first to understand the proton and neutron — hence the interest in hadron physics. There is the interesting possibility that the interactions and overlaps among closely packed nucleons in large or dense nuclei, or in high-energy collisions of heavy ions, may disturb the distributions of quarks — their spatial or momentum distributions — relative to their behavior when confined in isolated free nucleons. There are indeed hints of such behavior, e.g., the "EMC effect" where quarks in iron have a slightly different momentum distribution relative to that in the deuteron. Their mean momenta are reduced in iron and other heavy nuclei suggesting greater spatial freedom. Is this "liberation" a "cold" precursor to "hot" deconfinement? Are nucleons in nuclei physically "enlarged", or is this a manifestation of quarks exchanged between nucleons, tying the nucleus together and having more spatial mobility than when in a single free nucleon? Future experiments may help to answer these questions — questions raised, in large degree, by the underlying quark theory.

I would like now to draw some analogies between QED (electrical charges, atoms, and molecules) on the one hand, and on the other, QCD (color, hadrons, and nuclei). The similarity is such that one could rewrite Bjorken and Drell's QED text by inserting a traceless 3×3 matrix (λ of SU(3)) at the fermion gauge boson vertices and let QED become QCD (with α replaced by $\alpha_s \approx 1/10$). However, the gluons themselves have color and so mutually interact via the color forces (contrast the photon of QED which transmits but does not directly "feel" the electromagnetic force). These new intergluon interactions give rise to vertices involving three or four gluons at a point, and so a text on QCD requires more than just a coloring of Bjorken and Drell.

Now let's make a matrix to summarize how systems variously react to the forces.

Notice that the gluon and photon are in different slots. This small difference gives rise to the different long-range phenomena in QCD compared to QED (e.g. confinement versus ionization).

QED	QCD
electric charge	3 colors
attraction of opposites	attraction of unlike colors
electrically zero atoms	colorless hadrons
radiation photon	radiation gluons
magnetic effects	chromomagnetic effects
hyperfine splitting $^3S_1 - {}^1S_0$	color hyperfine $m_\rho - m_\pi$, $m_\Delta - m_N$
Fermi-Breit in hydrogen	Fermi-Breit splittings in hadron spectroscopy

		QED	QCD
Feel the force	Carry the charge	e^- Z^+ Na^+ Cl^-	Quarks Gluons
	Contain the charge	Atoms Molecules	Hadrons Nuclei
Do not feel the force Do not contain the charge		Neutrinos Photon	Leptons (ν,e)

Within atoms and hadrons one finds analogues. The Coulomb potential of hydrogen has an analogue in quark systems: as the quarks' relative separations $r \to o$, $V(r) \sim 1/r$, but at large r, $V(r) \sim r$, presumably due to the detailed self-interactions among the gluons that are transmitting the force. In the ground state of hydrogen the magnetic interaction ("one photon exchange") splits the $^3S_1$ and $^1S_0$ levels. In the ground state quark conglomerates, the high-J combinations have enhanced masses relative to their low-J counterparts due to "one gluon exchange"; thus the $\Delta(1230)$ resonance, with J = 3/2, has greater mass than the J = 1/2 nucleon.

Now move up a layer in complexity to the world of molecules (QED) and nuclei (QCD).

At the risk of being accused of oversimplification by the atomic experts, I will divide the interatomic forces into three broad classes, then make analogy in the QCD world with interhadronic forces at the level of the quarks and gluons.

	Covalent	van der Waals	Ionic
Atoms Molecules	e^- exchange	"two photon"	Na^+Cl^-
Hadrons Nuclei	quark exchange	"two gluon"	no analogue if color confined in neutral clusters

If this was the whole story, then nuclear physics from QCD would be a rerun of molecules from QED. However, confinement of color breaks the simple analogy. The quark exchange at large distances ($>$ 1 fm) is contained within the confined packages, dominantly pions. The confinement of gluons in glueballs also breaks the analogy with van der Waals' forces. The hope that QCD would predict observable color van der Waals' forces in nuclei is most probably flawed, as the gluons will be confined within colorless glueballs. Computer simulations of QCD suggest that the lightest glueballs have masses in excess of 1 GeV and so transmit forces over much less than a nucleon radius. Thus their presence is hidden in nuclear physics.

It is an open question whether analogues of ionic forces occur in dense or hot nuclear systems; whether multiquark clusters occur within nuclei; whether quark-gluon plasma may form in hot-dense systems.

If color attractions among quarks are the source of internucleon forces, then there could exist analogous clusters of mesons — "meson molecules". The instability of most mesons prevents formation of these systems, but π, K, η are stable on the time scales of the strong interactions and may have the chance to bind. Indeed Weinstein and Isgur find that such attractions occur in S-wave. The $\pi\pi$ system has a strong enhancement above $2m_\pi$ which may be manifested in the $\psi \to \omega\pi\pi$ dipion spectrum. The KK system binds forming nearly degenerate I = 0, 1 systems 10 MeV below $2m_K$. The S^*(975 MeV) and δ(980 MeV), scalar "mesons", thus appear to be meson molecules; meson analogues of the I=0 dueteron (whose I=1 partner is above $2m_N$).

The color attractions among quarks and gluons lead to the prediction of glueballs and hybrid hadrons — the latter where gluons play a dynamical role, attracted to quarks to form hybrid mesons and baryons.

The problem in predicting the masses of these states is that we have to simulate the effects of confinement. Perhaps the simplest way of doing this is to suppose that the constituent quarks or gluons are free until they hit an infinitely high wall. This is the essence of cavity or bag models. Confine a massless J=1/2 quark in a radius, R, and it gains an energy that scales as 1/R. This energy becomes of the order of 350 MeV if R is of the order of the proton radius, hence the proton mass may be modeled. For gluons, one solves the eigenvalue equations for J = 1 rather than J = 1/2 confined fields. There are electric or magnetic modes (actually TE and TM in the language of classical electrodynamics) with different eigenvalues. If R is the same as for quark systems, the typical confined-mass-scale is some 500 MeV per TE mode and 750 MeV per TM mode. Thus follows the prediction that the lightest systems consisting of at least two confined gluons weigh in at O(1 GeV) and that the lightest hybrid baryons weigh in at O(1.5 GeV). A problem is that as soon as the hyperfine shifts in energy are taken into account (this involves one first calculating the propagators of confined quarks and gluons), the lowest spin-J systems are pulled down significantly in mass. The lightest hybrid baryon might thus appear to have a mass near that of the proton which suggests either a profound rethink of baryon spectroscopy or that we have unearthed a naievity.

I suspect it is the latter. No one yet has convincingly set up a study of loop effects with renormalization within a cavity. These loop diagrams enter at the same order in perturbation theory to which the hyperfine shifts have been calculated and may alter the naive "effective" energies per confined gluon. In the case of quarks, their effects were subsumed in the MIT bag by an input mass parameter for the quark; this mass fitted to the overall mass scale of the spectroscopy. In the gluonic sector we have no mass scale to set the scale, and until we make sense of the (infinite!) self-energy diagrams, we cannot predict the absolute scale. So the mass separations among the various states may be reliable, but the absolute mass scale is beyond present analysis. To predict the masses of glueballs and hybrid hadrons, we have to resort to computer simulations — lattice QCD. This has proved to be a harder task than was originally thought.

The eventual discovery of the gluonic spectroscopy may give important insights into the nature of confinement of gluons. If lattice calculations, including quarks <u>and</u> gluons (to date, people work in the "quenched" approximation, which roughly translated means "ignore the quarks") merely point out masses of states that correspond to the particle data tables, we will confirm QCD but may still require much study to elucidate the analytic dynamics of confinement. The main outcome of such a success may be the advances that will have come in the art of computation and design of machines. Thus the significant questions posed by hadron physics are having a spinoff in the intellectual stimulation they provide to computational science and, in turn, the subsequent ability to encode problems in field theory, condensed matter, and other areas of science. I am reminded of the title of Tony Hey's talk at a recent meeting of the British Association for the Advancement of Science, and it provides an apt one-line summary of the multidisciplinary efforts flowing from computation at the nuclear-particle interface. It was: "Quarks, Supercomputers, and Oil Prospecting".

COLOR, THE PAULI PRINCIPLE, AND SPIN-FLAVOR CORRELATIONS

Color

If quarks possess a property called color, any quark being able to carry any one of three colors (say red, yellow, blue), then the Ω^- (and any baryon) can be built from distinguishable quarks:

$$\Omega^- \left(s_R^{\uparrow}\ s_Y^{\uparrow}\ s_B^{\uparrow}\right).$$

If quarks carry color but leptons do not, then it is natural to speculate that color may be the property that is the source of the strong interquark forces — absent for leptons.

Electric charges obey the rule "like repel, unlike attract" and cluster to net uncharged systems. Colors obey a similar rule: "like colors repel, unlike (can) attract". If the three colors form the basis of an SU(3) group, then they cluster to form "white" systems — viz. the singlets of SU(3). Given a random soup of colored quarks, the attractions gather them into white clusters, at which point the color forces are saturated.

The residual forces among these clusters are the nuclear forces whose origin will be mentioned later.

If quark (Q) and antiquark ($\bar{Q}$) are the $\underset{\sim}{3}$ and $\bar{\underset{\sim}{3}}$ of color SU(3), then combining up to three together gives SU(3) multiplets of dimensions as follows (see Ref. 3):

$$QQ = \underset{\sim}{3} \times \underset{\sim}{3} = \underset{\sim}{6} + \bar{\underset{\sim}{3}}$$

$$Q\bar{Q} = \underset{\sim}{3} \times \bar{\underset{\sim}{3}} = \underset{\sim}{8} + \underset{\sim}{1}$$

The $Q\bar{Q}$ contains a singlet — the physical mesons

$$QQ\bar{Q} = \underset{\sim}{15} + \underset{\sim}{6} + \underset{\sim}{3} + \underset{\sim}{3}$$

$$QQQ = \underset{\sim}{10} + \underset{\sim}{8} + \underset{\sim}{8} + \underset{\sim}{1}.$$

Note the singlet in QQQ — the physical baryons.

For clusters of three or less, only $Q\bar{Q}$ and QQQ contain color singlets and, moreover, these are the only states realized physically. Thus are we led to hypothesize that only color singlets can exist free in the laboratory; in particular, the quarks will not exist as free particles.

<u>Symmetries and Correlations in Baryons</u>

To have three quarks in color singlet:

$$\underset{\sim}{1} \equiv \frac{1}{\sqrt{6}} [(RB-BR)Y + (YR-RY)B + (BY-YB)R] \tag{1}$$

any pair is in the $\bar{\underset{\sim}{3}}$ and is antisymmetric. Note that $\underset{\sim}{3} \times \underset{\sim}{3} = \underset{\sim}{6} + \bar{\underset{\sim}{3}}$. These are explicitly

$\bar{\underset{\sim}{3}}_{anti}$	$\underset{\sim}{6}_{sym}$
RB-BR	RB+BR
RY-YR	RY+YR
BY-YB	BY+YB
	RR
	BB
	YY

(2)

Note well: <u>Any Pair is Color Antisymmetric</u>

The Pauli principle requires total antisymmetry and therefore any pair must be:

<u>Symmetric in all else</u>

("else" means "apart from color").

This is an important difference from nuclear clusters where the nucleons have no color (hence are trivially symmetric in color!). Hence for nucleons Pauli says

Nucleons are Antisymmetric in Pairs (3)

and for quarks

Quarks are Symmetric in Pairs (4)

If we forget about color (color has taken care of the antisymmetry and won't affect us again), then

(i) Two quarks can couple their spins as follows

$$\begin{Bmatrix} S = 1: & \text{symmetric} \\ S = 0: & \text{antisymmetric} \end{Bmatrix} \tag{5}$$

(ii) Two u,d quarks similarly form isospin states

$$\begin{Bmatrix} I = 1: & \text{symmetric} \\ I = 0: & \text{antisymmetric} \end{Bmatrix} \tag{6}$$

(iii) In the ground state L = 0 for all quarks; hence the orbital state is trivially symmetric. Thus for pairs in L = 0, we have

$$\begin{Bmatrix} S = 1 \text{ and } I = 1 & \text{correlate} \\ S = 0 \text{ and } I = 0 & \text{correlate} \end{Bmatrix}. \tag{7}$$

Thus the Σ^0 and Λ^0 which are distinguished by their u,d being I = 1 or 0 respectively also have the u,d pair in spin = 1 or 0 respectively:

$$\begin{Bmatrix} \Sigma^0 (u,d)_{I=1} s \leftrightarrow (u,d)_{S=1} S \\ \Lambda^0 (u,d)_{I=0} s \leftrightarrow (u,d)_{S=0} S \end{Bmatrix}. \tag{8}$$

Thus, the spin of the Λ^0 is carried entirely by the strange quark.

This is the source of the Σ-Λ mass difference. The $\vec{S}\cdot\vec{S}$ interaction acts between all possible pairs; thus

$$\Sigma^0 \ [(u,d)_1 s]: \quad \langle \vec{S}\cdot\vec{S} \rangle_1 + \langle \vec{S}\cdot\vec{S} \rangle_{s,1} \tag{9}$$

$$\Lambda^0 \ [(u,d)_0 s]: \quad \langle \vec{S}\cdot\vec{S} \rangle_0 \tag{10}$$

(note $\langle \vec{S}\cdot\vec{S} \rangle$ between a spinless diquark and anything vanishes; hence the absence of $\langle S\cdot S \rangle_{s,o}$).

Now

$$\langle \vec{S}\cdot\vec{S} \rangle_0 = -3 \ \langle \vec{S}\cdot\vec{S} \rangle_1, \tag{11}$$

(see p. 91 of Ref. 3). Further, if $m_s = m_{u,d}$, the Σ and Λ become mass degenerate, and so in this limit

$$\langle \vec{S}\cdot\vec{S} \rangle_{s,1} = -4 \ \langle \vec{S}\cdot\vec{S} \rangle_1. \tag{12}$$

For unequal masses of u and s, the magnetic interaction scales as the inverse mass. Hence finally

$$\Sigma^0 \sim \langle \vec{S}\cdot\vec{S}\rangle_1 \left\{1-4\,\frac{m_u}{m_s}\right\} \tag{13}$$

$$\Lambda^0 \sim \langle \vec{S}\cdot\vec{S}\rangle_0 \;\{-3 \qquad \}. \tag{14}$$

Then with $m_s > m_u$, we find $m_\Sigma > m_\Lambda$ as observed. Increasing m_s/m_u enhances the effect (e.g., for the charmed analogues $\Sigma_c[(u,d)c]$ and $\Lambda_c[(u,d)c]$ the splitting will be larger — again observed).

Color, the Pauli Principle, and Magnetic Moments

The electrical charge of a baryon is the sum of its constituent quark charges. The magnetic moment is an intimate probe of the correlations between the charges and spins of the constituents. Being wise, today we can say that the neutron magnetic moment was the first clue that the nucleons are not elementary particles. Conversely the fact that quarks appear to have $g \simeq 2$ suggests that they are elementary (or that new dynamics is at work if composite).

A very beautiful demonstration of symmetry at work is the magnetic moment of two similar sets of systems of three, viz.

$$\begin{Bmatrix} N & ; & P \\ ddu & ; & uud \end{Bmatrix} \qquad \mu_P/\mu_N = -3/2$$

and the nuclei

$$\begin{Bmatrix} H^3 & ; & He^3 \\ NNP & ; & PPN \end{Bmatrix} \qquad \mu_{He}/\mu_H = -2/3.$$

The Pauli principle for nucleons requires He^4 to have no magnetic moment:

$$\mu[He^4;\ P^\uparrow P^\downarrow N^\uparrow N^\downarrow] = 0.$$

Then

$$He^3 \equiv He^4 - N$$

$$H^3 \equiv He^4 - P$$

and so

$$\frac{\mu_{He3}}{\mu_{H^3}} = \frac{\mu_N}{\mu_P}$$

To get at this result in a way that will bring best comparison with the nucleon three-quark example, let's study the He^3 directly.

He^3 = ppn: pp are flavor symmetric; hence, spin antisymmetric; i.e., $S = 0$.

Thus

$$[He^3]^\uparrow \equiv (pp)_o\, n^\uparrow \tag{15}$$

and so the pp do not contribute to its magnetic moment. The magnetic moment (up to mass scale factors) is

$$\mu_{He^3} = 0 + \mu_N. \tag{16}$$

Similarly,

$$\mu_{H^3} = 0 + \mu_P. \tag{17}$$

Now let's study the nucleons in an analogous manner.

The proton contains u,u flavor symmetric and <u>color antisymmetric</u>; thus the spin of the "like" pair is symmetric (S = 1) in contrast to the nuclear example where this pair had S = 0. Thus coupling spin 1 and spin 1/2 together, the Clebsches yield

$$p^{\uparrow} = \frac{1}{\sqrt{3}}\,(u,u)_0 d^{\uparrow} + \sqrt{\frac{2}{3}}\,(u,u)_1 d^{\downarrow} \tag{18}$$

(contrast Eq. (15)), and (up to mass factors)

$$\mu_P = \frac{1}{3}\,(0{+}d) + \frac{2}{3}\,(2u-d). \tag{19}$$

Suppose that $\mu_{u,d} \propto e_{u,d}$, then

$$\mu_u = -2\,\mu_d \tag{20}$$

so

$$\frac{\mu_P}{\mu_N} = \frac{4u-d}{4d-u} = -\frac{3}{2} \tag{21}$$

(the neutron follows from proton by replacing $u \leftrightarrow d$).

I cannot overstress the crucial, hidden role that color played here in getting the flavor-spin correlation right.

We can extend this discussion to the full baryon octet. Six of these states contain two identical quark flavors (which by symmetry necessitates that this pair have total spin S = 1):

$$\left.\begin{array}{l} [P(uu)_1 d] \\ [N(dd)_1 u] \\ [\Sigma^+(uu)_1 s] \\ [\Sigma^-(dd)_1 s] \\ [\Xi^0(ss)_1 u] \\ [\Xi^-(ss)_1 d] \end{array}\right\} \tag{22}$$

The remaining pair are Σ^0 and Λ^0, both u,d's. In the former, the (ud) have I = 1 and hence S = 1. For the Λ^0, on the other hand, the (ud) have I = 0 and hence S = 0.

Thus the Λ^0 $[(u,d)_0 s]$ is analogous to the He^3 nuclear example. The magnetic moment is carried entirely by the third quark, namely s. The data yield[8]

$$\mu_\Lambda \to \mu_s \simeq \frac{3}{5}\mu_d. \tag{23}$$

The strange and down quarks have the same charge (-1/3) and so the datum fits with

$$m_d = \frac{3}{5} m_s \tag{24}$$

as already noted.

If we approximate $m_u = \frac{1}{3} m_p$ (thus the proton would have g = 3), then we can do a quick computation of baryon magnetic moments where the individual contributions to the g factors are

$$\left.\begin{aligned} u &= 2 \\ d &= -1 \quad (\text{ratio of } e_d/e_u) \\ s &= -3/5 \quad (\text{ratio of } m_d/m_s) \end{aligned}\right\}. \tag{25}$$

From the general spin structure of Eq. (18)

$$B^\uparrow = \frac{1}{\sqrt{3}}(q_1q_2)_\rightarrow q_3^\uparrow + \sqrt{\frac{2}{3}}(q_1q_2)_\uparrow q_3^\downarrow$$

we have

$$\mu = \frac{2}{3}(q_1+q_2) - \frac{1}{3} q_3 \tag{26}$$

into which the (25) are to be substituted as required. The resulting pattern is as follows

	Prediction	Data	
$P[(uu)_1 d]$	$\frac{1}{3}(4u-d) = 3$	2.79	(27)
$N[(dd)_1 u]$	$\frac{1}{3}(4d-u) = -2$	-1.9	(28)
$\Sigma^+[(uu)_1 s]$	$\frac{1}{3}(4u-s) = 2.8$	2.33 ± 0.13	(29)
$\Sigma^-[(dd)_1 s]$	$\frac{1}{3}(4d-s) = -1.1$	-1.41 ± 0.25	(30)
$\Xi^0[(ss)_1 u]$	$\frac{1}{3}(4s-u) = -1.5$	-1.25 ± 0.02	(31)
$\Xi^-[(ss)_1 d]$	$\frac{1}{3}(4s-d) = -0.5$	-0.75 ± 0.06	(32)

The trend is exceptionally well described. There are undeniably 20% effects not fully accounted for.

Encouraged by this success, we might look further at this problem since, after all, there are exchange effects in nuclei that cause 20% deviations from the naive additive approach analogous to that which we have used for quarks.

We can form contributions of u and d quarks from (27) and (28) for nucleon, (29,30) for Σ, (31,32) for Ξ, and the data yield

$$(u-d)_N = 2.9$$

$$(u-d)_\Sigma = 1.7 \pm 0.15$$

$$(u-d)_\Xi = 0.9 \pm 0.12.$$

As we go to systems with more strange quarks, the u,d quarks act as if their effective mass increases (by a factor of three??). There is a systematic trend but extremely dramatic.

Now let's study the strange quark. We can do this by supposing that the environmental dependence for u and d flavors is the same. Then

$$(s)_\Xi \equiv - \left(\Xi^0 + 2\Xi^-\right) = -0.69 \pm 0.03$$

$$(s)_\Sigma \quad \left(\Sigma^+ + 2\Sigma^-\right) \quad = -0.5 \pm 0.5$$

$$(s)_\Lambda = -0.6$$

So the strange quark gives its "canonical" contribution to baryons containing either one or two strange quarks.

THE PROTON'S SPIN: A QUARK MODEL PERSPECTIVE

Inelastic lepton scattering from nucleons at high momentum transfer measures the number densities of charged constituents, $q(x)$, $\bar{q}(x)$, as a function of the Bjorken variable x (essentially the ratio of the constituent and target longitudinal momenta in an infinite momentum frame). There is a weak dependence of these distributions on the momentum transfer, Q^2, but I shall suppress this in much of what follows.

If the beam and target are polarized, one can extract the helicity-dependent distributions for quarks or antiquarks polarized parallel $\left(q^\uparrow(x)\right)$ or antiparallel $\left(q^\downarrow(x)\right)$ to the target polarization. I shall define $\Delta q(x) \equiv q^\uparrow(x) - q^\downarrow(x)$; $q(x) \equiv q^\uparrow(x) + q^\downarrow(x)$, and similarly for antiquarks, $\bar{q}$.

Data are presented in two ways.[1-3] One is in terms of the polarization asymmetry

$$A(x) = \sum_i e_i^2 \left(\Delta q_i(x) + \Delta\bar{q}_i(x)\right)/\sum_i e_i^2 \left(q_i(x) + \bar{q}_i(x)\right), \tag{33}$$

(note that $-1 \leqslant A \leqslant +1$). The other involves the polarized structure function

$$g_1(x) = \frac{1}{2} \sum_i e_i^2 \left(\Delta q_i(x) + \Delta\bar{q}_i(x)\right), \tag{34}$$

thus

$$g_1(x) \equiv A(x)\, F_1(x). \tag{35}$$

In advance of the data, the expectations were that

(i) At $x > 0.2$ where valence quarks dominate, $A(x)$ should be large and positive.[3,4] This follows from intuition developed for constituent valence quarks in baryon spectroscopy where the Pauli principle requires $\Delta u > 0$, $\Delta d < 0$. As the charge-squared weighting of Δu is four times that of Δd in protons, so $A^p(x > 0.2) > 0$. Data confirm this brilliantly. For a neutron target, it is Δd that is weighted 4:1 relative to Δu, hence these tend to cancel and one predicts[4] a small (zero?) asymmetry on the neutron.

(ii) Form $g_1(x)$, which directly shows the charge weighted helicity-dependent distributions and integrate over all x.[5,6] If it were not for the charge weightings, this would measure the net $\Delta q + \Delta\bar{q}$ $\left(\Delta q \equiv \int_0^1 dx\Delta q(x)\right.$ = net quark polarization$\left.\right)$.

Explicitly, in the quark parton model

$$I^p \equiv \int dx\, g_1^p(x) = \frac{1}{2}\left\{\frac{3}{9}\,\Delta u + \frac{1}{9}\,(\Delta u+\Delta d+\Delta s)\right\} + \left(\Delta q_i \to \Delta\bar{q}_i\right). \tag{36}$$

The surprise[2] is that $I^p(\text{EMC}) \simeq 0.12$ and is almost saturated by[7] $\Delta u(\simeq 0.75)$ leaving

$$\sum_i (\Delta q + \Delta\bar{q})_i \simeq 0, \tag{37}$$

hence, the much-advertised claim that maybe "none of the proton's spin polarization is carried by quarks". This is a misinterpretation of Eq. (37). The valence quarks are highly polarized (point (i) above); thus, the interpretation of Eq. (37) is that something cancels or hides it. Candidates include a highly polarized sea spinning opposite to the valence quarks, orbital angular momentum, or gluon polarization.[8-10]

One can cancel out some charge weighting effects by looking at the difference of proton and neutron for which

$$I^p - I^n = \frac{1}{6}\,(\Delta u-\Delta d) \equiv \frac{1}{6}\left|\frac{g_A}{g_V}\right|, \tag{38}$$

which is Bjorken's sum rule.[5] The various g_A in the baryon octet give information on the differences of Δu, Δd, and Δs which are summarized by a measured parameter known as F/D. To extract the sum, Δq, we need the proton integral (Eq. (36)) or information on neutral current form factors

$$\tilde{g}_A(\nu p \to \nu p) = \Delta u - \Delta d - \Delta s. \tag{39}$$

I shall discuss this at the end of the talk. Preceding that, I shall discuss the question of Δs, since the measured F/D and the measured I^p can be combined to extract a value for Δs. This appears to be substantial; EMC claiming that

$$\Delta s = -0.23 \pm 0.08. \tag{40}$$

Implications and criticisms of this startling result will occupy the latter half of this talk. First, I will discuss what we know about the (constituent) quark polarization from static properties of the nucleon (magnetic moments, g_A/g_V) and review the extent to which the new insights do or do not require revision of this simple picture.

Spin Polarization of Valence (Constituent) Quarks

In the constituent quark model where $L_Z = 0$ the charges and the magnetic moments of neutron and proton place the following constraints on the probabilities for finding the flavors and spin correlations of "valence" quarks,

$$u_v = 2d_v \quad \frac{\mu_n}{\mu_p} = -\frac{2}{3} \rightarrow \Delta u_v = -4\Delta d_v. \tag{41}$$

The 56, $L_Z = 0$ wave function of the nonrelativistic quark model (NRQM) satisfies (41) but it is by no means unique. A hybrid state, where a gluon $(J_Z = \pm 1)$ is partnered by qqq in 70 (required by the Pauli principle for $qq\bar{q}$ in color 8) satisfies Eq. (41) for the coherent combination[11] $g(^2 8 + ^4 8)$ where the superscripts refer to the 2S+1 of the net spin of the qqq system. The "valence quarks" here are significantly depolarized relative to 56. One can also have significant polarized sea without destroying the magnetic moment relations. This is because

$$\frac{\mu_n}{\mu_p} = \frac{2\Delta d - \Delta u + (-2\Delta\bar{u} + \Delta\bar{d} + R\Delta\bar{s})}{2\Delta u - \Delta d + (-2\Delta\bar{u} + \Delta\bar{d} + R\Delta\bar{s})}, \tag{42}$$

where $R = m_d/m_s \simeq 3/5$. The electrical neutrality of the sea tends to shield its contribution. A detailed fit is made in Ref. 12.

The $(g_A g_V)$ for the octet of baryons also relate to the spin polarized probabilities such as

$$\left(\frac{g_A}{g_V}\right)_{np} = \Delta u_v - \Delta d_v \rightarrow -5\Delta d_v, \tag{43}$$

where we used Eq. (41). Thus immediately

$$\Delta d_v = -0.25; \; \Delta u_v = 1. \tag{44}$$

In the 56 NRQM one would have[3]

$$\Delta d_v = -1/3; \; \Delta u_v = 4/3; \; \Delta u_v + \Delta d_v = 1, \tag{45}$$

and the entire spin polarization comes from the quarks. However, from Eq. (44), we see that

$$\Delta u_v + \Delta d_v \simeq 3/4, \tag{46}$$

and so, in advance of the EMC data, only naive "quarkists" would have expected 100% for Δq_v. Anyone who worked with four-component spinors, of which the MIT bag is a specific model example, knew that the "orbital dilution" in the lower components played an essential role.[13] In fact, the Δq_v expectation is even less than Eq. (46). When one makes a best fit to all of the baryon octet g_A/g_V, one finds

$$\Delta q_v (\equiv 3F - D) = 0.55 \pm 0.10. \tag{47}$$

Note the appearance of F and D which summarizes the g_A/g_V. This parameter will appear later. Note that many analyses of the polarization data use[2,6,10,15] F/D = 0.63 (Ref. 16). However, this value fitted a value of the neutron lifetime that we now know to have been incorrect.[17,18] The correct current value[19-25] is lower than 0.63 and is dependent upon assumptions about SU(3) flavor breaking.

The earliest predictions for the deep inelastic polarization asymmetry in the valence-dominated region assumed that all Δq and q (valence) have the same x dependence. Thus (see Refs. 3 and 4 for origins of these formulae)

$$A^n(x) \simeq 4\Delta d + \Delta u \to 0,$$

(the zero following immediately from Eq. (41)) and

$$A^p(x) = \frac{5}{3}(-\Delta d) + \frac{1}{3}(g_A/g_V).$$

The prediction that $A^p > 0$ is non-trivial as a priori it could be anywhere in the range $-1 \leq A \leq +1$. The presence of a $q\bar{q}$ sea as $x \to 0$ was expected to cause $A(x \to 0) \to 0$. The other qualitative expectation[26,27] was that $A(x \to 1) \to 1$ as follows.

The valence picture above implicitly assumed that $u_v(x) = 2d_v(x)$ for all x. However, unpolarized data show this to be untrue in that it would require that

$$\frac{F_1^n(x)}{F_1^p(x)} = 2/3.$$

In practice, this ratio drops as $x \to 1$, suggesting that the $u(x \to 1) \gg d(x \to 1)$, a phenomenon which follows from spin dependence via single gluon exchange. Chromomagnetic hyperfine energy shifts split the Δ-N masses and elevate $u(x \to 1)$ over $d(x \to 1)$. They also cause $u^\uparrow(x \to 1)$ to dominate over $u^\downarrow(x \to 1)$, which the consequence that $A^{p,n}(x \to 1) \to 1$. Thus, a qualitative expectation for A^p emerged:

$$A^p(x \to 0) \to 0; \; A^p(x \simeq 1/3) \simeq 1/3 \left|\frac{g_A}{g_V}\right| ; \; A^p(x \to 1) \to 1.$$

These predictions turned out to be remarkably well verified and even agree with the latest EMC data.

Recently Close and Thomas[28] showed that, within the framework of the MIT bag model, one could relate the x-dependent distortion of the valence distributions to the measured chromomagnetic energy shift in the Δ-N masses. All of this suggests that the valence quark polarizations measured in polarized deep inelastic scattering are similar to the polarizations of the constituent quarks manifested in low-energy spectroscopy. This is an important constraint on model builders. The memory of the constituent quark spins is not lost as one proceeds to the deep inelastic: the valence quarks are highly polarized.

If, as is being claimed, the quarks and antiquarks contribute (within errors) nothing to the net spin polarization of the proton, then we must conclude that something is canceling the contribution of the valence quarks. Candidates include orbital angular momentum polarized gluons or a negatively polarized sea.

We already noted that in the constituent limit it is over naive to ignore orbital angular momentum. The presence of polarized gluons may be probed by studying the polarization dependence of direct photon production or spin dependence of heavy flavor production; a polarized sea may affect the inclusive production of hadrons[33,34] and fast $K^{-}(s\bar{u})$ production may be a tag for scattering from the sea.[34]

Dziembowski et al.[35] have studied the relation between constituent quarks and partons. They view the constituent quarks as being a conglomerate of partons-quarks, antiquarks, and gluons, thus

$$q_i^{\lambda_i}(x,Q^2) = \sum_{v,\lambda_v} \int_x^1 \frac{dy}{y}\, G_{v/N}^{\lambda_v}(y)\, q_{i/v}^{\lambda_i\lambda_v}\left(\frac{x}{y}, Q^2\right),$$

where the λ are helicity labels. The constituent quark distributions $G_{v/N}(y)$ reflect the dynamics that binds the quarks to form hadrons, and are determined by a light cone nucleon wavefunction. The constituent quark structure functions $q(x/y, Q^2)$ are adapted from Altarelli et al.[36] together with Carlitz and Kaur's ansatz[37] for the spin of soft valence partons within a polarized constituent quark.

This picture of partons convoluted within constituents generates some effective $L_z \neq 0$ but not enough to account for the spin deficit claimed by EMC. The data seem to fall below the model systematically for $x \leq 0.1$. If these small x data survive further experiments, then it seems that polarization of the sea (not included in Ref. 35) must be allowed for. This naturally leads to the question of whether there are polarized strange quarks in the proton.

Polarized Strange Quarks?

One exciting possibility is that the EMC data imply a large polarization of strange quarks and/or antiquarks within the proton. If true, this could have significant consequences. In particular, it could modify earlier analyses of electroweak parity violation in deuterium where Campbell et al. argue,[15] the polarized strange quarks could give contributions that dominate over electroweak radiative corrections. An extreme claim has appeared in the literature that the large value for Δs is in conflict with perturbative QCD. If true, this would be devastating. This claim comes about, in part, because an incorrect value of F/D has been used in the analyses. It is this parameter, and its implications for Δs, that I will now discuss.

Given the integral, I_p, of the polarized structure function $g_1^P(x,Q^2)$, one extracts Δs (including new QCD corrections)

$$I_p \equiv \int dx g_1^P(x,Q^2) = \frac{1}{18}\left(\frac{g_A}{g_V}\right)\left[\frac{9f-1}{f+1} - \frac{\alpha_s(Q^2)}{\pi}\,\frac{3f+1}{f+1}\right] + \frac{\Delta s}{3}\,, \tag{48}$$

where $f \equiv F/D$ with $\alpha_s(Q^2) = 0.27$, $g_A/g_V = 1.254 \pm 0.006$ and $I_p = 0.126 \pm 0.022$. A feeling for the sensitivity of Δs to f can be gauged from the approximate relation

$$\Delta s \simeq (f-0.40) \pm 0.07. \tag{49}$$

The widely used value, following the much-quoted fit of Ref. 16 has been

$$F/D = 0.63 \pm 0.02 \rightarrow \Delta s = -0.23 \pm 0.09. \tag{50}$$

If the sea is flavor-independent, then Eq. (50) summarizes the widely accepted interpretation of the EMC polarized structure function data where a significant negative polarization of the sea cancels out the positive polarization of the valence quarks.

This value was based on the original value for I_p quoted by EMC,[2] namely $I_p = 0.116 \pm 0.022$. However, the revised value,[29] $I_p = 0.126 \pm 0.022$, reduces the magnitude of Δs by 0.03, and so $\Delta s = -0.20 \pm 0.09$ should replace Eq. (50).

However, it does not seem to be widely appreciated that the F/D of Ref. 16 was much constrained by an outdated value of the neutron lifetime, and that Ref. 16 chose "to omit from (their) fit the neutron decay correlation (which yields) $g_A = 1.258 \pm 0.009$, which differs significantly from the result 1.239 ± 0.009 required by the neutron lifetime measurements". The value accepted as correct today[18] differs by some 3σ from the old value, and this, together with other data on hyperon beta decays,[16,18,19] shows that F/D is much smaller than the old value. Flavor symmetry breaking causes a spread in values of F/D, depending on which partial set of data one uses; indeed, the symmetry breaking even calls into question the utility of the F/D parameter,[20] and so Refs. 17 and 21 set up their analyses without direct reference to F/D. Translating their work into F/D, one finds that the value subsumed in Ref. 17 is F/D = 0.56 consistent with that implicit in Ref. 21 and, within errors, with the fitted value in Ref. 22. Reference 23 obtained an even smaller value of F/D = 0.545 ± 0.02. Recent improvements in the Σn beta decay data, in particular, may raise F/D to 0.58 (Ref. 24), but nowhere as high as the 0.63 used previously.

The magnitudes for Δs implied by these values for F/D are

$$F/D = 0.548 \pm 0.01 \rightarrow \Delta s = -0.12 \pm 0.06 \text{ Ref. 23} \tag{51}$$

$$F/D = 0.58 \pm 0.01 \rightarrow \Delta s = -0.15 \pm 0.08 \text{ Ref. 24} \tag{52}$$

Thus we see that the magnitude of the (negative) strange polarization may be only half as big as that previously assumed. The QCD-corrected value of the Ellis-Jaffe sum rule falls from 0.19 (the cited value when F/D = 0.63) to 0.17 if F/D = 0.56, thereby reducing the statistical significance of the much-advertised failure of this sum rule.

What independent information exists on Δs? Elastic neutrino-proton scattering can, in principle, probe this quantity,[32] and a fit to these data give

$$\Delta s = -0.15 \pm 0.09.$$

Note that this agrees with the revised value in the present paper arising from the smaller F/D and the revised EMC integral (Ref. 29).

One should also be aware that the neutrino experiment is also consistent with $\Delta s = 0$ which, in advance of the controversial EMC experiment, was the expectation.

Flavor-changing weak interactions, such as neutron beta decay, can yield

$$\frac{g_A}{g_V} \simeq 1.25 = \Delta u - \Delta d,$$

while the zero momentum limit of $\nu p \to \nu p$ can probe

$$\tilde{g}_A(0) = \Delta u - \Delta d - \Delta s \equiv \left(\frac{g_A}{g_V}\right)\left(1 - \frac{\Delta s}{1.25}\right),$$

and so a difference between $\tilde{g}_A(0)$ and g_A/g_V can, after radiative corrections, reveal nonzero Δs. (Our $\Delta s \equiv 1.25\eta$ of Ref. 32.)

A practical problem is that $\nu p \to \nu p$ is detected by proton recoil and so an extrapolation to $\vec{q} = 0$ is needed. One fits the $q^2 \neq 0$ data with a form factor, in essence

$$\frac{1-\Delta s/1.25}{(1 + Q^2/M_A^2)^2},$$

where M_A is a mass scale to be fitted. Other experiments have determined this to have the value $M_A = 1.032 \pm 0.036$ GeV. If one fixes M_A to equal the world average, then $\Delta s = -0.15 \pm 0.09$; hence the claim to support the nonzero strange polarization. However, Ref. 32 also makes another, less well-advertised, fit. They constrain $\Delta s = 0$ and find that in this case $\Delta s = 0$; $M_A = 1.06 \pm 0.05$ GeV. Thus, one sees that $\Delta s = 0$ yields M_A consistent with the world average and hence is equally acceptable as a solution. The crucial statement in Ref. 32 is that "M_A and $\eta(\Delta s)$ are strongly correlated". Thus, Ref. 32 does not require $\Delta s < 0$ and thereby does not necessarily lend support to those who desire $\Delta s \neq 0$. Thus the question of the magnitude of the (strange) sea polarization is open. It is likely to be significantly nearer to zero than is being assumed in much of the current literature. Some of the inferences claimed from the EMC polarization data may need re-evaluation therefore. In particular, there need be no conflict with perturbative quantum chromodynamics.[30]

Polarized Gluons?

It has recently been realized[8] that the perturbative QCD correction to the singlet part of $g_1^P(x)$ effectively scales (to $O(\alpha_s^2)$) and may be important. This may be incorporated by replacing the Δq in Section 1 by $\tilde{\Delta} q \equiv \Delta q - \alpha_s/2\pi\ \Delta G$, where $\Delta G \equiv \int_0^1 dx \Delta g(x)$ and $\Delta G(x) = g_\uparrow(x) - G_\downarrow(x)$ is the polarized gluon distribution. This modifies the polarized lepton analysis, but cancels out in the expressions for (g_A/g_V) and does not enter the magnetic moment (Section 2) analysis.

One consequence is that there may be a continuity between the low-energy polarization revealed in constituent quarks (magnetic moments and spin dependence of resonance excitation) and the deep inelastic polarization.

First of all, we summarize the data on the Δq (or equivalently $\tilde{\Delta} q$) from the various (g_A/g_V).

If we assume $SU(3)_F$ symmetry in the sense that $s(\Sigma^+) \equiv d(P)$, then we may write the various g_A in terms of F, D, or Δq as follows:

g_A	F,D	$\Delta q^{(P)}$	Data
np	F + D	$\Delta u - \Delta d$	1.26 ± 0.005
Λp	$F + \frac{1}{3} D$	$\frac{1}{3}(2\Delta u - \Delta d - \Delta s)$	0.72 ± 0.02
$\Xi\Lambda$	$F - \frac{1}{3} D$	$\frac{1}{3}(\Delta u + \Delta d - 2\Delta s)$	0.25 ± 0.05
Σn	F − D	$\Delta d - \Delta s$	-0.33 ± 0.02

Thus $F/D \equiv (\Delta u - \Delta s)/(\Delta u + \Delta s - 2\Delta d)$. Extracting the individual contributions involves a correlated fit. The EMC values, corrected for F/D, become $\tilde{\Delta} u = 0.80 \pm 0.06$, $\tilde{\Delta} d = -0.45 \pm 0.06$, and $\tilde{\Delta} s = -0.15 \pm 0.06$. One possibility is that $\Delta s = 0$, so that $\tilde{\Delta} s = -\alpha/2\pi \; \Delta G$. In this case, we obtain for

$$\Delta u \equiv \tilde{\Delta} u - \tilde{\Delta} s = 0.95 \pm 0.06 \tag{53}$$

$$\Delta d \equiv \tilde{\Delta} d - \tilde{\Delta} s = -0.30 \pm 0.06 \tag{54}$$

It is interesting to note that these values are consistent with those extracted from the magnetic moments (Eq. (4)) viz

$$\Delta u_v = 1, \; \Delta d_v = -0.25$$

The proton helicity is given by

$$\frac{1}{2} = \frac{1}{2} \Delta q + \left(\Delta G + L_Z\right) \tag{55}$$

Hence $\Delta G = 3.5$ and $L_Z = -3.35$ at $Q^2 \simeq 10$ GeV^2. As one devalues to lower Q^2, $d/dQ^2\left(\Delta G + L_Z\right) = 0$, and the individual contributions fall. It is an open question whether the "passive" L_Z in the constituent model (i.e., the dilution of S_Z due to relativistic spinors) at low Q^2 provides a consistent picture between constituent spin polarization and "parton" polarization.

Ellis _et al._[31] suggest that the modification to $\Delta q(x)$ be driven by evolution

$$\tilde{\Delta} q(x) = \Delta q(x) - \int_x^1 \frac{dy}{y} \, \Delta G(y) \sigma(x/y)$$

where $\sigma(Z)$ is the cross section for $\gamma^* g \to q\bar{q}$. If so, then $g_1^P(x \to 0) < 0$, the crossover from positive to negative moving to smaller x values as Q^2 increases.

It is tantalizing that such a picture may already be manifested at low Q^2 in the resonance region. It is well known that the prominent $D_{13}(1520)$ and $F_{15}(1690)$ resonances are excited dominantly in $\sigma_{3/2}$ when $Q^2 = 0$, but in $\sigma_{1/2}$ for $Q^2 \neq 0$. The change in helicity structure,[38] or change in sign of $g_1^P(Q^2)$, occurs at $Q^2 \simeq 0.4$ GeV^2 for D_{13} and $Q^2 \simeq 0.7$ GeV^2 for F_{15}. It is amusing that these correspond to $x \simeq 0.2$, and so Bloom-Gilman duality may approximately hold true even for polarized leptoproduction, with a Q^2 dependence to the x_c where $g_1^P(x_c) = 0$. The first resonance $P_{33}(1236)$ sits on top of an S-wave background; the relative Q^2 dependences are not well known. However, perturbative QCD applied to resonance excitation suggests that this excitation may also change its character with Q^2 such that its contribution to $g_1^P(x)$ changes sign at $x \to 0$. It will be interesting at CEBAF to verify if the resonance region indeed matches onto the deep inelastic, and at high energy labs to verify whether $g_1^P(z \to 0) < 0$.

References

1. G. Baum et al., Phys. Rev. Lett. 51:1135 (1983); V. W. Hughes and J. Kuti, Ann. Rev. Nucl. Part. Sci. 33:611 (1983).
2. J. Ashman et al. (EMC), Phys. Lett. B206:364 (1983; V. W. Hughes et al., Phys. Lett. B212:511 (1988).
3. F. E. Close, "Introduction to Quarks and Partons," Academic, New York (1979) Chap. 13.
4. J. Kuti and V. Weisskopf, Phys. Rev. D4:3418 (1971).
5. J. D. Bjorken, Phys. Rev. D1:1976 (1970).
6. J. Ellis and R. L. Jaffe, Phys. Rev. D9:1444 (1984).
7. J. Ellis, R. A. Flores, and S. Ritz, Phys. Lett. B194:493 (1987).
8. G. Altarelli and G. G. Ross, Phys. Lett. B212:391 (1988); R. Carlitz, J. Collins, and A. Mueller, Phys. Lett. B214:229 (1988).
9. L. M. Sehgal, Phys. Rev. D10:1663 (1974).
10. S. J. Brodsky, J. Ellis, and M. Karliner, Phys. Lett. B206:309 (1988).
11. T. Barnes and F. E. Close, Phys. Lett. 128B:277 (1983; F. Wagner, Proc. XVI Rencontre de Moriond (1982), ed. J. Tranthanhvan.
12. C. Carlson and J. Milana, College of William & Mary report, WM-89-101. The role of gluon exchange is discussed by H. Hogassen and F. Myhrer, Phys. Rev. D37:1950 (1988).
13. For example, p. 117 in Ref. 3.
14. This comes from combinations of g_A/g_V for np with that for Λp, Σn, and $\Xi\Lambda$; see Eq. (11) in Ref. 17.
15. B. A. Campbell, J. Ellis, and R. A. Flores, CERN-TH-5342/89.
16. M. Bourquin et al., Z. Phys. C21:27 (1983).
17. F. E. Close and R. G. Roberts, Phys. Rev. Lett. 60:1471 (1988).
18. M. Aguilar-Benitez et al. (Particle Data Group) Phys. Lett. B204:1 (1988).
19. S. Hsueh et al. (E715 collaboration) Phys. Rev. D38:2056 (1988).
20. H. J. Lipkin, Phys. Lett. 214B:429 (1988).
21. M. Anselmino, B. Ioffe, and E. Leader, Santa Barbara ITP report (1988) unpublished.
22. D. Kaplan and A. Manohar, Nucl. Phys. B310:527 (1988).
23. J. Donoghue, B. Holstein, and S. Klint, Phys. Rev. D35:934 (1987).
24. A. Beretvas, private communication; Z. Dziembowski and J. Franklin, Temple University, Philadelphia report TUHE-89-11 (1989).
25. R. Jaffe and A. Manohar, MIT report, MIT-CTP-1706 (1989) (but note that this inputs outdated neutron lifetime, which artificially increases the errors).
26. F. E. Close, Phys. Lett. 43B:422 (1973).
27. G. Farrar and D. Jackson, Phys. Rev. Lett. 35:1416 (1975).
28. F. E. Close and A. W. Thomas, Phys. Lett. B212:227 (1988).
29. EMC Collaboration, CERN preprint, CERN EP/89-73, June 1989.
30. F. E. Close, Phys. Rev. Letts. (in press).
31. J. Ellis, M. Karliner, and C. Sachrajda, CERN-TH-5471/89.
32. L. Ahrens et al., Phys. Rev. D35:785 (1987).
33. M. Frankfurt et al. (private communication).

34. F. E. Close and R. Milner, Oak Ridge National Laboratory report (1989).
35. Z. Dziembowski et al., Phys. Rev. D 39:3257 (1989).
36. G. Altarelli et al., Nucl. Phys. B69:531 (1974).
37. R. Carlitz and J. Kaur, Phys. Rev. Lett. 38:673 (1977).
38. See, e.g., V. Burkerdt, Research Program at CEBAF (Report of 1985 study group) ed. F. Gross (CEBAF, Newport News, 1986).

HADRON SPECTROSCOPY:
an overview with strings attached

Nathan Isgur

Department of Physics
University of Toronto
Toronto, Canada M5S 1A7

ABSTRACT

After recalling why most strong interaction phenomena cannot be studied with quark-gluon perturbation theory, I suggest that a phenomenological form of the old dual string theory may be a good model for the nonperturbative aspects of QCD. I begin by describing some recent progress within this framework in understanding the status of the quark model within QCD. The resulting picture of the ordinary hadrons, hybrids, and glueballs provides a simple framework for studying these elements of hadron spectroscopy. Within the same model, I discuss multiquark states, emphasizing their importance for understanding glue dynamics and, in a more practical context, nuclear physics.

PRELUDE: THE BREAKDOWN OF PERTURBATIVE QCD

We all know that asymptotic freedom guarantees that at sufficiently small distances Quantum Chromodynamics (QCD) becomes a weakly coupled quark-gluon theory which is amenable to a perturbative expansion in the running coupling constant $\alpha_s(M^2)$. However, the other side of this coin is that at large distances α_s becomes large so that quark-gluon perturbation theory may break down.

In fact, we now know from numerical studies that QCD predicts confinement: the potential energy between two static quarks grows linearly with their separation r with a constant of proportionality b, called the string tension, that is about 1 GeV/fm. I want to begin these lectures by reminding you that such a result rigorously implies the breakdown of perturbative QCD. Given that confinement is the central feature of strong interaction physics, we are therefore forced to seek new methods for the study of most strong interaction phenomena.

Since the pure gluon sector of QCD in which the static potential problem is posed has no dimensionful parameters, the equation for the string tension in QCD must take the form

$$b = \frac{1}{a^2} f[g(a)],$$

Hadrons and Hadronic Matter
Edited by D. Vautherin *et al.*
Plenum Press, New York, 1990

where $a \sim M^{-1}$ is the (spatial) ultraviolet cutoff placed on the calculation, and f is some function of the running strong coupling constant g(a) appropriate to the scale a. Since b is an observable, it cannot depend on a, so $\frac{db}{da^2}=0$, implying Regge theory itself, then, as the result of efforts to make this theory dual[1)]*, crossing symmetric, and analytic, the Veneziano formula[2)]. This formula, which was at first quite mysterious, was soon interpreted[3)] as containing the dynamics of quarks connected by strings, and the string theory of hadrons was born[4)]. It was eventually abandoned as a fundamental theory of hadrons, but has been revived, as you all know, as a fundamental theory of everything. I want to remind you that as an effective theory of hadrons the string theory had many virtues. I will also explain how this effective theory may be related to QCD (via the QCD flux tube model[5)]).

The string theory certainly seems to correspond very well with all of the qualitative features observed in the hadronic world: the confined quark spectrum, Regge phenomenology for cross sections, duality, etc. It was of course designed with these phenomena in mind. What is perhaps more impressive is that it also contains, at least qualitatively, many other features expected in QCD but not yet observed. To me the two most startling such examples are its predictions of hybrid mesons and glueballs. The old string theory had in addition to the known Regge trajectories others called "daughter trajectories". Some of these corresponded to states in which the string degrees of freedom were excited, in remarkably close correspondence to the flux tube model for hybrids[5)]. Glueballs were required in the old string theory for consistency: hadronic reactions were assumed to proceed by an elementary string vertex in which a string breaks, forming a $q\bar{q}$ pair (or the time reversed healing process). When the healing process occurred between the q and $\bar{q}$ of an ordinary meson, a closed loop of string would be formed[6)]. Such states can be associated with what we would today call glueballs and indeed correspond to the glueloops of the flux tube model[5)].

The exact relationship between QCD and string theory is not yet clear. There have been several attempts to make a rigorous connection[7)]. Perhaps because it is the only one I understand, I prefer the picture advocated in Ref. 5: QCD can be formulated as a theory of colour electric flux lines interacting with quarks and with each other. The full QCD theory differs from the old string theory in that, for a given quark sector like $q\bar{q}$, there are an infinite number of flux line topologies which can "string" the quarks together (instead of just one) and a corresponding infinity of vertices which convert a given quark and flux line configuration into another one (instead of just one basic vertex). However, it is plausible that at long wavelengths the simple string configurations and conversions between them are dominant; this is the essence of the flux tube model.

* Duality is the idea that a given amplitude can be viewed as arising either from t-channel exchanges or s-channel resonances. It is most simply visualized in terms of quark line diagrams: see below.

WHAT IS THE QUARK MODEL?

A. ADIABATIC SURFACES

The idea of adiabatic surfaces[5,8)] is central to our argument for the quark model approximation to QCD. Consider first QCD without dynamical fermions in the presence of fixed $q_1\bar{q}_2$ or $q_1q_2q_3$ sources. The ground state of QCD with these sources in place will be modified, as will be its excitation spectrum. For excitation energies below those required to produce a glueball, this spectrum will presumably be discrete and continuous as, for example, shown in Figure 1 as a function of the $q_1\bar{q}_2$ spatial separation $\vec{r}$. There will be analogous spectra for $q_1q_2q_3$ which are functions of the two relative coordinates $\vec{\rho} = \sqrt{\frac{1}{2}}(\vec{r}_1 - \vec{r}_2)$ and $\vec{\lambda} = \sqrt{1/6}(\vec{r}_1 + \vec{r}_2 - 2\vec{r}_3)$. We call the energy surface traced out by a given level of excitation as the positions of the sources are varied an adiabatic surface.

Let us now define the "quark model limit": the quark model limit obtains when the quark sources move along the lowest adiabatic surface in such a way that they are isolated from the effects of other (excited) surfaces.

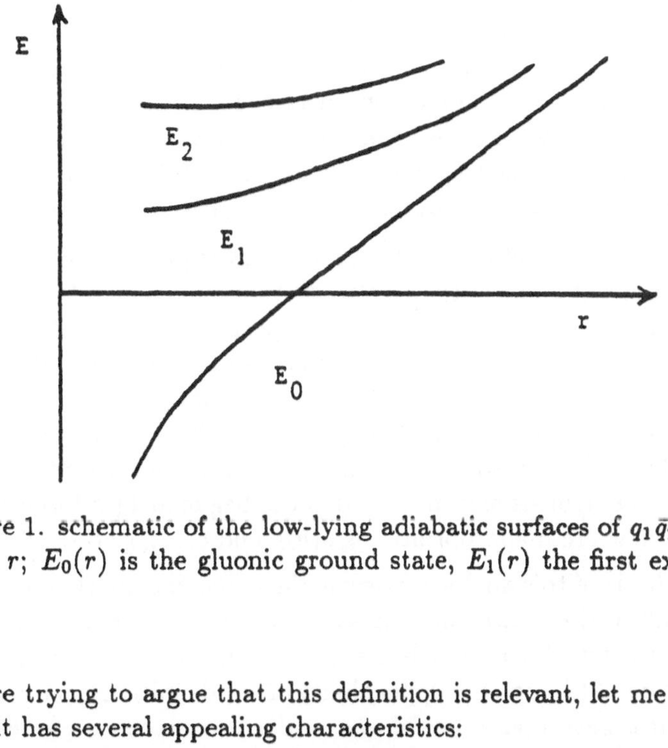

Figure 1. schematic of the low-lying adiabatic surfaces of $q_1\bar{q}_2$ at relative separation r; $E_0(r)$ is the gluonic ground state, $E_1(r)$ the first excited state, etc.

Before trying to argue that this definition is relevant, let me first simply note that it has several appealing characteristics:

1) One of the great "mysteries" of the quark model is that it describes the mesons and baryons in terms of a wavefunction which only gives the amplitude for the valence quark variables, even though in QCD the general state vector must also refer to the glue fields. Indeed, in QCD for fixed q_1 and $\bar{q}_2$, for example, there are an infinite number of possible states of the glue so that it is certainly not sufficient to simply specify the state of the quarks. In the "quark model limit", however, although there are an infinite number of possible glue states, for any fixed r there is one lowest-lying one. Moreover, although this lowest-lying state changes as r changes, it is completely determined by the quark coordinates. Thus we see the possibility that the quark model wavefunction had

a "secret suppressed subscript" describing the state of the glue: $\psi_0(\vec{r})$. We will argue below that there should be analogous (but as yet undiscovered) worlds $\psi_n(\vec{r})$ for $n > 0$ corresponding to hybrid mesons.

2) The "quark model limit" can easily be seen to be inapplicable to any systems more complicated that $q_1\bar{q}_2$ and $q_1q_2q_3$: such systems will always have adiabatic surfaces which cross so that the condition of isolation cannot be satisfied[5,8]. Figure 2 gives a simple illustration of this phenomenon in the $q_1q_2\bar{q}_3\bar{q}_4$ sector. We will return to this important distinction between the familiar mesons and baryons and multiquark systems below. For now we just note that the above definition allows us to begin to see why $q_1\bar{q}_2$ and $q_1q_2q_3$ may have a special status in QCD: only in these two cases is it possible that the state of the glue is (approximately) determined by the quark coordinates (see the point x=y in Figure 2).

With these attractions for motivation, we now proceed with the argument for the relvance of this definition.[5,8] We first recall a simple molecular physics analogy to this proposed approximation. Diatomic molecular spectra can be obtained in an adiabatic approximation by holding the two relvant atomic nuclei at fixed separation r and then solving the Schrödinger problem for the (mutually interacting) electrons moving in the static electric field of the nuclei. The electrons will, for fixed r, have a ground state and excited states which will eventually become a continuum above energies required to ionize the molecule. The resulting adiabatic surfaces then serve as effective internuclear potentials on which vibration-rotation spectra can be built. Molecular transitions can then take place within states built on a given surface or between surfaces.

In the "quark model limit" the quark sources play the rôle of the nuclei, and the glue plays the rôle of the electrons. From this point of view we cân see clearly that conventional meson and baryon spectroscopy has only scratched the surface of even $q_1\bar{q}_2$ and $q_1q_2q_3$ spectroscopy: so far we have only studied the vibration-rotation bands built on the lowest adiabatic surface corresponding to the gluonic ground state. We should expect to be able to build other "hadronic worlds" on the surfaces associated with excited gluonic states[5,9]: these states correspond to the hybrids first discussed in the bag model (in terms which from this point of view are inappropriate) as $q_1\bar{q}_2g$ and $q_1q_2q_3g$ states [10].

On the basis of this analogy it seems clear that the quark model limit will apply when all of the quarks in a meson or baryon are heavy. This certainly corresponds well with the established success of the "naive" quark model in the $c\bar{c}$ and $b\bar{b}$ sectors. Although the applicability of our definition thus seems assured in this limit, it is nevertheless useful to consider the corrections to the adiabatic approximation. These considerations will not only allow us to understand how to make more accurate predictions, but will also help us to understand why the quark model limit has a much wider range of validity than might have been expected. Of course, the simple argument for the applicability of the adiabatic approximation is just that the dynamics of the heavy quark systems $Q_1\bar{Q}_2$ and $Q_1Q_2Q_3$ are completely controlled by the asymptotically free colour Coulomb interaction and so have quark velocities of the order of $\alpha_s \to 0$ as $m_Q \to \infty$. The details are more interesting. We first note (using $Q\bar{Q}$ as our prototype) that as $m_Q \to \infty$ the low-lying states have radii $r_{Q\bar{Q}} \to (m_Q\alpha_s)^{-1} \to 0$ and

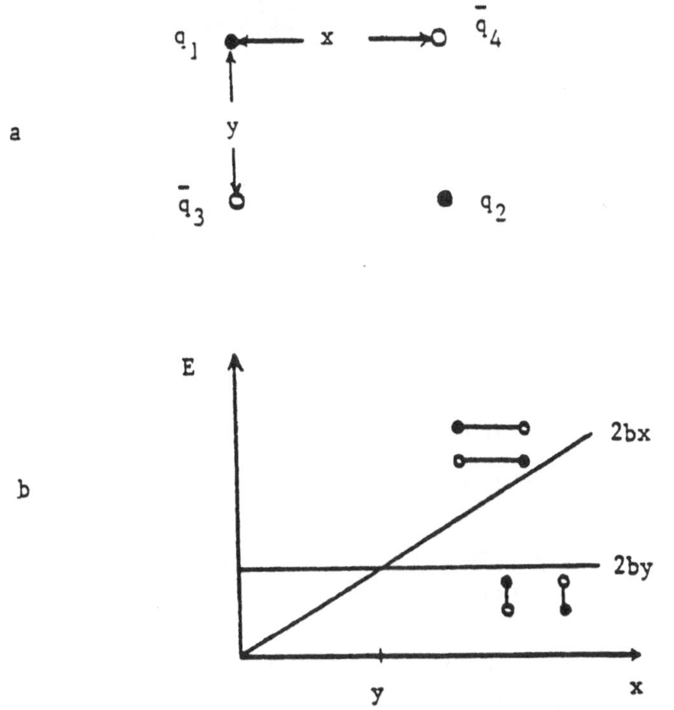

Figure 2.(a) a simple rectangular $q_1 q_2 \bar{q}_3 \bar{q}_4$ configuration with $x \equiv |\vec{r}_1 - \vec{r}_4| = |\vec{r}_3 - \vec{r}_2|$ and $y \equiv |\vec{r}_1 - \vec{r}_3| = |\vec{r}_4 - \vec{r}_3 - \vec{2}|$; (b) the two adiabatic surfaces corresponding to the "colour configurations" $(q_1\bar{q}_3)(q_2\bar{q}_4)$ and $(q_1\bar{q}_4)(q_2\bar{q}_3)$ shown for fixed y as a function of x, for x and y in the linear potential region; b is the $q\bar{q}$ string tension; the surfaces are labelled by their corresponding flux tube model states.

frequencies $\omega_{Q\bar{Q}} \sim m_Q \alpha_s^2 \to \infty$. Note that, although as previously stated $v \sim \omega r \sim \alpha_s \to 0$, the analogue of the condition usually quoted for the applicability of the adiabatic approximation in molecules ($\omega_{nuclei} << \omega_{electrons}$) does not apply. This is because $\omega_{Q\bar{Q}} \to \infty$ while ω_{glue}, corresponding to the gap between $E_0(r)$ and the first gluonic excitation, is never greater than the energy required to excite the glueball continuum. Even if we were to discount such continua, we would have to consider the other discrete surfaces which must lie no higher than about 1 GeV above $2m_Q$ corresponding to $\omega_{glue} \sim \omega_{quark}$. The physics of the decoupling of motion on the lowest adiabatic surface from excited surfaces is thus rather different than in the molecular case. It depends (in the language of second order perturbation theory) primarily on small matrix elements rather than on large energy denominators: in $Q\bar{Q}$ the decoupling occurs because the quarks produce very small oscillating colour dipole moments.

Now consider light quark systems. As the quark mass is decreased, the quark oscillations gradually approach a size governed by the QCD scale $b^{-\frac{1}{2}}$ (where b is the string tension, $b \sim \Lambda_{QCD}^2$). On the other hand, the quark frequencies are decreasing. For light quarks ω_{quark} also approaches the QCD scale, and we have $\omega_{quark} \leq \omega_{glue}$. (That ω_{quark} is somewhat less than ω_{glue} is indicated by the fact that the orbital excitations of the quark model are well known, but hybrids have not yet been discovered!) Nevertheless, given the scale of the perturbation in this case, I know of no simple argument for the vailidity of the quark model limit for lilght quark systems: they could go either way, and the issue can only be decided by explicit calculations. Since such explicit calculations necessarily involve non-perturbative gluon dynamics, any such discussion is at this time bound to be model dependent.

Having accepted that our conclusions will be model dependent, I will now introduce the flux tube model[5)] for gluon dynamics. We will then be in a position to study the validity of the adiabatic approximation for all values of the quark mass within this model.

B. THE FLUX TUBE MODEL

Even after successful numerical calculations within QCD are possible, it will still be useful, and in complex situations essential, to have models which summarise the very complex structure of this theory. The flux tube model[5)] is a model for QCD in the non-perturbative regime which emerges from considerations of Hamiltonian lattice QCD.

Hamiltonian Lattice QCD. In the Hamiltonian version of lattice QCD, space (but not time as in most numerical studies) is discretized. In this formulation the lattice spacing "a", without reference to a perturbative expansion, plays the rôle of the regulator mass M. Latticizing the theory also has another advantage: it allows us to set up a strong coupling perturbation expansion in which the expansion parameter for lattice QCD is 1/g instead of g. We may expect to be able to learn more about the strongly coupled regime of the theory in terms of such an expansion, and indeed this seems to be the case: for example, confinement is an automatic property of the $g \to \infty$ limit of lattice QCD. Moreover, the natural degres of freedom of the strong coupling regime are not quarks and gluons, but rather quarks and flux tubes, the latter being more in accord with various qualitative ideas on the nature of confinement in QCD. Of course, space is not coarse-grained (at least not on the scale of 10^{-15} metres), so that to relate lattice QCD to real QCD we must consider the limit $a \to 0$. In this limit $g \to 0$ so that a strong coupling expansion must fail; this is just the other side of the failure of the weak coupling expansion for small Q^2. If, however, it can be shown that the two regimes "match" around g=1, thereby proving that lattice QCD as $a \to 0$ is QCD, then one would nevertheless expect the strong coupling expansion to be useful in many situations where large scales dominate, just as the weak coupling expansion is useful for short distance physics.

A simple analogy may be useful. Consider approximating a continuous one dimensional harmonic oscillator by a particle hopping along a one dimensional lattice of points $x = na(n = ..., -2, -1, 0, 1, 2....)$ with lattice spacing "a". The lattice Hamiltonian can be chosen to be

$$H_{mn} = \left[\frac{1}{ma^2} + \frac{1}{2}ka^2n^2\right]\delta_{mn} - \frac{1}{2ma^2}\left[\delta_{m,n+1} + \delta_{m,n-1}\right] \tag{1}$$

since then the Schrödinger equation

$$i\frac{d\psi_m(t)}{dt} = H_{mn}\psi_n(t) \tag{2}$$

becomes

$$i\frac{\partial\psi(x,t)}{\partial t} = \left[-\frac{1}{2m}\frac{\partial^2}{\partial x^2} + \frac{1}{2}kx^2\right]\psi(x,t) \tag{3}$$

as $a \to 0$. Now for $a \to \infty$ with k and m fixed, the potential energy term $\frac{1}{2}ka^2n^2\delta_{mn}$ dominates and the eigenstates of (1) correspond to the particle sitting on single lattice sites; corrections to this limit are of relative order $\chi = 1/kma^4$ and one can systematically proceed to do perturbation theory in this hopping strength. Since the characteristic scale of the harmonic oscillator is $\alpha^{-1} = (km)^{-\frac{1}{4}}$, one will not get realistic wave functions or eigenergies for the harmonic oscillator for a $>> \alpha^{-1}$ where lowest order perturbation theory applies, but for $\chi \sim 1$ one will begin to get good approximations to the solutions of the continuum problem if one works to sufficiently high order in χ. By contrast, starting with free particle solutions to the continuum Hamiltonian and treating $\frac{1}{2}kx^2$ as a perturbation is hopeless. (The difference, of course, is that the hopping parameter expansion for the ground state, the example, will be accurate if a matrix of dimension of order $1/a\alpha$ is diagonalized.)

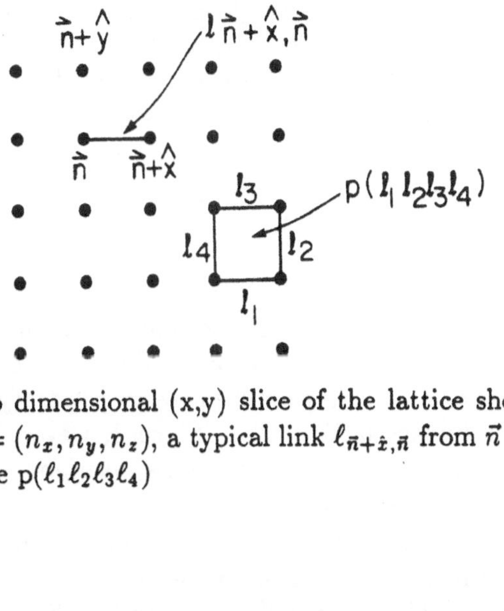

Figure 3. a two dimensional (x,y) slice of the lattice showing a typical lattice point $\vec{n} = (n_x, n_y, n_z)$, a typical link $\ell_{\vec{n}+\hat{x},\vec{n}}$ from $\vec{n}$ to $\vec{n}+\hat{x}$, and a typical plaquette $p(\ell_1\ell_2\ell_3\ell_4)$

We now turn to the formulation of QCD on a (cubic) spatial lattice. In this formulation the quark degrees of freedom of the theory "live" on the lattice sites while the gluonic degrees of freedom "live" on the links between these sites (see Figure 3). Let's consider first the theory without quarks: we describe this theory in terms of link variables U_ℓ which (before quantization) are 3×3 SU(3) group elements. The pure gauge field Hamiltonian is then the sum of two parts, one involving only the $U's$:

$$H_{glue} = \frac{g^2}{2a}\sum_{\ell} C_\ell^2 + \frac{1}{ag^2}\sum_{p} Tr[2-(U_{\ell_4}U_{\ell_3}U_{\ell_2}U_{\ell_1} + h.c.)] \tag{4}$$

with "a" the lattice spacing and g the corresponding coupling constant. Here C_ℓ^2 is defined in terms of the eight generators $E^a_{\ell\pm}$ of SU(3) transformations of U_ℓ at the beginning (-) or the end (+) of the link ℓ

$$[E^a_{\ell+}, U_\ell] = -\frac{\lambda^a}{2}U_\ell \tag{5}$$

$$[E^a_{\ell-}, U_\ell] = +U_\ell\frac{\lambda^a}{2} \tag{6}$$

by $C_\ell^2 = \sum_a (E^a_{\ell+})^2 = \sum_a (E^a_{\ell-})^2$. In the second term the product of the $U's$ is taken in order around the plaquette p. To complete lattice QCD one simply adds to (4) a lattice Hamiltonian H_{quark} for the quarks interacting with the glue. With the quark fields as site variables we have

$$H_{quark} = \sum_{flavour} m_q \sum_{\vec{n}} q^\dagger_{\vec{n}} q_{\vec{n}} + \frac{1}{a}\sum_{\substack{flavour \\ \ell_{ji}}} q_j^\dagger U_{\ell_{ji}} \alpha_{\ell_{ji}} q_i \tag{7}$$

where $\alpha_{\ell ji}$ is the Dirac matrix in the direction of the link ℓ_{ji}. Our complete Hamiltonian $H^{lattice}_{QCD} = H_{glue} + H_{quark}$ has H_{QCD} in $A^\circ = 0$ gauge as its naive continuum limit. Gauss' law takes the form of a constraint in the theory that the only physically relevant states are those which are gauge invariant.

We are now ready to consider the properties of $H^{lattice}_{QCD}$. We note first that in the strong coupling limit where "a" (and therefore g) is large

$$H^{lattice}_{QCD} \to H_{sc} = \frac{g^2}{2a}\sum_{\ell} C_\ell^2 + \sum_{\substack{flavour \\ \vec{n}}} m_q q^\dagger_{\vec{n}} q_{\vec{n}} \tag{8}$$

The eigenvalues of C_ℓ^2 are just those of the square Casimir of SU(3): 0 for the singlet, 4/3 for 3 or $\bar{3}$, 10/3 for 6 or $\bar{6}$, 3 for the octet, etc. The quark part of H_{sc} is, on the other hand, diagonalized by an arbitrary number of quarks and antiquarks at arbitrary lattice sites. Since, however, the only physically relevant eigenstates are those which are gauge invariant, the strong coupling eigenstates may be classified as follows:

1) <u>the strong coupled vacuum:</u> In this case all links are unoccupied ($C_\ell^2 = 0$) and there are no fermions: the total energy is zero.

2) <u>the pure glue sector:</u> There are still no quarks, but links are excited in such a way that gauge invariant states are produced. The simplest such pure glue states ("glueloops") have a closed path of links in the 3 (or $\bar{3}$) representation. These have energy $(2g^2L)/(3a^2)$ where L is the length of the path: the simplest such state just has the links around the perimeter of an elementary plaquette excited: $Tr[U_{\ell_4}U_{\ell_3}U_{\ell_2}U_{\ell_1}]|0>$, where $|0>$ is the vacuum. Of course, more complicated configurations are allowed, including those with non-triplet flux and those with more complicated topologies. For example, three flux links can emerge from a single lattice site since a gauge invariant combination can be formed with the ϵ_{ijk} invariant tensor. See Figure 4.

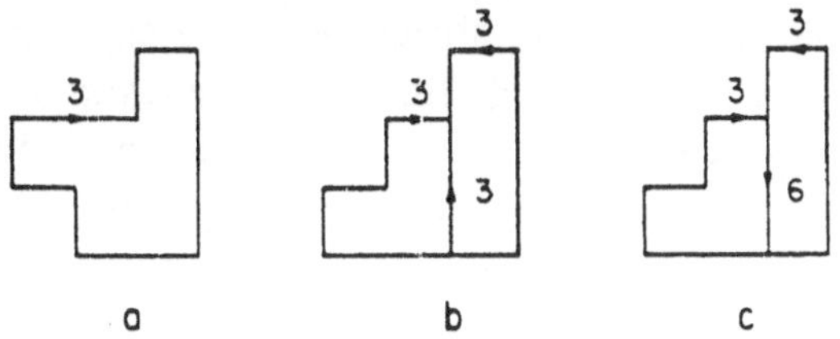

Figure 4. some primitive pure glue states.

3) <u>the meson sector:</u> The simplest quark-containing state consists of a quark and antiquark on the lattice jointed by a path of flux lines (for gauge invariance). These will have energy $m_q + m_{\bar{q}} + (2g^2L)/(3a^2)$ so that we automatically have quark confinement in strong coupling. See Figure 5.

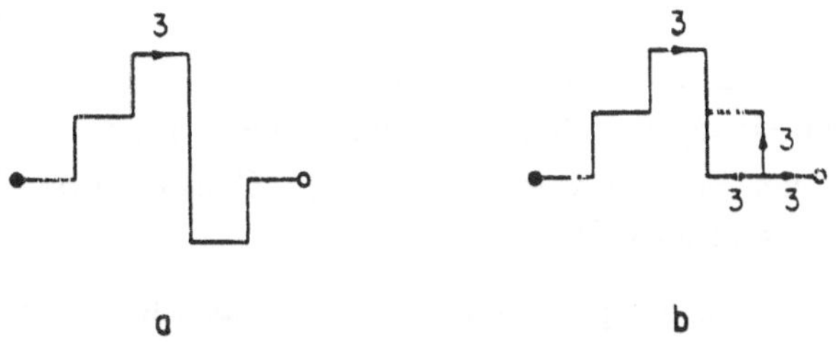

Figure 5. some primitive meson states.

4) <u>the baryon sector:</u> The next simplest quark-containing state consists of three quarks connected by an ϵ_{ijk}- type flux junction. Such quarks will also be confined. See Figure 6.

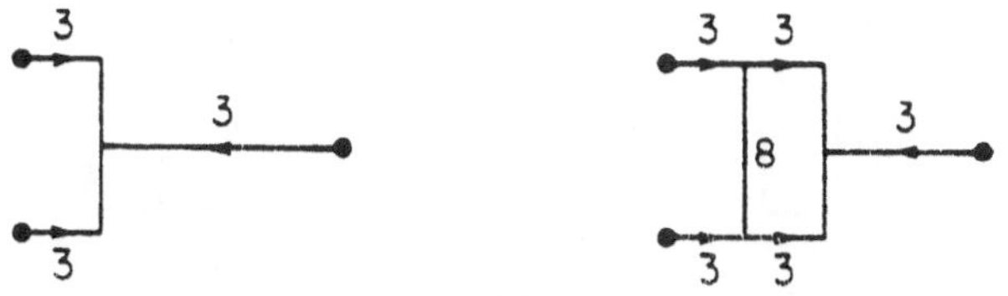

Figure 6. some primitive baryon states.

5) <u>multi-quark sectors:</u> When there are more quarks than those required for a meson or baryon, then in general the system will not be completely confined. The simplest such system consists of two quarks and two antiquarks. See Figure 7.

With these examples, the general structure of the eigenstates of the strong coupling limit is clear: it consists of "frozen" gauge invariant configurations of quarks and flux lines. Of course, these are not the eigenstates of QCD, but they do form a complete basis (in the limit a→ 0) for the expansion of the true strong interaction eigenstates.

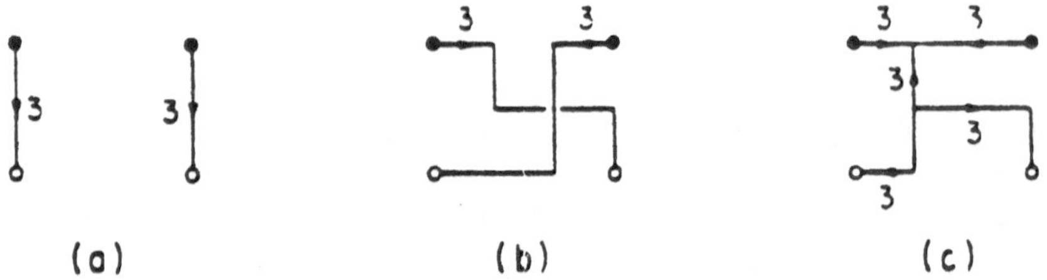

Figure 7. some primitive $qq\bar{q}\bar{q}$ states

The full eigenstates of QCD can be found (in principle!) by considering corrections to the strong coupling limit from the terms we have neglected so far. These terms can induce a variety of efffects. Consider first of all the $q^{\dagger}U\alpha q$ term. It can, among other things,

1) annihilate a quark at one point and recreate it at a neighbouring point with an appropriate flux link (Figure 8a);

2) break a 3-flux line and create a pair (Figure 8b).

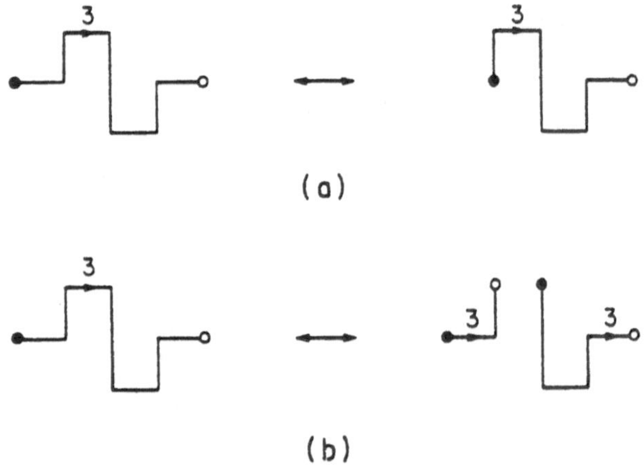

Figure 8. some effects of $q^{\dagger}U\alpha q$: (a) quark hopping, (b) flux breaking pair creation.

This term thus plays a rôle analogous to both the usual quark kinetic energy term and the quark-gluon coupling term of the weak coupled theory. Next consider the $Tr[2-(U_{\ell 4}U_{\ell 3}U_{\ell 2}U_{\ell 1}+h.c.)]$ term. It can, among other things,

1) allow flux to hop across plaquettes (Figure 9a);
2) change flux topology (Figure 9b).

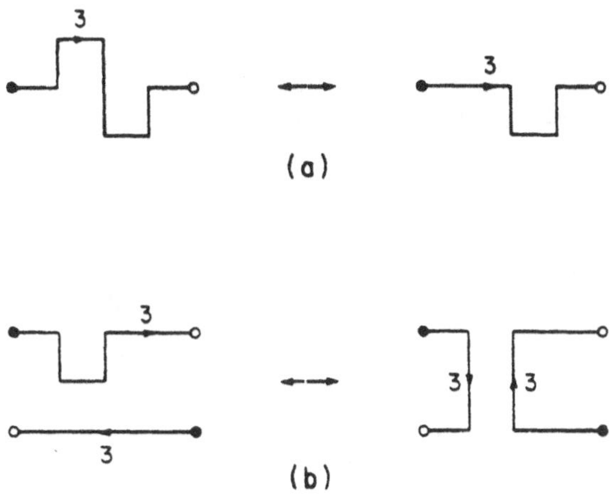

Figure 9. some effects of $Tr[2-(UUUU+h.c.)]$: (a) flux tube hopping, (b) flux tube topological mixing by rearrangement.

The full diagonalization of this Hamiltonian problem as a→0 would constitute an exact solution of QCD. Unfortunately, this diagonalization represents a numerical problem which is of a magnitude well beyond presently available computing capacity.

Extracting the Flux Tube Model. To illustrate the flux tube model in the simplest possible context, and to make direct contact with the previous discussion, we consider the sector of QCD which contains $Q_1\bar{Q}_2$. The complete set of gluonic base states formed by the eigenstates of the strong coupling limit for this system can be classified by their topology,* and as a first step in diagonalizing this problem we imagine organizing the Hamiltonian into blocks of fixed topology. Within each block the Hamiltonian then consists only of kinetic energy terms corresponding to Figures 8(a) and 9(a). Other perturbations either create $q\bar{q}$ pairs or new flux topologies and so are off-block-diagonal. Within each topological block QCD has thus been reduced to a (discrete) quantum string problem (in the generalized sense: these strings may have internal loops, various local string tensions depending on the colour representation excited, extra $q\bar{q}$ pairs, etc.). Since, in strong coupling, topological mixing is suppressed,

* By "topology" we mean the obvious classification into classes of flux excitations in which all the members of a given class could (in the absence of the lattice) be continuously transformed into one another.

the flux tube model assumes that the long-wavelength properties of QCD can be approximated by treating mixing between topological blocks as a perturbation. Thus in zeroth order the flux tube model treats QCD as a theory of non-interacting, discrete, multitopological strings. The lowest-lying eigenstates in this approximation will therefore be those associated with Q and $\bar{Q}$ connected by a single string of 3-flux, resembling the mesons of the old string model.

The flux tube model thus suggests a simple model for the adiabatic surfaces of Section 2: the glue between Q and $\bar{Q}$ behaves like a discrete quantum string. The lowest adiabatic surface then corresponds to the ground state of this string, while the low-lying excitations correspond (at least for large r) to the excitation of "phonons" in the string. (At higher mass "topologically excited" strings will create other families of adiabatic surfaces).

This extraordinarily simple picture of the low-lying adiabatic surfaces of QCD has recently received some direct support from Monte Carlo studies of QCD on a space-time lattice. First of all, there is now rather strong evidence for the existence of narrow flux tubes[11]. Measurements on the lattice of the chromoelectric field strengths in the neighbourhood of static $Q\bar{Q}$ colour charges show that this field is strongly aligned and collimated in a way consistent with a thin (discretized) string executing quantum zero-point fluctuations about the $Q\bar{Q}$ axis (see Fig. 10). There is also evidence for a chromomagnetic field

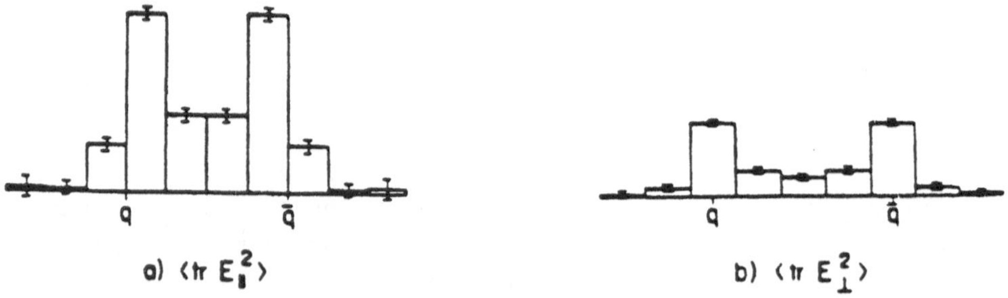

Figure 10. evidence for flux tubes from Euclidean lattice Monte Carlo studies. (Ref. 11).

(expected by the QCD analogue of Maxwell's law of magnetic field induction) circulating around the $Q\bar{Q}$ axis. At least as important are recent <u>dynamical</u> results[12)] on the excited adiabatic surfaces. These lattice studies find a first excited surface which not only has phonon quantum numbers, but also an energy gap over the ground state surface which is consistent with the $\frac{\pi}{r}$ expected at large r (see Fig. 11).

With this evidence in favour of the dynamical picture of the flux tube model for the low-lying adiabatic surfaces, it makes sense to push the model further by asking whether it would really lead to a "quark model limit". Merlin and Paton (Ref. 13) have thus studied the corrections to the adiabatic limit for quark masses ranging from $m_Q >> b^{\frac{1}{2}}$ to $m_q \sim b^{\frac{1}{2}}$. Such a calculation shows the behaviour one would hope to see: for the low-lying spectrum of the lowest adiabatic surface, non-adiabatic corrections are essentially negligible for

$m_Q > 1.5$ GeV, and they remain perturbations even when m_q is as small as 0.3 GeV. (The main correction is actually a "trivial" one: the glue contributes to the moment of inertia of the system. True mixing between adiabatic surfaces is a very small effect.)

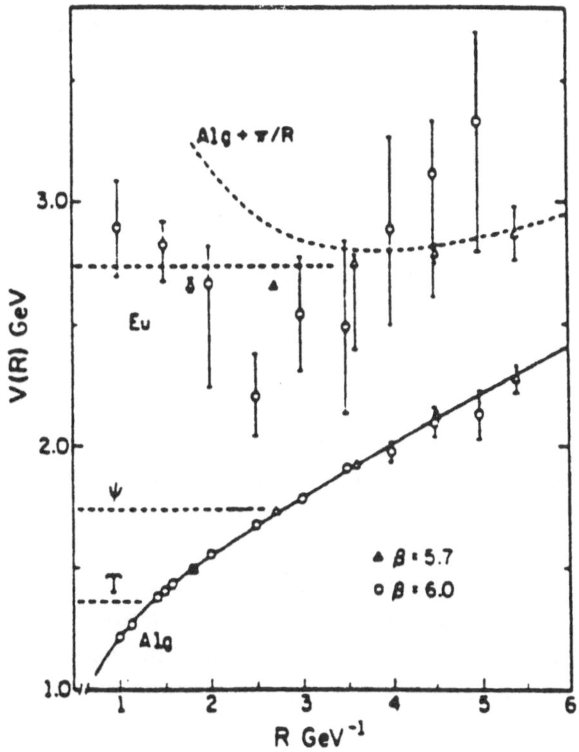

Figure 11. evidence for a phonon-like first excited adiabatic surface from Euclidean lattice Monte Carlo studies (Ref. 12).

Some Loose Ends. There are three potentially serious omissions in the "derivation" of the quark model presented so far: 1) the use of non-relativistic dynamics, 2) the neglect of dynamical $q\bar{q}$ pairs, and 3) the origin of the constituent quark mass.

I believe the use of non-relativistic dynamics (within the $q\bar{q}$ sector of Fock space) is a deficiency of no qualitative importance. If the constituent quark masses are $\gtrsim b^{\frac{1}{2}} \sim 300$ MeV, then a Schrodinger equation with suitable adjusted (i.e., non-fundamental) parameters can describe the spectrum and structure of the hadrons. Many workers have now shown that the non-relativistic quark model not only survives such a "relativization", but is even improved by it[8,14,15].

The argument that the coupling to dynamical $q\bar{q}$ pairs (which mixes sectors of Fock space) will not destroy the quark model is more phenomenological. The idea that such light pairs are created dominantly via flux tube breaking (see Figure 8) is not only attractive, but also remarkably successful in explaining the data on meson decay[16]. Assuming that the small pair creation constant of this model (which is an amplitude for the string to break to produce a constituent

$q\bar{q}$ pair) is realistic, the coupling between constituent quark Fock space sectors will also be weak and should be amenable to a perturbative treatment. This is of course the old argument of the dual string model that if Γ_{res}/m_{res} is small it can be used as a perturbative parameter of a narrow resonance approximation. The phenomenological support for this idea in the old days was just that if Γ/m were not small the linearity of the Regge trajectories would be spoiled[4].

The third "loose end" is, however, a real obstacle at the moment to the "derivation" of the quark model. What is required is to show that there exists a "quark model scale" a_{qm} at which QCD can be cut off with the properties that 1) $a_{qm} << r_{hadron} \sim 1fm$, 2) $r_{quark} << a_{qm}$, 3) $m(a_{qm}) = m_{constituent}$ and 4) the low-lying spectrum and other observables scale for $a \geq a_{qm}$. Naively, of course, $m_{constituent} \sim r_{quark}^{-1} \sim \Lambda_{QCD} \sim 1fm^{-1}$, but the conditions on a_{qm} are not impossible, and the lattice (for example) can tell us whether they are in fact realized in nature. My opinion is that they will emerge and that the physics behind them is connected with the failure of the naive argument that $m_{glueball}/m_{\rho} \sim 1$.

Is the Quark Model Derived? Despite its enormous phenomenological importance, the quark model thus cannot yet be derived from QCD. However, the flux tube model (a.k.a. "the string model") suggests a way of overcoming several difficulties in the path of such a derivation via the adiabatic approximation. This approach has some support from lattice Monte Carlo studies. Moreover, as we will see in the next Section, it makes some readily testable predictions. In any event, I would argue that the quark potential model is now on a sufficiently sound footing that we can drop the "naive" from its name!

STATUS OF CONVENTIONAL SPECTROSCOPY

Twenty five years after the birth of the quark model, light quark baryon and meson spectroscopy is in a deplorable state. For example, aside from the tensor mesons, of the remaining twelve states of the L=1 meson nonets, only four or five are reasonably well understood: the $J^{PC_n} = 0^{++}$ states are in a state of confusion which is probably related to the existence of $qq\bar{q}\bar{q}$ states (see below); the $1^{++} f_1(1420)$ does not seem to fit into its nonet, while the mass of the a_1 (the cornerstone of this nonet) has been called into doubt by recent measurements on $\tau \to v_\tau \pi\pi\pi$; and in 1^{+-} the h_1 has only recently been claimed while the $s\bar{s}$ state remains completely undiscovered. Matters are, as one would expect, even worse for the higher mass nonets, although there has been some recent progress.[17]

I am not of the opinion that we need to understand everything in low energy spectroscopy perfectly to be satisfied, but we are at present far from knowing enough to even verify our understanding of the general picture. Let me therefore provide one example of how we might go about hunting down some of these many missing states: the ω-like $^1D_2$ state with $J^{PC} = 2^{-+}$ (an "η_2"). As with any big game hunting, there are three essentials: 1) a map of the region, 2) a description of the beast, and 3) an appropriate trap. As a good general guide for this type of activity, I recommend Ref. (14), which provides spectroscopic predictions for this mass range based on a relativized version of the constituent quark model. In this case such a sophisticated map is not really required, however, since this η_2 is expected to be found within a few tens of MeV (the scale of OZI violation) from its isovector partner, the known π_2 (1680). The main characteristics of the beast which are needed to set a proper trap are, aside from its J^{PC}, its total width and prominent decay modes. The predictions of Ref. (16) for these properties are probably not far from the mark; they indicate a total width of about 400 MeV with $a_2\pi$ (70%) and $\rho\pi$ (10%) as the most prominent final states. As for the trap itself: this is not my department. I will nevertheless make one comment. These beasts are generally found hiding in a very dense jungle (i.e., background) populated by many other creatures (i.e., known states). It is very unlikely, therefore, that a simple trap will work. For example, it will almost certainly be necessary to have a J^{PC} filter on the trap (see Figure 12).

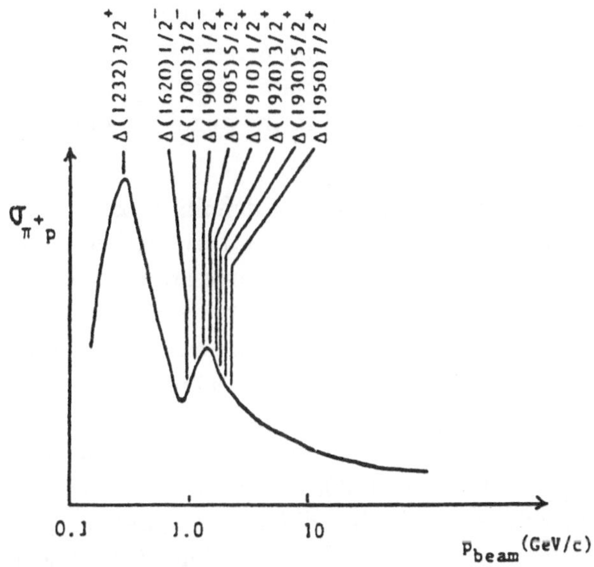

Figure 12. one of the lessons of baryon spectroscopy

HYBRIDS AND GLUEBALLS

Hybrids, which were originally discussed as $q\bar{q}g$ and $qqqg$ states in the bag model[10)], are still usually considered in that framework. The flux tube model provides an alternative nonperturbative picture of such states as, for example, $q\bar{q}$ states moving in the adiabatic potential of an excited flux tube. Since the bag model hybrid spectroscopy has been extensively discussed,[18)] and since my aim is to provide a theoretical *overview* as opposed to a *review*, I will only discuss the flux tube model predictions here. (Figure 11 makes this a possibly phenomenologically relevant exercise as well!)

The lowest order hybrid spectrum predictions of the flux tube model are very simple in character. Since the phonon has ± 1 units of angular momentum about the $q\bar{q}$ (body fixed) axis, good orbital angular momentum states must have total orbital angular momentum ≥ 1 and so have an effective angular momentum barrier in the $q\bar{q}$ relative coordinate. Such states therefore are "doubly excited" relative to the ordinary quarkonia: they have an excited string ($\omega_{gap} \sim 500$ MeV) and they have quarks which are required to have an r=0 node like an ordinary P-wave meson ($\omega_{orbit} \sim 500$ MeV). They are therefore predicted to lie far above the ρ meson at about 1.9 GeV. (Careful consideration of adiabatic corrections and spin-orbit effects produces splitting about this value; see Ref. 19). The ± 1 phonon helicity states combined with the four possible $q\bar{q}$ spin states produce eight nonets of such states with $J^{PC_n} = 2^{+-}_{-+}, 1^{+-}_{-+}, 0^{+-}_{-+}$, and 1^{++}_{--} (C_n is the conjugation of a neutral member of the nonet). Among these are three J^{PC_n} exotic nonets ($2^{+-}, 1^{-+}$, and 0^{+-}) whose existence would be unambiguous evidence for spectroscopy beyond the normal quark model. The crucial missing ingredient required in designing a search for these states is their hadronic couplings. These were supplied in Ref. 9 and are reproduced here as Table 1. One immediately sees a striking characteristic of these states (common to all the states of all eight nonets): they decouple from ordinary S-wave mesons. There is a simple reason for this[9]: kinematics forbid the phonon helecity from turning into relative angular momentum between the decay products, since the products' relative coordinate vector is parallel to the original $q\bar{q}$ axis. It must therefore be internalized in the angular momentum of one of the decay products.

There are other reasons why even the definitive exotic J^{PC} signals might have escaped detection so far. One is just their rather large masses. Another is that, of the nine candidate states, three are probably too broad to be seen with any clarity. When we turn to the six J^{PC} exotic hybrids which may be narrow enough to stand out as resonances [$y_2^{+-}(1900)$, $z_2^{+-}(2100)$, $x_1^{-+}(1900)$, $y_1^{-+}(1900)$, $z_1^{-+}(2100)$, and $y_0^{+-}(1900)$], we encounter further reasons why they may have escaped detection so far. The $y_1^{-+}(1900)$ decays mainly to $[a_1(1275)\pi]_S$ and $[\pi(1300)\pi]_P$: considering the notorious difficulty of seeing the a_1 and the large width of the $\pi(1300)$, these channels would probably not be conducive to finding the y_1^{-+}. Similar difficulties would seem likely to obscure the $z_1^{-+}(2100)$. The remaining four states, while still presenting formidible challenges, should be easier to see: $y_2^{+-}(1900)$ and $y_0^{+-}(1900)$ both decay dominantly to $[b_1(1235)\pi]_P$, $z_2^{+-}(2100)$ will decay much of the time to $[K^*(1420)\bar{K} + \text{c.c.}]_P$, and the x_1^{-+} (1900) will be found most of the time in $[b_1(1235)\pi]_S$.

The flux tube model then provides in a straightforward way an explanation of why meson hybrids have not yet been discovered and suggests definite ways in which to find them. Baryon hybrids in the model are predicted to be very massive ($m \gtrsim 2.2 GeV$); since they are undistinguished by any exotic quantum numbers, they will be exceptionally difficult to unravel from the very dense spectrum of ordinary baryons at such masses. Thus, at least with respect to hybrids, the flux tube model explains the hegemony of normal quark model states below 2 GeV.

The flux tube model predictions for the masses of hybrids are, as such things go, relatively straightforward: they are "normalized" by conventional $q\bar{q}$ spectroscopy, and involve no particularly risky approximations. These predic-

Table 1. the dominant decays of the low-lying exotic meson hybrids

hybrid state*	J^{PG}	(decay mode)$_L$ of decay	partial width (MeV)
$x_2^{+-}(1900)$	2^{++}	$(\pi a_2)_P$	450
		$(\pi a_1)_P$	100
		$(\pi h_1)_P$	150
$y_2^{+-}(1900)$	2^{+-}	$(\pi b_1)_P$	500
$z_2^{+-}(2100)$	2^{+-}	$(K\bar{K}^*(1420) + c.c.)_P$	250
		$(\bar{K}K_1(1400) + c.c.)_P$	200
$x_1^{-+}(1900)$	1^{--}	$(\pi b_1)_{S,D}$	100,30
		$(\pi f_1)_{S,D}$	30,20
$y_1^{-+}(1900)$	1^{-+}	$(\pi a_1)_{S,D}$	100,70
		$(\pi\pi(1300))_P$	100
		$(\bar{K}K_1(1400) + c.c.)_S$	$\sim$100
$z_1^{-+}(2100)$		$(\bar{K}K_1(1280) + c.c.)_D$	80
		$(\bar{K}K_1(1400) + c.c.)_S$	250
		$(\bar{K}K(1400) + c.c.)_P$	30
$x_0^{+-}(1900)$	0^{++}	$(\pi a_1)_P$	800
		$(\pi h_1)_P$	100
		$(\pi\pi(1300))_S$	900
$y_0^{+-}(1900)$	0^{+-}	$(\pi b_1)_P$	250
$z_0^{+-}(2100)$	0^{+-}	$(\bar{K}K_1(1280) + c.c.)_P$	800
		$(\bar{K}K_1(1400) + c.c.)_P$	50
		$(\bar{K}K(1400) + c.c.)_S$	800

*x,y, and z denote the flavour states $\frac{1}{\sqrt{2}}(u\bar{u} - d\bar{d})$, $\frac{1}{\sqrt{2}}(u\bar{u} + d\bar{d})$, and $s\bar{s}$. The subscript on a state is J, the superscripts are P and C_n.

tions of the masses of glueloops (also known as glueballs) is trickier: there is no known mass in this sector to set the "zero" and the approximations used to predict[5)] the spectrum are shakier. With these caveats (the first leads to an estimated ± 0.2 GeV uncertainty in the overall mass scale) the model predicts the lowest glueball to be a 0^{++} state at 1.5 GeV with all other states above 2 GeV and the lowest J^{PC} exotic around 2.5 GeV. These predictions used to be in conflict with results from lattice Monte Carlo calculations, but recent results[20,21)] are not so far away from them.

It is clearly too early to be sure, but it now seems possible that — apart from a 0^{++} glueball which will be very hard to disentangle from the very poorly understood ordinary scalar mesons — the "gluonic action" is all in the 2 GeV range. If so, it will be very difficult to see (against what will by that mass be a very dense spectrum of ordinary mesons) unless it has J^{PC} exotic quantum numbers. This suggests concentrating our efforts on finding the hybrids. Such a search will require high statistics to perform the isobar analyses required to prove that any bumps found have exotic J^{PC} (see Figure 12 again).

MULTIQUARK HADRONS AND QUARK EXCHANGE FORCES

I would like to begin the discussion of multiquark hadrons and quark exchange forces by returning to the string model history mentioned in the Introduction. This time my purpose is to show you that the idea of quark exchange forces is also a very old one with a fine pedigree. Figure 13 shows the quark line (duality) diagrams appropriate to meson-meson scattering in the old string theory. Figure 13(a) shows the classic duality diagram for meson-meson scattering. The diagram should be "read" by imagining a string stretched between the quark-antiquark pairs which sweeps out a membrane in space-time; also, as with a Feynman diagram, all possible time orderings of space-time events are implied. If one cuts this diagram vertically after the initial vertex one can expose a $q\bar{q}$ state in the t-channel; it can have any allowed quantum numbers so that this one diagram corresponds to the exchange of a whole tower of mesons with precise relationships between their masses and coupling constants. On the other hand, a horizontal cut can reveal a tower of $q\bar{q}$ meson resonances being formed in the s-channel. The physics is dual: it could be described in terms of *either* t-channel exchanges or s-channel resonances. (Note that it would be *incorrect* to sum both types of processes in this theory: a phenomenological model treating the mesons as the low-energy degrees of freedom coupled to each other in an effective Langrangian would give the *wrong answer*. Thus even if mesons were the only important low-energy degrees of freedom, it would be far from clear how to go about constructing an approximation to QCD based upon them.)

Now consider Figure 13(b), the analog to which is the diagram relevant to nucleon-nucleon scattering. There are certainly time-ordered parts of this diagram, where one of the exchanged quark lines zigs backward in time before zagging forward again, which correspond to meson exchange. However, there are also time orderings where there is never an additional $q\bar{q}$ pair (this is the case for the diagram as drawn): these are quark exchange and not meson exchange diagrams. If you imagine what is happening to the string in these diagrams, you will see that the strings (not the quarks) have touched, "broken", and rejoined

Figure 13. (a) the classic duality diagram for meson-meson scattering
(b) a duality diagram with a twist containing quark exchange

with a piece of the other string. A concrete picture for such processes will be discussed in the next Section; for now we simply note that the string picture expects them to exist.

I can perhaps make quark exchange seem more plausible by mentioning a system where such exchange forces are clearly dominant over meson exchange: the interaction between two hydrogen atoms has contributions from both electron exchange and positronium "meson" exchange. The dominance of electron exchange in atomic physics is due to both the suppression of positronium exchange by powers of α and its extremely short range. In hadronic physics neither factor particularly favours one type of force over the other (except for the pion's range) and we may expect them to be of comparable importance.

A. MULTIQUARK SYSTEMS IN QCD

It would be a mistake to assume that our successful phenomenological models for the $q\bar{q}$ (meson) and qqq (baryon) systems can be extended in a straightforward way to multiquark systems (i.e., ones with $n_q + n_{\bar{q}} > 3$). Although it is not clear that we understand why the quark model works as a low energy approximation to QCD, at least we can state a necessary condition for its existence: gluonic degrees of freedom must be "frozen out" at low energy, as discussed in a previous Section. However, this isolation of the lowest adiabatic surface which makes the quark model work for $q\bar{q}$ and qqq will fail for multiquark systems: they characteristically have many low-lying surfaces which cross each other[5)].

Let's use the $qq\bar{q}\bar{q}$ system as a "multiquark primer". It is difficult to visualize the full adiabatic surfaces associated with such a system, so in Figs. 14 (a) and 14(b) I show a slice through these surfaces associated with two simple rectangular configurations of the quarks for three low-lying flux tube configurations. It is immediately clear from these diagrams that the "quark model limit" does not apply: the state of these systems cannot be determined by the positions (and spins) of the quarks alone since there are places in this configuration space where two gluonic states are degenerate. The full wavefunction in such a situation must therefore include new discrete variables corresponding to the state of the glue. Of course these level crossings only occur in the extreme string limit; in reality the adiabatic surfaces will mix and repel as in molecular physics.

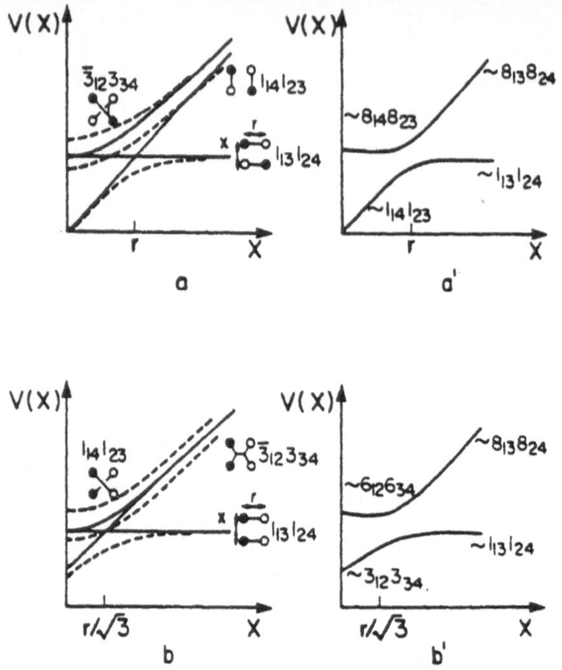

Figure 14.
the adiabatic potentials of the flux tube model and of the $\vec{F}_i \cdot \vec{F}_j$ potential model for two $qq\bar{q}\bar{q}$ geometries: (a) and (b) show the flux tube potentials before (solid curves) and after (shown schematically by dashed curves) topological mixing, while (a′) and (b′) show the related $\vec{F}_i \cdot \vec{F}_j$ potentials.

This mixing has its origin in the flux tube model in vertices of the QCD string picture which create transitions between different topologies: see Fig. 15 and compare to Fig. 9. Such an interaction should, on the long-wavelength scale appropriate to the flux tube model, be a contact interaction between the strings which then can act only when the string wavefunctions of the two asymptotic mesons overlap. (As we will discuss further below, this means that interhadron forces from quark exchange have no long range Van der Waals tail).

As a rough approximation we may say that topological mixing corresponds to the exchange of quarks between two interacting mesons, or that the "gluonic index" of our earlier discussion is really just a channel index corresponding to the two asymptotic states $(q_1\bar{q}_3)_1$ $(q_2\bar{q}_4)_1$ and $(q_1\bar{q}_4)_1(q_2\bar{q}_3)_1$, where $(q_i\bar{q}_j)_1$ denotes a colour singlet made of quark i and antiquark j. It should be noted from Fig. 14, however, that these two mesonic configurations cannot, alone, describe the low-lying adiabatic surfaces of $qq\bar{q}\bar{q}$. For example, in some regions of configuration space the "baryonium-like" topology with two baryon flux tube junctions is lowest-lying.

One of the lessons of this example is that the study of multiquark systems requires that we model gluon dynamics. (Ironically, this means that the mesons and baryons of traditional particle physics are less sensitive to gluon dynamics than the nuclei of traditional nuclear physics!) At short distances it seems clear that we can model gluon dynamics by simple one-gluon-exchange, and many of

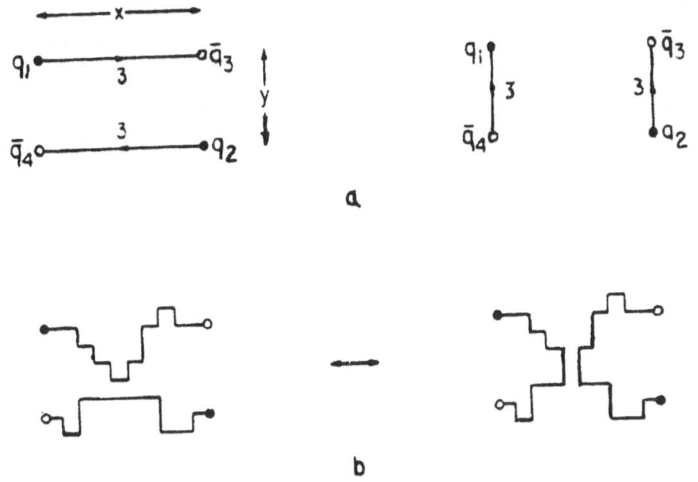

Figure 15.
(a) two low-lying $qq\bar{q}\bar{q}$ configurations, (b) topological mixing between these two configurations.

the most well-known results in multiquark physics (like the NN repulsive core) depend mainly on this region. At large distances we can be sure of little except that we don't yet know what to do, but let me mention two models:

1. the $\vec{F}_i \cdot \vec{F}_j$ model: One natural model for long range dynamics is to assume that the one-gluon-exchange model survives to all distances with only a change in radial dependence. This is certainly sensible given the way we visualize confinement as arising in the $q\bar{q}$ system from a modification of the short-distance dipole colour-electric field lines to a long-distance flux tube, and leads to the model

$$V = \sum_{i<j} V_{ij} \tag{9}$$

$$V_{ij} = -[\frac{\alpha_s}{r_{ij}} + c + \frac{3}{4} b r_{ij}] \vec{F}_i \cdot \vec{F}_j \tag{10}$$

where b is the usual mesonic string tension and $\vec{F}_i$ is the SU(3) colour matrix of quark i. This model has a venerable history. A variant was first suggested by Nambu[22] more than twenty years ago, and it has since been rediscovered by many others[23]. Because it reduces the gluonic degrees of freedom to a quark colour index degree of freedom, it is relatively easy to treat quantum mechanically and has been one of the most widely applied models. Since it reduces the infinite number of gluonic degrees of freedom to a finite number (in $qq\bar{q}\bar{q}$ there are only two distinct ways of coupling the four colour indices to a colour singlet so that this finite number is two), it is clearly not a complete model, but Figs. 14(a′), 14(b′) show that within this limitation it does a rather good job of reproducing the lowest-lying adiabatic surfaces of the flux tube model. The model has also been criticized for the fact that it predicts long range van der Waals forces between hadrons[24]. These arise because it produces mixings and repulsions between the pure string-like adiabatic surfaces of Figs. 14(a), 14(b) which are necessarily power law behaved while, as argued above, the true mixings are suppressed by the products of the overlaps of string wavefuntions. Since the predicted long range forces are in fact extremely weak[25)], while the mixings in the core of the wavefunctions (as opposed to the tail) presumably have strengths of the correct general magnitude, I don't consider this flaw to be a particularly deadly one.

2. the flux tube model: This model[5)] should give a more realistic picture of the long range forces in QCD, but at the moment the extreme string picture must be supplemented with a guess for the topological mixing amplitudes. For example:

a) In the strong coupling limit of Hamiltonian lattice QCD, topological mixing vanishes[5)]. If this limit were a good guide to the physics, long range quark exchange forces would vanish[26)]. This extreme possibility seems unlikely to me. In a relativistic string model, Fig. 13(b) is not suppressed by any small coupling (or $1/N_c$ factors) relative to the meson exchange diagram of Fig. 13(a) (to which it is related by crossing) so it should be of comparable strength.

b) Another extreme possibility is embodied in the string flip-flop model[27)]. In this model the scattering system always moves along the lowest adiabatic surface; it assumes, in some sense, maximal mixing. This scenario also seems unlikely: we have argued above that there will certainly be configurations where higher adiabatic surfaces are separated from the lowest surface by frequency gaps $\omega_{exc} << \omega_{quark}$, so that the extreme adiabatic approximation should not work.

Though it may be that neither of these extremes is realistic, they are both interesting to study since in some ways they should demonstrate the limits of possible quark exchange physics. If example a) were correct, non-meson-exchange interaction forces would be limited to short distances ($r \lesssim 1 fm$); in example b) there would be important interhadronic forces generated out to medium ranges ($r \lesssim 2 fm$) where, for example, nucleon-nucleon attraction takes place.

It will not be an especially easy task to decide what model of quark exchange forces is correct. Progress is being made in the theory: recent Monte Carlo simulations of lattice QCD seem to favour the existence of long range quark exchange[28)]. Nevertheless, this is an issue which will, in my opinion, best be resolved by experiment. In the next two Sections I will discuss the kinds of phenomena that we need to study to accomplish this resolution.

B. QUARK EXCHANGE IN NN: SOME CONFUSING REMARKS

Figure 16 shows a cartoon of two nucleons "exchanging" a meson. The importance of the cartoon is in its ability to remind us that the exchange of effective degrees of freedom at distance scales below which they can be treated as pointlike is a dubious procedure. It is not just that the particles will have form factors. There is also the fact that when two nucleons are separated by a nucleon diameter, they completely modify the structure of the vacuum between them so that it is very unlikely that anything resembling the asymptotic spectrum of states exists in this region.

It seems to me, therefore, to be completely appropriate to be dubious about the reality of meson exchange between hadrons unless the condition

$$r_{exchange} \sim m_{exchange}^{-1} >> r_{hadron} \tag{11}$$

is met. However, this condition is almost impossible in QCD: the QCD scale Λ_{QCD} determines all masses and radii, so they are all comparable. There is

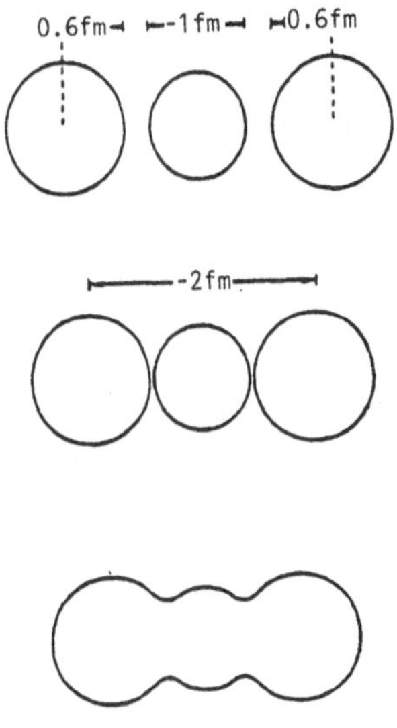

Figure 16.

A cartoon illustrating that for $r \lesssim 2fm$ meson exchange is unlikely to be appropriate to the description of the internucleon potential.

only one case in which this rule is evaded: chiral symmetry tells us that $m_\pi = 0 \cdot \Lambda_{QCD} = 0$ so that $m_\pi^{-1} >> \Lambda_{QCD}$ in the chiral limit. (That Λ_{QCD} is the only scale of QCD is only true in the chiral limit $m_q = 0$. The existence of small quark masses $m_u, m_d << \Lambda_{QCD}$ gives the pion a mass $m_\pi^2 \sim \frac{1}{2}(m_u + m_d)\Lambda_{QCD}$ but hardly changes other masses and radii.) It would therefore seem reasonable to adopt the point of view that meson exchange in the region $r > 2r_N$ (where most if not all of the tests of pion exchange take place, and where there are just very weak residual tails from the exchange of higher mass states) is legitimate, but that for $r < 2r_N$ it may be that other mechanisms (in particular quark exchange) are at work.

Indeed, the quark exchange calculations have had considerable successes in this region[29−31]. The most important and gold-plated is the ability of the quark exchange model to derive a strong NN repulsive core. Since these derivations rely almost exclusively on the one-gluon-exchange induced Breit-Fermi interactions, they do not have the uncertainties discussed above associated with long range quark exchange amplitudes. The basic physics is, moreover, rather simple: a combination of the Pauli principle, the Fermi spin-spin interaction, and the Heisenberg uncertainty principle (the "Pauli-Fermi-Heisenberg effect"[32]). Less certain, but potentially of equal importance, are calculations involving the long-range forces which claim to find significant nuclear attraction at intermediate distances based on both the $\vec{F}_i \cdot \vec{F}_j$ model and the string flip-flop model.

Unfortunately, there is a "confusion theorem" which makes it very difficult to decide by studying conventional nuclear physics which of these pictures (conventional meson exchange or quark exchange) is correct in the $r < 2r_N$ region. The theorem is based on two facts:

1) quark exchange has meson quantum numbers, produces exchange currents with these quantum numbers, and (except for the pion's quantum numbers) has the same range as meson exchange.

2) the repulsive core (whatever its origin) protects the integrity of the interacting nucleons so that they remain excellent effective degrees of freedom.

In my opinion this theorem means that we must go outside of the NN system to make progress in understanding it, and attempt to understand interhadron forces in a more general setting.

This raises a philosophical and sociological issue and I will depart from pure physics for a moment to discuss it. Both nuclear and particle physics are undergoing fission processes at the moment, and I like to visualize this process as in Figure 17. Neither nuclear structure physicists nor golden-Lagrangian-in-the-sky physicists[33] are too interested in the strong interactions anymore. In the case of the former group this is because they understand that their problem

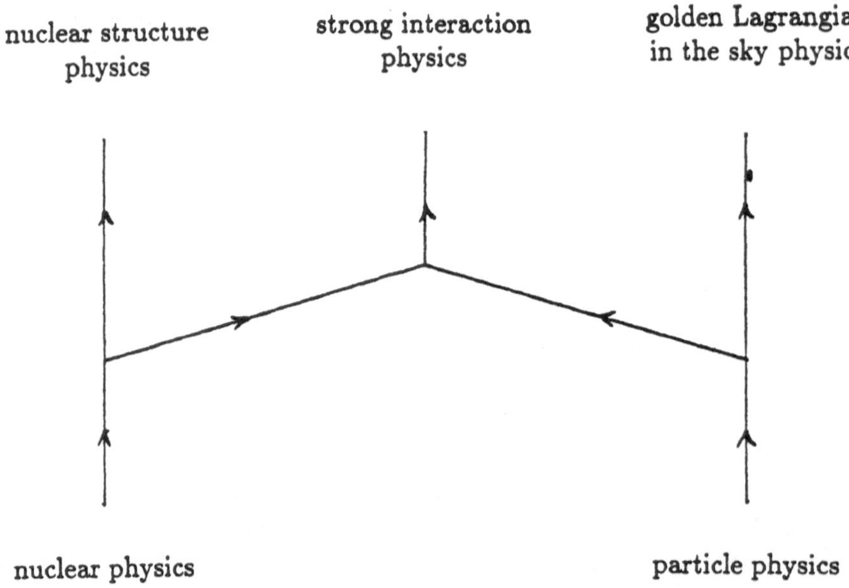

Figure 17. an interdisciplinary interaction.

is mainly a many body problem for which they need only a parameterization of the interparticle forces. In the case of the latter group it is because from their perspective the problem is solved: the QCD Lagrangian is known. The Figure indicates that, inevitably, the two strong interaction fission fragments must coalesce. Nuclear physics was, after all, born in an attempt to understand the strong interaction so there should be no problem with some fraction of the community continuing to pursue that goal in a context wider than the nucleus. It is also natural for particle physicists interested in strong interaction phenomena to expand their scope to include nuclear physics: as I am trying to make clear here, multiquark QCD is a very interesting and challenging part of the subject.

It is in the spirit of this fusion of interests that I will propose in the next Section that one of the most effective ways we can learn about the nature of the NN force and multiquark QCD is by studying meson-meson interactions.

C. K$\bar{K}$ MOLECULES AND ALL THAT

The Confusion Theorem of the preceeding Section is a real barrier to progress only in systems like NN where a repulsive core in the interhadron interaction prevents the hadrons from exploring freely their full multiquark structure. There are many possibilities, therefore, for avoiding it. Most closely akin to NN are the non-repulsive nonstrange dibaryons[34)]. The addition of strange quarks makes it simpler to avoid the Pauli exclusion principle and therefore to find nonrepulsive six quark systems: Jaffe's H dibaryon is the most famous example of this type[35)]. In qqqq$\bar{q}$, there are also some promising systems. However, the simplest such systems are in the qq$\bar{q}\bar{q}$ sector we used as a pedagogical example previously. Such "meson-meson" systems are not only simple; they also come in many readily accessible varieties (e.g., $\pi^+\pi^+$ and $\pi^+\pi^-$ have very different quark exchange characteristics) which will undoubtedly prove useful in unravelling their physics. Moreover, I will suggest below that we already have evidence for the existence of nontrivial multiquark states of this type.

Before presenting a phenomenological status report on the qq$\bar{q}\bar{q}$ system, which many of you might have met previously disguised as stuff called "baryonium", I will give a hysterical introduction to the subject. My reason for doing so is that the qq$\bar{q}\bar{q}$ system has a somewhat unfortunate history which should not be allowed to discourage us from giving it the attention it deserves. Indeed, a confluence of dubious experimental results and dubious theoretical models in the late 1970's and early 1980's created a multiquark fiasco. I am not competent to discuss what went wrong experimentally, but let me review the theoretical side of this fiasco in order to place it in perspective.

The story is basically one of throwing caution to the winds. Modelers from at least four different camps were, it seems to me, guilty:

1) <u>bag models</u>: The bag model posits the existence of a confining volume. Thus, by assumption, any number of quarks and antiquarks (so long as they are in an overall colour singlet), when placed in the bag, will produce a multiquark state that appears to be as legitimately confined as the q$\bar{q}$ and qqq states. On the other hand, it is obvious that the bag model should not be used in this

naive way: the complicated (and nontrivial) dynamics of confinement may be approximated by a bag for states in which this property is guaranteed (like $q\bar{q}$ and qqq), but it was not intended to be a viable approximation to this dynamics in all situations. Nevertheless, some practitioners published detailed spectra of multiquarks confined to a single bag which they interpreted as having essentially the same legitimacy as bag model predictions for, say, meson spectroscopy.

2) naive potential models: The non-relativistic analog of the error made by some bag modelers was to assume that, in a multiquark system, the interquark forces would be identical to those confining forces operative in mesons and baryons. This totally unjustifiable assumption will of course also automatically produce a rich spectrum of multiquark states.

3) colour chemistry: A more sophisticated mistake was made by the adherents of the "colour chemistry" school. They recognized that a given multiquark configuration could be connected by a variety of flux tube, or string, topologies. They went on to note that in some of the these topologies (see, for an example, Figure 14) all the quarks and antiquarks would be connected together in a single unit. Based on a false analogy with mesons and baryons, a rich spectrum of such states was then predicted. While closer to the truth, the fatal flaw of this school was, of course, its neglect of topological mixing.

4) $\vec{F}_i \cdot \vec{F}_j$ potential models: The most peculiar story of this set belongs to those modelers who tried to study multiquark systems in the context of the $\vec{F}_i \cdot \vec{F}_j$ potential model. Unlike all of the other modelers, these were actually using a picture for multiquark confinement which, as explained above, should be approximately valid. Yet they too predicted rich spectra of multiquark bound states! In this case the problem was not with the model, but rather that approximations used to solve the $\vec{F}_j \cdot \vec{F}_j$ model were illegitimate. The difficulty and its resolution are described below.

To summarize: the multiquark fiasco was created by a series of unfortunate circumstances. Theorists, making a variety of errors from a number of different points of view, supported each other (often with very similar spectra) in predicting multiquark states analogous in character to ordinary hadrons. This bad psychology/sociology was made even worse by the (temporary) experimental sighting of states. I suspect that the mirror image of this bad psychology/sociology operated on experimenters working in this field.

The basic problem with the multiquark theories which had predicted baryonium was pointed out in the context of the $\vec{F}_i \cdot \vec{F}_j$ model in Ref. 25. Spectra had been obtained in that model by using as a colour basis the states $|\bar{3}_{12}3_{34}>$ and $|6_{12}\bar{6}_{34}>$ (see Fig. 14; $C_{\alpha\beta}\bar{C}_{\rho\sigma}$ is a colour singlet state in which particles α and β have been coupled to colour C, while particles ρ and σ have been coupled to its conjugate representation), with $\bar{3}3 \leftrightarrow 6\bar{6}$ mixing treated as a perturbation. In Ref. 25 it is shown that this approximation is completely illegitimate. Indeed, since this procedure produces a purely discrete spectrum it must be wrong: not only does nature have a meson-meson continuum in $qq\bar{q}q$, but also one can easily show that the $\vec{F}_i \cdot \vec{F}_j$ model contains solutions corresponding to free mesons in situations where the system is split into two $q\bar{q}$ clusters in that mixture of $|\bar{3}_{12}3_{34}>$ and $|6_{12}\bar{6}_{34}>$ corresponding to having each of these mesonic clusters in colour singlets. We can thus prove that in at least one simple case the mixing terms cannot be treated perturbatively: at least an infinite number of the states of the (erroneously deduced) discrete tower of states must collapse into a continuum of free mesons. Indeed, matters are even worse: there are an infinite number of such continuous spectra, corresponding to the infinite number of possible internal excitations of the separated meson clusters.

From this perspective, one might expect that if non-trivial multiquark states exist in this system (i.e., states that are not essentially two mesons), then they are likely to correspond to either weakly bound states or to meson-meson resonances. In Ref. 25 the $J^{PC_n} = 0^{++}$ sector of the qq$\bar{q}\bar{q}$ system was studied in the $\vec{F}_i \cdot \vec{F}_j$ model to examine this conclusion. It was discovered that in this sector at least, it is correct: in almost all cases the rich discrete spectrum of the perturbative mixing solutions collapsed completely into meson-meson continua. The one exception to this conclusion (among u,d,s states) occurred for the K$\bar{K}$ system where in each of the isospin zero and one sectors the solutions indicated the existence of a single surviving discrete state in the form of a weakly bound K$\bar{K}$ "molecule" roughly analogous to the deuteron. It was suggested[25] that two bound states be identified with the long-known but mystifying states S^* and δ (now also known as the f_o (975) and a_0 (980)).

The qq$\bar{q}\bar{q}$ interpretation of these two states was first suggested by Jaffe in the context of the bag model[36]. The K$\bar{K}$ molecule interpretation is rather different:

1) In the K$\bar{K}$ molecule interpretation, the S* and δ are the only resonances; in the naive bag model, at least, one obtains a very rich spectrum.

2) In the K$\bar{K}$ molecule interpretation, the S* and δ are "molecular": they are weakly bound states of two kaon "atoms" separated by many "atomic" radii; in the bag model the quarks and antiquarks are in spherical orbits about the centre of the bag.

3) In the molecular picture, $m_{S^*} \simeq m_\delta \simeq 2m_K$ is as natural as $m_d \simeq 2M_N$; in the bag model it is accidental.

4) No SU(3) partners of the K$\bar{K}$ molecules are expected (although they have associated with them threshold effects in channels related by SU(3)) in contrast to the bag model interpretation.

5) In the K$\bar{K}$ molecule interpretation the S^* and δ naturally have narrow widths, while in the bag model it is difficult to understand why $\delta \to \eta\pi$ is not a strong "fall-apart" mode.

Let me now turn to a phenomenological status report on anomalous effects seen in low energy meson-meson interactions. First of all, as already mentioned, there is now a considerable body of evidence that the S* and δ are not simple q$\bar{q}$ $^3P_0$ mesons:

a) The equality of m_{S^*} and m_δ suggests an ideally mixed pair of mesons like ω and ρ but $\Gamma(S^* \to \pi\pi)/\Gamma(\delta \to \eta\pi) \simeq 1/4$ instead of four, the ratio one would expect for ideal mixing[37].

b) Both S^* and δ are very strongly coupled to $K\bar{K}$: $\Gamma(S^* \to K\bar{K})/p_K \geq 2\Gamma(S^* \to \pi\pi)/p_\pi$ while SU(3), which if anything would err on the side of making too large a prediction, would predict $\Gamma(S^* \to K\bar{K})/p_K = \frac{1}{3}\Gamma(S^* \to \pi\pi)/p_\pi$ for the ideally mixed partner of the δ[37].

c) Both S^* and δ are <u>much</u> narrower than expected in models of q$\bar{q}$ decay[37]: $\Gamma(S^* \to \pi\pi)/\Gamma(b_1 \to \omega\pi) \simeq 1/6$ versus a prediction of 9/2. I.e., $\Gamma(S^* \to \pi\pi)$ is predicted to be nearly 1 GeV if it is a q$\bar{q}$ state versus its observed partial width of 25 MeV!

d) Ratios of $S^* \to \gamma\gamma$ and $\delta \to \gamma\gamma$ to $f_2 \to \gamma\gamma$ and $a_2 \to \gamma\gamma$ in the $q\bar{q}$ model fail[38].

e) The decays $\psi \to \rho X, \omega X$, and ϕX obey the OZI rule if $X = {}^3P_2$ but fail to obey the rule if $X = S^*$ or δ in a way that suggests they have $s\bar{s}$ content[37].

There is evidence of anomalous behaviour in other low energy meson-meson systems as well. In the $\pi\pi$ system there seems to be some strongly attractive but non-resonant components. Enhancements in the production of low invariant mass $\pi\pi$ pairs have been reported in a variety of experiments, including the classic "ABC effect"[39], $\eta' \to \eta\pi\pi$, $\psi' \to \psi\pi\pi$, $\Upsilon(nS) \to \Upsilon(mS)\pi\pi$, $\gamma\gamma \to \pi^+\pi^-$, and $\psi \to \omega\pi\pi$. Similar-looking enhancements have been seen in some $K\pi$ channels in $\bar{p}p \to K\bar{K}\pi$[40]. What is needed, as a first step, is a systematic study of the low-energy meson-meson interactions. If we model our program on the classic $\pi^- d \to nn\gamma$ reaction used to study the low-energy nn interaction, then we would want to produce $\pi\pi, \pi K, \pi\eta, \pi\eta', \pi\rho, \pi\omega, \pi K^*, \pi\phi, ..., K\bar{K}, K\eta, ..., K\bar{K}^*$, ... as isolated as possible from other hadronic debris. However, we might be content at least initially to study these dimeson systems in whatever environment we find them: at first one would simply want to know which channels were attractive and which repulsive. At the moment, for example, we know nothing about the $\pi\eta$ interaction at threshold; we can worry later about knowing it with precision.

D. COMMENTS ON MULTIQUARK PHYSICS

There are sound theoretical reasons for believing that in many circumstances quark exchange forces dominate hadron-hadron interactions. While the nucleon-nucleon system is the most fundamental arena in which such forces may be in action, it is not a good testing ground for their presence. On the other hand, there is growing evidence that meson-meson systems have low-energy interactions which are dominated by quark exchange. The elucidation of the properties of such "exotic" systems may, somewhat paradoxically, be the most effective way of improving our understanding of the "mundane" nuclear systems of which our world is made.

SUMMARY AND CONCLUSIONS

With QCD as the link, all of hadronic spectroscopy is now an indivisable subject. In this discussion I have tried to present an overview of this subject from the perspective of the flux tube model (which may be viewed as the reincarnation of the old dual string model).

The outstanding issue in the field at the moment is certainly the search for gluonic states, but these searches are hindered by our still sketchy knowledge of the quarkonium spectrum. This plus the growing evidence that the gluonic degrees of freedom are not excited below the 2 GeV mass range suggests looking for J^{PC} exotic hybrids since all such gluonic hadrons will be broad and immersed in a continuum of other broad non-exotic resonances.

Finally I argued that multiquark systems contain new kinds of physics related to gluon dynamics (quark exchange). The $qq\bar{q}\bar{q}$ system may be the best route to understanding this physics and thereby the classic NN interaction problem where strong interaction physics was born.

ACKNOWLEDGEMENTS

This discussion is based almost entirely on my work in collaboration with Jack Paton, Simon Capstick, Steve Godfrey, Rick Kokoski, John Weinstein, and Kim Maltman, whose invaluable contributions I gratefully acknowledge.

REFERENCES

1. R. Dolen, D. Horn, and C. Schmid, Phys. Rev. Lett. **19**, 402 (1967); Phys. Rev. 166, 1768 (1968).
2. G. Veneziano, Nuovo Cimento **57A**, 190 (1968).
3. Y. Nambu, Proc. Int. Conf. on Symmetries and Quark Models, Wayne State University (1969); H.B. Nielsen, Proc. 15th Int. Cong. on High-Energy Physics, Kiev (1970); L. Susskind, Nuovo Cimento **69A**, 457 (1970).
4. For a review see S. Mandelstam, Phys. Rep. **13C**, 261 (1974).
5. N. Isgur and J. Paton, Phys. Lett. **124B**, 247 (1983); Phys. Rev. **D31**, 2910 (1985).
6. M.A. Virasoro, Phys. Rev. **177**, 2309 (1969); J.A. Shapiro, Phys. Lett. **33B**, 351 (1970); J. Schwartz, Phys. Rep. **89**, 223 (1982).
7. See, e.g., Y. Nambu, Phys. Lett **80B**, 372 (1979); A.M. Polyakov, Phys. Lett. **82B**, 247 (199); G.'t Hooft, Nucl. Phys. **B72**, 461 (1974); J.-L. Gervais and A. Neveu, Phys. Lett. **80B**, 255 (1979).
8. S. Capstick, S. Godfrey, N. Isgur, and J. Paton, Phys. Lett. **175B**, 457 (1986).
9. N. Isgur, R. Kokoski, and J. Paton, Phys. Rev. Lett. **54**, 869 (1985).
10. F.E. Barnes, California Institute of Technology Ph.D. thesis, 1977 (unpublished); Z. Phys. **C10**, 275 (1981); D. Horn and J. Mandula, Phys. Rev. **D17**, 898 (1978); P. Hasenfratz, R.R. Horgan, J. Kuti and J.M. Richard, Phys. Lett. **95B**, 299 (1980); F.E. Barnes, F.E. Close and S. Monaghan, Nucl. Phys. **B198**, 380 (1982); M. Chanowitz and S. Sharpe, Nucl. Phys. **B222**, 211 (1983); E. Golowich, E. Haqq and G. Karl, Phys. Rev. **D28**, 160 (1983).
11. See, e.g., J.W. Flowers and S.W. Otto, Phys. Lett **160B**, 128 (1985).
12. See, e.g., C. Michael, in "Lattice Gauge Theory: A Challenge in Large-Scale Couputing" (New York, Plenum, 1986), p. 227; A. Huntley and C. Michael, Nucl. Phys. **B286**, 211 (1987).
13. J. Merlin and J. Paton, J. Phys. **G11**, 439 (1985).
14. S. Godfrey and N. Isgur, Phys. Rev. **D32**, 189 (1985).
15. S. Capstick and N. Isgur, Phys. Rev. **D34**, 2809 (1986); D.P. Stanley and D. Robson, Phys. Rev. **D21**, 3180 (1980); J. Carlson, J. Kogut, and V.R. Pandaripande, Phys. Rev. **D27**, 233 (1983); **D28**, 2807 (1983); R.K. Bhaduri, L.E. Cohler, and Y. Nogami, Phys. Rev. Lett. **44**, 1369 (1980); and many others.
16. R. Kokoski and N. Isgur, Phys. Rev. **D35**, 907 (1987).

17. A. Donnachie and H. Mirzaie, Z. Phys. **C33**, 407 (1987); A. Donnachie and A.B. Clegg, Z. Phys. **C34**, 257 (1987).
18. See, e.g., F.E. Close at this conferece and references therein.
19. J. Merlin, D. Phil. thesis (University of Oxford, 1986).
20. C. Michael and M. Teper, Univ. of Oxford preprint (1988).
21. S. Sharpe in these proceedings.
22. Y. Nambu in Preludes in Theoretical Phsyics, ed. A. de Shalit, H. Feshbach, and L. van Hove, (North Holland, Amsterdam 1966), p. 133.
23. H.J. Lipkin, Phys. Lett. **45B**, 267 (1973) and in Common Problems in Low- and Medium- Energy Nuclear Physics, proceedings of the NATO Advanced Study Institute, Banff, 1978, ed. B. Castel, B. Gouland, and F.C. Khana (Plenum, New York, 1978), p. 173; R.S. Willey, Phys. Rev. **D18**, 270 (1978); M.B. Gavela *et al.*, Phys. Lett. **79B**, 459 (1979); and N. Isgur in The New Aspects of Subnuclear Physics, ed. A. Zichichi (Plenum, New York, 1980), p. 107.
24. O.W. Greenberg and H.J. Lipkin, Nucl. Phys. **A370**, 349 (1981) and refs. therein.
25. J. Weinstein and N. Isgur, Phys. Rev. **D27**, 588 (1983).
26. D. Robson, Phys. Rev. **D35**, 1018 (1987); G. Miller in proceedings of the Nuclear Chromodynamics Conference, Argonne National Laboratory, Argonne, Illinois, USA (1988).
27. F. Lenz *et al.*, Ann. Phys. **170**, 65 (1986).
28. C. Michael in proceedings of the Nuclear Chromodynamics Conference, Argonne National Laboratory, Argonne, Illinois, USA (1988).
29. K. Maltman and N. Isgur, Phys. Rev. **D29**, 952 (1984).
30. See, e.g., M. Oka and K. Yazaki, Phys. Lett. **90B**, 41 (1980); I. Bender and H.G. Dosch, Fortsch. Phys. **30**, 633 (1980).
31. See, e.g., D. Robson in Progress in Particle and Nuclear Physics, ed. D. Wilkinson (Pergamon, New York, 1982), vol. 8, p. 257; M. Rosina *et al.*, ibid., p. 417; A. Faessler *et al.*, Nucl. Phys. **A402**, 55 (1983); M. Cvetic *et al.*, Nucl. Phys. **A395**, 349 (1983).
32. H.J. Lipkin in Proceedings of the 2nd Conference on the Intersections of Particle and Nuclear Physics, AIP Conference Proceedings 150, ed. D.F. Geesman (AIP, New York, 1986), p. 657.
33. With acknowledgements to K. Johnson.
34. See, e.g., K. Maltman, Nucl. Phys. **A438**, 669 (1985); M. Oka and K. Yazaki, Phys. Lett. 90B, 41 (1980); M. Cvetic *et al.*, Phys. Lett. **93B**, 489 (1980).
35. R.L. Jaffe, Phys. Rev. Lett. **38**, 195 (1977).
36. R.L. Jaffe, Phys. Rev. **D15**, 267, 281 (1977); **D17**, 1444 (1978); R.L. Jaffe and K. Johnson, Phys. Lett. **60B**, 201 (1976); R.L. Jaffe and F.E. Low, Phys. Rev. **D19**, 2105 (1979).
37. J. Weinstein and N. Isgur, in preparation.
38. F.E. Barnes in Proceedings of the VIIth Int. Workshop on Photon-Photon Collisions, Paris, 1986.
39. A. Abashian, N.E. Booth, K.M. Crowe, R.E. Hill, and E.H. Rogers, Phys. Rev. **132**, 2296 (1963).

40. F. Feld-Dahme, Ludwig-Maximilians-Universität Ph.D. thesis, Munich (1987).

INTRODUCTION TO NAMBU JONA-LASINIO MODELS

APPLIED TO LOW ENERGY HADRONIC MATTER

Georges RIPKA

Service de Physique Théorique de Saclay
Laboratoire de l'Institut de Recherche Fondamentale du
Commissariat à l'Energie Atomique
F-91191 Gif-sur-Yvette Cedex, France
Martine JAMINON and Pierre STASSART
Université de Liège
Institut de Physique B5, B-4000 Liège, Belgium

1. Introduction

These lectures are an introduction to and not a survey of Nambu Jona-Lasinio theory applied to low energy (below 1 possibly 2 GeV) hadronic physics. The aim is to ease the reader's access to the literature and to help him make his own calculations. We have certainly not referred adequately to the ever increasing number of works on the subject and we stress more what people do than what people say.

The Nambu Jona-Lasinio model was originally introduced to explain the spontaneous chiral symmetry breaking of the physical vacuum [1]. It is a theory of Dirac particles with a local 4-fermion interaction and, as such, it belongs to the same class of effective theories as Landau's theory of Fermi liquids or the BCS theory of superconducting metals. Like the latter, it is not renormalisable and it requires the introduction of a momentum cut-off which is an important parameter of the theory.

The Nambu Jona-Lasinio model has been extensively applied to calculate the properties of low-lying mesons in terms of $q\bar{q}$ excitations of the vacuum. In Ref.[2,3] it is shown to give a unified description of soft pion theorems, PCAC, the Golberger Treiman relation [4], the KSFR relation [5] , the Weinberg relation [6], the the Gell-Mann Oakes Renner formula [8],quark-loop effects such as anomalies, anomalous decays and Wess-Zumino terms [7] , hidden symmetries [9] , and ρ meson dominance of electro-magnetic interactions [10] . Attempts have also been made to calculate the nucleon as a soliton consisting of a bound state of quarks [17,27,38] . Last but not least, the Nambu Jona-Lasinio model has very recently been applied to the standard model [15] thus relating the Higgs mass to the top quark mass.

Hadrons and Hadronic Matter
Edited by D. Vautherin *et al.*
Plenum Press, New York, 1990

In the present state of the art, the Nambu Jona-Lasinio model of hadronic matter stands in opposition (and is possibly complementary) to constituent quark models in which quarks are confined by string-like or flux-tube forces [18] or by color-dielectric [19] fields. These models require a roughly 300 MeV constituent quark mass as an input. The Nambu Jona-Lasinio model yields a constituent quark mass of this order. However it lacks the confinement properties of the constituent quark models. The merging of these two approaches would certainly be useful.

There have been several attempts to derive the Nambu Jona-Lasinio model from QCD. To quote but two: Diakonov and Petrov [16]relate the model to the instanton structure of the QCD vacuum; R.Ball [3] rewrites the QCD lagrangian in terms of new non-local fields which, at low energy, reduce to colorless mesons and glueballs.

We view the Nambu Jona-Lasinio theory to be part of an effective theory for the low energy (possibly 1-2 GeV) properties of hadronic matter. The ideal effective theory we are aiming at should account for the chiral symmetry breaking of the vacuum, its restoration at finite density and temperature, the structure, decays and electro-weak properties of low energy mesons ($\pi,\eta,K,\omega,\rho,\ldots$) the structure and low energy properties of baryons ($N,\Delta,\Sigma,\Lambda,\ldots$) together with their couplings to the mesons, the modification of hadrons propagating in normal and dense nuclear matter and it should form the basis for a calculation of the equation of state of dense matter which is encountered in the evolution and explosion of stars [14].

The effective lagrangian which describes low energy properties of hadronic matter may bear little ressemblance to the QCD lagrangian, just like the pairing hamiltonian used to explain superconductivity has little to do with QED but is, instead, closely related to the attractive potential between two electrons which exchange phonons. Low energy properties of hadronic matter may teach us nothing about the QCD lagrangian. They may, instead, teach us more about the as yet poorly understood structure of the QCD vacuum and its response to quark propagation. It is probably vain to expect much help or appreciation from experts in high energy perturbative QCD. It is not a small parameter we are trying to find. It is the relevant degrees of freedom. Some help may come from recent attempts to extrapolate between the low and high energy regimes of hadronic interactions [40].

2. The lagrangian

We use the method of background fields to write down the action. This method is convenient for discussing symmetries and for calculating mesons. We introduce a chiral field

$$U = e^{i\gamma_5\vec{\theta}.\vec{\tau}} \tag{2.1}$$

together with a scalar field φ. In the SU2 case we have 4 real fields (φ

and $\vec{\theta}$) which have a linear representation:

$$\varphi U \equiv \varphi\, e^{i\gamma_5 \vec{\theta}.\vec{\tau}} = \varphi^0 + i\gamma_5 \vec{\varphi}.\vec{\tau} \qquad (2.2)$$

where $\varphi^a \equiv (\varphi^a, \vec{\varphi})$ is a real chiral 4-vector. We introduce a vector field $V_\mu = (i\Phi, \vec{V})$ and we work in a Euclidean metric (see appendices A and B) where $\gamma_\mu = (i\beta, \vec{\gamma}) = -\gamma^{+}{}_{\mu}$ and:

$$\partial \equiv \gamma_\mu \partial_\mu = \beta\partial_\tau + \vec{\gamma}.\vec{\nabla} \qquad V \equiv \gamma_\mu V_\mu = -\beta\Phi + \vec{\gamma}.\vec{V} \qquad (2.3)$$

The action of the Nambu Jona-Lasinio lagrangian is:

$$I \equiv \int d_4x\, \bar{\psi}(-i\partial + \varphi U + V)\psi + \frac{a^2\varphi^2}{2} + \frac{b^2V_\mu^2}{2}$$

$$= \int d_4x\, \bar{\psi}\left(-i\partial + \varphi_0 + i\gamma_5\vec{\varphi}.\vec{\tau} + V\right)\psi + \frac{a^2\varphi_a^2}{2} + \frac{b^2V_\mu^2}{2} \qquad (2.4)$$

The notation of appendix A is used. ψ is a Dirac spinor representing the quarks. Notice that no kinetic (derivative) terms of the background fields φ and V occur. They appear in the action as constraints which can be eliminated using the equations of motion:

$$a^2\varphi^0 = -\bar{\psi}\psi \qquad a^2\varphi^a = -\bar{\psi}(i\gamma_5\tau_a)\psi \qquad b^2V_\mu = -\bar{\psi}\gamma_\mu\psi \qquad (2.5)$$

The action is then expressed entirely in terms of quark fields with 4-fermion interactions:

$$I = \int d_4x\, \bar{\psi}(-i\partial)\psi - \frac{(\bar{\psi}\psi)^2}{2a^2} + \frac{\left(\bar{\psi}\gamma_5\vec{\tau}\psi\right)^2}{2a^2} - \frac{\left(\bar{\psi}\gamma_\mu\psi\right)^2}{2b^2} \qquad (2.6)$$

The constants a^2 and b^2 are thus the inverse coupling constants of the 4-fermion interactions.

Other vector fields can and have been used [2,3,11,33]. In these lectures we will not cover the interesting subject of the iso-vector mesons ρ and A_1 [2,3,33,34].

Note that the background field which we used is arbitrary. We could have, for example, chosen another of the three channels of the 4-fermion interaction to define a background field. Other choices for the 4-fermion interaction have also been used [11].

The lagrangians (2.4) and (2.6) are equivalent. On the form (2.6) it is simple to check chiral symmetry, namely the invariance of the action with respect to <u>global</u> chiral rotations:

$$\psi \Rightarrow A\psi \qquad U \Rightarrow A^{-1}UA^{-1} \qquad A \equiv e^{-i\vec{\alpha}.\vec{\tau}\gamma_5/2} \qquad (2.7)$$

We shall see how the chiral symmetry of the lagrangian is spontaneosly broken in the physical vacuum. This will give rise to Goldstone bosons of zero mass which are identified to the pions (see section 6). In these lectures we will neglect the small current quark mass which breaks chiral symmetry explicitely and which gives mass to the pions.

The euclidean metric is useful for expressing the partition function as a functional integral [28]:

$$\mathrm{Tr}\; e^{-\beta H} = = \int \mathcal{D}(\varphi)\mathcal{D}(V)\mathcal{D}(\psi)\mathcal{D}(\bar\psi)\; e^{-I(\varphi,V,\psi,\bar\psi)}$$

$$= \int \mathcal{D}(\varphi)\mathcal{D}(V)\mathcal{D}(\psi)\mathcal{D}(\bar\psi)\; e^{-\int d_4x\, \bar\psi\left(-i\partial + \varphi_0 + i\gamma_5\vec{\varphi}.\vec{\tau} + V\right)\psi + \frac{a^2\varphi^2}{2} + \frac{b^2V_\mu^2}{2}} \tag{2.8}$$

The functional integral is gaussian in the chiral and vector fields φ and V. If we integrate out the meson fields, the partition function is expressed in terms of quark variables with the action (2.6). However, instead of integrating out the meson fields, we can integrate out the fermion fields $\bar\psi$ and ψ. We obtain then a partition function expressed in terms of the chiral and vector fields alone (this is loosely referred to as a "bosonization"):

$$\mathrm{Tr}\; e^{-\beta H} = \int \mathcal{D}(\varphi)\mathcal{D}(V)\; e^{-\int d_4x\left(-\,\mathrm{tr}\,\lg(-i\partial + \varphi U + V) + \frac{a^2\varphi^2}{2} + \frac{b^2V_\mu^2}{2}\right)} \tag{2.9}$$

where tr is over the internal indeces of the quark states (see appendix A). The form (2.9) defines an effective action:

$$I = \int d_4x \left(-\,\mathrm{tr}\,\lg(-i\partial + \varphi U + V) + \frac{a^2\varphi^2}{2} + \frac{b^2V_\mu^2}{2}\right) \tag{2.10}$$

which is convenient for the study of mesons.

3. The gap equation

In order to simplify some expressions and to conform to the literature, we make an analytic continuation of the vector field so as to make it real:

$$V_\mu \equiv (i\Phi,\vec{V}) = (\eta,\vec{V}) = V_\mu^* \tag{3.1}$$

This makes $V \equiv V_\mu\gamma_\mu = -V^\dagger$ anti-hermitian. We define a Dirac operator D:

$$D = -i\partial + \varphi U + V \qquad D^\dagger = i\partial + \varphi U^\dagger - V$$

$$D^\dagger D = (i\partial + \varphi U^\dagger - V)(-i\partial + \varphi U + V) = -\partial_\mu^2 + \varphi^2 + i(\partial\varphi U) + i\partial V + iV\partial + V_\mu^2 \tag{3.2}$$

We can write:

$$- \mathrm{Tr}\ \lg D = - \frac{1}{2} \mathrm{Tr}\ \lg D^{\dagger} D - \frac{1}{2} \mathrm{Tr}\ \lg D + \frac{1}{2} \mathrm{Tr}\ \lg D^{\dagger} \tag{3.3}$$

and the action (2.10) becomes:

$$I = - \frac{1}{2} \mathrm{Tr}\ \lg D^{\dagger} D - \frac{1}{2} \mathrm{Tr}\ \lg D + \frac{1}{2} \mathrm{Tr}\ \lg D^{\dagger} + \int d_4 x \left(\frac{a^2 \varphi^2}{2} + \frac{b^2 V_\mu^2}{2} \right) \tag{3.4}$$

In the equations above Tr means a trace over the Hilbert space of a single quark: $\mathrm{Tr}\ A \equiv \int d_4 x\ \mathrm{tr}\ A$. No confusion should arise with this notation and that of the partition function $\mathrm{Tr} e^{-\beta H}$ where Tr is a trace over the Fock space of 0,1,2,...quark states.

Let us check that a stationary point of the action can be found of the form:

$$\varphi = \varphi_0 \qquad (\partial_\mu \varphi_0) = 0 \qquad U = 1 \qquad V_\mu = 0 \tag{3.5}$$

This stationary point will represent the translationally invariant physical vacuum. It is convenient to subtract the value of the action at this point. The action becomes then:

$$I = - \frac{1}{2} \mathrm{Tr}\ \lg \left(1 + \frac{1}{-\partial_\mu^2 + \varphi_0^2} \left(\varphi^2 - \varphi_0^2 + i(\partial \varphi U) + i \partial V + i V \partial + V_\mu^2 \right) \right)$$

$$- \frac{1}{2} \mathrm{Tr}\ \lg \left(1 + \frac{1}{-i\partial + \varphi_0} (\varphi U - \varphi_0 + V) \right) + \frac{1}{2} \mathrm{Tr}\ \lg \left(1 + \frac{1}{i\partial + \varphi_0} \left(\varphi U^{\dagger} - \varphi_0 - V \right) \right)$$

$$+ \int d_4 x \ \frac{a^2 \left(\varphi^2 - \varphi_0^2 \right)}{2} + \frac{b^2 V_\mu^2}{2} \tag{3.6}$$

Consider the first order variation of the action around the point (3.5). It is easy to check that the first order terms in θ and V vanish. We set $\varphi = \varphi_0 + \tilde{\varphi}$. The first order variation of the action is:

$$I^{(1)} = - \frac{1}{2} \mathrm{Tr} \frac{1}{\partial_\mu^2 + \varphi_0^2} 2 \varphi \tilde{\varphi} + a^2 \int d_4 x\ \varphi_0 \tilde{\varphi} = \varphi_0 \left(\int d_4 x \tilde{\varphi} \right) \left(- \frac{1}{\beta \Omega} \mathrm{Tr} \frac{1}{-\partial_\mu^2 + \varphi_0^2} + a^2 \right) \tag{3.7}$$

The point (3.5) is a stationary point of the action if:

$$a^2 = \frac{1}{\beta \Omega} \mathrm{Tr} \frac{1}{-\partial_\mu^2 + \varphi_0^2} \equiv \frac{1}{\beta \Omega} \mathrm{Tr}\ G_0 \tag{3.8}$$

where we have introduced the "unperturbed propagator":

$$G_0 \equiv \frac{1}{-\partial_\mu^2 + \varphi_0^2} \tag{3.9}$$

Equation (3.8) is the "gap equation" so called for reasons which are explained in section 4. In Eq.(3.8) β is the inverse temperature (see appendix A). In these lectures we work in the zero temperature limit $\beta \Rightarrow \infty$.

It is useful to eliminate the parameter a^2 in favor of φ_0. The action becomes:

$$I = -\frac{1}{2}\,\mathrm{Tr}\;\lg\left(1 + \frac{1}{-\partial_\mu^2 + \varphi_0^2}\left(\varphi^2 - \varphi_0^2 + i(\partial\varphi U) + i\partial V + iV\partial + V_\mu^2\right)\right)$$

$$+ \mathrm{Tr}\;\frac{1}{2}\,\frac{1}{-\partial_\mu^2 + \varphi_0^2}\left(\varphi^2 - \varphi_0^2\right) + \int d_4x\;\frac{b^2 V_\mu^2}{2}$$

$$-\frac{1}{2}\,\mathrm{Tr}\;\lg\left(1 + \frac{1}{-i\partial + \varphi_0}(\varphi U - \varphi_0 + V)\right) + \frac{1}{2}\,\mathrm{Tr}\;\lg\left(1 + \frac{1}{i\partial + \varphi_0}\left(\varphi U^\dagger - \varphi_0 - V\right)\right) \tag{3.10}$$

We simplify the notation by defining an "interaction" $\mathcal{V}$:

$$\mathcal{V} \equiv \varphi^2 - \varphi_0^2 + i(\partial\varphi U) + i\partial V + iV\partial + V_\mu^2 \tag{3.11}$$

Since $\mathrm{Tr}\;G_0\mathcal{V} = \mathrm{Tr}G_0\left(\varphi^2 - \varphi_0^2\right) + \mathrm{Tr}\;G_0 V_\mu^2$ the action can be written in the more compact form:

$$I = -\frac{1}{2}\,\mathrm{Tr}\;\lg(1 + G_0\mathcal{V}) + \frac{1}{2}\,\mathrm{Tr}\;G_0\mathcal{V} - \frac{1}{2}\,\mathrm{Tr}\;G_0 V_\mu^2 + \int d_4x\;\frac{b^2 V_\mu^2}{2}$$

$$-\frac{1}{2}\,\mathrm{Tr}\;\lg\left(1 + \frac{1}{-i\partial + \varphi_0}(\varphi U - \varphi_0 + V)\right) + \frac{1}{2}\,\mathrm{Tr}\;\lg\left(1 + \frac{1}{i\partial + \varphi_0}\left(\varphi U^\dagger - \varphi_0 - V\right)\right) \tag{3.12}$$

We note that the action (2.10) has both quadratic and logarithmic divergences. By eliminating the inverse coupling constant a^2 in favor of φ_0 (using the quadratically divergent gap equation), we have constructed the action (3.12) which, as we shall see below, has only logarithmic divergences [13].

Several authors [2,27,34] regularize the real part $\frac{1}{2}\,\mathrm{Tr}\;\lg\;D^\dagger D$ of the action using Schwinger's proper time method [29]. This consists in writing:

$$-\frac{1}{2}\,\mathrm{Tr}\;\lg(1 + G_0\mathcal{V})$$

$$= -\frac{1}{2}\,\mathrm{Tr}\;\lg\left(-\partial_\mu^2 + \varphi^2 + i(\partial\varphi U) + i\partial V + iV\partial + V_\mu^2\right) + \frac{1}{2}\,\mathrm{Tr}\;\lg\left(-\partial_\mu^2 + \varphi_0^2\right)$$

$$= \mathrm{Tr}\int_0^\infty \frac{ds}{s}\,w(s)\left(e^{-s\left(-\partial_\mu^2 + \varphi_0^2\right)} - e^{\left(-\partial_\mu^2 + \varphi^2 + i(\partial\varphi U) + i\partial V + iV\partial + V_\mu^2\right)}\right) \tag{3.13}$$

The expression (3.13) is an identity if w(s)=1 and if the trace converges. The regularization consists in introducing a weight factor w(s) so as to cut off the low values of s which cause the divergence of the trace (at high momenta). Details on proper time regularization are given in Ref.[2]. We will discuss below other regularisations.

We shall see that the regularization of the Nambu Jona-Lasinio model introduces a cut-off Λ which has a low value, about 1 GeV or even less. It is an important parameter of the model and, as yet, its origin is poorly understood. In Ref.[16] the form of the weight factor is based on a study of the instanton structure of the vacuum. The regularization of the energy is discussed in sections 4 and 10 and meson propagator regularization is discussed in sections 7 and 9.

4. Dynamical chiral symmetry breaking, the quark constituent mass and finite baryonic density

Chiral symmetry is spontaneously broken when the stationary point (3.5) of the action has $\varphi_0 \neq 0$. The quarks then acquire a "constituent" mass equal to φ_0. To see this more clearly, we proceed to reformulate the gap equation in much simpler terms.

Spontaneous chiral symmetry breaking of the physical vacuum is easily calculated starting from the Hamiltonian derived from the action (2.3):

$$H = \int d_3 r \left(\psi^\dagger \left(\frac{\alpha.\nabla}{i} + \beta\varphi U - \Phi + \alpha.V \right) \psi + \frac{a^2\varphi^2}{2} + \frac{b^2 V_\mu^2}{2} \right) \tag{4.1}$$

The translationally invariant fields U and V, which we use to describe the physical vacuum, can be eliminated by defining rotated quark fields q:

$$q \equiv U^{1/2} e^{iV.r} \psi \tag{4.2}$$

Expressed in terms of the q fields, the hamiltonian (4.1) becomes:

$$H = \int d_3 r \left(q^\dagger \left(\frac{\alpha.\nabla}{i} + \beta\varphi - \Phi \right) q + \frac{a^2\varphi^2}{2} + \frac{b^2 (V^2 - \Phi^2)}{2} \right) \tag{4.3}$$

We note that the component Φ of the (constant) vector field acts as a chemical potential. A vacuum composed of filled negative energy Dirac sea orbitals can be constructed with $\Phi = 0$. We can then minimize the energy with zero space-like components of the vector field: $V = 0$. In this case, the hamiltonian (4.3) describes free quarks of mass φ which we call the constituent quark mass for two reasons. First because calculations made so far of the nucleon with the Nambu Jona-Lasinio model yield energies close to $N_c\varphi_0$ where φ_0 is the vacuum constituent quark mass (see section 10). Second because the values usually obtained for φ_0 are close to 300 MeV which is used in constituent quark models of the nucleon (see section 8). We assume the vacuum to be the ground state of the hamiltonian (4.3). With $\Phi = 0$ the vacuum is then a Dirac sea (Fig.1) with quarks filling negative energy plane wave orbits. The vacuum energy per unit volume is:

$$\frac{E_D}{\Omega} = - \frac{\nu}{\Omega} \sum_{\vec{k}} \sqrt{\vec{k}^2 + \varphi^2} + \frac{a^2\varphi^2}{2} \qquad (4.4)$$

where ν is the degeneracy of the quark plane wave orbits (see appendix A).

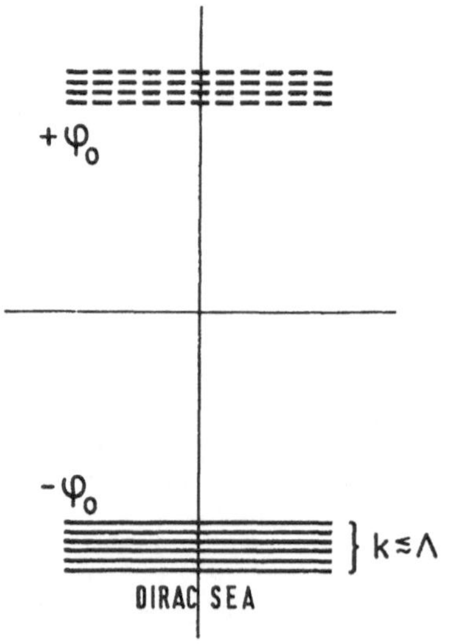

Figure 1. In the Nambu Jona-Lasinio model, the physical vacuum is described by quarks filling a Dirac sea with momenta $k \lesssim \Lambda$.

In the physical vacuum, the constituent quark mass φ takes the value φ_0 which makes the energy stationary:

$$\left.\frac{\partial E_D}{\partial \varphi}\right|_{\varphi=\varphi_0} = 0 \qquad a^2 = \frac{\nu}{\Omega} \sum_{\vec{k}} \frac{1}{\sqrt{\vec{k}^2 + \varphi_0^2}} \qquad (4.5)$$

This is the same φ_0, introduced in Eq.(3.5), which makes the action (2.3) stationary. The gap equations (4.5) and (3.8) are the same. To see this, we evaluate the trace in (3.8) using the plane wave basis of appendix A and we integrate over k_0:

$$a^2 = \frac{\nu}{\beta\Omega}\sum_k \frac{1}{k^2+\varphi_0^2} = \frac{\nu}{2\pi}\int_{-\infty}^{\infty} dk_0 \frac{1}{\Omega}\sum_{\vec{k}} \frac{1}{k_0^2+\vec{k}^2+\varphi_0^2} = \frac{\nu}{\Omega}\sum_{\vec{k}} \frac{1}{\sqrt{\vec{k}^2+\varphi_0^2}} \tag{4.7}$$

For a given inverse coupling constant a^2, Eq.(4.5) can be used to determine the constituent quark mass φ_0. Fig.1 shows that the energy gap in the quark spectrum is $2\varphi_0$. This is why Eqs.(4.5) or (3.8) are called gap equations. We shall use the gap equation to eliminate the unphysical coupling constant a^2 in favor of the more physical constituent quark mass φ_0. The vacuum energy per unit volume becomes thus:

$$\frac{E_D}{\Omega} = -\frac{\nu}{\Omega}\sum_k\left(\sqrt{k^2+\varphi^2} - \sqrt{k^2+\varphi_0^2} - \frac{\varphi^2-\varphi_0^2}{2\sqrt{k^2+\varphi_0^2}} + \frac{\left(\varphi^2-\varphi_0^2\right)^2}{8\left(k^2+\varphi_0^2\right)^{3/2}}\right)$$

$$-\frac{\nu}{\Omega}\sum_k\left(-\frac{\left(\varphi^2-\varphi_0^2\right)^2}{8\left(k^2+\varphi_0^2\right)^{3/2}} + \frac{\left(\varphi^2-\varphi_0^2\right)^2}{8\left(k^2+\Lambda^2\right)^{3/2}}\right)$$

$$= \frac{\nu}{16\pi^2}\left(\frac{1}{4}\left(\varphi^2-\varphi_0^2\right)\left(3\varphi^2-\varphi_0^2\right) - \frac{\varphi^4}{2}\,\mathrm{lg}\left(\frac{\varphi^2}{\varphi_0^2}\right) - \frac{\left(\varphi^2-\varphi_0^2\right)^2}{2}\,\mathrm{lg}\left(\frac{\varphi_0^2}{\Lambda^2}\right)\right) \tag{4.8}$$

To obtain this expression we have subtracted the energy at $\varphi = \varphi_0$ so as to set the vacuum energy to zero and we have added a Pauli-Villars type regularizing term so as to cut off momenta larger than Λ.

Figure 2 shows the form of the energy per unit volume of the vacuum, as given by (4.8). The curves are futher discussed in section 8. By construction, the minimum energy occurs at the point $\varphi = \varphi_0$. This point describes the physical vacuum. The point at $\varphi = 0$ is the energy per unit volume required to restore chiral symmetry in the vacuum. It is the MIT bag constant.

From Eq.(4.3), we see that, in the vacuum, the quarks acquire a mass φ_0. The vacuum energy depends only on φ and it is invariant with respect to chiral rotations (2.7) However the Dirac sea, occupied by the quarks in the vacuum, is not invariant with respect to chiral rotations of the quark wavefunctions. This is what is meant when we say that chiral symmetry is spontaneously broken in the physical vacuum. The non-vanishing quark mass φ_0 is a reflection of this. The ground state of the hamiltonian (4.1) describes a phase of the vacuum which is commonly measured by the order parameter $\langle\bar{\psi}\psi\rangle$ which is the expectation value, in the Dirac sea of quarks, of the operator $\bar{\psi}(\vec{r})\psi(\vec{r})$. This expectation value may be calculated by differentiating, with respect to φ, the quark constribution to the energy (4.4). In the vacuum, where $\varphi = \varphi_0$, the order parameter is equal to:

$$\langle\bar{\psi}\psi\rangle = -\frac{\varphi_0}{(2\pi)^3}\int d_3k\,\frac{1}{\sqrt{\vec{k}^2+\varphi_0^2}} \tag{4.9}$$

The regularization of this quadratically divergent expression is discussed in section 8. Note that a vanishing (or non-vanishing) constituent quark mass φ_0 implies a vanishing (or non-vanishing) order parameter $\langle\bar{\psi}\psi\rangle$.

The energy (4.8) depends on two parameters: the quark mass φ_0 (which replaces the inverse coupling constant a^2) and the cut-off Λ. We determine these in section 8.

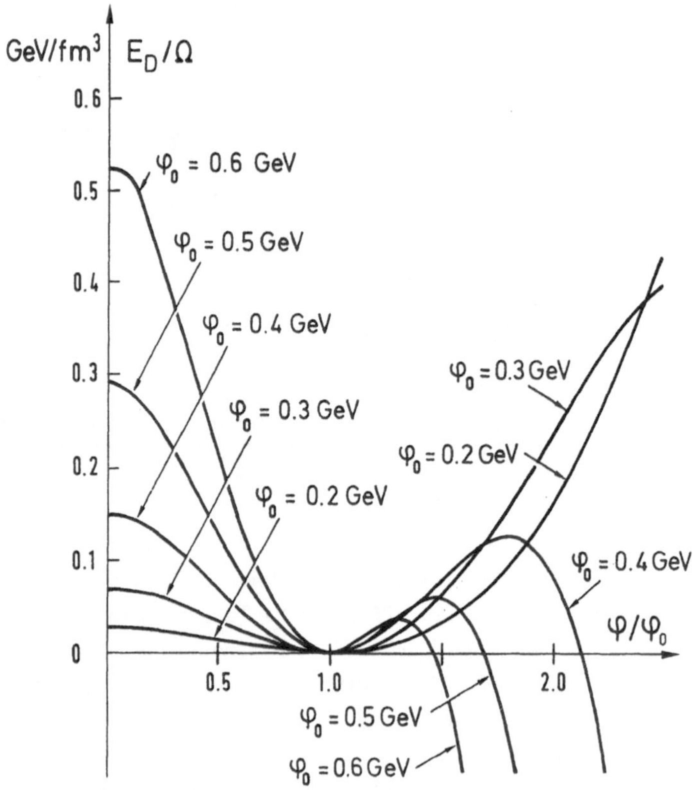

Figure 2. The energy per unit volume of the vacuum as given by Eq. (4.8). It is plotted as a function of the constituent quark mass φ in units of its vacuum value φ_0. The curves, which correspond to different values of φ_0 are discussed in section 8.

We now consider finite baryonic density. We construct baryonic matter at finite density by adding, to the Dirac sea, a Fermi sea of positive energy quarks or nucleons. Consider the simplest case where quarks fill positive energy orbits up to a Fermi momentum k_F. The latter is related to the baryonic density ρ_B (number of baryons per unit volume) by the expression:

$$\rho_B = \frac{N_q}{N_c \Omega} = \frac{\nu}{N_c} \frac{k_F^3}{6\pi^2} \tag{4.10}$$

where N_q is the number of quarks and N_c the number of colors. The Fermi momentum k_F is the same whether the Fermi sea contains quarks or nucleons (remember that quarks have a baryonic number equal to $1/N_c$). The case where quarks fill the Fermi sea is displayed on Fig.3.

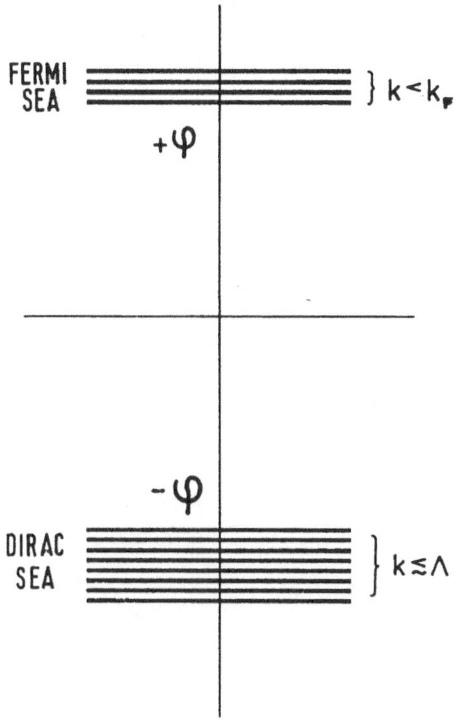

Figure 3. finite density baryonic matter is constructed by filling the positive energy orbits of a Fermi sea with quarks. As shown on Fig.4, the quark constituent mass φ is reduced relative to its vacuum value φ_0.

The energy per unit volume of the system is then:

$$\frac{E}{\Omega} = \frac{E_D}{\Omega} + \frac{\nu}{\Omega} \sum_{k < k_F} \sqrt{k^2 + \varphi^2} \tag{4.11}$$

where E_D is the Dirac sea energy (4.8). The second term is the contribution of the Fermi sea quarks and it is a roughly linearly increasing function of φ. Fig.4 shows that, at finite baryonic density, the minimum energy occurs at a value of φ which is smaller than φ_0. As the density increases, the equilibrium value of φ decreases and when it gets very small (typically of the size of the current quark mass neglected here) chiral symmetry of the vacuum is restored.

However, even at normal nuclear matter density ($k_F = 1.36\ \mathrm{fm}^{-1}$) the equilibrium value of φ is significantly reduced [11,12,20], the exact amount (10-30% approximately) depending on the parameters of the model [13,20]. (The parameters are discussed in section 8.) The decrease of φ is a partial restoration of chiral symmetry due to the presence of baryons in the Fermi sea. In sections 6 and 10 we shall see that the σ meson and the nucleon masses are proportional to φ so that they decrease in the nuclear medium in proportion to φ. The nucleon size increases in the same proportion. Such effects, for which experimental evidence exists, might significantly modify the equation of state of nuclear matter [26]. When finite baryonic matter is formed by putting nucleons instead of quarks into the Fermi sea, chiral symmetry is restored even at normal nuclear matter density [12,20] . This suggests that quantitative predictions of partial chiral symmetry restoration based on the mean field approximation (as done here) should not be taken too seriously.

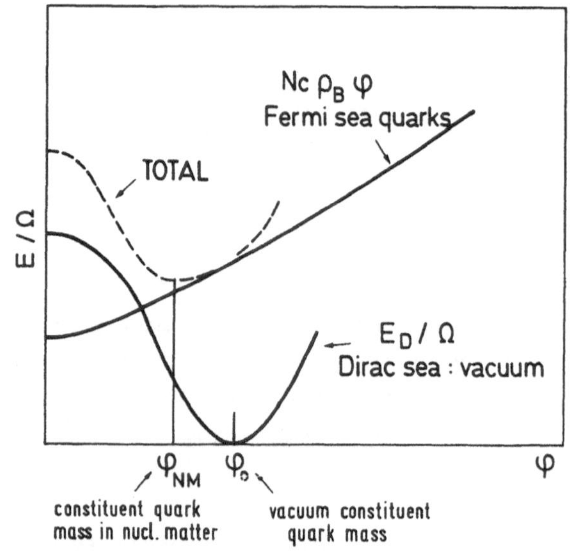

Figure 4. the Fermi sea quarks add a term to the energy per unit volume which is a roughly linearly increasing function of φ. The effect is to reduce the equilibrium value of φ relative to its vacuum vamue φ_0 . The σ meson and nucleon masses decrease (and their size increases) in the same ratio as φ.

5. Second order expansion of the action

In appendix B we show that the inverse field propagators are given by the second order derivatives of the action with respect to the fields.

The second order expansion of the action in powers of $\tilde{\varphi} \equiv \varphi - \varphi_0$, θ and V will therefore enable us to calculate the σ, π and ω propagators.

Consider the expansion of the action (3.12). No first order terms appear because (3.5) was chosen to be a stationary point. The last two terms of (3.12) contribute nothing to second order. Thus only the first order part of $\mathcal{V}$ is required:

$$\mathcal{V}^{(1)}= 2\varphi_0\tilde{\varphi} + i(\partial\tilde{\varphi}) - \varphi_0\gamma_5(\partial\theta) + i\partial V + iV\partial \tag{5.1}$$

where: $(\partial\theta) \equiv \gamma_\mu\left(\partial_\mu\vec{\theta}\right).\vec{\tau}$. The second order action is thus:

$$I^{(2)}= \frac{b^2}{2}\int d_4x\, V_\mu^2 - \frac{\mathrm{Tr}}{2}\, G_0 V_\mu^2$$
$$+ \frac{\mathrm{Tr}}{4} G_0\left(2\varphi_0\tilde{\varphi}+i(\partial\tilde{\varphi}) - \varphi_0\gamma_5(\partial\theta)+i\partial V+iV\partial\right)G_0\left(2\varphi_0\tilde{\varphi}+i(\partial\tilde{\varphi})-\varphi_0\gamma_5(\partial\theta)+i\partial V+iV\partial\right) \tag{5.2}$$

The trace over the γ_5 matrices kills the cross terms of the type $\theta\tilde{\varphi}$ or $V\vec{\theta}$. The $\tilde{\varphi}V$ cross terms also vanish (can you prove this?). The second order action thus separates into three terms:

$$I^{(2)}= I^{\varphi\varphi}+ I^{\theta\theta}+ I^{VV} \tag{5.3a}$$

where:

$$I^{\varphi\varphi}= \varphi_0^2 G_0\tilde{\varphi}G_0\tilde{\varphi} - \frac{1}{4} G_0(\partial\tilde{\varphi})G_0(\partial\tilde{\varphi}) \qquad I^{\theta\theta}= -\frac{\varphi_0^2}{4} G_0(\partial\theta)G_0(\partial\theta)$$

$$I^{VV}= \frac{1}{2} G_0 V^2+ \frac{1}{4} G_0(i\partial V + Vi\partial)G_0(i\partial V + Vi\partial) + \frac{b^2}{2}\int d_4x\, V_\mu^2 \tag{5.3b}$$

The three terms are calculated in sections 6 and 9.

6. Sigma and pi mesons

The sigma meson propagator is obtained by casting the term $I^{\varphi\varphi}$ of expression (5.3b) into the form (B19) of appendix B. Tracing out the Dirac matrices and using the plane wave basis (A2) we have:

$$I^{\varphi\varphi}= \mathrm{Tr}\ \varphi_0^2 G_0\tilde{\varphi}G_0\tilde{\varphi} + \frac{\mathrm{Tr}}{4} G_0\left(\partial_\mu\tilde{\varphi}\right)G_0\left(\partial_\mu\tilde{\varphi}\right)$$

$$= 2\nu \sum_{kk'}\frac{1}{\left(k^2+\varphi_0^2\right)\left(k'^2+\varphi_0^2\right)}\left(\varphi_0^2\langle k|\tilde{\varphi}|k'\rangle\langle k'|\tilde{\varphi}|k\rangle+ \frac{1}{4}\left(\langle k|\left(\partial_\mu\tilde{\varphi}\right)|k'\rangle\langle k'|\left(\partial_\mu\tilde{\varphi}\right)|k\rangle\right)\right) \tag{6.1}$$

The matrix elements can be written in the form:

$$\langle k|\tilde{\varphi}|k'\rangle = \frac{1}{\beta\Omega}\int d_4x\, e^{i(k'-k)}\tilde{\varphi}(x) \equiv \frac{1}{\beta\Omega}\tilde{\varphi}(q) \qquad \langle k|\left(\partial_\mu\tilde{\varphi}\right)|k'\rangle = \frac{-iq_\mu}{\beta\Omega}\tilde{\varphi}(q)$$

$$\frac{1}{\beta\Omega}\sum_q \tilde{\varphi}(q)\tilde{\varphi}(-q) = \int d_4x\ \tilde{\varphi}^2(x) \qquad \partial_\mu|k\rangle = ik_\mu|k\rangle \qquad \langle k|\partial_\mu = ik_\mu\langle k| \tag{6.2}$$

K and q are defined thus:

$$k + k' = 2K \qquad k' - k = q \qquad k = K - q/2 \cdot \qquad k' = K + q/2 \tag{6.3}$$

With these definitions the expression (6.1) becomes:

$$I^{\varphi\varphi} = \frac{1}{2\beta\Omega}\sum_q \tilde{\varphi}(q)\tilde{\varphi}(-q)\nu\left(4\varphi_0^2+q^2\right)f(q) \tag{6.4}$$

where f(q) is the function:

$$f(q) \equiv \frac{1}{\beta\Omega}\sum_K \frac{1}{\left(K^2+\frac{q^2}{4} - K.q + \varphi_0^2\right)\left(K^2+\frac{q^2}{4} + K.q + \varphi_0^2\right)} \tag{6.5}$$

The regularisation of this function is discussed in section 7.

Comparing (6.4) to the expression (B19) we see that the inverse propagator of the φ field is:

$$K_\varphi^{-1}(q) = \nu\left(4\varphi_0^2+q^2\right)f(q) \tag{6.6}$$

We identify the σ-meson with the pole occuring at $q^2 = -4\varphi_0^2$ or, equivalently, at $iq_0 = \pm\sqrt{\vec{q}^{\,2} + 4\varphi_0^2}$ in the complex q_0 plane. The σ-meson is therefore a particle of mass $m_\sigma = 2\varphi_0$ and momentum $\vec{q}$. Its on-shell mass is exactly equal to the quark mass gap (see Fig.1).

The pi meson propagator is calculated from the $I^{\theta\theta}$ term of expression (5.3b) in the same way. A straightforward caclculation yields:

$$I^{\theta\theta} = \frac{\varphi_0^2}{4}\,\mathrm{Tr}\ G_0(\partial_\mu\theta_a)G_0(\partial_\mu\theta_a)$$

$$= 2\nu\frac{\varphi_0^2}{4}\sum_{kk'}\frac{1}{\left(k^2+\varphi_0^2\right)\left(k'^2+\varphi_0^2\right)}\left(\langle k|(\partial_\mu\theta_a)|k'\rangle\langle k'|(\partial_\mu\theta_a)|k\rangle\right)$$

$$= \frac{\varphi_0^2}{2\beta\Omega}\sum_q \theta_a(q)\theta_a(-q)q^2\nu f(q) \tag{6.7}$$

The inverse θ propagator is thus:

$$K_\theta^{-1}(q) = \varphi_0^2 q^2 \nu f(q) \tag{6.8}$$

The pole occuring at $q^2 = 0$ describes a zero mass particle which is identified to the pion (check that the quantum numbers are right!). The zero mass is due to the spontaneously broken chiral symmetry of the vacuum. The three pions (a=1,2,3) are the corresponding Goldstone bosons.

Notice that the on-shell σ and π-meson masses do not depend on the function f(q) and nor on its regularization (for a given value of φ_0). However the field strengths do depend on the function f(q) as we now show. The σ and π fields can be defined to be proportional to the φ and θ fields respectively:

$$\sigma(x) \equiv \sqrt{Z_\sigma}\, \varphi(x) \qquad \pi_a(x) \equiv \sqrt{Z_\pi}\, \varphi_0 \theta_a(x)$$

$$K_\sigma(q) = Z_\sigma K_\varphi(q) \qquad K_\pi(q) = Z_\pi K_\theta(q) \tag{6.9}$$

The constants Z_σ and Z_π can be chosen such that the residues of the σ and π propagator poles are equal to 1. This yields the values:

$$Z_\sigma = \nu\, f(q)\Big|_{q^2 = -4\varphi_0^2} \qquad Z_\pi = \nu\, f(q)\Big|_{q^2 = 0} \tag{6.10}$$

Field theorists sometimes like to identify mesons by making an expansion of the action in powers of the derivatives of the fields (so-called gradient expansion) retaining only terms up to second order derivatives. The masses obtained this way (see Refs.[2,3] for example) are not on-shell mesons. Actually the mesons exchanged by nucleons in nuclear matter energy calculations are not on the energy shell. In the Hartree approximation, for example, they are q=0 mesons.

The calculation of zero momentum mesons is blind to another problem. Indeed there is nothing in this model to prevent the poles of the meson propagators from exceeding the mass gap. This can actually occur for some choices of the model parameters (the parameters are determined in section 8). Such on-shell mesons would decay into quark-antiquark pairs and they would acquire a width due to this process. This would be an unphysical width since quarks are expected to be bound by string-like forces. Such an occurence would be a failing of the model due to its unability to confine the quarks.

One can also calculate meson propagators and meson masses at finite baryonic density where the constituent quark mass is reduced [11,12] . One finds that the relation $m_\sigma = 2\varphi$ still holds at finite density so that the σ-meson mass is reduced in the same ratio as φ.

7. Regularisation of the function f(q)

We will find it useful to work with the following two identities:

$$f(q) \equiv \frac{1}{\beta\Omega}\sum_k \frac{1}{\left(k^2+\frac{q^2}{4}+k.q+\varphi^2\right)\left(k^2+\frac{q^2}{4}-k.q+\varphi^2\right)}$$

$$= \frac{\pi^2}{(2\pi)^4}\frac{1}{2}\int_{-1}^{1} du \int_0^\infty \frac{ds}{s} e^{-s\left(\varphi^2+\frac{q^2}{4}(1-u^2)\right)} \qquad (7.1)$$

and:

$$g_{\mu\alpha}(q) \equiv \frac{1}{\beta\Omega}\sum_k \frac{k_\mu k_\alpha}{\left(k^2+\frac{q^2}{4}+k.q+\varphi^2\right)\left(k^2+\frac{q^2}{4}-k.q+\varphi^2\right)}$$

$$= \frac{\pi^2}{(2\pi)^4}\frac{1}{2}\int_{-1}^{1} du \int_0^\infty \frac{ds}{s} e^{-s\left(\varphi^2+\frac{q^2}{4}(1-u^2)\right)} \left(\frac{\delta_{\mu\alpha}}{2s}+\frac{u^2 q_\alpha q_\mu}{4}\right) \qquad (7.2)$$

The function $g_{\mu\alpha}$ appears in the study of the vector meson (section 9). To prove this we apply the Feynman trick $\frac{1}{AB} = \int_0^1 dx \frac{1}{(B + x(A-B))^2}$ to:

$$g_{\mu\alpha}(q) = \frac{1}{\beta\Omega}\sum_k \frac{1}{2}\int_{-1}^{1} du \frac{k_\mu k_\alpha}{\left(k^2+\varphi^2+\frac{q^2}{4}+ukq\right)^2}$$

$$= \frac{1}{(2\pi)^4}\frac{1}{2}\int_{-1}^{1} du \int_0^\infty s\, ds \int d_4k\, k_\mu k_\alpha\, e^{-s\left(k^2+\varphi^2+\frac{q^2}{4}+ukq\right)} \qquad (7.3)$$

We now do the k integrals:

$$\int d_4k\, e^{-s\left(k^2+\frac{q^2}{4}\right) - sukq} = \frac{\pi^2}{s^2} e^{-\frac{sq^2}{4}(1-u^2)}$$

$$\int d_4k\, k_\mu k_\alpha\, e^{-s\left(k^2+\frac{q^2}{4}\right) - sukq} = \frac{e^{-\frac{sq^2}{4}}}{s^2u^2}\frac{\partial^2}{\partial q_\mu \partial q_\alpha}\int d_4k\, e^{-sk^2 - sukq}$$

$$= \frac{\pi^2}{s^2}\left(\frac{\delta_{\mu\alpha}}{2s}+\frac{u^2 q_\mu q_\alpha}{4}\right) e^{-\frac{sq^2}{4}(1-u^2)} \qquad (7.4)$$

The identities (7.1) and (7.2) follow.

The expression (7.1) for f(q) diverges logarithmically at large k. Several ways have been used to regularize it. The brute force method of cutting off the momenta k at some finite value Λ has the disadvantage of not yielding gauge invariant vector meson contributions to the quark loop (see section 9). The transformation of the k integral into a "proper-time" s integral is a trick introduced by Schwinger [29] in order to avoid this problem. The ultra-violet (high k) divergence of the momentum integral is now replaced by the infrared (low s) divergence of the proper time integral. The function $g_{\mu\alpha}$ has a quadratic divergence. In section 9 we shall find however that the quadratically divergent part does not contribute leaving only a logarithmic divergence.

One favorite regularization consists in cutting off the low values of s at a value $s = 1/\Lambda^2$ where Λ is a cut-off parameter. This effectively cuts off the momenta k which exceed Λ in a smooth way. The function f(q) is then equal to:

$$f(q) = \frac{\pi^2}{(2\pi)^4}\frac{1}{2}\int_{-1}^{1} du \int_{1/\Lambda^2}^{\infty} \frac{ds}{s}\, e^{-s\left(\varphi^2 + \frac{q^2}{4}(1-u^2)\right)} \tag{7.5}$$

More generally one can regularize the s integral (7.1) by introducing a cut-off profile function with a given shape. This is done in Ref. [27] with a profile function based on the study of the instanton structure of the QCD vacuum. Useful expansions of the exponential in (7.1), which has the form of a partition function, have also been developed [2].

Another method (used in these lectures) of regularising the function f(q) consists in subtracting off a term with φ replaced by a higher mass Λ. The function f(q) becomes then:

$$f(q) = \frac{1}{\beta\Omega}\sum_k \left(\frac{1}{\left(k^2 + \frac{q^2}{4} + k.q + \varphi^2\right)\left(k^2 + \frac{q^2}{4} - k.q + \varphi^2\right)} - \frac{1}{\left(k^2 + \frac{q^2}{4} + k.q + \Lambda^2\right)\left(k^2 + \frac{q^2}{4} - k.q + \Lambda^2\right)}\right)$$

$$= \frac{\pi^2}{(2\pi)^4}\frac{1}{2}\int_{-1}^{1} du \left(\lg\left(\Lambda^2 + \frac{q^2}{4}(1-u^2)\right) - \lg\left(\varphi^2 + \frac{q^2}{4}(1-u^2)\right)\right) \tag{7.6}$$

The integral over u can then be done analytically using the indefinite integral:

$$\int dx \, \lg(a^2 - x^2) = (a+x)\lg(a+x) - (a-x)\lg(a-x) - 2x \tag{7.7}$$

This method is akin to a Pauli-Villars regularization. It is not however the usual Pauli Villars regularization (see Ref. [31]) which preserves gauge invariance but not chiral symmetry [30]. We have preserved chiral symmetry because we have regularized only the term $D^{\dagger}D$ of the action (3.4).

The regularization (7.6) has the advantage of often yielding analytical results (see (8.9) for example). However it obviously makes no sense when $\Lambda < \varphi$. If the cut-off Λ turns out to be smaller than the quark constituent mass φ, then a proper-time regularization of the form (7.5) is preferable.

Notice that the regularized functions (7.5) and (7.6) are positive for real q^2 (only if $\Lambda > \varphi$ in the regularization (7.6)). The σ and π propagators (6.6) and (6.8) have no poles for real values of q^2. Thus the Nambu Jona-Lasinio model is free of Landau ghosts which imply an instability of the assumed translationally invariant vacuum. This constrasts the model with the σ-model in which such instabilities do occur [30].

In this section we have only discussed the methods which have been proposed to regularize the divergent function f(q). We have not adressed the all-important and as yet unsolved problem of justifying the regularization. The cut-off parameters used in low energy hadronic physics are often very low, less that a GeV, and results depend strongly on the value of Λ. The cut-off is therefore an important parameter of the theory and its physical meaning should be specified.

8. Determination and discussion of the model parameters

The action (2.4) or (2.10) depends on 3 parameters: the inverse coupling constants a^2 , b^2 and the cut-off Λ. The parameter b^2 will be determined in section 9 by fitting the omega meson mass. In section 3 we saw that the gap equation allows us to trade off the inverse coupling constant a^2 in favor of the vacuum constituent quark mass φ_0. This leaves us with two undetermined parameters: φ_0 and Λ. We require to fit the pion decay constant $f_\pi = 93$ MeV . This yields a relation, given below by Eq.(8.9), between φ_0 and Λ, and we are then left with a one parameter theory.

Remember that the pion decay constant f_π is determined by the pion lifetime. More precisely, the π^- decay determines the matrix element of the axial current between the vacuum and the pion initial state. The pion decay constant $f_\pi = 93$ MeV is <u>defined</u> by the equation:

$$\left\langle 0 \left| A_{\mu a}(x) \right| \pi_b(q) \right\rangle \equiv - i\delta_{ab} f_\pi \frac{q_\mu}{\sqrt{\omega_q}} \frac{e^{iqx}}{\sqrt{\Omega}} \tag{8.1}$$

with $\omega_q \equiv \sqrt{\vec{q}^{\,2} + m_\pi^2}$ and $q^2 = - m_\pi^2$. In the definition (8.1), $A_{\mu a}(x)$ is the Noether current associated to the chiral rotation (2.7), $|0\rangle$ is the physical vacuum and $|\pi_b(q)\rangle \equiv c^\dagger_{\vec{q},a} |0\rangle$ is a normalised state representing a pion on the mass shell of type a and momentum $\vec{q}$. The pion field is assumed to have the quantized form:

$$\pi_a(x) = \sum_{\vec{q}} \frac{1}{\sqrt{2\omega_q \Omega}} \left(e^{iqx} c_{\vec{q},a} + e^{-iqx} c^\dagger_{\vec{q},a} \right) \tag{8.2}$$

Let us calculate the axial current $A_{\mu a}(x)$ associated to the chiral rotation (2.7). By definition, it is given by the variation of the action which is first order in the derivative $(\partial_\mu \alpha_a)$:

$$\delta I \equiv \int d_4 x \; A_{\mu a} (\partial_\mu \alpha_a) \tag{8.3}$$

For the matrix element (8.1) we only require the terms which are linear in the pion (or θ) field. We start from the expression (3.12) of the action. The second line gives no contribution. The contribution of the first line turns out to be (check this for yourself!):

$$\delta I = \frac{\varphi_0^2}{2} \operatorname{Tr} G_0 (\partial_\mu \theta_a) G_0 (\partial_\mu \alpha_a) \tag{8.4}$$

Using the matrix elements (6.2) we deduce that the part of the axial current which is linear in the pion field is:

$$A_{\mu a}(x) = \frac{\nu\varphi_0^2}{\beta\Omega} \sum_q (-iq_\mu) e^{-iqx} f(q)\theta(q) = \frac{\nu\varphi_0}{\sqrt{Z_\pi}} \frac{1}{\beta\Omega} \sum_q (-iq_\mu) e^{-iqx} f(q)\pi_a(q) \tag{8.5}$$

Assuming the pion field to be quantized as in (8.2), the matrix element (8.1) is:

$$\langle 0 | A_{\mu a}(x) | \pi_b(q) \rangle \equiv -i\delta_{ab} \frac{\nu\varphi_0}{\sqrt{Z_\pi}} f(q) \frac{q_\mu}{\sqrt{\omega_q}} \frac{e^{iqx}}{\sqrt{\Omega}} \tag{8.6}$$

For a pion at rest and neglecting the pion mass, we can set $f(q) \approx f(0)$ and deduce the following expression for f_π:

$$f_\pi = \varphi_0 \sqrt{Z_\pi} = \varphi_0 \sqrt{\nu f(0)} \tag{8.7}$$

Notice that the Eq.(8.7) implies that the pion field is related to the θ field by the equation:

$$\pi_a(x) = f_\pi \theta_a(x) \tag{8.8}$$

which is assumed in non-linear σ-models. The Eq.(8.7) is slightly modified when the pion has mass because of on-shell extrapolations (See Refs. [11,12] for example). These are small effects which we neglect in these lectures for the sake of clarity.

Since $f_\pi = 93$ MeV is known, we use the Eq.(8.7) as a constraint on the parameters of the model. Using the expression (7.6) of the function f(q) with q = 0 we can rewrite the Eq.(8.7) in the form:

$$\Lambda = \varphi_0 \, e^{\frac{8\pi^2 f_\pi^2}{\nu\varphi_0^2}} \tag{8.9}$$

The pion decay constant fixes the mass scale and Eq.(8.9) relates the cut-off to the vacuum quark constituent mass. The relation is shown on Fig.5. If one assumes this relation to hold, the theory becomes a one-parameter model which we choose to be the quark constituant mass φ_0.

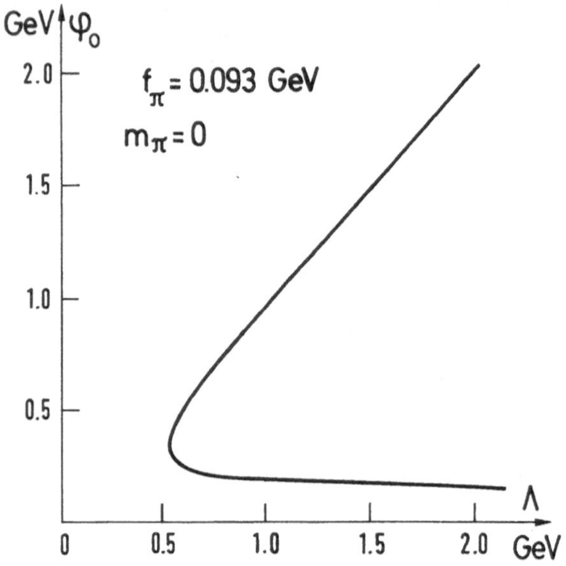

Figure 5. the relation (8.9) between the cut-off Λ and the constituent quark mass φ_0 obtained by fitting the pion decay constant f_π= 93 MeV.

Let us return to Fig.2 which shows the energy per unit volume of the vacuum as a function of φ. The curves correspond to different values of φ_0, the cut-off Λ being chosen so as to satisfy Eq.(8.9). Assume that φ_0 is small, say 0.2 GeV. The cut-off can then be quite large (see Fig.5). However the quark mass gap $2\varphi_0$ (see Fig.1) acquires then the uncomfortably low value of 0.4 GeV which would imply, for example, that a meson with a mass higher than 0.4 GeV would decay into real quark pairs! In general we would expect the theory to describe systems which are not excited above the mass gap $2\varphi_0$. So we don't want φ_0 to be too small. Assume now that φ_0 is large, say 0.6 GeV. We see from Eq.(8.9) that the cut-off Λ is then quite close to φ_0 so that the regularization (7.6) of the function f(q) becomes quite drastic although less so for the regularisation (7.5). Fig.2 shows that, for large values of φ_0, the vacuum becomes easily unstable against an increase of φ (a barrier no higher than 40 GeV/fm^3 would have to be overcome). It has also been reported that a high value of φ_0 tends to spoil the calculated electromagnetic decay width of the π_0 meson [35]. Other effects, not studied so far, might stabilize the vacuum, but at present large values of φ_0 also appear unattractive.

Most applications so far have used values close to the region where the cut-off is minimum: $\Lambda \simeq 0.6$ GeV. In this region Fig.3 shows that φ_0 is close to 0.3 GeV which is 1/3 of the nucleon mass and therefore close to the values used in constituent quark models. Notice however that this value is <u>not</u> arrived at by considering the nucleon. Most authors find that this is the region where best values are obtained for the quark condensate (4.9). The expression (4.9) can be regularized thus:

$$\langle\bar{\psi}\psi\rangle = -\frac{\varphi_0}{(2\pi)^4}\int d_4k\left(\frac{1}{k^2+\varphi_0^2} - \frac{1}{k^2+\Lambda_1^2} - \frac{\Lambda_1^2-\varphi_0^2}{(k^2+\Lambda_2^2)}\right) \tag{8.10}$$

This form involves two cut-off parameters Λ_1 and Λ_2. Indeed the quark condensate is a quadratically divergent quantity and therefore requires two subtractions to be regularized [13]. Alternatively one could regularise it by cutting off the low s values in the proper time representation [27]. For example:

$$\langle\bar{\psi}\psi\rangle = -\frac{\varphi_0}{(2\pi)^4}\int d_4k\int_{1/\Lambda^2}^{\infty} ds\; e^{-s\left(k^2+\varphi_0^2\right)} \tag{8.11}$$

The estimated value of the quark condensate $\langle\bar{\psi}\psi\rangle$ is $(250\ \text{MeV})^3$ [36]. This value is usually obtained in the minimum cut-off region of Fig.5. However the quark condensate is a quantity which is very sensitive to the cut-off procedure or to the profile function used to cut down the high quark-loop momenta (or the low proper-time values). This is because it has a quadratic divergence. In contrast, f_π which has only a logarithmic divergence is far less sensitive to the cut-off profile [13].

Summing up this discussion one can say that the parameters of the model are not ideal in the sense that the constituent quark mass is rather low (close to 0.3 GeV). However this value is remarquably close to the value required by constituent quark models to fit the nucleon mass. In the latter models the constituent quark mass is an input which the Nambu Jona-Lasinio model seems to explain. The low value of φ_0 would become more acceptable if one knew how to modify or extrapolate the Nambu Jona-Lasinio theory to energies higher than $2\varphi_0$. This is a subject of future development.

Let us note finally that $1/\sqrt{Z_\sigma}$ and $1/\sqrt{Z_\pi}$ represent the coupling constants of quarks to the σ and π fields. Indeed, for small values of θ, we have:

$$\bar{\psi}\varphi_0 U\psi = \frac{1}{\sqrt{Z_\sigma}}\bar{\psi}\sigma\psi + \frac{1}{\sqrt{Z_\pi}}\, i\bar{\psi}\,\vec{\pi}.\vec{\tau}\gamma_5\psi \tag{8.12}$$

Neglecting the q dependence of the function f(q), the coupling constants are equal to:

$$\frac{1}{\sqrt{Z_\sigma}} \simeq \frac{1}{\sqrt{Z_\pi}} = \frac{\varphi_0}{f_\pi} = 3.2 \qquad \text{for } \varphi_0 = 0.3 \text{ GeV} \tag{8.13}$$

We are dealing with a strongly coupled system.

9. The omega meson

We calculate the omega meson propagator starting from the third term I^{VV} of (5.3b). We begin by considering the first two terms of I^{VV} which are due to the quark loop:

$$I_Q^{VV} \equiv -\frac{\mathrm{Tr}}{2} G_0 V_\mu^2 - \frac{\mathrm{Tr}}{4} G_0 (\partial V + V\partial) G_0 (\partial V + V\partial) \quad \text{with} \qquad V \equiv \gamma_\mu V_\mu \quad \partial \equiv \gamma_\mu \partial_\mu \tag{9.1}$$

The quark loop contribution to the action (2.10) is gauge invariant, meaning that it is unchanged by the substitution $V_\mu \Rightarrow V_\mu + (\partial_\mu \alpha)$. We would like to maintain this property in spite of the non gauge invariant term $\frac{b2}{2} \int d_4 x\, V_\mu^2$ in the lagrangian. This may be achieved using the method of Schwinger [29]. As a bonus for maintaining gauge invariance, we will see that the quadratically divergent terms will drop out.

The first term of (9.1) may be cast into the form:

$$A \equiv -\frac{\mathrm{Tr}}{2} G_0 V_\mu^2 = -\frac{1}{2} 2\nu \sum_{kk'} \frac{1}{k^2 + \varphi_0^2} \langle k | V_\mu | k' \rangle \langle k' | V_\mu | k \rangle \tag{9.2}$$

We use the notation (6.2) for the matrix elements together with the proper time representation of the propagator:

$$\frac{1}{\beta\Omega} \sum_k \frac{1}{k^2 + \varphi_0^2} = \frac{1}{(2\pi)^4} \int d_4 k \int_0^\infty ds\, e^{-s(k^2 + \varphi_0^2)} = \frac{\pi^2}{(2\pi)^4} \int_0^\infty e^{-s\varphi_0^2} \frac{ds}{s^2} \tag{9.3}$$

We obtain thus:

$$A \equiv -\frac{\mathrm{Tr}}{2} G_0 V_\mu^2 = -\frac{1}{2} \frac{2\nu}{\beta\Omega} \sum_q V_\mu(q) V_\mu(-q) \frac{\pi^2}{(2\pi)^4} \int_0^\infty e^{-s\varphi_0^2} \frac{ds}{s^2} \tag{9.4}$$

The second term is:

$$-\frac{\mathrm{Tr}}{4} G_0 (\partial V + V\partial) G_0 (\partial V + V\partial) = -\frac{1}{4} \sum_{kk'} \frac{\mathrm{tr} \langle k | \partial V + V\partial | k' \rangle \langle k' | \partial V + V\partial | k \rangle}{\left(k^2 + \varphi_0^2\right)\left(k'^2 + \varphi_0^2\right)} \tag{9.5}$$

We trace the γ matrices in the numerator:

$$\mathrm{tr}\langle k|\partial V + V\partial|k'\rangle\langle k'|\partial V + V\partial|k\rangle = \mathrm{tr}(-kV_q k'V_{-q} - kV_q V_{-q} k - V_q k'k'V_{-q} - V_q k'V_{-q} k)$$

$$= -2(kV_q)(k'V_{-q}) - 2(kV_{-q})(k'V_q) + 2(kk')(V_qV_{-q}) - k^2(V_qV_{-q}) - k'^2(V_qV_{-q})$$

$$= -q^2(V_qV_{-q}) - 4(KV_q)(KV_{-q}) + (qV_q)(qV_{-q})$$

$$= -\frac{1}{2}F_{\mu\alpha}(q)F_{\mu\alpha}(-q) - 4(KV_q)(KV_{-q}) \tag{9.6}$$

To obtain the last line we wrote:

$$q^2(V_qV_{-q}) - (qV_q)(qV_{-q}) \equiv q^2V_\mu(q)V_\mu(-q) - q_\mu q_\alpha V_\mu(q)V_\alpha(-q) = \frac{1}{2}F_{\mu\alpha}(q)F_{\mu\alpha}(-q) \tag{9.7}$$

with $F_{\mu\alpha}(x) \equiv (\partial_\mu V_\alpha) - (\partial_\alpha V_\mu)$ a manifestly gauge invariant quantity.

Putting together the results (9.5) and (9.6) we obtain the second term in the form:

$$-\frac{\mathrm{Tr}}{4}G_0(\partial V+V\partial)G_0(\partial V+V\partial) =$$

$$= \frac{1}{4}\frac{2\nu}{\beta\Omega}\sum_q \frac{1}{2}F_{\mu\alpha}(q)F_{\mu\alpha}(-q)\ f(q) + \frac{1}{4}\frac{2\nu}{\beta\Omega}4\sum_q V_\mu(q)V_\alpha(-q)\ g_{\mu\alpha}(q) \equiv B+C \tag{9.8}$$

where f and g are the functions (7.1) and (7.2). We can extract the gauge invariant part of the term C by expliciting the function (7.2):

$$C = \frac{2\nu}{\beta\Omega}\sum_q V_\mu(q)V_\alpha(-q)\frac{\pi^2}{(2\pi)^4}\frac{1}{2}\int_{-1}^{1}du\int_0^\infty\frac{ds}{s}e^{-s\varphi_0^2 - sq^2(1-u^2)/4}$$

$$\left(\frac{\delta_{\mu\alpha}}{2s} + \frac{u^2}{4}\left(\delta_{\mu\alpha}q^2 - \delta_{\mu\alpha}q^2 + q_\mu q_\alpha\right)\right)$$

$$= \frac{2\nu}{\beta\Omega}\sum_q V_\mu(q)V_\mu(-q)\frac{\pi^2}{(2\pi)^4}\frac{1}{2}\int_{-1}^{1}du\int_0^\infty\frac{ds}{s}e^{-s\varphi_0^2 - sq^2(1-u^2)/4}\left(\frac{1}{2s} + \frac{u^2q^2}{4}\right)$$

$$-\frac{2\nu}{\beta\Omega}\sum_q \frac{1}{2}F_{\mu\alpha}(q)F_{\mu\alpha}(-q)\frac{\pi^2}{(2\pi)^4}\frac{1}{2}\int_{-1}^{1}du\int_0^\infty\frac{ds}{s}e^{-s\varphi_0^2 - sq^2(1-u^2)/4}\frac{u^2}{4}$$

$$\equiv D + E \tag{9.9}$$

The non gauge-invariant V_μ^2 terms are:

$$A + D = -\frac{1}{2}\frac{2\nu}{\beta\Omega}\sum_q V_\mu(q)V_\mu(-q)\frac{\pi^2}{(2\pi)^4}\int_0^\infty e^{-s\varphi_0^2}\frac{ds}{s^2}$$

$$+\frac{2\nu}{\beta\Omega}\sum_q V_\mu(q)V_\mu(-q)\frac{\pi^2}{(2\pi)^4}\frac{1}{2}\int_{-1}^{1}du\int_0^\infty\frac{ds}{s}e^{-s\varphi_0^2 - sq^2(1-u^2)/4}\left(\frac{1}{2s} + \frac{u^2q^2}{4}\right)$$

$$= \frac{1}{2}\frac{2\nu}{\beta\Omega}\sum_q V_\mu(q)V_\mu(-q)\ \frac{\pi^2}{(2\pi)^4}\int_0^\infty e^{-s\varphi_0^2}\frac{ds}{s}$$

$$\frac{1}{2}\int_{-1}^{1} du\left(-\frac{1}{s}+\frac{1}{s}e^{-sq^2(1-u^2)/4}+\frac{u^2q^2}{2}e^{-sq^2(1-u^2)/4}\right) = 0 \tag{9.10}$$

The reason it is zero is seen by integrating by parts:

$$\frac{1}{2s}\int_{-1}^{1} du\ e^{-sq^2(1-u^2)/4} = \frac{1}{2s}ue^{-sq^2(1-u^2)/4}\Big|_{-1}^{1} - \frac{1}{2s}\int_{-1}^{1}du\ \frac{sq^2u^2}{2}e^{-sq^2(1-u^2)/4}$$

$$= \frac{1}{2}\int_{-1}^{1}du\left(\frac{1}{s}-\frac{u^2q^2}{2}e^{-sq^2(1-u^2)/4}\right) \tag{9.11}$$

We are left with the gauge invariant terms B + E which can be written in the form:

$$I_Q^{VV} = \frac{1}{\beta\Omega}\sum_q \frac{1}{4}F_{\mu\alpha}(q)F_{\mu\alpha}(-q)\ \nu f_\omega(q) \tag{9.12}$$

$f_\omega(q)$ is given by the expression:

$$f_\omega(q) = \frac{\pi^2}{(2\pi)^4}\frac{1}{2}\int_{-1}^{1}du\int_0^\infty\frac{ds}{s}e^{-s\varphi_0^2 - sq^2(1-u^2)/4}(1-u^2) \tag{9.13}$$

which has only a logarithmic divergence and which may be regularized by one of the methods discussed in section 7.

Using (5.3b), (9.7) and (9.12) we obtain finally:

$$I^{VV} = \frac{1}{2\beta\Omega}\sum_q \nu f_\omega(q)V_\mu(q)\left(q^2\delta_{\mu\alpha} - q_\mu q_\alpha + \frac{b^2}{\nu f_\omega(q)}\delta_{\mu\alpha}\right)V_\alpha(-q) \tag{9.14}$$

We recognize, in parentheses, the inverse propagator of a massive vector meson. Let m_ω be its on-shell mass to which we fit the constant b^2:

$$b^2 = m_\omega^2 Z_\omega \qquad Z_\omega = \nu f_\omega(q)\big|_{q^2=-m_\omega^2} \tag{9.15}$$

The omega meson field is:

$$\omega_\mu(x) = \sqrt{Z_\omega}\ V_\mu(x) \tag{9.16}$$

10. Solitons.

Consider the possibility of forming bound states of quarks with the Nambu Jona-Lasinio lagrangian [27,17,38] . These are often called <u>solitons</u> because, if one intergrates out the quark degrees of freedom from the action, bound states of quarks appear as localized stationary solutions of the equations of motion of the meson fields. It is however simpler to deal with the quark degrees of freedom explicitely. One avoids this way Wess-Zumino terms and anomalies which are approximate quark loop effects and nothing more [32].

We begin by proving a useful property of the action which holds for stationary (time independent) systems. When the meson fields are time independent the Dirac operator (3.2) is:

$$D = \beta(\partial_\tau + h) \tag{10.1}$$

where h is the hermitian Dirac hamiltonian of a quark coupled to the meson fields. For the Dirac operator (3.2) for example, we have:

$$h = \frac{\vec{\alpha}.\vec{\nabla}}{i} + \beta\varphi U - \Phi + \vec{\gamma}.\vec{V} \tag{10.2}$$

The Dirac hamiltonian h is hermitian provided we do not make the analytic continuation $i\Phi \Rightarrow \eta$ as in (3.1) and in the following we shall assume the vector potential to be $V_\mu = (i\Phi,\vec{V})$ with Φ and $\vec{V}$ real. Let $|\lambda\rangle$ and e_λ be the eigenvectors and eigenvalues of the Dirac hamiltonian h:

$$h|\lambda\rangle = e_\lambda|\lambda\rangle \qquad \langle\lambda|\lambda\rangle = 1 \tag{10.3}$$

When h is time independent we can calculate traces in the basis $|k_0\rangle|\lambda\rangle$ where $\partial_\tau|k_0\rangle = ik_0|k_0\rangle$ so that:

$$\mathrm{Tr}\ \lg(\partial_\tau + h) = \sum_{k_0, e_\lambda} \lg(ik_0 + e_\lambda) = \sum_{k_0, e_\lambda} \lg(-ik_0 + e_\lambda) = \mathrm{Tr}\ \lg(-\partial_\tau + h) \tag{10.4}$$

It follows that:

$$\mathrm{Tr}\ \lg D = \mathrm{Tr}\ \lg D^\dagger = \frac{1}{2}\,\mathrm{Tr}\ \lg D^\dagger D \tag{10.5}$$

Therefore, provided one does not make the analytical continuation (3.1), one can calculate stationary states while omitting the imaginary part of the action (by dropping, for example, the last two terms of (3.12)). The latter however must be considered when dealing with rotating solitons.

We shall consider solitons in the approximation where the meson fields are classical (non quantized). In that case the hamiltonian H, given by the expression (4.1), describes independent quarks coupled to external meson fields. The eigenstates of H are Slater determinants composed of quark orbits which are the eigenstates of h as in (10.3). The energy of the system is:

$$E(\varphi,U,V) = \sum_{\lambda_{occ}} e_\lambda + \int d_3r \left(\frac{a^2\varphi^2}{2} + \frac{b^2V_\mu^2}{2}\right) \tag{10.6}$$

where λ_{occ} are the orbits which are occupied by the quarks.

In general we will be able to distinguish orbits which belong to the Dirac sea and a finite set of valence orbits (see Fig.6). For a localized soliton, the Dirac sea orbits are not the plane wave orbits considered in section 4 because the meson fields are not translationally invariant.

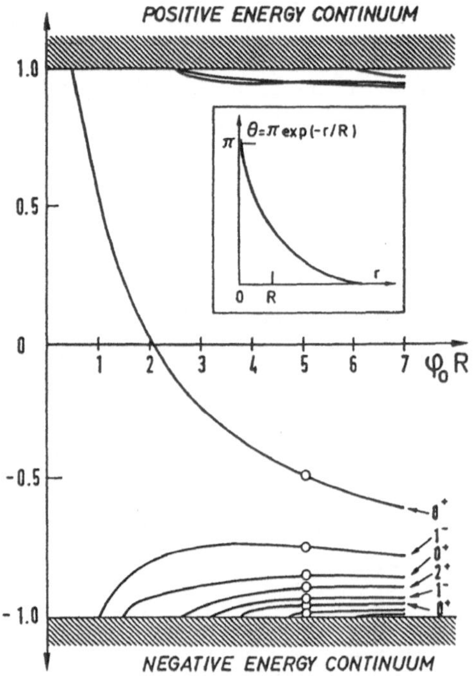

Figure 6. The spectrum of quark orbits in a hedgehog field taken from Ref.[37]. The orbits are plotted against the size parameter $R\varphi_0$ and they are in units of φ_0. The white dots show which orbits are filled for a baryon number 1 system like the nucleon. The quantum numbers are the parity and grand spin **K** = **I** + **T**.

Fig.6 shows that the valence orbits will not neccessarily be positive energy orbits. However, as long as an energy gap exists between them, Dirac sea and valence orbits can be distinguished without ambiguity. In this case we can separate the energy (10.6) into valence and Dirac sea contributions. Denoting by α the valence orbits we can write:

$$E(\varphi,U,V) = \sum_{\alpha} e_{\alpha} + E_D(\varphi,U,V) \tag{10.7}$$

The valence energies are calculated by solving the Dirac equation (10.3). The Dirac sea energy is the difficult part to calculate because it contains the ultra-violet divergencies of the theory. It is not proportional to the volume as in the translationally invariant case (4.8) because the fields differ from their vacuum values only in a localized region of space. Indeed the chiral field is assumed to acquire its vacuum value far away from the soliton:

$$U \underset{r \Rightarrow \infty}{\Rightarrow} 1 \qquad\qquad \varphi \underset{r \Rightarrow \infty}{\Rightarrow} \varphi_0 \tag{10.8}$$

We can calculate the energy from the action using the expression (B11). When the fields are treated classically, the second term can be neglected. The Dirac sea energy can be obtained from the action (3.12). As stated above, the second line does not contribute. For simplicity, consider the case where we neglect the contribution of the vector meson V_μ. A regularized form of the Dirac sea energy may be written in the form:

$$E_D(\varphi,U) = -\frac{1}{2}\,\mathrm{Tr}\,\lg(1+G_0\mathcal{V}) + \frac{1}{2}\,\mathrm{Tr}\,G_0\mathcal{V} - \frac{1}{4}\,\mathrm{Tr}\,G_0\mathcal{V}G_0\mathcal{V}$$

$$+\frac{1}{4}\,\mathrm{Tr}\,G_0\mathcal{V}G_0\mathcal{V} - \frac{1}{4}\,\mathrm{Tr}\,G_\Lambda\mathcal{V}G_\Lambda\mathcal{V} \tag{10.9}$$

where $G_\Lambda \equiv (-\partial^2 + \Lambda^2)^{-1}$. We have used the same Pauli-Villars type subtraction as in the translationally invariant case (4.8). If the cut-off Λ is close to or even smaller that φ_0 a proper-time regularisation may be better. The traces can be calculated quite simply in terms of the eigenvectors (10.3) of the Dirac hamiltonian (10.2). Indeed, when the fields are time independent, we have:

$$G_0^{-1} + \mathcal{V} \equiv -\partial^2 + \varphi^2 + i(\partial\varphi U) = -\partial_\tau^2 + h^2 \tag{10.10}$$

and since h commutes then with ∂_τ the basis $|k_0\rangle|\lambda\rangle$ may be used to calculate the traces in terms of the eigenvalues e_λ.

It is invariably found that a hegdehog shaped chiral field is energetically favored for the baryon number 1 system such as the nucleon or the delta. The method used to calculate the soliton is thus technically similar to the calculation of chiral solitons in the σ-model [21,33]. The main advantage of using the Nambu Jona-Lasinio model rather than the σ-model is that, as mentioned in section 7, the Nambu Jona-Lasinio model does not suffer from the σ-model instabilities due to high field gradients [30] thus allowing an unrestricted variation of the meson fields. Very recent soliton calculations (the last of Ref. [21]) show that an equilibrium size for the soliton can be achieved when the constituent quark mass exceeds about 350 MeV and that higher masses are required to bind the quarks. Further calculations will be required to determine the success or failure of the model to calculate the structure of the nucleon. Baryon structure using chiral models will be reviewed in forthcoming lectures [39].

By defining dimensionless distances and energies as follows:

$$\vec{x} \equiv \varphi_0 \vec{r} \qquad\qquad \epsilon_\lambda \equiv e_\lambda/\varphi_0 \tag{10.11}$$

it is simple to show that the soliton energy (10.7) is proportional to φ_0 and that its size is inversely proportional to φ_0. In a medium of finite baryonic density, the energy (and size) is proportional (and inversely proportional) to the asymptotic value of φ which decreases with baryonic density (see section 4). Thus the model explains, as other chiral models do, the swelling of nucleons in nuclear matter [26,24,17].

APPENDIX A

Plane-wave basis and traces in the Euclidian metric

We work with space-"time" coordinates in a Euclidian metric:

$$x_\mu = x^\mu = (\tau,\vec{r}) \qquad \int d_4x = \int_0^\beta d\tau \int d_3r \tag{A1}$$

where τ is the "time" variable and β the inverse temperature. Traces are evaluated in the plane wave basis:

$$\langle x|k\rangle = \frac{1}{\sqrt{\beta\Omega}}\, e^{ikx} \qquad kx = k_\mu x_\mu = k_0\tau + \vec{k}.\vec{r} \tag{A2}$$

where Ω is the integration volume. We assume periodic boundary conditions for the 4-momenta k_μ. In the limit $\beta \Rightarrow \infty$ of zero temperature and for infinite volume, we have:

$$\frac{1}{\beta\Omega}\sum_k = \frac{1}{(2\pi)^4}\int d_4k \qquad d_4k = dk_0\,dk_1\,dk_2\,dk_3 \tag{A3}$$

where each component k_μ is integrated from $-\infty$ to $+\infty$. The plane wave basis (A2) forms a complete orthonormal set for functions which are periodic in the interval $(0,\beta)$ and on the boundary of Ω:

$$\int d_4x\ |x\rangle\langle x| = \sum_k |k\rangle\langle k| = 1 \qquad \langle k|k'\rangle = \delta_{k,k'} \qquad \langle x|x'\rangle = \delta(x-x') \tag{A4}$$

The trace notation used in the text is:

$$\mathrm{Tr}\,A = \sum_k \mathrm{tr}\ \langle k|A|k\rangle = \sum_{ks}\langle ks|A|ks\rangle = \int d_4x \sum_s \langle xs|A|xs\rangle \tag{A5}$$

where $\sum_s$ is a sum over the internal (Dirac, isospin, color ...) variables. For Dirac particles the degeneracy of the plane wave states is ν. We take $\nu_q = 12$ for 2-flavor (u and d) quarks and $\nu_N = 4$ for the nucleons. With this definition:

$$\sum_s = 2\nu \tag{A6}$$

Calculation of the field propagators

This is a friendly reminder of the way to calclulate field propagators in the quadratic (small amplitude) approximation. We consider a system described by a Euclidean action:

$$I(\varphi) = \int d_4x \, \mathcal{L}(\varphi) \tag{B.1}$$

which is a functional of boson fields $\varphi_a(x)$. We work in the Euclidean metric (appendix A) with $x_\mu = x^\mu = (\tau, \vec{r})$ and $\int d_4x = \int_0^\beta d\tau \int_\Omega d_3r$ where β is the inverse temperature and Ω the volume of the system. The partition function is given by the path integral [28]:

$$Z \equiv \mathrm{Tr}\, e^{-\beta H} = \int \mathcal{D}(\varphi)\, e^{-I(\varphi)} \equiv \int \mathcal{D}(\varphi)\, e^{-\int_0^\beta d\tau \int_\Omega d_3r\, \mathcal{L}(\varphi)} \tag{B.2}$$

The energy of the system is $E = -\dfrac{\partial}{\partial\beta} \ln Z$.

To study the excited states of the system we introduce a source j_a for each field φ_a and we define the functional:

$$Z(j) = \mathrm{Tr}\, e^{-\beta\left(H - \int d_3r\, j.\varphi\right)} = \int \mathcal{D}(\varphi)\, e^{-I(\varphi) + \int d_4x\, j_a(x)\, \varphi_a(x)} \tag{B.3}$$

Using the T-product expansion of Dyson:

$$e^{-\beta\left(H - \int d_3r\, j.\varphi\right)} = e^{-\beta H}\, T\, e^{\int d_4x\, j_a(x)\, \varphi_a(x)} \tag{B.4}$$

with $\varphi_a(x) \equiv \varphi_a(\tau, \vec{r}) = e^{\tau H} \varphi_a(\vec{r}) e^{-\tau H}$, we see that expectation values of time-ordered products of fields are given by:

$$\langle \varphi_a(x) \rangle = \frac{\partial \ln Z}{\partial j_a(x)}$$

$$\langle T[\varphi_a(x)\varphi_b(y)] \rangle - \langle \varphi_a(x) \rangle \langle \varphi_b(y) \rangle = \frac{\partial^2 \ln Z}{\partial j_a(x) \partial j_b(y)} \tag{B.5}$$

where the source is set to zero after the derivatives are taken.

To evaluate the generating functional (B.3) in the small amplitude approximation we set $\varphi = \varphi_c + \tilde{\varphi}$ and we choose the "classical value" φ_c of the field, to be a stationary point of the action:

$$\left.\frac{\partial I}{\partial \varphi_a(x)}\right|_{\varphi = \varphi_c} = 0 \tag{B.6}$$

We then expand the action up to second order in the fluctuating part $\tilde{\varphi}$:

$$I(\varphi) = I\left(\varphi_c + \tilde{\varphi}\right) = I(\varphi_c) + \frac{1}{2}\int d_4x\, d_4y\, \tilde{\varphi}_a(x) \left.\frac{\partial^2 I}{\partial\varphi_a(x)\partial\varphi_b(y)}\right|_{\varphi=\varphi_c} \tilde{\varphi}_b(y)$$

$$\equiv I_c + I^{(2)} \tag{B.7}$$

The quadratic (small amplitude) approximation consists in evaluating the generating functional (B.3) with the action expanded only up to second order, as in (B.7):

$$Z(j) = e^{-I_c + j.\varphi_c} \int \mathcal{D}(\tilde{\varphi})\, e^{-\frac{1}{2}\tilde{\varphi}K^{-1}\tilde{\varphi} + j.\tilde{\varphi}} \tag{B.8}$$

where we use the abbreviated notation:

$$I^{(2)} = \frac{1}{2}\tilde{\varphi}K^{-1}\tilde{\varphi} \equiv \frac{1}{2}\int d_4x\, d_4y\, \tilde{\varphi}_a(x)\, \langle xa|K^{-1}|yb\rangle\, \tilde{\varphi}_b(y)$$

$$\langle xa|K^{-1}|yb\rangle \equiv \left.\frac{\partial^2 I}{\partial\varphi_a(x)\partial\varphi_b(y)}\right|_{\varphi=\varphi_c}$$

$$j.\varphi \equiv \int d_4x\, j_a(x)\, \varphi_a(x) \qquad\qquad I_c \equiv I(\varphi_c) \tag{B.9}$$

The integral (B.8) is gaussian and it is easily evaluated:

$$\ln Z(j) = -I_c + j.\varphi_c - \frac{1}{2}\mathrm{Tr}\ln K^{-1} + \frac{1}{2}jKj \tag{B.10}$$

When the classical field φ_c is time independent, the energy of the system is, in this approximation:

$$E = \frac{1}{\beta} I_c + \frac{1}{2\beta}\mathrm{Tr}\ln K^{-1} \tag{B.11}$$

The first term is the "classical energy" and the second term is the "one loop" contribution. It is due to the zero-point fluctuations of the normal modes of the system (also referred to as the ground state correlation energy in the random phase approximation). The expression (B.11) is only valid in the zero temperature limit ($\beta \Rightarrow \infty$) when lnZ is strictly proportional to β.

The expression (B.10) can also be used in conjunction with (B.5) in order to calculate expectation values of time-ordered products in the small amplitude approximation. We thus find that the "classical value" φ_c, which makes the action stationary, is equal to the expectation value of the field:

$$\langle \varphi_a(x) \rangle = \frac{\partial \ln Z}{\partial j_a(x)} = \varphi_c(x) \tag{B.12}$$

We also find that expectation values of time-ordered products of fields are equal to the matrix elements of K:

$$\langle T[\varphi_a(x)\varphi_b(y)]\rangle - \langle\varphi_a(x)\rangle\langle\varphi_b(y)\rangle \equiv \langle T[\varphi_a(x)\varphi_b(y)]\rangle_c$$

$$= \frac{\partial^2 \ln Z}{\partial j_a(x)\partial j_b(y)} = \langle xa|K|yb\rangle \tag{B.13}$$

We see that K is the propagator of the fields, also called the Green's function.

We consider a translationally invariant system in the limit ($\beta\Rightarrow\infty$) of zero temperature. The propagators of such a system acquire a simple form. Let $P_\mu = \left[i(H-E_0),\vec{P}\right]$ be the Euclidean energy-momentum vector, where H is the Hamiltonian, E_0 the ground state energy and $\vec{P}$ the momentum operator. We can write:

$$\varphi(x) = e^{-ixP}\,\varphi\,e^{ixP} \equiv e^{\tau H - i\vec{r}.\vec{P}}\,\varphi\,e^{-\tau H + i\vec{r}.\vec{P}} \tag{B.14}$$

In the zero temperature limit, expectation values become ground state expectation values [28] . Let $|\Phi_0\rangle$ be the ground state and assume it has zero momentum:

$$H|\Phi_0\rangle = E_0|\Phi_0\rangle \qquad \vec{P}|\Phi_0\rangle = 0 \qquad P_\mu|\Phi_0\rangle = 0 \tag{B.15}$$

The propagator (B.13) is then equal to:

$$K_{ab}(x_1 - x_2) \equiv \langle x_1 a|K|x_2 b\rangle$$

$$= \langle\Phi_0|\varphi_a e^{i(x_1-x_2)P}\varphi_b|\Phi_0\rangle_c\,\theta(\tau_1-\tau_2) + \langle\Phi_0|\varphi_b e^{-i(x_1-x_2)P}\varphi_a|\Phi_0\rangle_c\,\theta(\tau_2-\tau_1) \tag{B.16}$$

It follows that the propagator K is diagonal in the momentum representation:

$$\langle q_1 a|K|q_2 b\rangle = \delta_{q_1 q_2}K_{ab}(q_1) \qquad K_{ab}(q) = \int d_4x\, e^{-iqx}K_{ab}(x) \tag{B17}$$

The excited states $|\Phi_\alpha\rangle$ of the translationally invariant system may be labelled by their enrer E_α and momentum $\vec{k}_\alpha$::

$$|\Phi_\alpha\rangle \equiv \left|\alpha\, E_\alpha\, \vec{k}_\alpha\right\rangle \qquad H|\Phi_\alpha\rangle = E_\alpha|\Phi_\alpha\rangle \qquad \vec{P}|\Phi_\alpha\rangle = \vec{k}_\alpha|\Phi_\alpha\rangle$$

$$P_\mu|\Phi_\alpha\rangle = k_{(\alpha)\mu}|\Phi_\alpha\rangle \qquad k_{(\alpha)\mu} = \left(i(E_\alpha - E_0),\, \vec{k}_\alpha\right) \tag{B.18}$$

Inserting the complete set of states $|\Phi_\alpha\rangle$ into the expression (B.16) of the propagator and using (B.17), we obtain the spectral decomposition of K(q):

$K_{ab}(q) =$

$$(2\pi)^3 \sum_{\alpha\neq 0} \left(\delta(\vec{k}_\alpha - \vec{q}) \frac{\langle \Phi_0 | \varphi_a | \Phi_\alpha \rangle \langle \Phi_\alpha | \varphi_b | \Phi_0 \rangle}{E_\alpha - E_0 + i q_0} + \delta(\vec{k}_\alpha + \vec{q}) \frac{\langle \Phi_0 | \varphi_b | \Phi_\alpha \rangle \langle \Phi_\alpha | \varphi_a | \Phi_0 \rangle}{E_\alpha - E_0 - i q_0} \right) \tag{B.18}$$

where q is the 4-vector $q_\mu = (q_0, \vec{q})$.

The spectral decomposition (B.18) allows us to give a physical interpretation of the propagator. The poles of the propagator in the complex q_0 plane occur on the imaginary axis. In view of the spectral decomposition (B.18) they can be interpreted as excitation energies of states with momentum $\vec{q}$ and energy $iq_0 = \pm (E_\alpha - E_0)$. Propagators calculated at real q_0 occur in the calculation of the equation of state of dense matter. The calculation of on-shell mesons requires imaginary q_0.

Since K is diagonal in the momentum representation, we can write the second-order variation of the action in the form:

$$I^{(2)} = \frac{1}{2} \tilde{\varphi} K^{-1} \tilde{\varphi} = \frac{1}{2\beta\Omega} \sum_q \tilde{\varphi}_a(q) \, K^{-1}_{ab}(q) \, \tilde{\varphi}_b(-q) \tag{B.19}$$

where $\tilde{\varphi}_a(q) = \int d_4x \, e^{iqx} \, \tilde{\varphi}_a(x)$.

ACKNOWLEDGMENTS

We thank R. Mendez Galain for a critical reading of the manuscript and for many suggested improvements. We are also grateful to R.Alkofer, V.Bernard, W.Broniowski, D.Diakonov, S.Kahana, M.Praszalowicz and I.Zahed for useful discussions.

REFERENCES

[1] Y.Nambu and G.Jona-Lasinio, Phys.Rev. 122 (1961) 354.
[2] D.Ebert and H.Reinhardt, Nucl.Phys. B271 (1986) 188.
[3] R.D.Ball, "Desperately seeking mesons", Crakow workshop on Skyrmions and anomalies, 1987 (World Scientific).
[4] M.Goldberger and S.Treiman, Phys.Rev.110 (1958) 996.
[5] K.Kawarabayashi and M.Suzuki, Phys.Rev.Lett. 16 (1966) 255. Riazuddin and Fayazuddin, Phys.Rev. 147 (1966) 1071.
[6] S.Weinberg, Phys.Rev.Lett.18 (1984) 142.
[7] J.Wess and B.Zumino, Phys.Lett. 18 (1971) 95.
[8] M.Gell-Mann, R.J.Oakes and B.Renner, Phys.Rev. 175 (1968) 2195.
[9] M.Bando, T.Kugo and Y.Yamazaki, Phys.Rev.Lett. 54 (1985) 1215; Nucl.Phys. B159 (1985) 493; Progr.Theor.Phys. 73 (1985) 1541. O.Kayamakcala, S.Rajeev and J.Schechter, Phys.Rev. D20 (1984) 2345.
[10] N.Kroll, T.D.Lee and B.Zumino, Phys.Rev. 157 (1967) 1376.
[11] V.Bernard, Ulf G.Meissner and I.Zahed, Phys.Rev.Lett. 59 (1987) 966. V.Bernard, R.L.Jaffe and Ulf G.Meissner, Nucl.Phys. B308 (1988) 753.

V. Bernard and Ulf G.Meissner, Phys.Rev. D38 (1988) 1551; Nucl.Phys. A489 (1988) 647.
E.Ruiz Arriola, P.Alberto and K.Goecke, Zeits.Phys. A333 (1989) 203.
M.Wakamatsu and W.Weise, Zeitschr.Phys. A331 (1988) 173.

[12] M.Jaminon, G.Ripka and P.Stassart, "Chiral mesons in dense nuclear matter", Saclay preprint SPhT/88-207.

[13] M. Jaminon, G.Ripka and P.Stassart, Phys.Lett. B227 (1989) 191.

[14] Phys.Reports 163 (1988).
S.H.Kahana, J.Cooperstein and E.Baron, Phys.Lett. B196 (1987) 259.

[15] W.A.Bardeen, C.T.Hill and M.Lindner, "Minimal dynamical symmetry breaking of the standard model", Fermilab preprint FERMI-PUB-89/127-T.

[16] D.I.Diakonov and V.Yu Petrov, Nucl.Phys. B245 (1984) 259; Nucl.Phys. B272 (1985) 457; Sov.Phys.JETP 62 (1985) 204.

[17] E.Ruiz, C.V.Christov and K.Goeke, "Medium effects on nucleon properties", Bochum preprint (1989).
W.Weise, invited talk at the International Nuclear Physics Conference, Sao Paolo, Brazil, 1989.

[18] See the lectures of N.Isgur in the present volume.
R.Kokoski and N.Isgur, Phys.Rev. D35 (1987) 907.
S.Capstick and N.Isgur, Phys.Rev. D34 (1986) 2809.
A.B.Migdal, S.B.Khokhlachev and L.N.Scur, Sov.Phys.JETP 64 (1986) 441.
M.Baker, J.S.Ball and F.Zachariasen, Phys.Rev. D37 (1988) 1036; Phys.Rev. D34 (1986) 3894.

[19] W.Broniowski, Crakow workshop on skyrmions and anomalies, (World Scientific) 1987.
A.Schuh and H.J.Pirner, Phys.Lett.B173 (1986) 19.
G.Chanfray and H.J.Pirner, Phys.Rev. C35 (1987) 760.

[20] G.Ripka, "Nambu Jona-Lasinio models applied to dense hadronic matter", XII Workshop on Nuclear Physics, Iguazu Falls, Argentina, 1989 (to be published by World Scientific).

[21] S.Kahana, G.Ripka and V.Soni, Nucl.Phys. A415 (1984) 351.
M.C.Birse and M.K.Banerjee, Phys.Rev.D31 (1985) 118.
Th.Meissner, E.R.Arriola, F.Grummer, H.Mavromatis and K.Goeke, Phys.Lett. B214 (1988) 312.

[22] A.Chodos and C.B.Thorn, Phys.Rev. D12 (1975) 2733.

[23] T.H.R.Skyrme, Nucl.Phys. 31 (1962) 556.
G.S.Adkins, C.R.Nappi and E.Witten, Nucl.Phys. B228 (1983) 552.

[24] G.Ripka, "Can we control the proliferation of nucleon models?", in Windsurfing the Fermi sea, Volume 2, (North Holland) 1987.

[25] B.D.Serot and J.D.Walecka, Advances in Phys. 16 (1986) Plenum Press.

[26] G.E.Brown, Lecture delivered at the XII Workshop on Nuclear Physics, Iguazu Falls, Argentina, 1989 (to be published by World Scientific).
G.E.Brown, A.Sethi and N.M.Hintz, "Proton-nucleus scattering and the swelling of the nucleon", Stony-Brook preprint, 1989.

[27] D.I.Diakonov, V.Yu Petrov and P.V.Pobylitsa, Nucl.Phys. B306 (1988) 809.

D.I. Diakonov, V.Yu Petrov and M.Praszalowicz, Nucl.Phys. B323 (1989) 53.
M.Praszalowicz, "Solitons in the Chiral Quark Model", Brookhaven National Laboratory preprint (1989).

[28] J.P.Blaizot and G.Ripka "Quantum Theory of Finite systems" MIT Press (1987)

[29] J.S.Schwinger, Phys.Rev. 82 (1951) 664

[30] G.Ripka and S.Kahana, Phys.Rev. D36 (1987) 1233.
V.Soni, Phys.Lett. B195 (1987) 569.

[31] C.Itzykson and J.B.Zuber, Quantum Field Theory, McGraw-Hill, 1980.

[32] I.A.R.Aitchison and C.W.Fraser, Phys.Rev. D31 (1985) 2605.

[33] W.Broniowski and M.K.Banerjee, Phys.Lett. B167 (1986) 21.

[34] L.N.Epele, H.Fanchiotti, C.A.Garcia Canal and R.Mendez Galain, Phys.Rev. D39 (1989) 1473;"Isospin violation in chiral effective models", La Plata preprint, 1989.

[35] A.H.Blin, B.Hiller and M.Schaden, Zeitschr.Phys. A331 (1988) 75.

[36] J.Gasser and H.Leutwyler, Ann.Phys. 158 (1984) 142.
M.A.Schifman, A.I.Vainshtein and V.I.Zakharov, Nucl.Phys. B147 (1979) 385,448,519.

[37] G.Ripka and S.Kahana, Phys.Lett. B155 (1985) 327.

[38] H.Reinhardt and R.Wunsch, Phys.Lett. B215 (1988) 577.

[39] G.Ripka, "Soliton models of baryons", Les Houches School on Hadronic Physics with Multi-GeV Electrons, Febuary 1990.

[40] W.A.Bardeen, J.Bijnens and J.-M.Gérard, Phys.Rev.Lett. 62 (1989) 1343; J.-M.Gérard, "Physics below one GeV", MPI München preprint (1989).

[41] K.Goecke, J.N.Urbano, M.Fiolhais and M.Harvey, Phys.Lett. B164 (1985) 249.
J.P.Blaizot and G.Ripka, Phys.Rev. D38 (1988) 1556.

THE QCD VACUUM AND CHIRAL SYMMETRY [1]

Jürg Gasser

Institut für Theoretische Physik
Universität Bern
Sidlerstrasse 5, CH-3012 Bern, Switzerland

1 INTRODUCTION

These lectures review low-energy processes in the framework of QCD. In particular, methods and results of chiral perturbation theory (χPT) will be discussed.

At low energies $E << M_W$ the interactions of leptons and hadrons are described by QCD + QED up to corrections of order (E/M_W). In the following I shall disregard electromagnetic contributions and correspondingly I set $e = 0$. We are then left with QCD, whose Lagrangian $\mathcal{L}_{QCD}$ contains only a few parameters: the renormalization group invariant scale Λ and the running quark masses $m_u, m_d, m_s \ldots$, whose absolute values are [1,2]

$$\begin{aligned} m_u(1\text{GeV}) &= 5.1 \pm 1.5\text{MeV} \\ m_d(1\text{GeV}) &= 8.9 \pm 2.6\text{MeV} \\ m_s(1\text{GeV}) &= 175 \pm 55\ \text{MeV} \\ m_c(1\text{GeV}) &= 1.35 \pm .05\text{GeV} \\ m_b(1\text{GeV}) &= 5.3 \pm .1\text{GeV}\,. \end{aligned} \tag{1.1}$$

The quark masses m_u, m_d and m_s are small on a typical hadronic scale like the mass of the rho or of the proton,

$$\frac{m_u}{M_p} = .005\,, \qquad \frac{m_d}{M_p} = .01\,, \qquad \frac{m_s}{M_p} = .18\,. \tag{1.2}$$

It therefore makes sense to consider the limit where these masses are set equal to zero, $m_u = m_d = m_s = 0$ (*chiral limit*). The remaining quarks $c, b, \ldots$ are not

[1]Work supported in part by Schweizerischer Nationalfonds

Hadrons and Hadronic Matter
Edited by D. Vautherin *et al.*
Plenum Press, New York, 1990

light: although one may of course study the theoretical limit in which these masses also vanish, it does not seem to be possible to recover the actual mass values by an expansion around that limiting case. At low energies a better approximation is obtained if the quarks $c, b, \ldots$ are instead treated as infinitely heavy. In this limit the degrees of freedom associated with these quarks freeze and may be ignored in the effective low-energy theory.

In the chiral limit QCD thus contains only one parameter, the scale Λ. The mass of the proton is some pure number multiplying Λ and likewise for all the other states of the theory - the numbers $M_\rho/M_p, M_\Delta/M_p, \ldots$ are determined in a parameter free manner. In this sense the chiral limit of QCD may be called a theory without any adjustable parameters: QCD is of course unable to predict the value of M_p in GeV units, but it determines all dimensionless hadronic quantities in a parameter free manner. The elastic cross section for pp scattering e.g. is some fixed function of the variables s/M_p^2 and t/M_p^2 , multiplying the square of the Compton wavelength of the proton.

It is well known that it is unfortunately very difficult to actually *calculate* masses, cross sections and decay amplitudes in this beautiful theory, because $\mathcal{L}_{QCD}$ is formulated in terms of quark and gluon-fields which do not create asymptotically observed particles (confinement). Several methods have therefore been devised in the past to cope with this problem in different regimes of the energy scale:

i) Processes at high energies. At high energies, the effective coupling constant α_{QCD} becomes small, and conventional perturbation theory in α_{QCD} is applicable.

ii) Lattice calculations. This is the only method known today which leads directly from the QCD Lagrangian to the mass spectrum, decay matrix elements, cross sections ... (see the lectures by J. Negele [3]). Unfortunately the CPU time needed for full fledged QCD calculations is enormous, and I believe that one will have to wait still a long time before this program achieves the accuracy one is aiming at in χPT. [In order to reproduce e.g. the asymmetries in the mass spectrum, the lattice must extend to distances which are large in comparison to M_π^{-1}.]

iii) Chiral perturbation theory. This method exploits the symmetry of the QCD Lagrangian and its ground state: one solves in a perturbative manner the constraints imposed by chiral symmetry and unitarity by expanding the Green functions in powers of the external momenta and of the quark masses m_u, m_d and m_s. To illustrate the idea consider elastic pion-nucleon forward scattering at low energies in the chiral limit. Chiral symmetry implies that $T^{\pi^- p}$ has the following form near threshold [4]

$$T^{\pi^- p} = \frac{q}{2F_\pi^2}(1 - g_A^2)(1 + \cdots). \tag{1.3}$$

I have denoted by q the pion momentum in the center of mass frame. $F_\pi = 93.3$MeV ($g_A = -1.259$) is the pion (neutron) decay constant, and the dots denote terms of order (q/M_p).

The result (1.3) displays the first order term in a systematic expansion of the amplitude $T^{\pi^- p}$ in powers of q/M_p. This term algebraically dominates the remainder for sufficiently small values of q and thus provides an accurate parametrization of the full amplitude near threshold. Symmetry considerations alone do determine neither F_π nor g_A - they do pin down however the leading term in the low-energy expansion as a function of these two parameters.

As one goes away from threshold, the higher order terms (denoted by dots in eq. (1.3)) come into play. I will show in the following sections that χPT is a method which allows to determine these corrections in a systematic manner, including terms which are due to nonvanishing quark masses [5,6,7,8,9,10,11,12]. (The use of χPT is of course not restricted to the pion-nucleon amplitude, it applies to the low-energy expansion of any Green function in QCD. The method extends also to the weak interactions [13] and to the evaluation of quark mass and finite size effects in lattice calculations [14].)

The remaining sections are organized as follows. In section 2 chiral symmetry and its spontaneous breakdown are reviewed. In section 3 I discuss various properties of the QCD ground state. Sections 4, 5 and 6 contain an outline of χPT, and in sections 7 and 8 I present some results obtained in this framework.

2 CHIRAL SYMMETRY AND ITS SPONTANEOUS BREAKDOWN

To simplify the discussion I consider the chiral limit throughout this section (unless stated otherwise) and I correspondingly set $m_u = m_d = m_s = 0$. In this limit the QCD Lagrangian for 3 flavours is

$$\mathcal{L}^{\circ}_{QCD} = -\frac{1}{2g^2} < G_{\mu\nu}G^{\mu\nu} > + \bar{q}i\gamma^{\mu}D_{\mu}q \tag{2.1}$$

where $G_{\mu\nu}$ is the field strength associated with the gauge-field A_μ, and $D_\mu = \partial_\mu - iA_\mu$ is the covariant derivative. (Note that I include the coupling constant g in the gauge-field A_μ.) The symbol $< A >$ stands for the trace of the matrix A. The column vector q contains the three light quark fields u, d, s

$$q = \begin{pmatrix} u \\ d \\ s \end{pmatrix} . \tag{2.2}$$

The index o in eq. (2.1) reminds us that $m_u = m_d = m_s = 0$. $\mathcal{L}^{\circ}_{QCD}$ is a gauge-field theory: it is invariant under local gauge transformations which act on the *colour* indices,

$$\mathcal{L}^{\circ}_{QCD}(G'_{\mu\nu}, q', D'_{\mu}q') = \mathcal{L}^{\circ}_{QCD}(G_{\mu\nu}, q, D_{\mu}q) \tag{2.3}$$

where

$$\begin{aligned} A'_\mu &= UA_\mu U^\dagger + iU\partial_\mu U^\dagger \\ G'_{\mu\nu} &= UG_{\mu\nu}U^\dagger \\ \begin{pmatrix} q'_1 \\ q'_2 \\ q'_3 \end{pmatrix} &= U \begin{pmatrix} q_1 \\ q_2 \\ q_3 \end{pmatrix} ; \qquad q = u, d, s \\ D'_\mu &= \partial_\mu - iA'_\mu . \end{aligned} \tag{2.4}$$

2.1 Chiral symmetry

In addition to the gauge transformation (2.4), $\mathcal{L}^{\circ}_{QCD}$ is also invariant under the following group of (space-time independent) transformations which act on the *flavour* indices:

$$\mathcal{L}^{\circ}_{QCD}(G_{\mu\nu}, q', D_\mu q') = \mathcal{L}^{\circ}_{QCD}(G_{\mu\nu}, q, D_\mu q) \tag{2.5}$$

where

$$q' = g_R \frac{1}{2}(1+\gamma_5)q + g_L \frac{1}{2}(1-\gamma_5)q$$
$$g_I g_I^\dagger = \mathbf{1}, \qquad \det g_I = 1; \qquad I = L, R. \tag{2.6}$$

The gauge-field A_μ is not touched by these transformations. The group of transformations (2.6) is called $SU(3)_R \times SU(3)_L$, and the relation (2.5) goes under the name 'chiral symmetry of QCD'. There are $2 \times (3^2 - 1) = 16$ conserved Noether currents associated with the symmetry (2.5),

$$\begin{aligned} J_I^{\mu a} &= \bar{q}_I \gamma^\mu \frac{\lambda^a}{2} q_I \\ \partial_\mu J_I^{\mu a} &= 0; \qquad I = L, R; \qquad a = 1, \ldots, 8 \\ q_{R,L} &\doteq \frac{1}{2}(1 \pm \gamma_5) q \, . \end{aligned} \tag{2.7}$$

(λ^a denote the Gell-Mann matrices.)

The associated 16 conserved charges

$$\begin{aligned} Q_I^a &\doteq \int_{x^\circ = const} J_I^{oa} d^3x \\ \frac{dQ_I^a}{dx^0} &= 0 \end{aligned} \tag{2.8}$$

generate the algebra of $SU(3)_R \times SU(3)_L$,

$$\begin{aligned} \left[Q_I^a, Q_I^b\right] &= i f^{abc} Q_I^c \\ \left[Q_L^a, Q_R^b\right] &= 0 \, . \end{aligned} \tag{2.9}$$

Applications of the commutation relations (2.9) to hadronic production and decay processes are called "current algebra calculations", "PCAC relations", ... They were very fashionable in the late sixties and early seventies [15]. Manipulations with these commutators is replaced in our days by chiral perturbation theory.

2.2 Spontaneous breakdown of chiral symmetry

If massless QCD is a good approximation to our world, the ground state of the theory cannot be symmetric with respect to $SU(3)_R \times SU(3)_L$ (in contrast to the Lagrangian $\mathcal{L}^{\circ}_{QCD}$), i.e.,

$$Q_L^a \mid 0 >= Q_R^a \mid 0 >= 0 \qquad \text{(symmetric ground state)} \tag{2.10}$$

is impossible. Indeed if (2.10) was true, then the two point functions of the vector and axial currents

$$\begin{aligned} V^{\mu a} &= J_R^{\mu a} + J_L^{\mu a} \\ A^{\mu a} &= J_R^{\mu a} - J_L^{\mu a} \end{aligned} \tag{2.11}$$

would be the same,

$$< 0 \mid A^a_\mu(x) A^b_\nu(y) \mid 0 > = < 0 \mid V^a_\mu(x) V^b_\nu(y) \mid 0 > , \tag{2.12}$$

because $< 0 \mid J_R^{a\mu} J_L^{b\nu} \mid 0 >$ vanishes if (2.10) holds. The Fourier transform of the two point functions $V^{ab}_{\mu\nu} =< 0 \mid V^a_\mu V^b_\nu \mid 0 >$ and $A^{ab}_{\mu\nu} =< 0 \mid A^a_\mu A^b_\nu \mid 0 >$ have been determined using experimental data on the semileptonic weak decays $\tau \to \nu_\tau + n\pi$ [16]. They turn out to be very different from each other: $V^{ab}_{\mu\nu}$ is peaked at M_ρ = 770 MeV where $A^{ab}_{\mu\nu}$ is smooth. The axial two point function $A^{ab}_{\mu\nu}$ on the other hand contains a peak at $M_{a_1} \simeq$ 1100 MeV where $V^{ab}_{\mu\nu}$ is smooth. This result is in contradiction to the relation (2.12), and one concludes that the physical ground state must be asymmetric under $SU(3)_R \times SU(3)_L$ [17],

$$\text{Symmetry of } \mathcal{L}^\circ_{QCD} \neq \text{Symmetry of ground state}. \tag{2.13}$$

This fundamental fact is called "spontaneous breakdown of chiral symmetry".

A spontaneously broken symmetry calls for Goldstone bosons: if some generator Q^a of $SU(3)_R \times SU(3)_L$ does not leave the vacuum invariant, $Q^a \mid 0 > \neq 0$, then we must have a physical state $Q^a \mid 0 >$ with the same energy eigenvalue as $\mid 0 >$, because H_{QCD} commutes with Q^a. If Q^a is a vector charge then the state $Q^a \mid 0 >$ describes a massless scalar, if Q^a is an axial charge then we get a massless pseudoscalar [18].

The eight lightest hadrons (π, K, η) are pseudoscalars. They are the Goldstone particles associated with the axial charges. That these particles are not exactly massless as it would be appropriate for Goldstone particles is attributed to the fact that $\mathcal{L}^\circ_{QCD}$ is not really chirally invariant - the masses of the light quarks break the symmetry $SU(3)_R \times SU(3)_L$. If we were able to vary the quark mass parameters in the real world then we could see that π, K and η indeed become massless as we let m_u, m_d and m_s tend to zero.

The vector symmetry is not broken spontaneously: i) there is no trace of light scalars in the spectrum of mesons (the lightest scalars are heavier than the proton) ii) there is very good evidence for the corresponding symmetry group $SU(3)_V$ to be a normal symmetry of the spectrum: hadrons do occur in nearly degenerate multiplets that consitute representation of $SU(3)_V$.

To summarize, the observed hadron spectrum suggests that the ground state of QCD has the property

$$\begin{aligned} Q^a_V \mid 0 > &= 0 \\ Q^a_A \mid 0 > &\neq 0; \qquad a = 1, \ldots, 8 \end{aligned} \tag{2.14}$$

where $Q^a_V (Q^a_A)$ denotes the conserved charge associated with the vector (axial) current in eq. (2.11).

2.3 The order parameter $< 0|\bar{q}q|0 >$

Gell-Mann, Oakes and Renner [19] identified a long time ago the vacuum expectation value of the scalar operators $\bar{u}u, \bar{d}d$ and $\bar{s}s$ as a quantitative measure of spontaneous chiral symmetry breakdown,

$$< 0 \mid \bar{u}u \mid 0 > = < 0 \mid \bar{d}d \mid 0 > = < 0 \mid \bar{s}s \mid 0 > \neq 0 . \tag{2.15}$$

(Recall that we consider here the chiral limit $m_u = m_d = m_s = 0$. In the real world where all the quark masses are different from zero, the order parameters in eq. (2.15) experience isospin - and $SU(3)$ - splitting.)

Why do these quark bilinears play the role of an order parameter ? The formal reasoning goes as follows. Since the vector charges annihilate the vacuum, we have

$$< 0 \mid [Q_V^a, \bar{q}\lambda^b q] \mid 0 >= 0; \qquad a, b = 1, \ldots, 8. \tag{2.16}$$

By choosing appropriate values a, b in

$$[Q_V^a, \bar{q}\lambda^b q] = i f^{abc} \bar{q}\lambda^c q \tag{2.17}$$

one finds from (2.16) that all nondiagonal bilinears have zero vacuum expectation value,

$$< 0 \mid \bar{u}d \mid 0 >=< 0 \mid \bar{u}s \mid 0 >= \cdots = 0. \tag{2.18}$$

Furthermore, the choice $a = 1, b = 2$ and $a = 6, b = 7$ leads to

$$< 0 \mid \bar{u}u \mid 0 >=< 0 \mid \bar{d}d \mid 0 >=< 0 \mid \bar{s}s \mid 0 > . \tag{2.19}$$

Next we use

$$[Q_A^1, \bar{q}\gamma_5\lambda^1 q] = -(\bar{u}u + \bar{d}d) \tag{2.20}$$

and sandwich this relation between the vacuum state $\mid 0 >$. Since $Q_A^a \mid 0 > \neq 0$ according to (2.14), it is then consistent to assume that

$$< 0 \mid \bar{u}u \mid 0 >=< 0 \mid \bar{d}d \mid 0 >=< 0 \mid \bar{s}s \mid 0 > \neq 0. \tag{2.21}$$

On the other hand, the Goldstone theorem can again be used to show that (2.21) implies the existence of 8 massless pseudoscalar particles whose coupling to the currents A_μ^a is nonvanishing,

$$< 0 \mid A_\mu^a(x) \mid \pi^b(p) >= i p_\mu F_0 \delta^{ab} e^{-ipx} . \tag{2.22}$$

The constant F_0 is related to the physical pion decay constant $F_\pi = 93.3$ MeV by

$$F_\pi = F_0(1 + O(m_{quark})) . \tag{2.23}$$

F_0 is referred to as "the pion decay constant in the chiral limit".

As was noted above, the Goldstone bosons become massive as the quark masses are turned on. In particular, the Gell-Mann-Oakes-Renner relation for the $(\text{mass})^2$ of the charged pion reads

$$M_{\pi^+}^2 = -\frac{1}{F_0^2}(m_u + m_d) < 0 \mid \bar{u}u \mid 0 > + O(m_{quark}^2) . \tag{2.24}$$

Neglecting the terms of order m_{quark}^2 in this equation, we obtain from $(m_u + m_d)$ (1 GeV) $\simeq$ 14 MeV

$$< 0 \mid \bar{u}u \mid 0 >= -(225\text{MeV})^3 \tag{2.25}$$

at a scale of 1 GeV.

3 THE QCD VACUUM

Let me begin this section with an important remark: I will not provide an up to date review of what is known about the QCD vacuum, because I am not an expert in this field. Readers who are interested in such a review may consult e.g. the comprehensive book by E.V. Shuryak [20]. Here I shall content myself with a discussion of a few of the many remarkable properties which the QCD vacuum seems to enjoy. (The material in this section is not needed for an understanding of the following exposition of χPT, and readers interested mainly in the latter may wish to go directly to section four.)

I shall use in the following the name 'vacuum' for the ground state of the theory, $H_{QCD} \mid 0 >= 0$. According to the standard wisdom, this state has the following properties (among many others):

1. $\mid 0 >$ spontaneously breaks chiral $SU(3)_R \times SU(3)_L$ down to diagonal $SU(3)_V$. The order parameters associated with this symmetry breakdown are given by the vacuum expectation values of quark bilinears. In the chiral limit $m_u = m_d = m_s = 0$,

$$< 0 \mid \bar{u}u \mid 0 >=< 0 \mid \bar{d}d \mid 0 >=< 0 \mid \bar{s}s \mid 0 >\simeq -(225\mathrm{MeV})^3 . \tag{3.1}$$

2. The quark condensates $< 0 \mid \bar{q}q \mid 0 >$ melt at T $=\mathrm{T}_c \simeq 200$MeV,

$$< \bar{q}q >_T= \left(Tr\, \mathrm{e}^{-H_{QCD}/T}\bar{q}q\right) / Tr\, \mathrm{e}^{-H_{QCD}/T} \quad \stackrel{T \to T_c}{\to} \quad 0 . \tag{3.2}$$

3. According to Shifman, Vainshtein and Zakharov [21], various nonperturbative aspects of QCD can be parametrized by additional condensates,

$$\begin{aligned} < 0 \mid G^a_{\mu\nu} G^{a\mu\nu} \mid 0 > &\sim 0.470\mathrm{GeV}^4 \\ < 0 \mid f_{abc} G^a_{\mu\nu} G^{b\nu\rho} G^c_{\rho\sigma} g^{\sigma\mu} \mid 0 > &\sim 0.045\mathrm{GeV}^6 \\ &\vdots \end{aligned} \tag{3.3}$$

(Recall that the coupling constant g is included in the gauge-field.) None of these properties have been rigorously shown to hold within QCD. Yet there exist many interesting nonperturbative approximation schemes in which they can be examined. In the following short discussion I concentrate on

- Models which relate $\mid 0 >$ to the properties of a (dual) superconductor (the vacuum as a dielectric medium).
- Articles related to the condensates: Is $< 0 \mid \bar{q}q \mid 0 >, < 0 \mid G^2 \mid 0 > \neq 0$? Why? Consequences?
- Investigations concerning the properties of the vacuum at finite temperatures; phase transitions.

3.1 The vacuum as a dielectric medium

i) Classical Electrodynamics. Consider the electric field of a charge $e > 0$ located at the origin. In empty space, its Coulomb potential is

$$V(r) = \frac{e}{4\pi r}. \tag{3.4}$$

If the charge is immersed in a dielectric medium, one has

$$V_d(r) = \frac{e}{4\pi r}\frac{1}{1+\chi_e}. \tag{3.5}$$

The electric suszeptibility χ_e is positive on general grounds [22], and the elctric field E_d in the dielectric medium is therefore smaller than in empty space, $E_d < E$: the charge e is screend.

ii) Quantumelectrodynamics. Charge-screening also takes place in QED. Indeed consider the modification of the Coulomb potential (3.4) due to the vacuum polarization at lowest order. One finds

$$\begin{aligned} \frac{e}{4\pi r} &\rightarrow \frac{e}{4\pi r}Q(r) \\ Q(r) &= 1 + \frac{\alpha}{4\pi^{\frac{1}{2}}(mr)^{\frac{3}{2}}}e^{-2mr} + \cdots \qquad mr \gg 1 \\ \frac{dQ}{dr} &< 0. \end{aligned} \tag{3.6}$$

where $\gamma = 0.5772\ldots$ is Eulers constant. (The explicit expression for the effective charge $eQ(r)$ may be found e.g. in ref. [23].) We see from this result that indeed $eQ(r)$ decreases with increasing distance: The vacuum fluctuations in QED act in the same way as an ordinary dielectric medium in classical electrodynamics (at least to lowest order in an expansion in powers of α).

iii) Quantumchromodynamics. Consider the potential between two static coloured charges at distance r. The quantum corrections at lowest order modify the Coulombic form of the potential according to

$$\frac{g^2}{4\pi r} \rightarrow \frac{g^2(r)}{4\pi r} \tag{3.7}$$

where

$$\frac{dg^2}{dr} > 0, \tag{3.8}$$

see ref. [24] for the explicit expression of $g^2(r)$.

According to this result, coloured charges in QCD are not screened by quantum fluctuations: $g^2(r)$ is an *increasing* function of the distance r between the two charges. Vacuum fluctuations in QCD thus appear to act as a medium with *negative* suszeptibility $\chi_g < 0$ (antiscreening of charge).

iv) Antiscreening and confinement. It was suggested a long time ago that antiscreening in QCD may be at the origin of confinement [25]. In ref. [26,27] a (numerically) solvable model is discussed. It exhibits antiscreening and indeed leads to charge confinement. In the simplest version of the model, one considers 2 coloured static charges in a static (abelian) chromoelectric field. The dielectric 'constant' is a function of the electric field (leading logarithmic (abelian) model):

$$\vec{\nabla}\cdot(\epsilon(E)\vec{E}) = j^0$$
$$\epsilon(E) = \frac{b_0}{4}\ln\frac{E^2}{\kappa^2}; \qquad E = \mid \vec{E} \mid \tag{3.9}$$

where the current j^0 describes two charges of opposite sign located at x=y=0,z=±a,

$$j^0 = Q\delta(x)\delta(y)\left[\delta(z-a) - \delta(z+a)\right] \tag{3.10}$$

and where b_0 and κ are two parameters of the model. The form of the dielectric constant $\epsilon(E)$ in (3.9) is motivated by the QCD effective potential.

The static potential between the two charges has been worked out analytically from (3.9) for small and large distances with the result [26,27]

$$V \to \begin{cases} c(r\ln r)^{-1}, & r \to 0 \\ d_1 r + d_2 \ln r, & r \to \infty\,. \end{cases} \tag{3.11}$$

This shows that indeed the charges are confined. The partial differential equations for the electric field $\vec{E}$ have furthermore been solved numerically [27], and I invite the reader to enjoy the numerous graphic representations of various physical quantities which were evaluated in the course of these calculations (flux function, energy density,dielectric constant $\epsilon(E)$,. . .). In particular the result of these investigations show that the dielectric suszeptibility $\chi_e = \epsilon - 1$ is negative and close to -1 almost everywhere, except in a thin tube which connects the two charges at $z=\pm a$. In this tube $\chi_e \gg 1$ (normal region). The tube furthermore coincides with the region where the electric flux is nonvanishing (see figs.(13,14) in ref. [27]). This result is similar to the situation where a magnetic monopole-antimonopole pair is immersed in a superconductor, except that the role of electric and magnetic quantities is interchanged: hence the name *dual* superconductor for this phenomenological picture of confinement.

It was suggested a long time ago [28] that colour confinement is the dual analog of confinement of magnetic monopoles in an ordinary superconductor. This idea was taken up and elaborated on by many authors, both in the framework of abelian and nonabelian field-theoretic models [29].

3.2 The order parameter

It is vital to know which of the global symmetries of $\mathcal{L}_{QCD}$ are spontaneously broken by the ground state and which remain intact. There are various investigations of this question using lattice regularized QFT, considerations of anomalies and $1/N_c$ expansions. For an early review of the topic see [30]. Here I wish to touch upon the order parameter $< 0 \mid \bar{q}q \mid 0 >$ as discussed in [31,32].

The condensate $< 0 \mid \bar{u}u \mid 0 >$ can be obtained from the fermion propagator in the external gauge-field A_μ,

$$< 0 \mid \bar{u}u \mid 0 > = -\int [\mathrm{d}A]\, \mathrm{Tr} S_F(x,x \mid A) \mathrm{e}^{-S_{YM}}$$
$$(\not{D} + m) S_F(x,y \mid A) = \delta^4(x-y)\,. \tag{3.12}$$

Here S_{YM} is the action of QCD with fermions (mass m) integrated out. The trace is taken over colour and Dirac indices. Representing the propagator through its spectral density

$$S_F(x,y \mid A) = \int_{-\infty}^{\infty} \frac{\mathrm{d}\lambda}{m - i\lambda} \rho(\lambda, x, y \mid A) \tag{3.13}$$

one finds that the condensate in the chiral limit is given by

$$< 0 \mid \bar{u}u \mid 0 >_{|m=0} = -\pi \int [\mathrm{d}A] \operatorname{Tr} \rho(0, x, x \mid A) \mathrm{e}^{-S_{YM}} \tag{3.14}$$

Floratos and Stern [32] showed that there are gluon-field configurations for which ρ remains finite and non-zero as $\lambda \to 0$. For such field configurations the quark propagator indeed spontaneously breaks chiral symmetry according to (3.14).

However, it has not yet been proven that these chiral symmetry breaking effects, which are produced by a particular set of gauge-field configurations, survive the averaging in eq. (3.12).

3.3 The condensate $< 0 \mid G^2 \mid 0 >$

According to ref. [21], the vacuum condensate $< 0 \mid G^2 \mid 0 >$ induces the leading power correction to the perturbative short distance behaviour of various Green functions. The numerical value for the condensate obtained in [21] is

$$\begin{aligned} < 0 \mid G^a_{\mu\nu} G^a_{\rho\sigma} \mid 0 > &= \frac{1}{3} A (g_{\mu\rho} g_{\nu\sigma} - g_{\mu\sigma} g_{\nu\rho}) \\ A &\sim 0.12 \mathrm{GeV}^4 . \end{aligned} \tag{3.15}$$

(The constant A must indeed be positive if the true ground state $\mid 0 >$ has lower energy than the perturbative vacuum $\mid 0)$ where $(0 \mid G^2 \mid 0) = 0$).

There is a vast literature on the sum rule technique devised to obtain the value of various condensates from data [33]. (Some of these condensates have now become accessible to evaluation on the lattice [34].) Here I wish to recall the influence of a nonvanishing condensate $< 0 \mid G^2 \mid 0 >$ on the energy levels of heavy $q\bar{q}$-pairs [35,36]. In ref. [36] the quadratic stark effect (quarks immersed in a constant gluon background-field) has been calculated with the result

$$\begin{aligned} M_{nl} &= 2m - \frac{m\beta^2}{4n^2} + \frac{m < B^2 >}{(m\beta)^4} n^6 \epsilon_{nl} \\ < B^2 > &= \frac{1}{4} < 0 \mid G^a_{\mu\nu} G^{a\mu\nu} \mid 0 > \\ \beta &= \frac{4}{3} \alpha_s, \qquad \varepsilon_{nl} = \mathrm{O}(1) \end{aligned} \tag{3.16}$$

where n is the principal quantum number and l the orbital angular momentum of the heavy quark-antiquark pair (quark mass m). The dimensionless quantity ε_{nl} is of order one [36]. The formula (3.16) is accurate up to terms of order $m\beta^4$, $m < D_\mu G^2_{\rho\sigma} > (m\beta)^{-6}, m < G^3_{\rho\sigma} > (m\beta)^{-6}$ and up to effects due to the pair density of light quarks.

The main feature of the result (3.16) is the extreme sensitivity of the correction on the principal quantum number, $\delta M_{nl} \sim n^6$. As a result of this, even for very heavy quarks the nonperturbative effects are large on excited levels. Using the estimate $< B^2 > \simeq (600\mathrm{MeV})^4$, one finds that in the ground state of the $b\bar{b}$ system the nonperturbative correction relative to the Schrödinger binding energy $m\beta^2/(4n^2)$ is 13% (α_s

$= 0.35$), 34% ($\alpha_s = 0.3$) and 100% ($\alpha_s = 0.25$). In the $c\bar{c}$-system the nonperturbative effects completely overwhelm the Coulomb force even in the ground state. I refer to [36] for further discussions. In particular, in this reference the influence of the condensate on the wave function of the $q\bar{q}$-system is worked out as well.

The calculations of refs. [35,36] have been extended to include colour fields which stochastically fluctuate in (Euclidean) space-time with a finite correlation time [37]. From fits to the $b\bar{b}$ system the authors of ref. [38] find a gluon correlation time $T_G = 0.2. - 2.2 fm$.

3.4 The order parameter at finite temperature

Asymptotic freedom predicts that at high temperatures, the interaction among the quarks and gluons is weak - colour is liberated and chiral symmetry is restored [39,20]. It may be that the restoration of chiral symmetry and the liberation of colours occurs in a single phase transition from the low-temperature hadron phase to the high-temperature quark-gluon-plasma phase. Estimates of the critical temperature from various sources (Monte-Carlo calculations, studies of correlation functions or using simplified models) lead to values of T_c = 100-300 MeV [40].

Here I wish to comment on the condensate $< \bar{u}u >_T$ defined in eq. (3.2). Below the critical temperature, i.e. in the hadronic phase, χPT allows to work out its low-temperature expansion [14,41]. As an example I quote the following low-energy theorem which is valid in the chiral limit $m_u = m_d = 0$ (chiral $SU(2) \times SU(2)$) as $T \to 0$,

$$< \bar{u}u >_T = < 0 \mid \bar{u}u \mid 0 > \left\{ 1 - \frac{3T^2}{24F^2} - \frac{3}{8}\left(\frac{T^2}{12F^2}\right)^2 + O(T^6) \right\} . \qquad (3.17)$$

The constant F denotes the pion decay constant in the chiral limit.

The reason why χPT allows to make predictions for $< \bar{u}u >_T$ is based on the following observations:

1. At temperature T, only states with energy $E \leq T$ contribute in the sum (3.2). At sufficiently low temperature, the contribution from massive states is therefore suppressed, and the sum extends only over $n-$ pion states $\mid \pi^{a_1}(p_1) \ldots \pi^{a_n}(p_n) >$.
2. The pions in the these states have energy $E_i^\pi \leq T$.
3. At low temperature, one therefore needs to evaluate the matrix elements $< \pi^{a_1}(p_1) \ldots \pi^{a_n}(p_n) \mid \bar{u}u \mid \pi^{a_1}(p_1) \ldots \pi^{a_n}(p_n) >$ for low energetic pions.
   This is a case for χPT, see following sections.

The result (3.17) indeed predicts that the condensate decreases at increasing temperatures. In ref. [41], the calculation of $< \bar{u}u >_T$ has been carried further to the three-loop level, including quark mass effects and contributions from massive hadrons in the dilute gas approximation. The authors of ref. [41] estimate the value of the critical temperature T_c at which the chiral phase transition takes place and obtain the result $T_c \sim 170\text{MeV}(\sim 190\text{MeV})$ in the chiral limit $m_u = m_d = 0$ (in the real world).

4 OUTLINE OF χPT

Consider the Green functions built from vector and axial currents in the chiral limit,

$$G_{m,n} =< 0 \mid TV_{\mu_1} \ldots V_{\mu_m} A_{\sigma_1} \ldots A_{\sigma_n} \mid 0 > . \tag{4.1}$$

(I have suppressed $SU(3)$ indices and spacetime coordinates.) From these Green functions one may evaluate e.g. scattering matrix elements and form factors via the LSZ reduction formula - hence it is useful to know them.

Chiral symmetry and its spontaneous breakdown leave their mark in $G_{m,n}$:

1. $SU(3)_R \times SU(3)_L$ -symmetry induces relations ("Ward identities") between the Green functions,

$$G_{m,n} \quad \Leftrightarrow \quad \begin{pmatrix} G_{m-1,n} \\ G_{m+1,n-2} \end{pmatrix} . \tag{4.2}$$

2. The presence of massless pseudoscalars which couple to the axial currents according to (2.22) produces one-particle singularities in $G_{m,n}$. As an example consider the two point function of the axial current,

$$i \int \mathrm{d}x\, e^{ip(x-y)} < 0 \mid TA_\mu^a(x) A_\nu^b(y) \mid 0 > = \delta^{ab} \frac{p_\mu p_\nu}{-p^2} F_0^2 + \cdots . \tag{4.3}$$

The first term on the right hand side displays the pole which is due to the presence of Goldstone particles.

These two properties suffice to work out the interactions among the 8 Goldstone bosons at low energies in the chiral limit. Indeed, the following results have been established more than 20 years ago [42,4,5]:

1. The scattering amplitude for the process $\pi^a(p_1)+\pi^b(p_2) \to \pi^{a_1}(p_1')+\ldots+\pi^{a_n}(p_n')$ starts with a term of order p^2 for any n,

$$T^{ab;a_1\ldots a_n} = O(p^2) . \tag{4.4}$$

2. The leading term in eq. (4.4) is completely fixed in terms of the fourmomenta $p_1, p_2, p_1', \ldots, p_n'$ and of the pion decay constant F_0. In particular, the amplitude for elastic meson-meson scattering

$$\pi^a(p_1) + \pi^b(p_2) \to \pi^c(p_1') + \pi^d(p_2') \tag{4.5}$$

becomes

$$T^{ab;cd} = \frac{1}{3F_0^2} \left[(t-u) f^{abr} f^{cdr} + (s-u) f^{acr} f^{bdr} + (s-t) f^{adr} f^{bcr} \right] + O(p^4) \tag{4.6}$$

where

$$s = (p_1 + p_2)^2, \quad t = (p_1 - p_1')^2, \quad u = (p_1 - p_2')^2 \tag{4.7}$$

are the usual Mandelstam variables. The symbol $O(p^4)$ denotes terms which are at least of fourth order in the pion momenta.

From (4.4) one concludes that the interaction among the Goldstone bosons is weak at low energies. The leading term (4.6) should therefore be a good approximation to the full amplitude near threshold. To check, one may e.g. evaluate the (I = 1, J = 1) scattering length a_1^1 for pions. The result is

$$a_1^1 = \frac{1}{24F_0^2} = 0.030 M_{\pi^+}^{-2} . \quad (4.8)$$

(Here I have used $F_0 = 93.3$ MeV. To express the scattering length in powers of M_{π^+} = 139.57 MeV is pure convention - this can of course be done even in the chiral limit which is considered here.). The experimental value is [43]

$$a_1^1 = (0.038 \pm 0.002) M_{\pi^+}^{-2} . \quad (4.9)$$

The low-energy representation (4.6) indeed is amazingly good.

Of course one ought to be able to evaluate the corrections to the leading order result (4.8). They arise from higher order terms in the low-energy expansion (4.6) of the scattering amplitude and from quark mass effects. Chiral perturbation theory (χPT) is a systematic procedure to calculate these corrections. It consists in constructing the most general form of the Green functions which correspond to a field theory with spontaneous chiral symmetry breakdown

$$SU(3)_R \times SU(3)_L \to SU(3)_{R+L} . \quad (4.10)$$

Quark mass effects are automatically included in the procedure. The construction is done by expanding the Green functions in powers of the external momenta and of the quark masses masses m_u, m_d and m_s. The Green functions so obtained obey all Ward identities associated with (4.10).

Several remarks are in order at this place.

1. Green functions exhibit threshold factors of the square root or logarithmic type. In χPT such factors are never expanded, and the low-energy representation is perfectly valid also above threshold.

2. Contrary to statements made in the literature [44], Green functions calculated in χPT satisfy unitarity (at each order in the low-energy expansion).

3. What is the energy range in which χPT can be trusted? It is first of all clear that in the region where resonances appear ($\rho, K^*, a_0, \ldots$) one encounters a natural barrier which cannot be surpassed without additional effort. I do not want to enter this matter here and instead refer to the article by Donoghue et al. [45]. In this reference partial waves obtained in χPT at the one-loop level are compared with the available data. The authors conclude that χPT gives a valid representation of the partial waves up to center of mass energies $\sqrt{s} \sim 700$MeV. In ref. [44], lowest order Padé approximants are furthermore worked out for the scalar and vector form factor of the pion. The author finds that this procedure extends the validity of χPT. In my opinion, this is a very interesting point which deserves further investigation.

4. Ultimately one compares Green functions - evaluated within χPT - with the data. What can one conclude from such a comparison about the underlying theory? The answer is simple: The only ingredients in the calculation are symmetry properties of QCD and assumptions about its spectrum in the chiral limit. If physical quantities evaluated within χPT do not agree with the data, one has to conclude that QCD is not able to describe these data either.

5. Last but not least, one may use χPT to pin down some of the parameters which occur in $\mathcal{L}_{QCD}$ (e.g. by analyzing the quark mass expansion of the hadron spectrum). While one cannot determine the absolute value of the quark masses in this way, one may pin down the two ratios $m_u : m_d : m_s$ within rather small errors [1].

5 THE NONLINEAR σ-MODEL AT TREE LEVEL

In this section I consider the interaction among the Goldstone bosons. Massive particles (baryons) are treated later.

Every field theory $\mathcal{L}_0 + \mathcal{L}_I$ where

1. $\mathcal{L}_0$ is $SU(3)_R \times SU(3)_L$ symmetric
2. the ground state is only symmetric under $SU(3)_V$
3. $\mathcal{L}_0$ conserves 8 axial currents with normalization
   $< 0 \mid A^a_\mu(0) \mid \pi^b(p) >= ip_\mu\ F_0\delta^{ab}$
4. $\mathcal{L}_I$ has the same one particle matrix elements as the quark mass term in $\mathcal{L}_{QCD}$
5. the only massless particles in the theory are the 8 Goldstone bosons associated with the spontaneous symmetry breakdown

leads to the same leading terms in the low-energy expansion as $\mathcal{L}_{QCD}$ [5].

The simplest theory with these properties is the nonlinear σ-model,

$$\mathcal{L}_M^{(2)} = \frac{F_0^2}{4} < \partial_\mu U \partial^\mu U^\dagger + 2B_0\mathcal{M}(\mathcal{U} + \mathcal{U}^\dagger) > \tag{5.1}$$

where

$$U(x) \in SU(3), \qquad \mathcal{M} = \text{diag}(m_u, m_d, m_s) \tag{5.2}$$

and $< A >$ denotes the trace of the matrix A. F_0, B_0 are two arbitrary real parameters which are not fixed by chiral symmetry alone. Explicit calculations show [11] that F_0 coincides with the pion decay constant introduced before - hence I used the same symbol in (5.1). On the other hand, B_0 is related to the order parameter by $< 0 \mid \bar{u}u \mid 0 >= -F_0^2 B_0(1 + 0(\mathcal{M}))$. Finally the index (2) in eq. (5.1) reminds us that the nonlinear σ - model Lagrangian contains two derivatives.

For $\mathcal{M} = 0, \mathcal{L}_M^{(2)}$ is invariant under the global $SU(3)_R \times SU(3)_L$ transformations

$$U \stackrel{G}{\rightarrow} g_R U g_L^\dagger, \qquad G = SU(3)_R \times SU(3)_L . \tag{5.3}$$

The corresponding 16 conserved currents are

$$A^a_\mu = R^a_\mu - L^a_\mu, \qquad V^a_\mu = R^a_\mu + L^a_\mu \tag{5.4}$$

with

$$R_\mu^a = \frac{F_0^2}{2i} < \frac{\lambda^a}{2} U \partial_\mu U^\dagger >, \quad L_\mu^a = \frac{F_0^2}{2i} < \frac{\lambda^a}{2} U^\dagger \partial_\mu U > . \tag{5.5}$$

The ground state $U = \mathbf{1}$ is invariant only under diagonal $SU(3)_R \times SU(3)_L$ rotations $g_R = g_L$ and $\mathcal{L}_M^{(2)}$ thus indeed exhibits spontaneous breakdown of chiral symmetry as was required in point 2 above.

5.1 Calculations with $\mathcal{L}_M^{(2)}$

In order to calculate masses and scattering amplitudes, we parametrize the $SU(3)$ matrix $U(x)$ by

$$U = \exp \frac{i\phi}{F_0}, \qquad < \phi >= 0 \tag{5.6}$$

and expand $\mathcal{L}_M^{(2)}$ in powers of ϕ,

$$\begin{aligned} \mathcal{L}_M^{(2)} &= \frac{1}{4} < \partial_\mu \phi \partial^\mu \phi - 2B_0 \mathcal{M} \phi^2 > \\ &+ \frac{1}{48F_0^2} < -4\partial_\mu \phi \partial^\mu \phi^3 + 3\partial_\mu \phi^2 \partial^\mu \phi^2 + 2B_0 \mathcal{M} \phi^4 > + \cdots \end{aligned} \tag{5.7}$$

I have dropped a constant term which is independent of the field ϕ. $\mathcal{L}_M^{(2)}$ contains an infinite number of terms -indicated by dots in eq. (5.7) - which lead to interactions among the the Goldstone bosons: terms of order ϕ^4 induce elastic meson - meson scattering which was discussed before, terms of order ϕ^6 describe the production process $\pi^a(p_1) + \pi^b(p_2) \to \pi^{a_1}(p_1') + \ldots \pi^{a_4}(p_4')$, and so on.

Consider now the *kinetic term* in (5.7). Expanding ϕ in Gell-Mann matrices and evaluating the trace one can readily read off the meson masses (I put $m_u = m_d$ for simplicity):

$$\begin{aligned} \overset{\circ}{M}{}_\pi^2 &= 2\hat{m} B_0 \\ \overset{\circ}{M}{}_K^2 &= (\hat{m} + m_s) B_0 \\ \overset{\circ}{M}{}_\eta^2 &= \frac{2}{3}(\hat{m} + 2m_s) B_0 . \end{aligned} \tag{5.8}$$

<u>Remarks</u>

1. The physical pion mass - which I denote by M_π^2 throughout - differs from the quantity $\overset{\circ}{M}{}_\pi^2$ in (5.8) by terms of order m_{quark}^2,

$$M_P^2 = \overset{\circ}{M}{}_P^2 (1 + O(\mathcal{M})); \qquad P = \pi, K, \eta . \tag{5.9}$$

   I will show in the next section how these corrections can be calculated.

2. The (meson mass)^2 is proportional to the quark mass and thus vanishes as $m_{quark} \to 0$, in agreement with Goldstone's theorem.

3. The masses $\overset{\circ}{M}{}_P^2$ exhibit the Gell-Mann-Okubo relation

$$\overset{\circ}{M}{}_\eta^2 = \frac{1}{3}(4 \overset{\circ}{M}{}_K^2 - \overset{\circ}{M}{}_\pi^2) . \tag{5.10}$$

4. Neglecting the higher order terms in (5.9), one can evaluate the ratio $m_s/\hat{m}$ from eq. (5.8) with the result [2]

$$\frac{m_s}{\hat{m}} = \frac{2\,\overset{\circ}{M}{}_K^2 - \overset{\circ}{M}{}_\pi^2}{\overset{\circ}{M}{}_\pi^2} = 25.9\,. \tag{5.11}$$

The correction to this result - due to the higher order terms in (5.9) - is small, see ref. [1].

Next we come to the *terms of order* ϕ^4 and first consider elastic pion-pion scattering,

$$\begin{aligned}
&\pi^a(p_1) + \pi^b(p_2) \to \pi^c(p_1') + \pi^d(p_2') \\
T^{ab;cd} &= \delta^{ab}\delta^{cd}A(s,t,u) + \delta^{ac}\delta^{bd}A(t,s,u) + \delta^{ad}\delta^{bc}A(u,t,s) \\
s &= (p_1+p_2)^2, \quad t = (p_1 - p_1')^2, \quad u = (p_1 - p_2')^2\,.
\end{aligned} \tag{5.12}$$

Evaluating the tree graph contributions from the terms of order ϕ^4 in (5.7) one finds

$$A^{(2)}(s,t,u) = \frac{s - \overset{\circ}{M}{}_\pi^2}{F_0^2}\,. \tag{5.13}$$

The index (2) reminds that this result is the first order term in the low-energy expansion of the full amplitude which I denote by $A(s,t,u)$. The result (5.13) agrees in the chiral limit with the leading term which was displayed earlier in eq. (4.6). For the scattering lengths a_0^0 and a_1^1 one obtains from (5.13)

$$\begin{aligned}
a_0^0 &= \frac{7\,\overset{\circ}{M}{}_\pi^2}{32\pi F_0^2} = 0.16 \qquad (0.26 \pm 0.05) \\
a_1^1 &= \frac{1}{24\pi F_0^2} = 0.030 M_{\pi^+}^{-2} \qquad (0.038 \pm 0.002) M_{\pi^+}^{-2}
\end{aligned} \tag{5.14}$$

in reasonable agreement with the experimental values [43] which are given in brackets.

Finally, I come the the decay $\eta \to \pi^+\pi^-\pi^0$ which is also determined by the terms of order ϕ^4 at leading order (tree graphs). One finds for the width

$$\Gamma_{\eta\to\pi^+\pi^-\pi^0} = 66\text{eV}\,, \tag{5.15}$$

i.e., nearly a factor of five smaller than its experimental value [3] $\Gamma^{exp}_{\pi^+\pi^-\pi^0} = 317 \pm 20$ eV! This is obviously not a particularly successful prediction of the lowest order calculation.

5.2 Where are we?

In the preceding paragraph we have used the nonlinear σ-model in the tree approximation to calculate the leading order term of various physical quantities. Whereas the threshold parameters a_0^0 and a_1^1 come out reasonably well, the prediction for $\eta \to 3\pi$ is far from its experimental value. The need for evaluating the corrections to these tree calculations is obvious.

---

[2] I use F_0=93.3 MeV , $\overset{\circ}{M}_K = 495$ MeV and $\overset{\circ}{M}_\pi = 135$ MeV in all numerical evaluations.
[3] Here I use $\Gamma_{\eta\to 2\gamma} = 520 \pm 30$eV [46].

What about graphs with loops? On the one hand, if we simply disregard these graphs the theory violates unitarity. On the other hand, since the nonlinear σ-model is not renormalizable, the contributions from graphs involving loops are not well defined. This apparent dilemma has a remarkably simple solution [9]: Consider for definiteness graphs containing one loop. The infinities which arise if one calculates these graphs require counter terms. Using dimensional regularization (which preserves chiral symmetry) one finds that the counter terms necessary to renormalize one-loop graphs are of order p^4. Since the regularization is consistent with Lorentz invariance, parity and chiral symmetry, the counter terms have the structure of the most general chiral invariant Lagrangian $\mathcal{L}_M^{(4)}$ of order p^4. This Lagrangian will again contain constants, similar to F_0 and B_0 in $\mathcal{L}_M^{(2)}$, which are not fixed by chiral symmetry alone. With a suitable renormalization of the constants which occur in $\mathcal{L}_M^{(4)}$ one thus gets finite results for all Green functions to one-loop order. The fact that the nonlinear σ-model requires counter terms which do not have the structure of the Lagrangian $\mathcal{L}_M^{(2)}$ one starts with is characteristic of the low-energy expansion: one needs only two constants F_0, B_0 to specify the low-energy behaviour of the Green functions to leading order, and one needs additional constants to characterize the behaviour at next-to-leading order. (There are 12 of them at order p^4, see below.) One needs counter terms of increasing complexity as one evaluates the σ-model to higher orders, and one finds an increasing number of low-energy constants if one carries the low-energy expansion to higher orders. (In principle, all of these low-energy constants are determined by the parameters $\Lambda, m_c, m_b, \ldots$ of the underlying renormalizable theory.)

General power counting arguments [9] show that graphs containing n loops are suppressed by p^{2n} in comparison to tree graphs. The loop graphs therefore do not modify the leading low-energy behaviour which is given by the tree graphs of $\mathcal{L}_M^{(2)}$. The one-loop graphs do, however, contribute at first nonleading order in the low-energy expansion.

The evaluation of loops is taken up in the next section.

6 LOOPS

In the following I shall elaborate on the external field technique to evaluate loops in the nonlinear σ-model [47,10].

I denote by Z the generating functional for the Green functions of vector, axial vector, scalar and pseudoscalar quark currents:

$$e^{iZ[v,a,s,p]} = < 0 \mid T\exp\{i \int dx \bar{q}[\gamma_\mu(v^\mu + \gamma_5 a^\mu) - (s - ip\gamma_5)]q\} \mid 0 > . \qquad (6.1)$$

The external fields v_μ, a_μ, s and p are hermitean 3×3 matrices in flavour space, with tr a_μ = tr $v_\mu = 0$. The scalar field s contains the quark mass matrix, $s = \mathcal{M} + \delta s$, and $\mid 0 >$ denotes the QCD ground state in the chiral limit $m_u = m_d = m_s = 0$. The functional derivatives of Z with respect to v_μ, a_μ, s and p generate the relevant Green functions in the usual way.

As I explained in the last section, chiral symmetry implies [10,11] that Z admits a low-energy representation of the form

$$e^{iZ(v,a,s,p)} = \int [dU] \exp[i \int dx \mathcal{L}_{eff}(U, v, a, s, p)] . \tag{6.2}$$

The right hand side is a meson field theory, the meson fields being described by a unitary 3×3 matrix U as before. U transforms under *local* chiral rotations as

$$U \xrightarrow{G} g_R U g_L^\dagger, \qquad G = SU(3)_R \times SU(3)_L . \tag{6.3}$$

The effective Lagrangian $\mathcal{L}_{eff}$ consists of a string of terms

$$\mathcal{L}_{eff} = \mathcal{L}^{(2)} + \mathcal{L}^{(4)} + \mathcal{L}^{(6)} + \cdots . \tag{6.4}$$

I refer to $\mathcal{L}^{(2n)}$ as a term of order p^{2n}: The field U counts as a quantity of $O(p^0)$, the derivative ∂_μ and the external fields v_μ, a_μ are booked as $O(p)$, and the fields s, p count as $O(p^2)$.

The first term $\mathcal{L}^{(2)}$ is the nonlinear σ-model Lagrangian in the presence of external fields:

$$\begin{aligned} \mathcal{L}^{(2)} &= \frac{F_0^2}{4} < D_\mu U D^\mu U^\dagger + \chi U^\dagger + \chi^\dagger U > \\ D_\mu U &= \partial_\mu U - i(v_\mu + a_\mu)U + iU(v_\mu - a_\mu); \qquad \chi = 2B_0(s + ip) . \end{aligned} \tag{6.5}$$

For $v_\mu = a_\mu = p = 0$, $s =$ diag (m_u, m_d, m_s), $\mathcal{L}^{(2)}$ reduces to $\mathcal{L}_M^{(2)}$ discussed in the last section. The low-energy expansion of Z is given by the expansion of the effective meson field theory in the number of loops,

$$Z = Z^{(2)} + Z^{(4)} + Z^{(6)} + \cdots . \tag{6.6}$$

The leading term $Z^{(2)}$ coincides with the classical action associated with $\mathcal{L}^{(2)}$. This term reproduces the results discussed in the last section. The next to leading term $Z^{(4)}$ contains two types of contributions: one-loop graphs generated by $\mathcal{L}^{(2)}$ and tree graphs involving one vertex from $\mathcal{L}^{(4)}$. In addition to the most general chiral invariant Lagrangian $\mathcal{L}^4$ of $O(p^4)$, $\mathcal{L}^{(4)}$ also contains a piece $\mathcal{L}^{WZ}$ to account for the chiral anomaly [48]:

$$\mathcal{L}^{(4)} = \mathcal{L}^4 + \mathcal{L}^{WZ}$$

$\mathcal{L}^4 = \sum_{i=1}^{10} L_i P_i + 2$ contact terms in the external fields

$$\begin{aligned} &P_1 = < D_\mu U^\dagger D^\mu U >^2 && P_2 = < D_\mu U^\dagger D^\nu U >< D^\mu U^\dagger D^\nu U > \\ &P_3 = < D_\mu U^\dagger D^\mu U D_\nu U^\dagger D^\nu U > && P_4 = < D_\mu U^\dagger D^\mu U >< \chi^\dagger U + \chi U^\dagger > \\ &P_5 = < D_\mu U^\dagger D^\mu U(\chi^\dagger U + U^\dagger \chi) > && P_6 = < \chi^\dagger U + \chi U^\dagger >^2 \\ &P_7 = < \chi^\dagger U - \chi U^\dagger >^2 && P_8 = < \chi^\dagger U \chi^\dagger U + \chi U^\dagger \chi U^\dagger > \\ &P_9 = -i < F_R^{\mu\nu} D_\mu U D_\nu U^\dagger + F_L^{\mu\nu} D_\mu U^\dagger D_\nu U > && P_{10} = < U^\dagger F_R^{\mu\nu} U F_{L\mu\nu} > . \end{aligned} \tag{6.7}$$

The quantity $F_R^{\mu\nu}(F_L^{\mu\nu})$ stands for the field strength associated with the nonabelian external field $v_\mu + a_\mu (v_\mu - a_\mu)$. $L_1, \ldots, L_{10}$ are ten real coupling constants which absorb the divergences of the one-loop graphs generated by $\mathcal{L}^{(2)}$. The corresponding renormalized, scale dependent couplings will be denoted by $L_i^r(\mu)$ in the following.

In [10,11] it is suggested that the prescription just outlined gives the most general solution to the constraints imposed by chiral symmetry and unitarity at order p^4. The generating functional $Z^{(2)} + Z^{(4)}$ contains 12 real low-energy couplings

$$F_0, B_0; \qquad L_1^r, \ldots, L_{10}^r \tag{6.8}$$

which are not determined by chiral symmetry alone. (There are 2 additional real couplings H_1, H_2 associated with the 2 contact terms indicated in (6.7). They are not measurable quantities.)

As was mentioned before, the couplings (6.8) are fixed by the dynamics of the underlying theory through the renormalization group invariant scale Λ and the heavy quark masses $m_c, m_b, \ldots$. With present techniques it is, however, not possible to calculate them directly from the QCD Lagrangian (for several attempts see [49]). In the absence of such a calculational scheme they have been determined by comparison with experimental low-energy information and by using large-N_c arguments (see table1 in ref. [11]). It has furthermore been shown that the couplings L_i^r are nearly completely dominated by resonance exchange [10,50].

In the next section I will review some of the results which were obtained from the evaluation of the next to leading order term $Z^{(4)}$ in the loop expansion (6.6).

7 MESONS

7.1 Nonanalyticity

Due to the presence of Goldstone particles, the low-energy expansions are in general not pure Taylor series [6,7,8]. As an example consider the quark mass expansion of the pion decay constant [7,11] in the isospin symmetry limit $m_u = m_d = 0$,

$$F_\pi = F_0 \left\{ 1 - 2\mu_\pi - \mu_K + \frac{4M_\pi^2}{F_0^2} L_5^r + \frac{4(2M_K^2 + M_\pi^2)}{F_0^2} L_4^r \right\} + \mathrm{O}(\mathcal{M}^2) \tag{7.1}$$

where

$$\mu_P = (32\pi^2)^{-1} M_P^2 F_0^{-2} \ln \frac{M_P^2}{\mu^2} \tag{7.2}$$

and μ denotes the scale introduced by the dimensional regularization scheme. (F_π is of course independent of the scale μ: the scale dependence of μ_K, μ_π, L_4^r and L_5^r in eq. (7.1) cancel. This is not obvious here, because I have not provided the scale dependence of the couplings L_i^r.)

The leading quark mass term in (7.1) is of the logarithmic type $m_q \ln m_q$ and algebraically dominates over the terms $\mathrm{O}(m_q)$ for sufficiently small values of m_q. Such nonanalytic contributions potentially give large corrections to the leading order term in the low-energy expansion, see e.g. the logarithmic contribution in the quark mass expansion of the $\pi\pi$-scattering length a_0^0 which is displayed in (7.4,7.5).

7.2 Elastic $\pi\pi$-scattering

In [10] the chiral $SU(2) \times SU(2)$ expansion of the elastic $\pi\pi$-scattering amplitude $A(s,t,u)$ had been worked out to one loop. The result takes the form

$$A(s,t,u) = \frac{1}{F_\pi^2}(s - M_\pi^2) + B(s,t,u) + C(s,t,u) + O(p^6). \qquad (7.3)$$

The function $B(s,t,u)$ contains the (renormalized) contribution from the loops, and $C(s,t,u)$ is a polynomial of order p^4 in the variables s, t and M_π^2. It contains four low-energy constants $\bar{l}_1, \ldots, \bar{l}_4$, whose relation to the couplings L_i^r introduced in the last section has been worked out in [11]. The explicit form of the loop and polynomial contributions $B(s,t,u)$, $C(s,t,u)$ may be found in [10].

To work out the consequences of the representation (7.3) we consider first the isoscalar S-wave $\pi\pi$ - scattering length discussed before. From the result (7.3) one obtains

$$a_0^0 = \frac{7M_\pi^2}{32\pi F_\pi^2}\left\{1 + \frac{5M_\pi^2}{84\pi^2 F_\pi^2}\left(\bar{l}_1 + 2\bar{l}_2 - \frac{3}{8}\bar{l}_3 + \frac{21}{10}\bar{l}_4 + \frac{21}{8}\right)\right\} + O(M_\pi^6). \qquad (7.4)$$

As $M_\pi \to 0$, the low-energy constants $\bar{l}_1, \ldots, \bar{l}_4$ all tend to infinity like - ln M_π^2.The quark mass expansion of a_0^0 therefore takes the form

$$a_0^0 = \frac{7M_\pi^2}{32\pi F_\pi^2}\left\{1 - \frac{9M_\pi^2}{32\pi F_\pi^2}\ln\frac{M_\pi^2}{\mu^2} + O(M_\pi^2)\right\}. \qquad (7.5)$$

The correction to the soft pion theorem (5.14) is not of order M_π^2, it is of order M_π^2 ln M_π^2. Taking the scale of the logarithm at $\mu = 1$ GeV we get a correction of 25%. Note that the nonanalytic correction goes in the right direction to decrease the discrepancy between the soft pion prediction $a_0^0 = 0.16$ and the observed value $a_0^0 = 0.26 \pm 0.05$.

The numerical value of the correction depends on the choice of the scale μ in eq. (7.5), i.e., on the size of the analytic terms of order M_π^2. In contrast to the representation (7.5) which only exhibits the nonanalytic term the full expression (7.4) is scale independent.

In [51] we have worked out the prediction of various other threshold parameters on the basis of (7.3). The low-energy constants which occur in the polynomial $C(s,t,u)$ were pinned down from the "experimental" value of the D-wave scattering lengths [43], from $SU(3)$ mass formulae and from $F_K/F_\pi = 1.22$ [52]. The calculated values of the threshold parameters are within $1\frac{1}{2}$ standard deviation of the measured ones in all cases. In particular, we obtained

$$a_0^0 = .20 \pm 0.01, \qquad (7.6)$$

which amounts to a 25% correction to the leading order result $a_0^0 = 0.16$. The error quoted in (7.6) only measures the accuracy to which the first order correction can be calculated; it does not include an estimate of contributions due to higher order terms.

7.3 $\eta \to 3\pi$

We have seen in section 5 that the leading order prediction for the decay width $\Gamma_{\eta\to 3\pi}$ fails miserably, see eq. (5.15). In [53] we have worked out the one-loop correction to that result and found

$$\Gamma_{\eta\to\pi^+\pi^-\pi^0} = (160 \pm 50)\text{eV}. \qquad (7.7)$$

The correction to the leading order result is large - $\sim 55\%$ in amplitude - yet the result (7.7) is still a factor of 2 from the data. Various facets of this intriguing result are discussed in [54]. I refer to this reference for details and just quote its conclusion: the decay $\eta \to 3\pi$ still remains to be understood.

7.4 Vector and scalar form factors

The prediction of the vector form factors is in good agreement with the data [55]. We have also calculated [55] the scalar form factor $< K^+ \mid \bar{u}s \mid \pi^0 >$ which is measured in K_{l_3} decays. We obtain a low-energy theorem which relates the slope λ_0 to the ratio F_K/F_π (the theorem is a modified version of the Dashen-Weinstein relation [56]). With $F_K/F_\pi = 1.22$ as input, χPT predicts

$$\lambda_0 = 0.017 \pm 0.004 . \tag{7.8}$$

The experimental situation concerning the value of λ_0 is far from clear. The old high statistics SLAC experiment [57] confirmed the theoretical expectations with the value $\lambda_0 = 0.019 \pm 0.004$. More recent experiments however report substantially higher values of λ_0:

$$\lambda_0 = \begin{cases} 0.039 \pm 0.010 & [58] \\ 0.046 \pm 0.006 & [59] \\ 0.0341 \pm 0.0067 & [60] \end{cases} . \tag{7.9}$$

I see no way to reconcile these findings with chiral symmetry.

7.5 K_{l_4}-decays

Rosselet et al. [61] have measured $3 \cdot 10^4$ $K^+ \to \pi^+\pi^-e^+\nu_e$ decays. This decay is particularly interesting for several reasons. First it allows to measure in a rather direct way the $\pi\pi$-phase shift difference $\delta_0^0 - \delta_1^1$ at low energies via the Watson final state theorem. Secondly, the low-energy constants $L_1^r, \ldots, L_5^r$ and L_9^r enter the decay amplitude at the one-loop level in χPT. The constants L_1^r, L_3^r and L_4^r were pinned down in [11] by using large-N_c arguments. Data on K_{l_4}-decays can thus be used to test these predictions. Thirdly, it will be interesting to see whether in these decays a puzzle similar to $\eta \to 3\pi$ occurs: The calculation of both of these decays involves an expansion in m_s. Because M_K is only slightly lighter than M_η, one might expect that a failure of predicting $\eta \to 3\pi$ implies a failure of the prediction for K_{l_4}-decays.

K^+-decays involve the axial and vector current matrix elements

$$\begin{aligned} &< \pi^+(p_1)\pi^-(p_2) \mid A_\mu^{4-i5}(0) \mid K^+(p_K) >= \\ &\frac{f}{M_K}(p_1+p_2)_\mu + \frac{g}{M_K}(p_1-p_2)_\mu + \frac{r}{M_K}(p_K-p_1-p_2)_\mu \end{aligned} \tag{7.10}$$

and

$$< \pi^+(p_1)\pi^-(p_2) \mid V_\mu^{4-i5}(0) \mid K^+(p_K) >= \frac{ih}{M_K^3}\varepsilon_{\mu\nu\rho\sigma}p_K^\nu(p_1+p_2)^\rho(p_1-p_2)^\sigma , \tag{7.11}$$

Table 1. K_{l_4} decays.

	Experiment	χPT (leading order)
$K^+ \to \pi^+\pi^-e^+\nu_e$		
Γ	$(3.26 \pm 0.15)10^3\text{sec}^{-1}$	$1.29 \cdot 10^3\text{sec}^{-1}$
$\mid V_{us}f \mid$	1.23 ± 0.03	0.82
$\mid V_{us}g \mid$	1.05 ± 0.06	0.82
$\mid V_{us}h \mid$	0.59 ± 0.15	0
$K^+ \to \pi^0\pi^0e^+\nu_e$		
Γ	$\left(1.61 \begin{array}{c} +0.39 \\ -0.27 \end{array}\right) \cdot 10^3\text{sec}^{-1}$	$0.54 \cdot 10^3\text{sec}^{-1}$
$K_L^0 \to \pi^0\pi^\pm e^\mp\nu_e$		
Γ	$(1.2 \pm 0.4) \cdot 10^3\text{sec}^{-1}$	$0.42 \cdot 10^3\text{sec}^{-1}$

The data are taken from ref. [61] for $K^+ \to \pi^+\pi^-e^+\nu_e$ and from the PDG [62] for $K^+ \to \pi^0\pi^0e^+\nu_e$ and $K_L^0 \to \pi^0\pi^\pm e^\mp\nu_e$. For $\mid V_{us} \mid$ we use $\mid V_{us} \mid = 0.2197$ [52,64,62].

where f, g, r and h are Lorentz invariant form factors which depend on the 3 variables $p_1 \cdot p_2, p_1 \cdot p_K$ and $p_2 \cdot p_K$. The contribution from the form factor r to the decay rate is multiplied by m_e^2 and is thus completely negligible. This form factor will therefore be dismissed in the following discussion.

The experimental value for the form factors together with the prediction of lowest order χPT is given in the table. The experimental values in the channel $K^+ \to \pi^+\pi^-e^+\nu_e$ are taken from [61], whereas the widths for the other two decays are taken from the PDG [62]. (The PDG quotes $\Gamma_{K^+\to\pi^+\pi^-e^+\nu_e} = (3.15 \pm 0.11) \cdot 10^3\text{sec}^{-1}$ for the weighted average over 5 experiments.) Finally the last column displays the lowest order result of χPT [63], obtained with $\mid V_{us} \mid = 0.2197$ [52,64,62]. The leading order result $f = 0.82\,(g = 0.82)$ is smaller than the experimental value by $\sim 30\%$ ($\sim 20\%$). (Part of the discrepancy in f is due to the I =0 S-wave $\pi\pi$ final state interactions which are not present in g.) Finally the form factor h is zero at leading order in χPT. It will receive a contribution at order p^4 from the Wess-Zumino Lagrangian $\mathcal{L}^{WZ}$.

Of course it is now very interesting to check whether agreement with the experimental results can be obtained at next order in χPT. The evaluation of the loop correction to the form factors f and g has been done and the comparison with the data is nearly completed [65]. The results will be published elsewhere.

In ref. [66] a different approach is pursued: the form factors f and g are represented as Omnès functions, using a phenomenological parametrization of the $\pi\pi$-phase shifts which is consistent with the data up to the resonance region. A good agreement between the calculated and measured form factors f and g at threshold is obtained in this way. The slope of the form factor f at threshold also agrees with the value quoted in ref. [61].

8 BARYONS

We have seen in the last sections that the chiral Lagrangian for the strong interactions in the meson sector is completely known up to and including terms of order p^4. The framework has been extended recently to include strong interaction processes with one external nucleon [12]. The leading order term in the elastic πN scattering amplitude is e.g. obtained by evaluating tree graphs with the nonlinear σ-model Lagrangian including baryon fields (I consider chiral $SU(2) \times SU(2)$):

$$\begin{aligned} \mathcal{L}_1 &= \mathcal{L}_{\pi N}^{(1)} + \mathcal{L}_{\pi\pi}^{(2)} \\ \mathcal{L}_{\pi N}^{(1)} &= \bar{\psi}(i\gamma^\mu \nabla_\mu - \overset{\circ}{M} + i \overset{\circ}{g}_A \gamma^\mu \gamma_5 \Delta_\mu)\psi \\ \mathcal{L}_{\pi\pi}^{(2)} &= \frac{F^2}{4} < \partial_\mu U \partial^\mu U^\dagger + \chi U^\dagger + \chi^\dagger U >, \end{aligned} \tag{8.1}$$

where

$$\begin{aligned} \nabla_\mu &= \partial_\mu \psi + \Gamma_\mu \psi \\ \Gamma_\mu &= \frac{1}{2}\left[u^\dagger, \partial_\mu u\right] \\ \Delta_\mu &= \frac{1}{2} u^\dagger \partial_\mu U u^\dagger \\ u^2 &= U = \exp i\phi^a \tau^a / F \\ \chi &= 2B \begin{pmatrix} m_u & 0 \\ 0 & m_d \end{pmatrix}. \end{aligned} \tag{8.2}$$

$\mathcal{L}_1$ contains four real parameters $F, B, \overset{\circ}{M}$ and $\overset{\circ}{g}_A$. They are related to the pion decay constant $F_\pi = 93.3$ MeV, the nucleon mass $M_N = 938$MeV, the neutron decay constant $g_A = -1.259$ and the condensate $< 0 \mid \bar{u}u \mid 0 >$ by

$$\begin{aligned} \left(F, \overset{\circ}{M}, \overset{\circ}{g}_A\right) &= (F_\pi, M_N, g_A)_{|m_u = m_d = 0} \\ F^2 B &= - < 0 \mid \bar{u}u \mid 0 >_{|m_u = m_d = 0} . \end{aligned} \tag{8.3}$$

The constants F, B introduced in eqs. (8.1,8.2) differ from the analogous constants F_0, B_0 used in the previous sections by terms of order m_s: $(F_0, B_0) = (F, B)_{|m_s = 0}$.

The elastic pion-nucleon amplitude $T_{\pi N}$ can be expressed in terms of four Lorentz invariant functions $A^\pm, B^\pm$. The leading order term in the low-energy expansion is

$$\begin{aligned} A^\pm_{tree} &= A^\pm_{pv} = \frac{g^2}{\overset{\circ}{M}} \begin{pmatrix} 1 \\ 0 \end{pmatrix} \\ B^\pm_{tree} &= B^\pm_{pv} + \frac{1}{2F^2} \begin{pmatrix} 0 \\ 1 \end{pmatrix}; \qquad g^2 = \frac{\overset{\circ}{M} \overset{\circ}{g}_A}{F} \end{aligned}$$

$$B^{\pm}_{pv} = g^2 \left(\frac{1}{\overset{\circ}{M}{}^2 - s} \mp \frac{1}{\overset{\circ}{M}{}^2 - u} \right) - \frac{g^2}{2\,\overset{\circ}{M}{}^2} \begin{pmatrix} 0 \\ 1 \end{pmatrix} . \tag{8.4}$$

The resulting amplitude $T_{\pi N}$ is of order p at threshold [4].

For a detailed discussion of χPT in the baryon sector I refer the reader to ref. [12]. In the following I concentrate on the pion-nucleon sigma term

$$\sigma = \frac{\hat{m}}{2M_N} < p \mid \bar{u}u + \bar{d}d \mid p > . \tag{8.5}$$

($| p >$ denotes a physical one-proton state with normalization $< p' \mid p >= 2(2\pi)^3 p^\circ \delta^3(\vec{p'} - \vec{p})$.) For the following discussion it is usful to introduce the quantity

$$\Sigma \doteq F_\pi^2 \bar{D}^+(\nu = 0, t = 2M_\pi^2); \qquad \nu = \frac{s-u}{4M_N} \tag{8.6}$$

where $\bar{D}^+$ denotes the amplitude $D^+ = A^+ + \nu B^+$ with pseudovector Born term subtracted [4]. Chiral symmetry relates the sigmaterm σ to the amplitude Σ in a manner which will be specified below. Using the quark mass expansion of the energy levels in the baryon octet, one can on the other hand express σ in terms of physical masses and an additional parameter which one may identify with the matrix element of the operator $\bar{s}s$ in the proton. In principle, elastic πN-scattering data can therefore be used to measure σ and to determine the strange quark content of the proton [68,69,1,70].

The actual evaluation of the matrix element $< p \mid \bar{s}s \mid p >$ and of σ involves several nontrivial steps which are visualized in the figure. I review briefly steps 1 and 2. In step 3 the amplitude at $\nu = t = 0$ is reconstructed from the data which are available above [5] $T_\pi \simeq 30$MeV. Chiral perturbation theory then relates the amplitude $T_{\pi N}$ at $\nu = t = 0$ to Σ (step 4). The remarkable work of the Karlsruhe group [71] has shown that this extrapolation of the data to the Cheng-Dashen-point can be done in a meaningful manner. The result for the Σ-term is $\Sigma = 64 \pm 8$MeV [72]. (See also [73,74], where a method is proposed which allows to relate experimental errors to the errors in Σ.)

Step 1: Using the quark mass expansion of the baryon octet at leading order, one finds

$$\sigma = \frac{\hat{m}}{m_s - \hat{m}} \frac{M_\Xi + M_\Sigma - 2M_N}{1-y} + O(\mathcal{M}^{\frac{3}{2}}) \tag{8.7}$$

where

$$y = \frac{2 < p \mid \bar{s}s \mid p >}{< p \mid \bar{u}u + \bar{d}d \mid p >} \tag{8.8}$$

measures the strange quark content of the proton. One may instead express σ in terms of M_N, M_Ξ, M_Λ and the nucleon mass $\overset{\circ}{M}$ in the chiral limit $m_u = m_d = m_s = 0$,

$$\begin{aligned} \sigma &= \frac{2\hat{m}}{2\hat{m} + m_s} \left\{ M_N - \overset{\circ}{M} + \frac{3}{2} \frac{m_s}{m_s - \hat{m}} (M_\Xi - M_\Lambda) \right\} + O(\mathcal{M}^{\frac{3}{2}}) \\ \overset{\circ}{M} &= M_N\big|_{m_u = m_d = m_s = 0} . \end{aligned} \tag{8.9}$$

[4]The point $\nu = 0, t = 2M_\pi^2$ is called 'Cheng-Dashen-point' in the literature[67].
[5]$T_\pi = \sqrt{M_\pi^2 + k^2} - M_\pi$ where k is the pion LAB momentum.

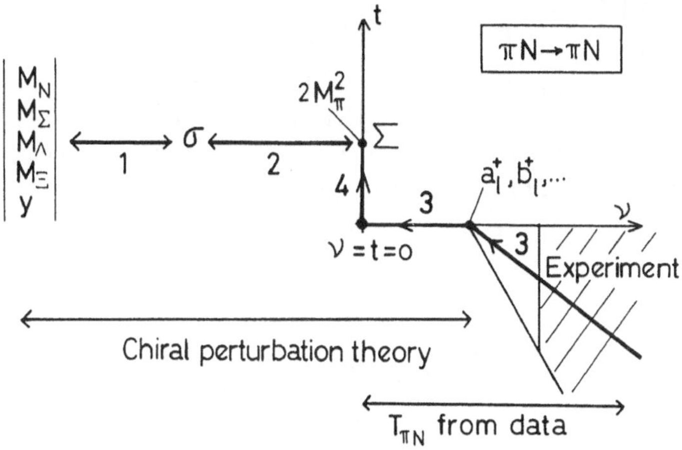

Fig 1. Relation between σ and the amplitude $T_{\pi N}$.
Step 1: σ is related to M_N, M_Σ, M_Ξ and the parameter y in eq. (8.8).
Step 2: Chiral symmetry relates σ to the πN amplitude at the Cheng-Dashen-point. Step 3: The amplitude at $\nu = t = 0$ is reconstructed from the data. Step 4: χPT relates Σ to $T_{\pi N}$ at $\nu = t = 0$.

Numerically, using [1] $m_s/\hat{m} = 25$, one finds

$$\sigma = \frac{26.4\text{MeV}}{1-y} + O(\mathcal{M}^{\frac{3}{2}}) \tag{8.10}$$

and

$$\sigma = (92.7 - 0.074 \overset{\circ}{M})\text{MeV} + O(\mathcal{M}^{\frac{3}{2}}) \tag{8.11}$$

respectively from eqs. (8.7,8.9). With $y = 0$ one has $\sigma = 26.4$MeV. This value corresponds to $\overset{\circ}{M} = 896$MeV according to (8.11).

At the one-loop level in χPT, the expression for σ becomes [69,1]

$$\sigma = \frac{\hat{\sigma}}{1-y}; \qquad \hat{\sigma} = (35 \pm 5)\text{MeV}. \tag{8.12}$$

This result includes all effects of order $\mathcal{M}^{3/2}$ in the quark mass expansion and incorporates an estimate of the terms of order $\mathcal{M}^2$.

Step 2: To relate σ to the amplitude Σ, one introduces the scalar form factor

$$\hat{m} < p' \mid \bar{u}u + \bar{d}d \mid p >= \bar{u}'u\sigma(t); \qquad t = (p'-p)^2. \tag{8.13}$$

At $t = 0, \sigma(t)$ coincides with the sigma term, $\sigma(0) = \sigma$. The Ward-Takahashi identity satisfied by $T_{\pi N}$ on the other hand relates the amplitude $\bar{D}^+$ to the form factor $\sigma(t)$

and a remainder [75],

$$\bar{D}^+(\nu,t) = F_\pi^{-2}\sigma(t) + q'^\mu q^\nu r_{\mu\nu}\,. \tag{8.14}$$

The remainder $q'^\mu q^\nu r_{\mu\nu}$ is not fixed by chiral symmetry requirements alone. It is in general not a good approximation to neglect this term, because it is of the same order in the quark mass expansion as the sigma term. However, at the Cheng-Dashen point one has $q'^\mu q^\nu r_{\mu\nu} = O(M_\pi^4 \ln M_\pi^2)$ and thus [67,75]

$$\begin{aligned} \Sigma &= \sigma + \Delta_\sigma + \Delta_R \\ \Delta_\sigma &= \sigma(2M_\pi^2) - \sigma(0) \\ \Delta_R &= O(M_\pi^4 \ln M_\pi^2)\,. \end{aligned} \tag{8.15}$$

The quantity Δ_R was evaluated for the first time in ref. [12] with the result

$$\Delta_R = 0.35\text{MeV}\,. \tag{8.16}$$

To estimate Δ_σ one needs to study the t-dependence of the form factor $\sigma(t)$. At the one-loop level in χPT one finds [1,12]

$$\Delta_\sigma = 4.6\text{MeV}\,. \tag{8.17}$$

(The result quoted in [76] is algebraically wrong by a factor of 2.) From eqs. (8.15 - 8.17) we obtain finally the one-loop result

$$\Sigma - \sigma = 5\text{MeV}\,. \tag{8.18}$$

The value $\Sigma = 64 \pm 8$MeV quoted above gives therefore $\sigma \sim 60$MeV. A σ-term of this size calls for rather drastic changes in the standard picture[68,69,1,70], as it implies that the matrix element $< p \mid \bar{s}s \mid p >$ is large, $y \sim 0.40$, and that half of the nucleon mass is generated by the mass of the strange quark [69,1].

Interestingly enough, it appears that the one-loop result (8.17) underestimates the actual value of Δ_σ substantially, leading in particular to an astonishingly large square radius of the nucleon scalar form factor. Of course, these higher order effects will also affect the relation between the πN-amplitude $T_{\pi N}$ (extracted from the data) and the sigma-term (steps 3,4 in the figure). The problem is under study, and we hope to provide these higher loop corrections soon [77].

9 SUMMARY AND OUTLOOK

Unitarity, chiral symmetry and its spontaneous breakdown constrain the Green functions very strongly. I have shown how one can reconstruct these Green functions at low energies in terms of a few coupling constants which are not fixed by symmetry considerations alone. Once these constants are pinned down, one can evaluate other physical quantities in a parameter free manner. Many calculations have already been done - yet there are also many places where the power of χPT has not yet been fully explored, e.g., weak and strong interactions in the baryon sector. Moreover the challenge to get a handle on both the low energy constants in the weak sector and on the higher loop corrections is still with us.

ACKNOWLEDGEMENTS

I thank the organizers of this school for providing us with such a marvellous place where we have had the opportunity to discuss about many fascinating topics. In addition I thank P. Hasenfratz, J. Hošek, M. Maggiore and S. Pokorski for enlightening discussions. Furthermore I am very greatful to M. von Ins and L. Gasser for their invaluable help in my struggle with LATEX#¿%.

REFERENCES

[1] J. Gasser and H. Leutwyler, Phys. Rep. **87C** (1982) 77.

[2] C.A. Dominguez and E. de Rafael, Ann. Phys. (N.Y.) **174** (1987) 372.

[3] J.W. Negele, QCD and hadrons on a lattice. Lectures given at this school.

[4] S. Weinberg, Phys. Rev. Lett. **18** (1967) 188.

[5] R. Dashen, Phys. Rev. **183** (1969) 1245;
R. Dashen and M. Weinstein, Phys. Rev. **183** (1969) 1261; **188** (1969) 2330.

[6] L.-F. Li and H. Pagels, Phys. Rev. Lett. **26** (1971) 1204; **27** (1971) 1089; Phys. Rev. **D5** (1972) 1509.

[7] P. Langacker and H. Pagels, Phys. Rev. **D8** (1973) 4595; **D10** (1974) 2904.

[8] H. Pagels, Phys. Rep. **16C** (1975) 219.

[9] S. Weinberg, Physica **96A** (1979) 327.

[10] J. Gasser and H. Leutwyler, Ann. Phys. (N.Y.) **158** (1984) 142.

[11] J. Gasser and H. Leutwyler, Nucl. Phys. **B250** (1985) 465.

[12] J. Gasser, M.E. Sainio and A. Švarc, Nucl. Phys. **B307** (1988) 779.

[13] J. Bijnens and M.B. Wise, Phys. Lett. **137B** (1984) 245;

J. Bijnens, H. Sonoda and M.B. Wise, Phys. Rev. Lett. **53** (1984) 2367;

G. D'Ambrosio and D. Espriu,Phys. Lett. **175B** (1986) 237;

J. F. Donoghue et al., Phys. Lett. **179B** (1986) 361;

J. L. Goity, Z. Phys. **C34** (1987) 341;

G. Ecker, A. Pich and E. de Rafael, Nucl. Phys. **B291** (1987) 692; Phys. Lett. **189B** (1987) 363; Nucl. Phys. **B303** (1988) 665;

J. Kambor, J. Missimer and D. Wyler, The chiral loop expansion of the nonleptonic weak interactions of mesons, Preprint ETH Zürich, August 1989.

[14] J. Gasser and H. Leutwyler, Phys. Lett. **184B** (1987) 83; **188B** (1987) 477; Nucl. Phys. **B307** (1988) 763.

[15] S.L. Adler and R.F. Dashen, 'Current' Algebras and Applications to Particle Physics', Benjamin, New York (1968).

[16] R.D. Peccei and J. Solà, Nucl. Phys. **B281** (1987) 1;
C.A. Dominguez and J. Solà, Z. Phys. **C40** (1988) 63.

[17] Y. Nambu, Phys. Rev. Lett. **4** (1960) 380; Phys. Rev. **117** (1960) 648;
Y. Nambu and G. Jona-Lasinio, Phys. Rev. **122** (1961) 345; Phys. Rev. **124** (1961) 246.

[18] J. Goldstone, Nuovo Cim. **19** (1961) 154;
J. Goldstone, A. Salam and S. Weinberg, Phys. Rev. **127** (1962) 965.

[19] M. Gell-Mann, R.J. Oakes and B. Renner, Phys. Rev. **175** (1968) 2195.

[20] E.V. Shuryak, 'The QCD vacuum, hadrons and the superdense matter', World Scientific Publishing Company, Singapore (1988).

[21] M.A. Shifman, A.I. Vainshtein and V.I. Zakharov, Nucl. Phys. **B147** (1979) 385,448,519.

[22] L.D. Landau and E.M. Lifshitz, 'Electrodynamics of continuous media', Pergamon Press, New York (1960), section 14.

[23] C. Itzykson and J.-B. Zuber, 'Quantum Field Theory', McGraw-Hill Inc., New York (1980), p.328.

[24] J.B. Kogut, A review of the lattice gauge theory approach to quantum chromodynamics, in: 'Recent advances in field theory and statistical mechanics', Les Houches, Aug.2 - Sept.10, 1982 (Session XXXIX) . North-Holland Physics Publishing, Amsterdam (1984).

[25] J.B. Kogut and L. Susskind, Phys. Rev. **D9** (1974) 3501.

[26] H. Lehmann and T.T. Wu, Nucl. Phys. **B237** (1984) 205.

[27] S. Adler and T. Piran, Rev. Mod. Phys. **56** (1984) 1.

[28] S. Mandelstam, Phys. Rev. **D19** (1979) 2391;
G. 't Hooft, Nucl. Phys. **B190 [FS3]** (1981) 455.

[29] J.W. Alcock, M.J. Burfitt and W.N. Cottingham, Nucl. Phys. **B266** (1983) 299;
V.P. Nair and C. Rosenzweig, Phys. Rev. **D31** (1985) 401;
M. Backer, J.S. Ball and F. Zachariasen, Phys. Rev. **D31** (1985) 2575;
F. Zachariasen, Comm. Nucl. Part. Phys. **17** (1987) 135;
J.S. Ball and A. Caticha, Phys. Rev. **D37** (1988) 524;
M. Baker, J.S. Ball and F. Zachariasen, Phys. Rev. **D37** (1988) 1036;
J. Hošek, Phys. Lett. **226B** (1989) 377.

[30] M.E. Peskin, Chiral symmetry and chiral symmetry breaking, in: 'Recent advances in field theory and statistical mechanics', Les Houches, Aug.2 - Sept.10, 1982 (Session XXXIX) . North-Holland Physics Publishing, Amsterdam (1984).

[31] T. Banks and A. Casher, Nucl. Phys. **B169** (1980) 103;
E. Marinari, G. Parisi and C. Rebbi, Phys. Rev. Lett. **47** (1981) 1795;
C. Vafa and E. Witten, Nucl. Phys. **B234** (1984) 173.

[32] E. Floratos and J.Stern, Phys. Lett. **119B** (1982) 419.

[33] M.A. Shifman, Ann. Rev. Nucl. Part. Sci. **33** (1983) 199;
L.J. Reinders, H. Rubinstein and S. Yazaki, Phys. Rep. **127C** (1985) 1;
S. Narison, Riv. Nuovo Cimento **10** (1987) 1;
R.A. Bertlmann, QCD Sum Rules: Selected Topics. Plenary Talk given at the conference 'Quarks - 88', May 1988, Tbilisi, USSR (Preprint UWThPh-1988-26).

[34] E.M. Ilgenfritz and M. Müller-Preussker, Phys. Lett. **119B** (1982) 395;
M. Campostrini, G. Curci, A. Di Giacomo and G. Paffuti, Z. Phys. **32** (1986) 377;
M. Campostrini, A. Di Giacomo and Y. Gündüc, Gluon Condensation in $SU(3)$ Lattice Gauge Theory, Preprint Università di Pisa, IFUP-TH 13/89.

[35] M.B. Voloshin, Nucl. Phys. **B154** (1979) 365; Sov. J. Nucl. Phys. **35** (1982) 592.

[36] H. Leutwyler, Phys. Lett. **98B** (1981) 447.

[37] The literature on the subject may be traced from ref. [38].

[38] A. Krämer, H.G. Dosch and R.A. Bertlmann, Estimate of the Background Gluon Correlation Time from Bottonium, Preprint UWThPh-1989-10.

[39] D. Gross, R. Pisarski and L. Yaffe, Rev. Mod. Phys. **53** (1981) 43;
O.K. Kalshnikov, Fortschr. Phys. **32** (1984) 525;
L. van Hove, 'The Quark-Gluon Plasma - A Progress Report', CERN TH 5069/88.

[40] The literature on the subject may be traced from ref. [41].

[41] P. Gerber and H. Leutwyler, Nucl. Phys. **B321** (1989) 387.

[42] S. Weinberg, Phys. Rev. Lett. **17** (1966) 616.

[43] M.M. Nagels et al., Nucl. Phys. **B147** (1979) 189.

[44] T.N. Truong, Chiral Perturbation Theory and Final State Theorem, Preprint Ecole Polytechnique, Palaiseau (France), March 1988.

[45] J.F. Donoghue, C. Ramirez and G. Valencia, Phys. Rev. **D38** (1988) 2195.

[46] A.W. Nilsson, Report given at the XXIV Int. Conf. on High Energy Physics, Munich (1988).

[47] D.G. Boulware and L.S. Brown, Ann. Phys. (N.Y.) **138** (1982) 392.

[48] J. Wess and B. Zumino, Phys. Lett. **37B** (1971) 95.

[49] J. Balog, Phys. Lett. **149B** (1984) 197;
A.A. Andrianov, Phys. Lett. **157B** (1985) 425;
N.I. Karchev and A.A. Slavnov, Theor. Math. Phys. **65** (1985) 1099 [Teor. Mat. Fiz. **65** (1985) 192];
L.-H. Chan, Phys. Rev. Lett. **55** (1985) 21;
A. Zaks, Nucl. Phys. **B260** (1985) 241;
P. Simic, Phys. Rev. **D34** (1986) 1903;
A.A. Andrianov et al., Phys. Lett. **186B** (1987) 401.

[50] G. Ecker, J. Gasser, A. Pich and E. de Rafael, Nucl. Phys. **B321** (1989) 311; J.F. Donoghue, C. Ramirez and G. Valencia, Phys. Rev. **D39** (1989) 1947; G. Ecker, J. Gasser, H. Leutwyler, A. Pich and E. de Rafael, Phys. Lett. **B223** (1989) 425.

[51] J. Gasser and H. Leutwyler, Phys. Lett. **125B** (1983) 325.

[52] H. Leutwyler and M. Roos, Z. Phys. **C25** (1984) 91.

[53] J. Gasser and H. Leutwyler, Nucl. Phys. **B250** (1985) 539.

[54] H. Leutwyler, Chiral perturbation theory - strong and semileptonic weak interactions, in: Proceedings of the Ringberg Workshop, Nucl. Phys. **B7A** (Proc.Suppl.) (1989) 42.

[55] J. Gasser and H. Leutwyler, Nucl. Phys. **B250** (1985) 517.

[56] R. Dashen and M. Weinstein, Phys. Rev. Lett. **22** (1969) 1337.

[57] G. Donaldson et al., Phys. Rev. **D9** (1974) 2960.

[58] D.G. Hill et al., Nucl. Phys. **B153** (1979) 39.

[59] Y. Cho et al., Phys. Rev. **D22** (1980) 2688.

[60] V.K. Birulev et al., Nucl. Phys. **B182** (1981) 1.

[61] L. Rosselet et al., Phys. Rev. **D15** (1977) 574.

[62] Particle Data Group: G.P. Yost et al., Phys. Lett. **204B** (1988) 1.

[63] J.A. Cronin, Phys. Rev. **161** (1967) 1483; L.M. Chounet et al., Phys. Rep. **C4** (1972) 199; C. Riggenbach, Chirale Störungstheorie und K_{l_4} - Zerfälle. Lizentiatsarbeit, Institut für theoretische Physik, Universität Bern, Bern (1988).

[64] J.F. Donoghue, B.R. Holstein and S.W. Klimt, Phys. Rev. **D35** (1987) 934.

[65] J.F. Donoghue, J. Gasser, B.R. Holstein and C. Riggenbach, work in progress.

[66] T.N. Truong, Modern application of dispersion relation: chiral perturbation vs dispersion technique, in: Festschrift for Professor K. Nishijima of the University of Tokyo and the Research Institute of Fundamental Physics, Kyoto University (Preprint CERN-TH4748/87).

[67] T.P. Cheng and R. Dashen, Phys. Rev. Lett. **26** (1971) 594.

[68] T.P. Cheng, Phys. Rev. **D13** (1976) 2161; **D38** (1988) 2869.

C.A. Dominguez and P. Langacker, Phys. Rev. **D24** (1981) 1905;

R.L. Jaffe, Phys. Rev. **D21** (1980) 3215;

J.F. Donoghue and C.R. Nappi, Phys. Lett. **168B** (1986) 105;

J.F. Donoghue, Proc. 2nd Intern. Workshop on πN physics (Los Alamos,1987), eds. W.R. Gibbs and B.M.K. Nefkens, Los Alamos report LA-11184-C (1987), p. 283.

[69] J. Gasser, Ann. Phys. (N.Y.) **136** (1981) 62.

[70] J. Gasser, Nucl. Phys. **B279** (1987) 65;

J. Gasser, Proc. 2nd Intern. Workshop on πN physics (Los Alamos,1987), eds. W.R. Gibbs and B.M.K. Nefkens, Los Alamos report LA-11184-C (1987), p. 266.

[71] G. Höhler, in: Landoldt-Börnstein, ed. H. Schopper (Springer, Berlin, 1983) Vol. 9 b2.

[72] R. Koch, Z. Phys. **C15** (1982) 161.

[73] J. Gasser, H. Leutwyler, M.P. Locher and M.E. Sainio, Phys. Lett. **213B** (1988) 85.

[74] M.E. Sainio, How to extract the pion-nucleon sigma term from data. Invited talk given at the 3rd International Symposium on Pion-Nucleon and Nucleon-Nucleon Physics, Gatchina, Leningrad, April 17-22, 1989. To be published in the proceedings.

[75] L.S. Brown, W.J. Pardee and R.D. Peccei, Phys. Rev. **D4** (1971) 2801.

[76] H. Pagels and W.J. Pardee, Phys. Rev. **D4** (1971) 3335.

[77] J. Gasser, H. Leutwyler and M.E. Sainio, work in progress.

HAMILTONIAN FORMULATION OF TWO-DIMENSIONAL GAUGE THEORIES ON THE LIGHT-CONE *

F. Lenz
Institute for Theoretical Physics
University of Erlangen-Nürnberg,
D-8520 Erlangen, Fed. Rep. Germany

Introduction

In Quantum Chromodynamics, the theory of strong interactions, gluons and quarks are the microscopic degrees of freedom. Hadrons constitute the effective degrees of freedom in terms of which hadronic and nuclear reactions as well as nuclear structure are traditionally described. The fundamental problem of low energy strong interaction theory is to understand, within the framework of QCD, the transition from microscopic to phenomenological degrees of freedom. Quark models have been important in clarifying this relation between microscopic and effective degrees of freedom as far as the structure of single hadrons is concerned. As the historical development already indicates, compositeness together with a simple picture about the underlying confining dynamics is sufficient for a qualitative understanding of hadronic properties in terms of "constituent" quarks.

The attempt to generalize quark models to the description of hadronic interactions has been less successful. From the formal point of view satisfactory non relativistic quantum mechanical models have been developed to describe the dynamics of multiquark systems in the presence of confining interactions. However here, compositeness together with simple confinement dynamics is apparently not sufficient for a qualitative understanding of hadronic interactions. In particular, medium range attraction and short range repulsion of the N-N interaction although conspicuously similar to effective interactions of other composite systems in physics, so far have not found a simple, unified explanation in terms of the underlying constituent quark dynamics. To a large extent, the theoretical attempts to understand the N-N interactions within the quark model have degenerated into fitting algorithms of the N-N phase shifts like other theoretical attempts before. Furthermore, there are obvious intrinsic limitations of quark models. The pion as the lightest particle plays a prominent role in low-energy hadronic interactions, its quark-model description however cannot be justified theoret-

*Supported by the Bundesministerium für Forschung und Technologie

Hadrons and Hadronic Matter
Edited by D. Vautherin *et al.*
Plenum Press, New York, 1990

ically. More generally, phenomena in hadronic interactions in which chiral symmetry is relevant, (e. g. low energy pion-nucleon scattering) are outside the scope of quark models. Similar remarks apply to those features of the strong interaction which are related to the axial U(1) anomaly. The large value of the η' mass for instance cannot be understood in quark models and there are indications from deep inelastic scattering that spin properties of the nucleon are severely affected by the anomaly. Thus it appears that on the one hand the quarkmodel is too remote from the fundamental QCD and on the other hand its theoretical framework too restrictive to allow for a qualitatively correct description of the relation between microscopic and effective degrees of freedom in hadronic interactions. For understanding of low-energy hadronic interactions, field-theoretic aspects of the strong interaction seem to be more essential than for many questions in hadron spectroscopy. Application of field-theoretical methods and investigations of field theoretical models are necessary in order to enlarge the theoretical framework for studies of strong interaction phenomena and to broaden the basis for construction of models which can be related to QCD. Here I describe such field-theoretic investigations of the strong interaction in the context of two-dimensional models, QED_2 and QCD_2.

QCD_2 as well as QED_2 both exhibit confinement of the microscopic degrees of freedom, and therefore constitute relevant and well defined theoretical models for analytic studies of hadronic interaction in terms of quark degrees of freedom. Unlike QCD_4, in these two dimensional gauge theories complete analytical solutions exist for certain limiting cases which provide a sufficient basis for a general understanding of these models. Clearly, these merits of two dimensional theories are intimately related to their shortcomings. Trivially, in two dimensions there are no transverse degrees of freedom; there is no spin for Fermions and there are (almost) no "real" photons and gluons. Confinement is not generated by the gluon dynamics; rather since infrared fluctuations in one space dimension are not damped, the $1/k^2$ Coulombpropagator coupled to either ordinary or color charges gives rise to a linearly confining potential. Despite these simplifying and unrealistic features, the two dimensional gauge theories constitute non-trivial relativistic theories of self- interacting Fermions which, as QCD_4 exhibit non trivial phenomena in addition to the confinement of the elementary Fermions, like chiral symmetry breaking, infinitely degenerate vacua and the axial anomaly. In detail, the aim of the investigations to be presented is to derive a Hamiltonian formulation of two-dimensional gauge theories, to construct on this basis the Hilbert space of physical states and to establish thereby a well defined framework accessible to both analytical and numerical studies of systems of strongly interacting Fermions in two dimensions.

Since the pioneering paper by 't Hooft [1] in (1974), QCD_2 has been the subject of many investigations. Meson and baryon spectra, properties of the hadrons, as well as hadronic interactions have been studied [2–11]. Essentially all the analytical tools of quantum field theory as well as numerical methods [12], [13], [14] have been employed in these investigations. Despite the many insights into strong interaction physics in two dimensions a coherent picture of the different states and phases, and the various phenomena of QCD_2 is missing; claims of inconsistencies [15], [16], [17] and proposals of alternative formulations [18] have appeared in the literature. The origin of this confused situation lies mainly in the not well understood infrared properties of QCD_2. These serious problems will be resolved in the formulation of QCD_2 in a finite interval which

I will present. In this way none of the ambiguities arises which plague the continuum formulations. However, the technically convenient axial gauges have to be abandoned and QCD_2 has to be solved in the Coulomb-gauge. This generalizes the finite interval treatment of the Schwinger model [19], [20]. The result of these investigations will be a properly defined Hamiltonian containing none of the redundant variables.

This Hamiltonan formulation will be derived in light-cone coordinates. As the original study by 't Hooft [1], most of the investigations of QCD_2 make use of the great technical simplifications offered by the light-cone dynamics. However this great technical simplicity is connected with serious conceptual difficulties. In particular, it is not clear whether and how on the light- cone non perturbative phenomena associated with nontrivial phases can be described . For instance for QCD_2 with a large number of colors, it is known from general arguments [21], and calculations [8] using ordinary coordinate quantization [7], [9] that the vacuum is non trivial and for massless Fermions characterized by a non-vanishing value of the quark condensate. Straightforward application of light cone dynamics yields the perturbative vacuum but on the other hand reproduces correctly the spectrum of elementary excitations. Similarly, the light-cone treatment of QED_2 yields the correct meson spectrum [22] but is not able to account directly for the axial anomaly and the Fermion condensate (cf. [23], [30]). These discrepancies in formulations of quantum field theories on the light- cone and in ordinary coordinates respectively are not well understood and shed serious doubts on the equivalence of these two formulations. (In the context of QCD_2 for instance, this equivalence has been demonstrated so far only numerically (cf. refs. [1], [9]) in studies of the meson spectrum.) These conceptual difficulties have to be resolved in formulating a well defined Hamiltonian theory of light-cone dynamics. I consider such an attempt as unavoidable if field theoretical descriptions are to be applied to complex multi-quark systems relevant for the study of hadronic interactions. In such investigations, the great technical simplicity of the light-cone formulation seems to be indispensible.

The first part of these investigations [24] is therefore devoted to the study of light-cone dynamics. Most of the difficulties are encountered already on the level of non-interacting theories. Formulation of the theory in finite light-cone intervals will be important for understanding light-cone theories as effective theories with well defined properties. These finite interval light-cone techniques will be used in the second Chapter to investigate the well known non perturbative phenomena of QED_2. The treatment here follows the finite interval description of ordinary coordinate quantization. In the third part the light- cone Hamiltonian for QCD_2 will be derived, the structure of the vacuum will be discussed as well as the elementary excitations.

1 Dynamics on the light-cone

1.1 The continuum light-cone formalism

In this first paragraph a brief introduction into light-cone dynamics is presented (cf. refs. [25], [26], [27]). The basic differences and simplifications of light-cone in comparison to ordinary coordinate quantization are emphasized and the related difficulties are discussed. Based on this discussion, a finite light-cone interval technique is developed which allows to interpret light-cone theories as effective theories. Here only Fermion-fields are discussed. Appendix I describes light-cone quantization of free Bosons in the finite interval.

Starting point of the formal developments is the Lagrangean of 2-dimensional electrodynamics.

$$\mathcal{L} = \overline{\chi}(i\partial_\mu + gA_\mu)\gamma^\mu\chi - m\overline{\chi}\chi - \frac{1}{4}F^{\mu\nu}F_{\mu\nu} \tag{1}$$

The electromagnetic field is described by the vector potential $A_\mu, \mu = 0,1$ which in turn determines the field strength tensor

$$F_{\mu\nu} = \partial_\mu A_\nu - \partial_\nu A_\mu \tag{2}$$

The matter field in two dimensions is described by a spinor with two components which we write as

$$\chi = \begin{pmatrix} \phi \\ \psi \end{pmatrix}$$

The γ^μ are 2 x 2 Dirac matrices satisfying

$$\{\gamma^\mu, \gamma^\nu\} = 2g^{\mu\nu} \qquad \mu, \nu = 0,1 \tag{3a}$$

and can be respresented by

$$\gamma^0 = \begin{pmatrix} 0 & 1 \\ 1 & 0 \end{pmatrix} \quad \gamma^1 = \begin{pmatrix} 0 & -1 \\ 1 & 0 \end{pmatrix} \tag{3b}$$

In addition we introduce

$$\gamma^5 = \gamma^0\gamma^1 = \begin{pmatrix} 1 & 0 \\ 0 & -1 \end{pmatrix} \tag{3c}$$

This choice of the γ matrices is particularly convenient for the description of massless fermions, since in this representation it is only the mass term which couples the two components ϕ, ψ as is read off directly from the Lagrangean

$$\begin{aligned} \mathcal{L} \;=\; & \phi^+(i\partial_0 + gA_0)\phi \;+\; \psi^+(i\partial_0 + gA_0)\psi \\ +\; & \phi^+(i\partial_1 + gA_1)\phi \;-\; \psi^+(i\partial_1 + gA_1)\psi \\ -\; & m(\phi^+\psi + \psi^+\phi) \;+\; \tfrac{1}{2}(\partial_0 A_1 - \partial_1 A_0)^2 \end{aligned} \tag{4}$$

Clearly the two spinor components ϕ, ψ can be obtained with the help of the projection operator on "right" and "left-handed" fermions

$$P_{R,L} = \frac{1}{2}(1 \pm \gamma_5) = \begin{matrix} \begin{pmatrix} 1 & 0 \\ 0 & 0 \end{pmatrix} \\ \\ \begin{pmatrix} 0 & 0 \\ 0 & 1 \end{pmatrix} \end{matrix} \tag{3d}$$

and we refer in the following to ϕ and ψ as right and left handed components respectively. To simplify further the Lagrangean, new coordinates are introduced

$$x^+ = \tfrac{1}{\sqrt{2}}(x^0 + x^1) \quad \partial_+ = \tfrac{1}{\sqrt{2}}(\partial_0 + \partial_1)$$
$$x^- = \tfrac{1}{\sqrt{2}}(x^0 - x^1) \quad \partial_- = \tfrac{1}{\sqrt{2}}(\partial_0 - \partial_1) \tag{5}$$

and new combinations of the gauge fields

$$A_\pm = \frac{1}{\sqrt{2}}\,(A_0 \pm A_1) \tag{6}$$

in terms of which the Lagrangean becomes:

$$\mathcal{L} = \sqrt{2}\phi^+(i\partial_+ + gA_+)\phi + \sqrt{2}\psi^+(i\partial_- + gA_-)\psi$$
$$-m(\phi^+\psi + \psi^+\phi) + \tfrac{1}{2}(\partial_+ A_- - \partial_- A_+)^2 \tag{7}$$

We first solve the free theory

$$A_\pm = 0$$

which proceeds in the standard way. The Euler Lagrange equations

$$\frac{\delta\mathcal{L}}{\delta\phi^+} = 0 = i\sqrt{2}\partial_+\phi - m\psi \tag{8a}$$

$$\frac{\delta\mathcal{L}}{\delta\psi^+} = 0 = i\sqrt{2}\partial_-\psi - m\phi \tag{8b}$$

are used to determine the normal modes of the fermion field. The equations of motion of non interacting fields are solved by plane waves with periodic time dependence

$$\begin{pmatrix} \phi \\ \psi \end{pmatrix} = \begin{pmatrix} \phi_0 \\ \psi_0 \end{pmatrix} e^{-i(p_+x^+ + p_-x^-)} \tag{9a}$$

The Euler Lagrange equations relate p_+ with p_-

$$p_+p_- = \frac{1}{\sqrt{2}}\,(p_0 - p_1)\,\frac{1}{\sqrt{2}}\,(p_0 + p_1) = \frac{1}{2}\,(p_0^2 - p_1^2) = \frac{m^2}{2} \tag{9b}$$

and relate the two spin components with each other

$$\psi_0 = \frac{m}{\sqrt{2}\,p_-}\,\phi_0 \tag{9c}$$

Eq. (9b) is nothing else than the standard dispersion relation between energy and momentum. I emphasize that up to this point the choice of coordinates is of no physical relevance. We clearly recognize in the plane wave the standard $p_\mu x^\mu$ term which we could have obtained equally easily in ordinary coordinates $x_{0,1}$. The distinction between standard dynamics and light-cone dynamics arises classically by the choice of the initial value problem and quantum mechanically by the basic commutation relations. The crucial step in both the classical and quantum theory is the selection of either t or x^+ as time variable.

We decide to interpret in the following x^+ as time variable and develop now the light-cone dynamics within the canonical formalism. We realize that only equation (8 a) is an equation of motion, while (8 b) does not involve a time derivative. As a consequence, given the right handed spinor component ϕ at a given light-cone time x^+ and for all x^- the left-handed ψ is (up to a constant) determined and given by

$$\psi(x^+,x^-) = \frac{m}{2i\sqrt{2}} \int dz \epsilon\,(x^- - z)\, \phi\,(x^+,z) \tag{10}$$

with

$$\epsilon(x) = \frac{x}{\mid x \mid}$$

Eliminating this dependent degree of freedom, the Lagrangean (7) (with $A_{\pm} = 0$) is written in terms of the true degree of freedom, the right-handed spinor component ϕ as

$$\mathcal{L} = \sqrt{2}\phi^+ i\partial_+\phi - \frac{m^2}{2i\sqrt{2}}\phi^+(x) \int dz \epsilon(x-z)\phi(z) \tag{11}$$

We can now apply the standard procedure of canonical quantization. We define the conjugate momentum

$$\pi(x) = \frac{\delta\mathcal{L}}{\delta\partial_+\phi} = i\sqrt{2}\phi^+ \tag{12}$$

the Hamiltonian

$$H = \int dx(\pi\partial_+\phi - \mathcal{L}) = \frac{m^2}{2i\sqrt{2}} \int dxdz\phi^+(x)\epsilon(x-z)\phi(z) \tag{13}$$

and impose equal (light-cone) time anti-commutation relations

$$\{\pi(x^+,x^-),\ \phi\,(x^+,y^-)\} = i\sqrt{2}\,\{\phi^+(x^+,x^-), \phi(x^+,y^-)\} = i\delta(x^- - y^-) \tag{14}$$

The theory is solved by Fourier-transforming the field operators

$$\phi\,(x) = \frac{1}{2^{\frac{1}{4}}}\,(\frac{1}{2\pi})^{\frac{1}{2}} \int dp\; e^{-ipx}\phi(p) \tag{15}$$

In terms of the annihilation and creation operators satisfying standard anti-commutation relations (as follows from (14))

$$\{\phi^+\,(p)\,,\phi\,(p')\} \;=\; \delta\,(p-p')$$

the Hamiltonian is diagonal

$$\begin{aligned} H &= \tfrac{m^2}{4i}\,(\tfrac{1}{2\pi}) \int dpdq\phi^+(p)\phi(q) \int dxdz\; e^{i(px-qz)}\epsilon(x-z) \\ &= \int dp \tfrac{m^2}{2p}\phi^+(p)\phi(p) \end{aligned} \tag{16}$$

and the construction of ground and excited states follows as in the usual quantization of Fermion-fields. According to (16) the light cone energy of an occupied momentum state p is $m^2/2p$ (cf. (9b)) and therefore the state of lowest energy, the vacuum $\mid \Omega >$ is obtained by occupying all states of negative energy, which are the states of negative momentum

$$\phi^+(p) \mid \Omega > \; = \; 0 \qquad \text{for} \;\; p < 0$$

$$\phi(p) \mid \Omega > \; = \; 0 \qquad \text{for} \;\; p > 0$$

The excited states are obtained by applying creation and annihilation operators. In particular, the single particle states $\phi^+(p) \mid \Omega > (p > 0)$ and $\phi(p) \mid \Omega > (p < 0)$ describe right moving fermions with (excitation) energies $m^2/2p$.

The characteristic features of light- cone quantization implying great simplifications also for interacting theories are consequences of one elementary property: the uniqueness in the dispersion relation between light-cone momentum and light-cone energy. This is schematically shown in Figure 1. While for a given p_1 single particle states of positive and negative energy exist, for a given p_- there is a unique state. This uniqueness implies independence of the vacuum from the value of the mass [25] and most importantly independence of the interaction. In ordinary coordinates, interactions in general mix positive and negative energy states and thereby "non- perturbative" vacua are generated while on the light cone such a mixing is not possible.

The absence of non-trivial vacua implies great technical simplifications but also poses serious conceptual problems. In particular it is not clear whether this implies a failure of the light-cone description if applied to theories with non trivial vacua. We will investigate this problem in the later Chapters. A related difficulty which can be discussed already within the free theory is that of the regularization of the total energy. We note that the divergence of the vacuum energy – the energy of the filled Dirac sea

$$E = \int_{-\infty}^{0} \frac{m^2}{2p} \, dp$$

arises on the one hand as in ordinary quantization from large momenta, on the other hand this expression also has an infrared divergence. This is a consequence of the light-cone dispersion relation in which the long wave-length particles have high-energy. In ordinary coordinates this divergence can be cured e. g. by splitting the space components of the arguments in the Fermion-operators defining the Hamiltonian. On the light-cone such a splitting of the space components is of no help since the distance between two events

$$ds^2 = x_0^2 - x_1^2 = 2x^+x^- \tag{17}$$

is in general spacelike for equal ordinary times ($x_0 = 0$) while always light like for equal light-cone times ($x^+ = 0$) . Thus regularization of the light-cone Hamiltonian requires splitting in the time-components of the arguments of $\phi^+(x) \cdot \phi(z)$ in (13) which in turn however requires knowledge of the time evolution to be determined by the Hamiltonian.

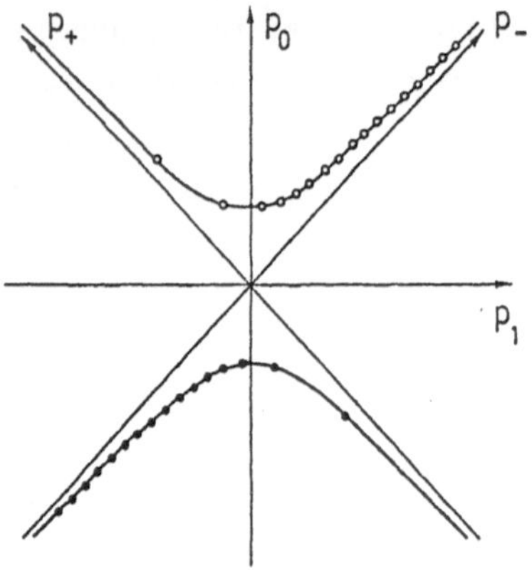

Fig. 1. Energy as function of the momentum. The axes correspond to standard p_0, p_1 and light-cone p_+, p_- choice of coordinates. The occupied negative energy states and the unoccupied positive energy states for equally spaced discrete light-cone momenta (p_-) are shown.

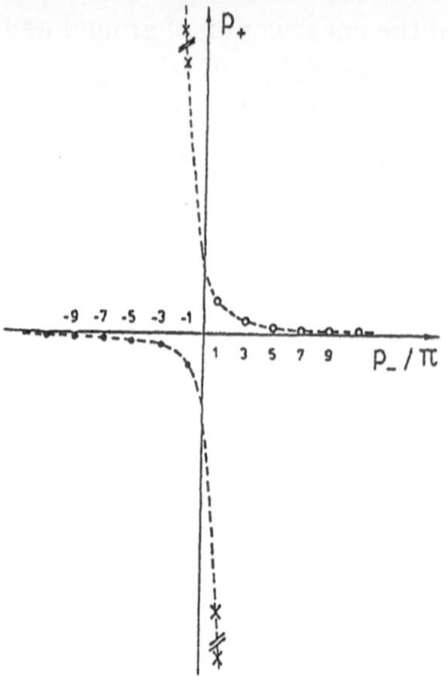

Fig. 2. Solution of the dispersion relation (21b) for the values $L = 1$, $\varphi = \frac{\pi}{2}$ in (21a). Note the order of magnitude difference in energy between the two solution (circles and croses) for a given momentum.

1.2 Finite light-cone intervals

We will clarify the physical origin of the formal problems discussed above by formulating the light-cone dynamics in a finite interval. The extension of the interval is L. For a finite interval, boundary conditions have to be specified. If f denotes either of the two spinor components

$$f(x^+, x^-) = \begin{matrix} \phi \\ \psi \end{matrix} \tag{18a}$$

we require

$$f(x^+ - \epsilon, L) = e^{-i\varphi} f(x^+, 0) \tag{18b}$$

This form of the boundary condition is unusual since the light-cone times on the two sides of equation (18 b) are different by the amount $\epsilon > 0$. This choice guarantees that the two space-time points connected via the boundary condition are spacelike separated

$$ds^2 = 2\cdot(x^+ - \epsilon - x^+)\cdot(L-0) = -2\epsilon\cdot L \tag{18c}$$

For applying the canonical formalism we must get rid of the time-dependence in the boundary condition by redefinition of the fields

$$\tilde{f}(x^+,x^-) = f(x^+ - \epsilon\frac{x^-}{L}, x^-) \rightarrow \tilde{f}(x^+,L) = e^{-i\varphi}\tilde{f}(x^+,0) \tag{19a}$$

Changing variables and further rescaling the fields

$$\begin{array}{lcl} x^{-\prime} = x^- & \rightarrow & \partial_- = \partial_{-\prime} + \frac{\epsilon}{L}\partial'_+ \\ x^{+\prime} = x^+ + \epsilon\cdot\frac{x^-}{L} & & \partial_+ = \partial_{+\prime} \end{array} \tag{19b}$$

$$\tilde{\phi} \rightarrow 2^{-\frac{1}{4}}\phi; \tilde{\psi} \rightarrow 2^{-\frac{1}{4}}\sqrt{\frac{L}{\epsilon}}\psi; \tilde{A}_+ \rightarrow A_+; \tilde{A}_- \rightarrow A_- + \frac{\epsilon}{L}A_+ \tag{19c}$$

the Lagrangean (7) in its final form reads

$$\begin{aligned} \mathcal{L} = \; & \phi^+(i\partial_+ + gA_+)\phi + \psi^+(i\partial_+ + gA_+ + \tfrac{L}{\epsilon}(i\partial_- + gA_-))\psi \\ & -m\sqrt{\tfrac{L}{2\epsilon}}(\psi^+\phi + \phi^+\psi) + \tfrac{1}{2}(\partial_+A_- - \partial_-A_+)^2 \end{aligned} \tag{20}$$

I emphasize for the following two important properties of this Lagrangean. As in the continuum light-cone Lagrangean (7) only a spatial derivative of the left handed Fermion fields appears. Unlike the continuum version , the Lagrangean (20) contains time derivatives of both right and left handed Fermion fields.

Again we first discuss in detail the free Fermion- theory ($A_\pm = 0$). The plane wave Ansatz (9 a) solves the corresponding Euler Lagrange equation with the following condition for light-cone energy and momenta

$$0 = \begin{vmatrix} p_+, & -m\sqrt{\frac{L}{2\epsilon}} \\ -m\sqrt{\frac{L}{2\epsilon}}, & p_+ + \frac{L}{\epsilon}p_n^\varphi \end{vmatrix} \quad ; \qquad p_n^\varphi = \frac{2n\pi}{L} + \varphi/L \tag{21a}$$

In contrast to the derivation above we now again find two solutions as in ordinary space-time description

$$p_+ = -\frac{1}{2\epsilon}\left((2n\pi+\varphi) \pm \sqrt{(2n\pi+\varphi)^2 + 2m^2\epsilon L}\right) \tag{21b}$$

In Figure 2 the solution of the dispersion relation is shown. The existence of two solutions for given light-cone momentum reflects the presence of time derivatives of both spinor components. Therefore both right and left-handed spinors represent independent degrees of freedom and consequently momenta conjugate to both spinor components

$$\pi_\varphi = \frac{\delta\mathcal{L}}{\delta\partial_+\phi}, \qquad \pi_\psi = \frac{\delta\mathcal{L}}{\delta\partial_+\psi} \tag{22a}$$

can be defined. The resulting Hamiltonian reads

$$H = -\frac{L}{\epsilon}\int dx \psi^+(x) i\partial_- \psi(x) + m\sqrt{\frac{L}{2\epsilon}}\int dx(\psi^+(x)\phi(x) + \phi^+(x)\psi(x)) \tag{22b}$$

We perfom a normal mode expansion of the field operators

$$\begin{matrix} \phi\,(x) \\ \psi\,(x) \end{matrix} = \frac{1}{L^{1/2}} \sum_n \begin{matrix} a_n \\ b_n \end{matrix} e^{-ip_n^{\varphi}\cdot x} \tag{23a}$$

$$\{a_n^+, a_m\} = \{b_n^+, b_m\} = \delta_{n,m} \tag{23b}$$

The Hamiltonian expressed in terms of the normal mode operators is decomposed

$$H = H_0 + H_1 \tag{24a}$$

$$H_0 = -\frac{L}{\epsilon}\sum_n b_n^+ p_n^{\varphi} b_n = \frac{-1}{\epsilon}\sum_n b_n^+(2\pi n + \varphi) b_n \tag{24b}$$

$$H_1 = m\sqrt{\frac{L}{2\epsilon}}\sum_n (a_n^+ b_n + b_n^+ a_n) \tag{24c}$$

into H_0 acting only in the Hilbert space describing the left-handed b-modes and a coupling term H_1 between a and b modes. We also note the absence of an operator which is diagonal in the right handed modes. For a free theory the Hamiltonian can be exactly diagonalized by a Bogoliubov transformation which mixes right and left-handed Fermions and the eigenvalues (21 b) are obtained. For interacting fields this is obviously not possible and we develop here a more general approximation scheme. Comparison of H_0 with H_1 shows that the relative coupling strength between a and b modes is controlled by the dimensionless parameter

$$\eta = m^2 \epsilon L \tag{25}$$

which we assume – by appropriate choice of ϵ – to satisfy

$$\eta << 1 \tag{26}$$

In this case, the b-modes are much higher in energy than the a-modes (cf. eq. 21 b and Figure 2) and therefore will only be little affected by the coupling H_1. This suggests to eliminate these almost frozen b-modes and thereby define an effective low-energy theory of the right-handed fermions. The smallness of η allows to carry through this elimination procedure perturbatively.

The ground state $|\,0>$ of the b-modes corresponding to H_0 is obtained by occupying all negative energy b-mode states thus ($\varphi > 0$):

$$|\,0>: \quad b_n^+\,|\,0>= 0 \quad n \geq 0; \quad b_n\,|\,0>= 0 \quad n < 0 \tag{27}$$

To first order in the pertubation H_1, this state is modified

$$
\begin{aligned}
\mid \tilde{0} > &= \mid 0 > + m\sqrt{\tfrac{L}{2\epsilon}} \textstyle\sum_n \frac{1}{\frac{1}{\epsilon}(2\pi n+\varphi)} (b_n^+ a_n - a_n^+ b_n) \mid 0 > \\
&= \mid 0 > + \mid \delta >
\end{aligned} \tag{28}
$$

Note that so far we have not specified the quantum state of the a-modes which to lowest order is not determined (i.e. infinitely degenerate). The effective Hamiltonian is obtained in second order perturbation theory (the first order contribution of H_1 vanishes) and taking the expectation value with respect to the b modes only

$$
\begin{aligned}
H_a &=: \; < 0 \mid H_1 \frac{1}{E_0 - H_0} H_1 \mid 0 >_b = \\
&= [m\sqrt{\tfrac{L}{2\epsilon}}]^2 \textstyle\sum_n < 0 \mid (a_n^+ b_n + b_n^+ a_n) \frac{1}{\frac{1}{\epsilon}(2\pi n+\varphi)} (b_n^+ a_n - a_n^+ b_n) \mid 0 >_b \\
&= \textstyle\sum_n \frac{m^2}{2(2\pi n+\varphi)/L} [a_n^+ a_n \theta(-n) - a_n a_n^+ \theta(n)]
\end{aligned} \tag{29}
$$

which up to a constant yields

$$
H_a = \sum_n \frac{m^2}{2(2\pi n + \varphi)/L} a_n^+ a_n \tag{30}
$$

the effective a-mode Hamiltonian. H_a is the discrete version of the continuous light-cone Hamiltonian (16).

The free continuum light-cone theory thus is derived as an effective theory in which the left handed high energy b-modes have not disappeared simply but have been eliminated within a well defined procedure yielding an effective low energy theory for the right-handed b-modes. This elimination procedure allows to properly define effective operators. As an example, important for later applications we consider the condensate operator $\Gamma(x)$ defined in terms of the original spinors (eqs. 1, 2) and expressed in terms of the rescaled variables (19 c) as

$$
\Gamma(x) = \overline{\chi}(x)\chi(x) = \sqrt{\frac{L}{2\epsilon}} (\phi^+(x)\psi(x) + \psi^+(x)\phi(x)) \tag{31}
$$

For translationally invariant states $\mid \psi >$ we have

$$
< \psi \mid \Gamma(x) \mid \psi > = \frac{1}{L} < \psi \mid \Gamma_0 \mid \psi > = \frac{1}{L}\sqrt{\frac{L}{2\epsilon}} < \psi \mid \sum_n (a_n^+ b_n + b_n^+ a_n) \mid \psi > \tag{32}
$$

We eliminate the b-modes by taking matrix-elements of Γ_0 in the b-ground state $\mid \tilde{0} >$ (eq. 28)

$$
\begin{aligned}
\Gamma_0^a &= < \tilde{0} \mid \Gamma_0^a \mid \tilde{0} >_b = < 0 \mid \Gamma_0^a \mid \delta >_b + < \delta \mid \Gamma_0^a \mid 0 >_b \\
&= \textstyle\sum_n \frac{m}{2(2\pi n+\varphi)/L} 2 \cdot [a_n^+ a_n \theta(-n) - a_n a_n^+ \theta(n)] \\
&= \textstyle\sum_n \frac{2m}{2(2\pi n+\varphi)/L} a_n^+ a_n = \frac{\delta H_a}{\delta m}
\end{aligned} \tag{33}
$$

and thus the effective condensate operator in the Hilbert space of the a modes is obtained.

The light-cone free fermion theory is however not yet complete, the various physical quantities like the total energy or the Fermion condensate have to be regularized; the a-mode vacuum is determined by H_a and obviously obtained by filling all negative (a-mode) momentum states. This yields for the total energy

$$E = \sum_{n<0} \frac{m^2 \cdot L}{2(2\pi n + \varphi)} \tag{34a}$$

This expression has to be regularized by imposing for instance a cutoff in momentum i. e. we require

$$| p_n | = \frac{| 2\pi n + \varphi |}{L} < \frac{N}{L} = \Lambda \tag{34b}$$

and obtain

$$E = -\sum_{n>0}^{\Lambda \cdot L} \frac{m^2 L}{2(2\pi n - \varphi)} \sim -\frac{m^2}{4\pi} L \, ln \Lambda L + 0(L) \tag{34c}$$

For fixed length L of the light-cone interval, this represents the total vacuum energy. However as the logarithmic L-dependence of the energy density indicates, these finite interval vacua cannot be used to define a proper continuum limit, a further modification of the theory is needed. The low momentum states have to be removed from the Hilbert space as well.

Thus we allow only momentum states satisfying

$$\lambda << | \frac{2\pi n}{L} | << \Lambda \tag{35a}$$

and therefore, by construction an energy density independent of L

$$E/L = -\sum_{\lambda L}^{\Lambda \cdot L} \frac{m^2}{2(2\pi n - \varphi)} = -\frac{m^2}{4\pi} \, ln \frac{\Lambda}{\lambda} \tag{35b}$$

results. This complication of defining a proper continuum limit is inherent to the light-cone description and we will encounter it again in the interacting theories. The difficulty can be completely avoided at the expense of solving a dynamically more complicated problem and choosing the parameter η (25) large. For

$$\eta = m^2 \epsilon L >> 1 \tag{36a}$$

perturbation theory cannot be applied anymore. Calculation of the total energy yields after introducing the ultraviolet cutoff Λ

$$E = -\sum_{0}^{\Lambda L} \frac{m^2}{2\sqrt{(2\pi n - \varphi)^2 + 2\epsilon m^2 L}} = -\frac{m^2}{4\pi} L \, ln \frac{\Lambda}{\epsilon m^2 / 4\pi^2} \tag{36b}$$

i. e. ϵm^2 acts as infrared momentum cutoff. Thus the whole effect of the non-perturbative mixing of a and b modes is to provide an infrared momentum cutoff for performing the continuum limit.

2 QED$_2$ on the light cone

2.1 Lagrangean and global symmetries

I present here a study of two-dimensional Quantum Electrodynamics within the light-cone framework as developed in the first Chapter. Here results are rederived within this new context which are well-known from investigations [19], [20] of QED$_2$ in ordinary space time quantization. The aim of this study is to display how within light-cone dynamics non perturbative phenomena are derived. Particular attention will be paid therefore to the axial anomaly and the appearance of an infinitely degenerate vacuum.The tool is the canonical formalism and therefore a careful discussion of the true degrees of freedom will be an important step for construction of the Hamiltonian. Techniques and insights developed in this Chapter will prepare the grounds for the technically more involved study of QCD$_2$ in the third Chapter.

Starting point is the light-cone Lagrangean (20)

$$\begin{aligned}\mathcal{L} = \; & \phi^+(i\partial_+ + gA_+)\phi + \psi^+(i\partial_+ + gA_+ + \tfrac{L}{\epsilon}(i\partial_- + gA_-))\psi \\ & + \tfrac{1}{2}(\partial_+ A_- - \partial_- A_+)^2 - m\sqrt{\tfrac{L}{2\epsilon}}\,(\phi^+\psi + \psi^+\phi)\end{aligned} \tag{37}$$

which has been derived in the first chapter for the choice (5, 19 b) of the coordinates.

As seen be inspection this Lagrangean is invariant under the global transformation

$$\begin{aligned}\phi(x^+,x^-) &\rightarrow \phi'(x^+,x^-) = e^{i\chi}\phi(x^+,x^-) \\ \psi(x^+,x^-) &\rightarrow \psi'(x^+,x^-) = e^{i\chi}\psi(x^+,x^-)\end{aligned} \tag{38}$$

in which right and left handed fermion-fields are changed by a common x^+,x^-–independent phase. This invariance leads to current conservation which in the light-cone variables has the form

$$\partial_+(\phi^+(x)\phi(x) + \psi^+(x)\psi(x)) + \frac{L}{\epsilon}\partial_-(\psi^+(x)\psi(x)) = 0 \tag{39}$$

We note that only left-handed Fermions (4) contribute to the space component of the current. In the Lagrangean (37) no spatial derivative of the right handed Fermion fields appears.

Charge conservation is a consequence of the continuity equation (39) which we write as

$$\partial_+ Q = \partial_+ Q_a + \partial_+ Q_b = 0 \tag{40a}$$

where we have defined separate right and left handed charges

$$\begin{aligned}Q_a &= \int dx^- \phi^+(x^+,x^-)\phi(x^+,x^-) \\ Q_b &= \int dx^- \psi^+(x^+,x^-)\psi(x^+,x^-)\end{aligned} \tag{40b}$$

In massless QED$_2$, – the Schwinger model – the Lagrangean is in addition invariant under chiral rotations, i. e. phases of right and left handed Fermions are changed differently

$$\phi(x^+,x^-) \to \phi'(x^+,x^-) = e^{+i\chi_5}\,\phi(x^+,x^-)$$
$$\psi(x^+,x^-) \to \psi'(x^+,x^-) = e^{-i\chi_5}\,\psi(x^+,x^-) \tag{41a}$$

The continuity equation for the axial current results

$$\partial_+(\phi^+(x)\phi(x) - \psi^+(x)\psi(x)) - \frac{L}{\epsilon}\partial_-(\psi^+(x)\psi(x)) = 0 \qquad (m=0) \tag{41b}$$

and the axial charge defined by

$$Q_5 = Q_a - Q_b \tag{42a}$$

is conserved

$$\partial_+ Q_5 = 0 \qquad (m=0) \tag{42b}$$

It is interesting to perform at this point the formal limit to the light-cone continuum. For this purpose we have to use the original fields (19 c), i. e. to replace ψ by $\tilde{\psi}$

$$\psi^+(x)\psi(x) = \frac{\sqrt{2}\epsilon}{L}\,\tilde{\psi}^+(x)\,\tilde{\psi}(x) \tag{43a}$$

and drop in the continuity equations (39, 41) the ϵ/L suppressed terms with the result

$$\partial_+(\phi^+(x)\phi(x)) \pm \partial_-(\tilde{\psi}^+(x)\,\tilde{\psi}(x)) = 0 \tag{43b}$$

i. e. in this formal limit only the charge of the dynamical right-handed fermions occurs and superficially conservation of vector and axial charge are seen to be identical. This is a direct consequence of the absence of the time derivative of left handed Fermions in the continuum light-cone Lagrangean (7).

2.2 Choice of the gauge

The Lagrangean (37) is invariant under the local gauge transformations

$$\begin{matrix}\psi(x^+,x^-)\\ \phi(x^+,x^-)\end{matrix} \to \begin{matrix}\psi'(x^+,x^-)\\ \phi'(x^+,x^-)\end{matrix} = e^{ig\theta(x^+,x^-)}\cdot \begin{matrix}\psi(x^+,x^-)\\ \phi(x^+,x^-)\end{matrix} \tag{44a}$$

$$A_\pm(x^+,x^-) \to A'_\pm(x^+,x^-) = A_\pm + \partial_\pm\theta(x^+,x^-) \tag{44b}$$

i. e. under this combined space-time dependent change of the Fermion-phases and the gauge fields, the Lagrangean (37) does not change

$$\mathcal{L}(\psi,\phi,A_\pm) \to \mathcal{L}(\psi',\phi',A'_\pm) = \mathcal{L}(\psi,\phi,A_\pm) \tag{44c}$$

This invariance can be interpreted as redundance in the number of the degrees of freedom and is therefore used to eliminate some of the gauge fields by "gauge fixing". For instance, in the continuum light-cone formulation the A_- - degree of freedom is in general eliminated ($A_- = 0$).

The freedom of gauge transformations is restricted in the finite interval to those which leave the boundary conditions for the Fermions (cf. 19 a)

$$
\begin{aligned}
\psi(x^+, L) &= e^{-i\varphi}\,\psi(x^+, 0) \\
\phi(x^+, L) &= e^{-i\varphi}\,\psi(x^+, 0)
\end{aligned}
\tag{45}
$$

invariant. Therefore gauge fixing is more subtle and is performed in two steps. Starting from a general gauge $A_\pm\,(x^+, x^-)$, we define the gauge function θ (eq. 44) by

$$
\theta(x^+, x^-) = -\int_0^{x^-} dz A_-(x^+, z) \tag{46a}
$$

which yields a vanishing space component of the gauge potential.

$$
A'_-(x^+, x^-) = A_-(x^+, x^-) + \partial_-\theta(x^+, x^-) = 0 \tag{46b}
$$

However the transformed Fermion fields

$$
\psi'(x^+, x^-) = e^{-ig\int_0^{x^-} dz A_-(x^+, z)}\,\psi(x^+, x^-) \tag{46c}
$$

in general do not satisfy the boundary conditions (45)

$$
\psi'(x^+, L) = e^{-ig\int_0^L dz A_-(x^+, z)} e^{-i\varphi} \cdot \psi'(x^+, 0) \tag{46d}
$$

To restore the (time independent) boundary conditions (45), another gauge transformation given by the gauge function $\vartheta(x^+, x^-)$

$$
\vartheta(x^+, x^-) = \frac{x^-}{L}\int_0^L dz A_-(x^+, z) \tag{47a}
$$

is performed

$$
\begin{matrix} \psi"(x^+, x^-) \\ \phi"(x^+, x^-) \end{matrix} = e^{ig\frac{x^-}{L}\int_0^L dz A_-(x^+, z)} \cdot e^{-ig\int_0^{x^-} dz A_-(x^+, z)} \cdot \left\{ \begin{matrix} \psi(x^+, x^-) \\ \phi(x^+, x^-) \end{matrix} \right. \tag{47b}
$$

with the desired results for the boundary values of the Fermions

$$
\begin{matrix} \psi"(x^+, L) \\ \phi"(x^+, L) \end{matrix} = \begin{matrix} \psi(x^+, L) \\ \phi(x^+, L) \end{matrix} = e^{-i\varphi} \cdot \begin{matrix} \psi(x^+, 0) \\ \phi(x^+, 0) \end{matrix} = e^{-i\varphi} \cdot \begin{matrix} \psi"(x^+, 0) \\ \phi"(x^+, 0) \end{matrix} \tag{47c}
$$

The space component of the gauge potential is changed by these two transformations

$$
\begin{aligned}
A"_-(x^+, x^-) &= A_-(x^+, x^-) + \partial_-\theta(x^+, x^-) + \partial_-\vartheta(x^+, x^-) \\
&= \tfrac{1}{L}\int_0^L dz A_-(x^+, z)
\end{aligned}
\tag{48}
$$

into an x^-– independent but light-cone time dependent variable, i. e. all but the zero mode of A can be gauged away . Thus we end up in the (light-cone) Coulomb-gauge rather than the light-cone axial gauge

$$
\partial_- A"_-(x^+, x^-) = 0 \tag{49}
$$

2.3 The Poisson-equation

The Lagrangean (37) does not contain a time derivative of A_+ and therefore the Euler Lagrange equation for A_+

$$\partial_- \frac{\delta \mathcal{L}}{\delta \partial_- A_+} = \frac{\delta \mathcal{L}}{\delta A_+} \tag{50a}$$

$$\partial_-^2 A_+ = g[\phi^+(x)\phi(x) + \psi^+(x)\psi(x)] \tag{50b}$$

is not an equation of motion. Given the Fermion fields (50 b) can be integrated

$$\begin{aligned} A_+(x^+, x^-) &= g \int_0^L dz G(x^-, z)[\phi^+(x^+, z)\phi(x^+, z) + \psi^+(x^+, z)\psi(x^+, z)] \\ &\quad + x^- E(x^+) + B(x^+) \end{aligned} \tag{51b}$$

with the help of some Greensfunction G which satisfies

$$\partial_-^2 G(x^-, y^-) = \delta(x^- - y^-) \qquad \text{e. g.} \quad G(x^-, y^-) = \frac{1}{2} \mid x^- - y^- \mid \tag{51b}$$

The field strength is given by

$$F = \partial_+ A_- - \partial_- A_+ = \partial_+ A_-(x^+) - E(x^+) - g \int dz G'(x^-, z)[\phi^+\phi + \psi^+\psi] \tag{51c}$$

$E(x^+)$, $B(x^+)$ are integration constants. $B(x^+)$ can be transformed to zero with a purely time dependent gauge transformation while the time dependent background field strength $E(x^+)$ can be absorbed into the definition of the x-independent vector potential of the Coulomb gauge as is clear from the expression for F. Note that this elimination of the background field is not possible in the axial gauge. We finally impose periodic boundary conditions for A_+ which specifies the Greensfunction G

$$G(x, y) = -\frac{L}{4\pi^2} \sum_{n \neq 0} \frac{1}{n^2} e^{-ip_n(x-y)}; \quad p_n = \frac{2n\pi}{L} \tag{52}$$

Periodicity in A_+ is compatible with the quasi-periodic boundary conditions (45) for the Fermions. (The summation restriction $n \neq 0$ will be discussed and justified later.) In this way the time evolution of the A_+ field is completely determined by the time evolution of the matter fields. The connection between the two is most easily established in momentum space. With the following definitions of the Fourier-transformed operators

$$A(x) = \frac{1}{L} \sum_n A(n) e^{-ip_n x^-} \tag{53}$$

$$j_\phi(x) = \phi^+(x)\phi(x) = \frac{1}{L} \sum_n j_\phi(n) e^{-ip_n x^-} \tag{54a}$$

$$j_\psi(x) = \psi^+(x)\psi(x) = \frac{1}{L} \sum_n j_\psi(n) e^{+ip_n x^-} \tag{54b}$$

$$j_\phi(n) = \sum_r a_r^+ a_{r+n}; \quad j_\psi(n) = \sum_r b_{r+n}^+ b_r; \quad j(n) = j_\phi(n) + j_\psi^+(n) \tag{54c}$$

the solution of the Poisson equation (50 b) is

$$A(n) = -\frac{gL^2}{4\pi^2} \frac{1}{n^2} j(n) \qquad (n \neq 0) \tag{55}$$

2.4 The Hamiltonian

With this result we have now specified the true degrees of freedom of QED_2. These are the right and left handed Fermion fields and a single quantum-mechanical degree of freedom, the x^- independent A_- field of the Coulomb-gauge. The Fermion conjugate momenta are defined as in the free theory (eq. 22), the momentum Π conjugate to A_- is defined by

$$\Pi = \frac{\delta \int_0^L dx\mathcal{L}}{\delta A_-} = L\,\partial_+ A_- \tag{56}$$

With these definitions and using the Fourier-transformed operators of eq. (54) the Hamiltonian is

$$H = H_0 + C + H_m \tag{57a}$$

$$H_0 = \frac{\Pi^2}{2L} - \frac{1}{\epsilon}\sum_n b_n^+(2\pi n + \varphi + gLA_-)b_n \tag{57b}$$

$$C = \sum_{n\neq 0} \frac{g^2 L}{8\pi^2 n^2}\,(j_\phi^+(n) + j_\psi(n))(j_\phi(n) + j_\psi^+(n)) \tag{57c}$$

$$H_m = m\sqrt{\frac{L}{2\epsilon}}\sum_n (b_n^+ a_n + a_n^+ b_n) \tag{57d}$$

The Fermion operators satisfy the standard anticommutation relations (23 b) and furthermore we have

$$[\Pi, A_-] = \frac{1}{i} \tag{57e}$$

H_0 is determined by the kinetic energies of the left-handed Fermion normal modes, the kinetic energy of the single A_- mode and their coupling C represents the Coulomb interaction which does not distinguish between right and left handed Fermions and finally H_m is the coupling of the left and right handed fermions via the mass-term. To fully specify the Hamiltonian we need to comment on the summation restriction in the Coulomb interaction. The expression (53) and (55) for $A_+(x)$ represents a solution to the Poisson equation only if $j(0) = 0$, i. e. if the total charge (cf. eq. (54, 40)) vanishes. This condition is imposed as a constraint on the physical states $\mid\varphi>$,

$$Q \mid \varphi >= j(0) \mid \varphi >= (j_\phi(0) + j_\psi(0)) \mid \varphi >= \sum_n (a_n^+ a_n + b_n^+ b_n) \mid \varphi >= 0 \tag{58}$$

The charge operator commutes with the Hamiltonian and therefore time evolution does not lead out of the space of physical states $\mid\varphi>$.

Interpretation of the Hamiltonian (57) and solution for the eigenstates requires regularization of H_0 as well as of other physical observables like the charge operators. In general the eigenvalues of H_0 and $Q_{a,b}$ will be infinite due to infinite contributions from the filled Dirac sea. The expressions are regularized by considering the action of these operators on reference states $\mid N_b>$ which as suggested by the free theory (27) are defined by

$$b_n^+ \mid N_b >= 0 \quad n \geq N_b ; \qquad b_n \mid N_b >= 0 \quad n < N_b \tag{59}$$

i. e. $\mid N_b >$ generalizes the free vacuum state by shifting the Fermi-surface from 0 to N_b. The (heat-kernel) regularization of the operator of the left handed charge Q_b is defined by

$$\begin{aligned} Q_b \mid N_b > &= \textstyle\sum_n b_n^+ b_n \mid N_b >= \sum_{N_b}^{\infty} 1 \cdot \mid N_b > \\ &=: \textstyle\sum_{n=N_b}^{\infty} e^{-\lambda(2\pi n + \varphi + gL\cdot A_-)} \mid N_b > \\ &= \left[\left(\frac{1}{2\pi\lambda}\right) + \left(-N_b + \frac{1}{2} - \frac{\varphi}{2\pi} - \frac{gLA_-}{2\pi}\right) \right] \mid N_b > \end{aligned} \tag{60a}$$

Obviously the exponential cutoff depending on n regularizes the contribution from large momenta. Addition of the A_- dependent term follows from gauge invariance which allows momenta to appear only together with the vector potential. (A more formal argument connecting this heat kernel regularization to point splitting is given in Appendix II.) As a result the regularized left-handed charge depends on the dynamic variable A_-. The infinite ($\lambda \to 0$) contribution to the charge is independent of both A_- and N_b and is therefore dropped. The same method is used to regularize the Hamiltonian with the result

$$H_b \mid N_b >= -\frac{1}{\epsilon} \sum_n b_n^+ (2\pi n + \varphi + gLA_-) b_n \mid N_b >= \frac{2\pi}{\epsilon} \left(\frac{Q_b^2}{2} - \frac{1}{24} \right) \mid N_b > \tag{60b}$$

The same effect which generates the A_- dependence in Q_b is responsible for the change from the linear to the quadratic A_- dependence of H_0. Thus after regularization, the quantum mechanical degree of freedom A_- is described by the harmonic oscillator Hamiltonian

$$\begin{aligned} H_A &= \frac{\Pi^2}{2L} + \frac{\pi}{\epsilon} \left[\frac{gLA_-}{2\pi} + \frac{\varphi}{2\pi} + N_b - \frac{1}{2} \right]^2 = \left(\frac{g^2 \cdot L}{2\pi\epsilon} \right)^{1/2} A_0^+ A_0 \\ [A_0, A_0^+] &= 1 \end{aligned} \tag{61a}$$

where we have introduced the appropriate creation and annihilation operators

$$A_0 = \frac{1}{\sqrt{2}} \left(-\frac{\Pi}{u} + i\, u (A_- + \frac{1}{gL} (\varphi - \pi + 2\pi N_b)) \right) \tag{61b}$$

with

$$u = \left(\frac{g^2 L^3}{2\pi\epsilon} \right)^{1/4} \tag{61c}$$

and therefore the total Hamiltonian after regularization reads

$$H = \left(\frac{g^2 L}{2\pi\epsilon} \right)^{1/2} A_0^+ A_0 - \frac{2\pi}{\epsilon} \sum_n n : b_n^+ b_n : + C + \tilde{H}_m \tag{62a}$$

$$\tilde{H}_m = m \sqrt{\frac{L}{2\epsilon}} \sum_n (b_n^+ a_n + a_n^+ b_n) - \frac{1}{\epsilon} \sum_n : b_n^+ b_n : (\varphi + gLA_-) \tag{62b}$$

The normal ordering in this expression refers to the reference state N_b. We have included in $\tilde{H}_m$ the coupling between left handed Fermions and the A_- degree of freedom. This term only contributes if the number of left handed Fermions is not conserved which in turn requires a non vanishing mass.

Most of the following discussion is concerned with $m = 0$, the Schwinger model. Since the number of b-particles is conserved all the b-mode excitations for fixed N_b can be generated by functions of the bilinear current operator j_ψ (54). Furthermore since the current operators satisfy bosonic commutation relations (as shown in A III).

$$[j^+_{\psi,\phi}(n), j_{\psi,\phi}(m)] = -n\,\delta_{n,m} \tag{63a}$$

we have (cf. A 31)

$$[-\sum_n n : b^+_n b_n :, \sum_r b^+_{r+m} b_r] = [\sum_p j^+_\psi(p) j_\psi(p), j_\psi(m)] \tag{63b}$$

and therefore the final form of the Hamiltonian of the Schwinger model (m = 0) is

$$\begin{aligned} H &= \left(\tfrac{g^2 L}{2\pi\epsilon}\right)^{1/2} A^+_0 A_0 + \tfrac{2\pi}{\epsilon} \textstyle\sum_{p>0} j^+_\psi(p) j_\psi(p) \\ &+ \textstyle\sum_{n\neq 0} \tfrac{g^2 L}{8\pi^2 n^2} \left(j^+_\phi(n) + j_\psi(n)\right)\left(j_\phi(n) + j^+_\psi(n)\right) \end{aligned} \tag{64}$$

In order to implement the neutrality condition (58) we introduce reference states $| N_a >$ for the right-handed Fermions

$$a^+_n \mid N_a >= 0 \quad n < N_a \quad a_n \mid N_a >= 0 \quad n \geq N_a \tag{65}$$

i. e. N_a is the Fermi-surface of the Dirac sea of right handed fermions filled from $-\infty$ to $N_a - 1$. In analogy to the regularization (60) we now have

$$Q_a = \sum_{n=-\infty}^{N_a-1} e^{\lambda(2\pi n+\varphi+gLA_-)} = N_a - \frac{1}{2} + \frac{\varphi}{2\pi} + \frac{gLA_-}{2\pi} \tag{66}$$

and the neutrality requirement (58) is

$$Q = Q_a + Q_b = 0 = N_a - N_b \tag{67}$$

i. e. in a neutral system, the Fermi-surfaces of a and b modes concide.

The Hamiltonian of the Schwinger model contains the decoupled A_- oscillator, the kinetic energy of the b modes written in bosonized form and the Coulomb-interaction. The relative importance of the various contributions is controlled by a parameter η' analogous to η introduced within the free theory in (25)

$$\eta' = g^2 \epsilon L \tag{68}$$

As in the free theory, the light cone description is the weak coupling limit of the theory (64)

$$\eta' << 1.$$

in which the theory can be solved perturbatively in the Coulomb-interaction. To leading order the ground state of the j_ψ Hamiltonian has to be determined. The energy is minimal for

$$j_\psi(p) \mid 0 >_{\psi,A_-} = 0 \tag{69a}$$

i. e. in the absence of ψ exciations and A_- excitations

$$A_0 \mid 0 >_{\psi,A} = 0 \tag{69b}$$

The effective Hamiltonian of the right handed Fermions is (up to a constant) given by

$$\begin{aligned} H_a &= < 0 \mid H \mid 0 >_{\psi,A_-} \\ &= \textstyle\sum_{n\neq 0} \frac{g^2 L}{8\pi^2 n^2} j_\phi^+(n) j_\phi(n) \end{aligned} \tag{70}$$

Here no second order contribution to H_a as in the free theory arises. This effective Hamiltonian defined by operators satisfying the commutation relations (63a) is nothing else than the Hamiltonian of free Bosons in the weak coupling limit as the comparison with (A16b) shows. The following identifications have to be made

$$A(n) = \frac{1}{\sqrt{n}} j_\phi(n) \qquad m^2 = g^2/\pi \tag{71}$$

Thus we have established in the finite light-cone interval theory the well known equivalence of massless QED_2 to a free Boson theory. Clearly the equivalence can be demonstrated also in the strong coupling limit $\eta' >> 1$. However in this case, the strong coupling between right and left handed Fermions requires a non perturbative dynamical calculation.

2.5 Anomaly and vacuum structure

At the same time, this derivation also demonstrates that none of the non- perturbative phenomena which the Schwinger model exhibits are lost in this light-cone formulation of the theory. Only at the level of the effective theory where the left-handed degrees of freedom are eliminated the framework is too restrictive for a description of such phenomena.

As in ordinary coordinate quantization, the "axial anomaly" i. e. the break-down of chiral symmetry is an effect of the Dirac sea and the requirement of gauge invariance of the regularization procedure. Left and right handed charges are given by eqs. (60a) and (66). They both depend on the gauge potential A_-. This dependence cancels in the total charge Q (67) and thereby the necessary requirement of neutrality could be implemented. The axial charge (42a) however remains A_- dependent

$$Q_5 = Q_a - Q_b = N_a + N_b + \frac{\varphi - \pi}{2\pi} + \frac{gL}{2\pi} A_- \tag{72}$$

As a consequence, the axial charge is not a constant of motion

$$[Q_5, H] = \left[\frac{gL}{2\pi} A_-, \frac{1}{2L}\Pi^2\right] \neq 0 \tag{73}$$

Clearly at the level of the effective theory with the Hilbert space of the b- and the A-modes eliminated only one charge, that associated with the a-modes can be defined, i. e.

$$q_a = j_\phi(0) \tag{74a}$$

and appropriately regularized. In this effective theory, with the A_- degree of freedom eliminated, gauge invariance of the regularization is not an issue. We have

$$[H_a, q_a] = 0 \tag{74b}$$

i. e. the possibility of a conserved vector current and an anomalous axial current does not exist at this level of theoretical description (cf. ref. [30] where an infinity of new degrees of freedom in addition to the a modes are introduced to replace the eliminated b modes).

The other important non perturbative phenomenon of the Schwinger model is the appearence of an infinitely degenerate vacuum. In construction of the vacuum we have introduced the Fermi surface of a and b modes which by the neutrality condition have to coincide (eq. 67). The different groundstates constructed are thus characterized by N_b and to each such vacuum state is associated a different zero momentum photon state, the ground state of H_{A_-}. Although the eigenenergies of H_{A_-} do not depend on N_b, the wave-functions do. N_b determines the equilibrium position of the A_- -oscillator. The ground state wave function is given by

$$\psi_{N_b}(A_-) = \psi_0 \, e^{-\frac{1}{2}(\frac{g^2L^3}{2\pi\epsilon})^{1/2}(A_- + \frac{2\pi}{gL}(N_b + \frac{\varphi-\pi}{2\pi}))^2} \tag{75}$$

I emphasize the different behaviour of the wave function in the weak and strong coupling limits. The overlap of wave-functions corresponding to neighbouring Fermi-surfaces is given by

$$\int \psi_{N_b+1}(A_-)\psi_{N_b}(A_-)dA_- = e^{-(\pi^3/2q^2L\epsilon)^{1/2}} \begin{array}{ll} \nearrow & 0 \quad g^2L\epsilon << 1 \\ \searrow & 1 \quad g^2L\epsilon >> 1 \end{array} \tag{76}$$

which shows that from weak to strong coupling limit the wave functions change from completely localized to completely extended ones.

The most general ground state can be written as linear superposition of the groundstates with fixed number of right and left- handed Fermions.

$$| \Omega >= \sum_N C_N \mid N, \psi_N(A_-) > \tag{77}$$

I will now discuss a particular class of linear combinations which are simultaneous eigenstates of the residual gauge transformations. The most general gauge transformations have been defined in eq. (44) via the space-time gauge-functions $\theta(x^+, x^-)$. Relevant for the canonical formalism are time independent gauge transformations which preserve the Coulomb gauge

$$\theta = \alpha x^- + \beta = \theta(x^-) \tag{78a}$$

and do not change the boundary conditions (45) for the Fermions, i. e.

$$ig\,(\theta\,(L) - \theta(0)) = 2i\pi m; \quad m = 0, \pm 1, \pm 2, \ldots \tag{78b}$$

and therefore (dropping the irrelevant constant β)

$$\theta(x^-) = \frac{2\pi m}{g}\,\frac{x}{L} \tag{78c}$$

Quantum mechanically, these residual gauge transformations are described by the unitary operator e^{iG} with

$$G = G(m) = \frac{2\pi m}{g \cdot L}\,\Pi - \int_0^L dx 2\pi m \cdot \frac{x}{L}(\phi^+(x)\phi(x) + \psi^+(x)\psi(x)) \tag{79a}$$

It is straightforward to prove

$$e^{iG} A_- e^{-iG} = A_- + \frac{2\pi m}{gL} \tag{79b}$$

$$e^{iG} \overset{\phi}{\psi}\, e^{-iG} = e^{2i\pi m \cdot \frac{x}{L}}\, \overset{\phi}{\psi} \tag{79c}$$

which imply the transformation properties of the creation and annihilation operators

$$\begin{aligned} e^{iG(m)}\phi(x^+, x^-)e^{-iG(m)} &= L^{-1/2} \textstyle\sum_n e^{iG(m)} a_n e^{-iG(m)} e^{-i(2n\pi+\varphi)\frac{x}{L}} \\ = e^{2i\pi m \frac{x^-}{L}} \phi(x^+, x^-) &= L^{-1/2} \textstyle\sum_n a_n e^{2i\pi m \frac{x}{L}} e^{-i(2n\pi+\varphi)\frac{x^-}{L}} \end{aligned} \tag{80a}$$

$$e^{iG(m)} a_n e^{-iG(m)} = a_{n+m}, \quad e^{iG(m)} b_n\, e^{iG(m)} = b_{n+m} \tag{80b}$$

Thus the operator e^{iG} shifts both Fermi surfaces and the value of the A_- mode and therefore connects different vacuum states with each other according to

$$e^{iG(m)} \mid N, \psi_N(A_-) > = \mid N + m, \psi_{N+m}(A_-) > \tag{81}$$

Thus the degeneracy of the vacuum is expressed as a consequence of the translational invariance of the theory with respect to the shift operator e^{iG}. The eigenstates to this shift operator are the Bloch-states

$$\mid \theta > = \sum_N e^{-iN\theta} \mid N, \psi_N(A_-) > \tag{82}$$

$$e^{iG(m)} \mid \theta > = e^{im\theta} \mid \theta > \tag{83}$$

2.6 The Fermion condensate

As a consequence of the infinite degeneracy of the vacua, it is possible to construct states in which the breakdown of chiral symmetry shows up in a non-vanishing value of the Fermion condensate. This is true for the θ-vacua of eq. (82). According to eqs. (31), (32) the condensate is given by

$$\gamma = <\theta \mid \Gamma(x) \mid \theta > / < \theta \mid \theta >=$$

$$= g\sqrt{\tfrac{1}{2g^2\epsilon L}} \sum_N e^{i\theta} < N+1, \psi_{N+1}(A_-) \mid a_N^+ b_N \mid N, \psi_N(A_-) > / < \theta \mid \theta >$$

$$+h.c. \tag{84}$$

$$= 2g\sqrt{\tfrac{1}{2g^2\epsilon L}} cos\theta < N+1 \mid a_N^+ b_N \mid N >< \psi_{N+1}(A_-) \mid \psi_N(A_-) >$$

$$= 2g\sqrt{\tfrac{1}{2g^2\epsilon L}} cos\theta e^{-(\pi^3/2g^2\epsilon L)^{1/2}}$$

and is indeed in general non-vanishing. However according to this expression, the condensate value vanishes with $\eta' \to 0$ as a consequence of the vanishing overlap of the A_- wavefunctions for different N_b (eq. 76). This result is in conflict with that obtained in ordinary coordinate quantization. After our discussion of Chapter I concerning the continuum limit, this does not come as a surprise. The theory has to be first properly regularized to allow for a reasonable continuum limit. The easiest way to perform here this regularization is to transform via the unitary transformation of the free Boson theory constructed in Appendix I from the $\eta' << 1$ to the $\eta' >> 1$ ground state. Comparison of the Hamiltonian (64) with (A 16 b) shows that the unitary transformation (A 20, A 21)

$$\mid \theta, \epsilon' >= U(\epsilon', \epsilon) \mid \theta, \epsilon > \tag{85a}$$

is given by

$$U(\epsilon', \epsilon) = \prod_{m>0} e^{-\alpha_m B_m/m}\; e^{-\alpha_0 (A_0^{+2} - A_0^2)} \tag{85b}$$

with

$$B_m = j_\phi^+(m)\, j_\psi^+(m) - j_\phi(m)\, j_\psi(m) \tag{85c}$$

and

$$e^{2\alpha_m} = \left(\frac{\pi^2 m^2 + \epsilon' L g^2/2\pi}{\pi^2 m^2 + \epsilon L g^2/2\pi} \right)^{1/2} \tag{85d}$$

We therefore define the regularized value of the condensate by

$$\gamma(\epsilon') = \frac{1}{L}\sqrt{\frac{L}{2\epsilon'}} < \theta \mid U^+(\epsilon', \epsilon) \sum_n (a_n^+ b_n + b_n^+ a_n) U(\epsilon', \epsilon) \mid \theta > / < \theta \mid \theta > \tag{86}$$

and obtain using the result (A 41) of Appendix IV

$$\gamma(\epsilon') = 2g\sqrt{\tfrac{1}{2g^2\epsilon' L}} \cos\theta < N+1 \mid U^+(\epsilon', \epsilon)\, a_N^+ b_N\, U(\epsilon', \epsilon) \mid N > \cdot$$

$$\cdot < \tilde{\psi}_{N+1}(A_-) \mid \tilde{\psi}_{N+1}(A) > \tag{87}$$

$$= 2g\sqrt{\tfrac{1}{2g^2\epsilon' L}} \cos\theta\, e^{-\sum_{m>0} \frac{1}{m}(e^{-2\alpha_m} - 1)} \cdot e^{-(\pi^3/2g^2\epsilon' L)^{1/2}}$$

where we have used the fact, that the operator containing the A_- mode in (85b) simply rescales the A_- and N_b dependences (cf. 75)

$$\begin{aligned} |\tilde{\psi}_N(A_-)> &= e^{-\alpha_0(A_0^{+2}-A_0^2)} \,|\, \psi_N(A_-)> \\ &= e^{-\alpha_0(A_0^{+2}-A_0^2)} \,|\, \psi(A_- + \tfrac{2\pi}{gL}(N_b + \tfrac{\varphi-\pi}{L\pi}))> \\ &= \lambda^{1/2} \,|\, \psi(\lambda A_- + \tfrac{2\pi}{gL}\lambda(N_b + \tfrac{\varphi-\pi}{2\pi}))>; \quad \lambda = e^{-2\alpha_0} = \sqrt{\tfrac{\epsilon}{\epsilon'}} \end{aligned}$$

In the expression (87), we take the limit $\epsilon g^2 L \to 0$ and therefore have for arbitrary ϵ'

$$\gamma(\epsilon') = 2g\sqrt{\frac{1}{2g^2\epsilon' L}}\, e^{S(\epsilon')-(\pi^3/2g^2\epsilon' L)^{1/2}} \cos\theta \tag{88a}$$

with

$$S(\epsilon') = \sum_{m>0}\left(\frac{1}{m} - \frac{1}{\sqrt{m^2+m_0^2}}\right), \quad m_0^2 = \epsilon' L g^2/2\pi^3 \tag{88b}$$

Eq. (88) represents the general result for the Fermion condensate. In the weak coupling limit, $g^2\epsilon' L << 1$ as discussed above, the condensate is suppressed by the essential singularity of γ as a function of this coupling parameter. The continuum limit cannot be reached on the basis of these light-cone finite interval ground states, i. e. in the weak coupling limit. In the strong coupling limit we have

$$g^2\epsilon L >> 1 \qquad S(\epsilon') = C + \ln\frac{m_0}{2}, \quad C - Euler constant$$

and a finite condensate value results

$$\gamma(\epsilon') = \frac{g}{2\pi^{3/2}}\, e^c \cos\theta \tag{89}$$

3 QCD$_2$ on the Light Cone

3.1 Lagrangean of QCD_2

The following discussion of 2 dimensional quantum chromodynamics is based on the finite interval light cone description developed in the first Chapter and generalizes the methods of solutions presented in the discussion of QED$_2$. The emphasis will be as in Chapter II on the explicit construction of the Hamiltonian and the investigation of the structure of the resulting ground state. A brief discussion of the elementary excitations for massless Fermions and within the large N_c (Number of colors) approximation will be given. However for most of the following discussion no use is made of the large N_c approximation and SU(N) and U(N) theories are distinguished. In the coordinates defined by eqs. (7, 19 b) the QCD-Lagrangean has the following form

$$\begin{aligned} \mathcal{L} &= \phi_i^+(i\partial_+\delta_{ik} + gA_{+ik})\phi_k + \psi_i^+(i\partial_+\delta_{ik} + gA_{+ik})\psi_k \\ &+ \tfrac{L}{\epsilon}\psi_i^+(i\partial_-\delta_{ik} + gA_{-ik})\psi_k + F_{ik}F_{ki} - m\sqrt{\tfrac{L}{2\epsilon}}(\phi_k^+\psi_k + \psi_k^+\phi_k) \end{aligned} \tag{90}$$

Here, the gauge fields $A_{\pm ik}$ are SU(N) matrices and can be expressed as linear combinations of the SU(N) Gell-Mann matrices λ^α

$$A_{\pm ik} = \sum_\alpha A_\pm^\alpha \lambda_{ik}^\alpha / 2 \tag{91}$$

(The connection between the notation in terms of components and in terms of λ - matrices is given in Appendix V.) The field strength (omitting space time indices) is given by

$$F_{kl} = \partial_+ A_{-kl} - \partial_- A_{+kl} - ig(A_{+km}A_{-ml} - A_{+ml}A_{-km}) \tag{92}$$

The Fermion spinors ϕ, ψ carry a color index $(i, k = 1, \ldots N)$ and in eq. (90), (92) as in the following summation over indices occuring twice is implied. Finally boundary conditions have to be imposed in the finite light-cone interval; we choose

$$\begin{aligned} \phi(x^+, L) &= e^{-i\varphi}\phi(x^+, 0) \\ \psi(x^+, L) &= e^{-i\varphi}\psi(x^+, 0) \end{aligned} \tag{93}$$

3.2 Choice of the Gauges

QCD is constructed such to be invariant under the transformations

$$\begin{aligned} \phi(x^+, x^-) &\to \phi'(x^+, x^-) = U\phi(x^+, x^-) \\ \psi(x^+, x^-) &\to \psi'(x^+, x^-) = U\psi(x^+, x^-) \\ A_\pm(x^+, x^-) &\to A'_\pm(x^+, x^-) = UA_\pm U^+ - \tfrac{1}{ig} U\partial_\pm U^+ \end{aligned} \tag{94}$$

where U is an arbitrary space-time dependent unitary matrix. While these gauge transformations leave invariant the Lagrangean (90) and therefore the equations of motion, in general the transformation (94) modify the boundary conditions (93). As in electrodynamics not all the gauge degrees of freedom can be eliminated in the presence of boundary conditions and gauge fixing has to be performed by generalizing the method developed in Chapter II. In the first step, we attempt to gauge away an arbitrarily given $A_-(x^+, x^-)$ by the unitary transformation

$$U^+ = P\, e^{+ig\int_0^{x^-} dz A_-(x^+, z)} \tag{95a}$$

Here P denotes "path ordering", which for the 1dimensional integral is the same as the time ordering of ordinary time dependent perturbation theory. Here it is necessary due to the non-commutativity in the color indices of A_-. U^+ is by construction solution of the differential equation

$$\partial_- U^+ = igA_-(x^+, x^-)U^+ \tag{95b}$$

and therefore we have achieved

$$A'_-(x^+, x^-) = 0 \tag{96}$$

This tranformation changes in general the boundary conditions (93)

$$\begin{matrix}\phi'(x^+,x^-)\\ \psi'(x^+,x^-)\end{matrix} = Pe^{-ig\int_0^{x^-} dzA_-(x^+,z)} \begin{matrix}\phi(x^+,x^-)\\ \psi(x^+,x^-)\end{matrix}$$

This change is compensated by another gauge transformation defined by

$$e^{i\vartheta} = Pe^{+ig\int_0^L dzA_-(x^+,z)} \tag{97a}$$

$$V = e^{i\vartheta\cdot x^-/L} \tag{97b}$$

The transformed spinors are given by

$$\begin{matrix}\phi''(x^+,x^-)\\ \psi''(x^+,x^-)\end{matrix} = V \begin{matrix}\phi'(x^+,x^-)\\ \psi'(x^+,x^-)\end{matrix} = V(x^+,x^-)U(x^+,x^-) \begin{matrix}\phi(x^+,x^-)\\ \psi(x^+,x^-)\end{matrix} \tag{98}$$

and since

$$U(x^+,x^-=0) = V(x^+,x^-=0) \tag{99a}$$

$$V(x^+,L)U(x^+,L) = 1 \tag{99b}$$

the boundary conditions are restored. Under these transformations, the original $A^-(x^+,x^-)$ has been changed into $A''_-(x^+,x^-)$

$$A''_- = VA'_-V^+ - \frac{1}{ig}V\partial_-V^+ = \frac{1}{gL}\vartheta \tag{100a}$$

i. e. A''_- is a gauge field in the Coulomb-gauge.

$$\partial_-A''_- = 0 \tag{100b}$$

In the last step we can diagonalize the hermitean x^- independent matrix A''_- by an x^- independent final gauge transformation $W(x^+)$

$$A'''_- = WA''_-W^+ - \frac{1}{ig}W(x^+)\partial_-W^+(x^+) \tag{101a}$$

which being x^- independent does not affect the boundary conditions (93). As a result, the dynamical gauge fields A_- can be written as

$$A'''_-(x^+) = \textstyle\sum'_\alpha \lambda^\alpha/2\, A^\alpha_-(x^+) \tag{101b}$$

where the summation is restricted to values of α corresponding to diagonal SU(N) λ - matrices. This result implies that in the finite interval the dynamically independent gauge fields are described by N-1 degrees of freedom.

3.3 The Poisson equation

In the next step we eliminate the dynamically dependent $A_+(x^+,x^-)$ gauge degrees of freedom by explicit solution of the Poisson equation. The gauge field dependent terms of the Lagrangean (90) are

$$\mathcal{L}_{YM} = \frac{1}{2} F^{\alpha} F^{\alpha} + g\,(A_+^{\alpha} j_+^{\alpha} + \frac{L}{e} A_-^{\alpha} j_-^{\alpha}) \tag{102}$$

with the colored matter currents

$$j_+^{\alpha} = \phi_i^+ \frac{\lambda_{ik}^{\alpha}}{2} \phi_k + \psi_i^+ \frac{\lambda_{ik}^{\alpha}}{2} \psi_k \tag{103a}$$

$$j_-^{\alpha} = \psi_i^+ \frac{\lambda_{ik}^{\alpha}}{2} \psi_k \tag{103b}$$

Note that as in QED the space components of the currents are given by the left-handed quarks only. The field strength is decomposed as

$$F^{\alpha} = \partial_+ A_-^{\alpha} - \partial_- A_+^{\alpha} + g f^{\alpha\beta\gamma} A_+^{\beta} A_-^{\gamma} = \partial_+ A_-^{\alpha} - g^{\alpha}(x) \tag{104}$$

The Euler Lagrange equation for A_+

$$\partial_- \frac{\delta \mathcal{L}_{YM}}{\delta \partial_- A_+^{\alpha}} = \frac{\delta \mathcal{L}_{YM}}{\delta A_+^{\alpha}} \tag{105b}$$

is

$$\partial_-(-\partial_+ A_-^{\alpha} + g^{\alpha}) = g j_+^{\alpha} - g f^{\alpha\beta\gamma}(\partial_+ A_-^{\beta} - g^{\beta}) A_-^{\gamma} \tag{105b}$$

and constitutes a constraint equation for g^{α} which in terms of its color components reads

$$\partial_- g_{ij}(x) = g\, j_{ij} - i g g_{ij}(x)\,(A_{-j} - A_{-i}) \tag{106a}$$

Similarly the defining equation for g^{α} (104) is an equation of constraint for A_+

$$\partial_- A_{ij}^+(x) = g_{ij}(x) - i g A_{ij}^+(x)(A_{-j} - A_{-i}) \tag{106b}$$

Here the j_{ij} are the color components of j_+^{α} and are given by

$$j_{ij}(x) = \frac{1}{2}\left[\phi_j^+ \phi_i - \frac{\delta_{ij}}{N} \sum_l \phi_l^+ \phi_l\right] + \frac{1}{2}\left[\psi_j^+ \psi_i - \frac{\delta_{ij}}{N} \sum_l \psi_l^+ \psi_l\right] \tag{106c}$$

Thus given the currents j_{ij} determined by the Fermion fields, and the dynamical gauge field degrees of freedom (eq. 101 b)

$$A_{-i} = \textstyle\sum_{\alpha}' \lambda_{ii}^{\alpha}/2\, A_-^{\alpha} \tag{106d}$$

g^{α} and subsequently A_+^{α} can be determined. The integration of the differential equations (106 a, b) is carried out with periodic boundary conditions for both A_{ij}^+ and g_{ij}. Defining the Fourier-transformed variables

$$\begin{matrix} g_{ij}\,(x) \\ A_{ij}\,(x) \\ j_{ij}\,(x) \end{matrix} = \frac{1}{L}\sum_n \begin{matrix} g_{ij}\,(n) \\ A_{ij}\,(n) \\ j_{ij}\,(n) \end{matrix} \quad e^{-i\frac{2n\pi}{L}\cdot x} \tag{107}$$

the solution of the Poisson equation (106) reads

$$g_{ij}(n) = \frac{ig}{\frac{2n\pi}{L} - g(A_{-j} - A_{-i})} j_{ij}(n) \tag{108a}$$

$$A^+_{ij}(n) = \frac{-g}{\left[\frac{2n\pi}{L} - g(A_{-j} - A_{-i})\right]^2} j_{ij}(n) \tag{108b}$$

3.4 The Hamiltonian.

By choosing properly the gauge and representing A_+ in terms of the independent dynamical variables we now can proceed to construct the Hamiltonian. We first rewrite the Lagrangean by splitting the field energy into off-diagonal and diagonal contributions

$$\begin{aligned} \sum_\alpha \int_0^L dx F^\alpha F^\alpha &= \sum''_\alpha \int_0^L dx g^\alpha(x) g^\alpha(x) + \sum'_\alpha \int_0^L dx (\partial_+ A^\alpha_- - g^\alpha(x))^2 \\ &= \sum_\alpha \int_0^L dx\ g^\alpha(x) g^\alpha(x) + \sum'_\alpha L(\partial_+ A^\alpha_-)^2 \end{aligned} \tag{109a}$$

where we have used that for α referring to a diagonal λ-matrix we have

$$\int_0^L dx\ g^\alpha(x) = 0 \qquad \alpha - \text{diag.} \tag{109b}$$

as follows from (106 b) and the required periodicity of A_+. The N–1 momentum variables conjugate to the quantum mechanical gauge variables A^α_- are given by

$$\Pi^\alpha = \frac{\delta \int_0^L \mathcal{L}_{YM} dx}{\delta \partial_+ A^\alpha_-} = L\, \partial_+\, A^\alpha_- \tag{110}$$

and thus the Hamiltonian after expressing g_{ij} and A^+_{ij} in terms of the currents reads

$$\begin{aligned} H &= \tfrac{1}{2L} \sum'_\alpha \Pi^{\alpha^2} - \tfrac{2\pi}{\epsilon} \sum_{ni} b^+_i(n) \left(n + \varphi/2\pi + \sum_\alpha c_\alpha\, \lambda^\alpha_{ii}/2\right) b_i(n) + \\ &\quad + C + H_m \qquad (111a) \\ &= \tfrac{1}{2L} \sum'_\alpha \Pi^{\alpha^2} + H_b + C + H_m \end{aligned}$$

where the Coulomb interaction C is given by

$$C = \frac{g^2 L}{4\pi^2} \sum'^{n}_{ij} \frac{j^{ij}(n) j^{ji}(-n)}{[n - \sum_\alpha c_\alpha (\lambda^\alpha_{jj} - \lambda^\alpha_{ii})/2]^2} \tag{111b}$$

In the sum, the n = 0 term has to be omitted for i = j. The c_α are the rescaled quantum mechanical gauge degrees of freedom.

$$c_\alpha = \frac{gL}{2\pi} A^\alpha_- \tag{111c}$$

and the mass term

$$H_m = m\sqrt{\frac{L}{2\epsilon}} \sum_{in} (a^+_i(n)\, b_i(n) + b^+_i(n)\, a_i(n)) \tag{111d}$$

Finally the current operators in momentum space are

$$\begin{aligned} j^{ij}(n) &= \tfrac{1}{2} \sum_r \Big[a^+_j(r) a_i(r+n) - \tfrac{\delta_{ij}}{N} \sum_l a^+_l(r) a_l(r+n) + \\ &\qquad + b^+_j(r) b_i(r+n) - \tfrac{\delta_{ij}}{N} \sum_l b^+_l(r) b_l(r+n) \Big] \qquad (112) \\ &= j^{ij}_\phi(n) + j^{ij}_\psi(n) \end{aligned}$$

where as in (23 a) $a_i(n), b_i(n)$ are the annihilation operators of right and left handed Fermions respectively of color i.

The comparison with the QED Hamiltonian (57) at the same level of the analysis displays the general similarity and important differences. In both cases, the gauge degrees of freedom couple to the left handed charges, the $U(1)$ charge in QED and the $SU(N)$ left handed charges in QCD_2. In QCD_2 an additional coupling of Fermions and gauge degrees of freedom is present through the "gluon-propagator" in the Coulomb-interaction which is modified as a result of the gluon self-interaction in the basic Lagrangean (90). In coordinate space and using for illustrative purpose the standard linear propagator, the transition from electrodynamics to chromodynamics corresponds to a phase modulation

$$| x - y | \rightarrow | x - y | \ e^{-ig(A_{-j} - A_{-i})(x-y)} \tag{113}$$

for a source carrying the color $i\bar{j}$.

To proceed further, regularization of the Hamiltonian (111) is required which is performed in close analogy to that of the previous Chapter. For the regularization we introduce reference states

$$| N_b >=| N_b^1, \ldots N_b^N > \tag{114a}$$

defined by

$$b_i^{(+)}(n) \mid N_b >= 0 \qquad n < N_b^i \qquad (n \geq N_b^i) \tag{114b}$$

i. e. all the left-handed Fermions occupy states starting from some finite momentum up to infinity. Note that these (finite momentum) limits of the occupation, the Fermi surfaces in general are different for different colors. The (gauge invariantly) regularized left handed charges Q_b^i are defined by

$$Q_b^i \mid N_b >= \sum_n b_i^+(n) b_i(n) e^{-\lambda(n + \frac{\varphi}{2\pi} + \sum_\alpha c_\alpha \lambda_{ii}^\alpha/2)} \mid N_b > \tag{115a}$$

$$Q_b^i = c_{ib}^0 - \sum_\alpha c_\alpha \ \lambda_{ii}^\alpha/2 \tag{115b}$$

$$c_{ib}^0 = -N_b^i - \frac{\varphi - \pi}{2\pi} \tag{115c}$$

Similarly we introduce reference states for the right-handed Fermions

$$| N_a >=| N_a^1, \cdots N_a^N > \tag{116a}$$

defined by

$$a_i^{(+)} \mid N_a >= 0 \qquad n \geq N_a^i \quad (n < N_a^i) \tag{116b}$$

and regularize the right handed color charges Q_a^i

$$\begin{aligned} Q_a^i \mid N_a > &= \sum_n a_i^+(n) a_i(n) \ e^{\lambda(n + \frac{\varphi}{2\pi} + \sum_\alpha c_\alpha \lambda_{ii}^\alpha/2)} \mid N_a > \\ &= (N_a^i + \tfrac{\varphi - \pi}{2\pi} + \sum_\alpha c_\alpha \ \lambda_{ii}^\alpha/2) \mid N_a > \end{aligned} \tag{117a}$$

which yields

$$Q_a^i = -c_{ia}^0 + \sum_\alpha c_\alpha \lambda_{ii}^\alpha / 2 \tag{117b}$$

with

$$c_{ia}^0 = -N_a^i - \frac{\varphi - \pi}{2\pi} \tag{117c}$$

As in QED (eq. 60) the Hamiltonian H_b (eq. 111) acting on $| N_b >$ is determined by the left handed charges

$$H_b \mid N_b >= \frac{2\pi}{\epsilon} \sum_i \left(\frac{Q_b^{i2}}{2} - \frac{1}{24} \right) \mid N_b > \tag{118a}$$

and therefore H_b can (up to a constant) be written as

$$\begin{aligned} H_b &= \tfrac{\pi}{\epsilon} \sum_{i=1}^N (c_{ib}^0 - \sum_\alpha c_\alpha \lambda_{ii}^\alpha/2)^2 + \ :H_b: \\ &= \tfrac{\pi}{2\epsilon} \sum_\alpha (c_\alpha - c_\alpha^0)^2 + \tfrac{\pi}{\epsilon N} (\sum_{i=1}^N \ c_{ib}^0)^2 + \ :H_b: \end{aligned} \tag{118b}$$

where in the last step we have performed the i- sum and introduced

$$c_\alpha^0 = \sum_i \lambda_{ii}^\alpha \, c_{ib}^0 \tag{119}$$

The total Hamiltonian therefore becomes

$$\begin{aligned} H = \ & -\tfrac{g^2 L}{8\pi^2} \sum_\alpha' \tfrac{\partial^2}{\partial c_\alpha^2} + \tfrac{\pi}{2\epsilon} \sum_\alpha' (c_\alpha - c_\alpha^0)^2 + \tfrac{\pi}{\epsilon N} (\sum_{i=1}^N \ c_{ib}^0)^2 \\ & + :H_b: + C + H_m \end{aligned} \tag{120}$$

As in QED, the gauge degrees of freedom are described by harmonic oscillators

$$H_c = \left(\tfrac{g^2 L}{4\pi\epsilon} \right)^{1/2} \sum_\alpha' a_\alpha^+ \, a_\alpha \tag{121a}$$

with

$$a_\alpha = \frac{1}{\sqrt{2}} \left[-\frac{1}{iu} \frac{\partial}{\partial c_\alpha} + i\, u\, c_\alpha \right] \tag{121b}$$

$$u = \left(\frac{4\pi^3}{g^2 \epsilon L} \right)^{1/4} \tag{121c}$$

$$\psi(c_\alpha) = \psi_0 \, e^{-(4\pi^3/g^2 L\epsilon)^{1/2} \, (c_\alpha - c_\alpha^0)^2} \tag{121d}$$

The quantities c_α^0 denote the equilibrium positions of the N - 1 gauge degrees of freedom and are determined by the Fermi surfaces of the left handed Fermions. The ground state wave functions are given by eq. (121 d) and we note that again the motion becomes classical in the weak coupling limit $\eta = g^2 \epsilon L << 1$, i. e. the wave functions are localized at the equilibrium positions c_α^0. In contradistinction to QED, the dynamics of the gauge degrees of freedom is (in the massless limit) not described by those unperturbed oscillators. Rather, these oscillators are coupled via the Coulomb-propagator (111 b) to the Fermions; furthermore the normal ordered $: H_b :$ in general contributes, since the

Coulomb interaction (and trivially the mass -term) change the number of b-particles of a given color.

As in electrodynamics a summation restriction had to be introduced for the definition of the Coulomb-propagator (eq. (111 b). Thus the Poisson equation can only be integrated with periodic boundary conditions if the following neutrality condition, interpreted as a constraint on physical states $\mid \varphi >$, is satisfied

$$j^{ii}(0) \mid \varphi >= 0 \qquad i = 1, \dots N \tag{122}$$

Note that the requirement is not that of vanishing total color charge of the Fermions; since color charge is associated with the zero momentum gluons, this charge can be exchanged between Fermions and gluons. However, since the gluons are neutral (only diagonal components, $i\bar{i}$ charges) neutrality of the Fermions is necessary for total singlet states. Finally the neutrality requirement is consistent with the Hamiltonian

$$[H, j^{ii}(0)] = 0$$

exactly because of the neutrality of the gluons. The neutrality requirement (122) constrains the choice of the Fermi surfaces. We have

$$\begin{aligned} j^{ii}(0) \mid N_b, N_a >&= \tfrac{1}{2} \textstyle\sum_n \Big\{ \big[a_i^+(n) a_i(n) - \tfrac{1}{N} \sum_s a_s^+(n) a_s(n)\big] e^{+\lambda(n+\xi)} + \\ &+ e^{-\lambda(n+\xi)} \big[b_i^+(n) b_i(n) - \tfrac{1}{N} \textstyle\sum_s b_s^+(n) b_s(n)\big]\Big\} \mid N_b, N_a > \\ &= \tfrac{1}{2}\big[N_a^i - \tfrac{1}{N} \textstyle\sum_s N_a^s - (N_b^i - \tfrac{1}{N} \sum_s N_b^s)\big] \mid N_b, N_a >= 0, \; i = 1, 2, \dots N \end{aligned} \tag{123}$$

with $\xi = \frac{\varphi}{2\pi} + \sum_\alpha c_\alpha \lambda_{ii}^\alpha / 2$ (cf. 115 a, 117 a). As in electrodynamics, the gauge field dependence of the regularized left and right handed color charges cancels in their sum and therefore the constraint relates Fermi-surfaces of right and left handed Fermions

$$N_b^i = N_a^i + K \tag{124}$$

i. e. right and left handed Fermi surfaces in general do not concide. They can differ however only by a color independent constant K. A final change of variables is particularly useful for the massless limit. We shift the momentum variables such that all the Fermi-surfaces are identified and have the value 0, i. e. we introduce

$$\tilde{a}_i(s) = a_i(N_a^i + s); \qquad \tilde{b}_i(s) = b_i(N_b^i + s) \tag{125a}$$

and shift correspondingly the gauge oscillators to 0 value equilibrium positions

$$\tilde{c}_\alpha = c_\alpha - c_\alpha^0 \tag{125b}$$

The Hamiltonian in these shifted variables is given by

$$\begin{aligned} H \;=\; & -\tfrac{g^2 L}{8\pi^2} \textstyle\sum_\alpha' \frac{\partial^2}{\partial \tilde{c}_\alpha^2} + \frac{\pi}{2\epsilon} \sum_\alpha' \tilde{c}_\alpha^2 + \frac{\pi}{\epsilon N} (\sum_{i=1}^N c_{ib}^0)^2 \\ & - \tfrac{2\pi}{\epsilon} \textstyle\sum_{in} : \tilde{b}_i^+(n) \tilde{b}_i(n) : (n + \frac{1}{2} - \frac{1}{N} \sum_{l=1}^N c_{lb}^0 + \sum_\alpha \tilde{c}_\alpha \lambda_{ii}^\alpha / 2) \\ & + \tfrac{g^2 L}{4\pi^2} \textstyle\sum_{ij}'^{\,n} \frac{\tilde{j}^{ij}(n) \tilde{j}^{ji}(-n)}{[n - \sum_\alpha \tilde{c}_\alpha (\lambda_{jj}^\alpha - \lambda_{ii}^\alpha)/2]^2} \\ & + m \sqrt{\tfrac{L}{2\epsilon}} \textstyle\sum_{in} (\tilde{a}_i^+(n+K) \tilde{b}_i(n) + \tilde{b}_i^+(n) \tilde{a}_i(n+K)) \end{aligned} \tag{126}$$

(the current operators $\tilde{j}^{ij}$ are obtained by replacing in the expression (112) for j^{ij} the original Fermion operators by the shifted ones (125 a)).

It is instructive in particular for the following symmetry considerations and the large N limit to compare the Hamiltonian (126) with the corresponding expression for the U(N) gauge theory. All the steps of the above derivation can be almost literally repeated. Here we choose the N components of the diagonal gauge fields as dynamical variables

$$\tilde{c}_i = \frac{gL}{2\pi} A_{-i} - c^0_{ib} \tag{127}$$

As above shifted Fermion operators (125a) are introduced in terms of which the (Fourier-transformed) U(N) currents are

$$j^U_{ij}(n) = \sum_r (\tilde{a}^+_j(r)\tilde{a}_i(r+n) + \tilde{b}^+_j(r)\tilde{b}_i(r+n)) \tag{128}$$

and the Hamiltonian is

$$\begin{aligned} H_{U(N)} &= -\tfrac{g^2L}{16\pi^2} \textstyle\sum_{i=1}^N \frac{\partial^2}{\partial \tilde{c}_i^2} + \frac{\pi}{\epsilon} \sum_{i=1}^N \tilde{c}_i^2 \\ &\quad -\tfrac{2\pi}{\epsilon} \textstyle\sum_{in} : \tilde{b}^+_i(n)\, \tilde{b}_i(n) : (n + \frac{1}{2} + \tilde{c}_i) \\ &\quad +\tfrac{g^2L}{4\pi^2} \textstyle\sum'_{n,ij} \frac{j^u_{ij}(n) j^u_{ji}(-n)}{(n-(\tilde{c}_j-\tilde{c}_i))^2} + m\sqrt{\frac{L}{2\epsilon}} \sum_{in}(\tilde{a}^+_i(n)\, \tilde{b}_i(n) + \tilde{b}^+_i(n)\, \tilde{a}_i(n)) \end{aligned} \tag{129}$$

The neutrality condition is formally the same as in the SU(N) theory eq. (122) however due to the different structure of the current (128) it yields here

$$N^u_{bi} = N^u_{ai} \tag{130}$$

In comparison, the following differences between SU(N) and U(N) theories are important. The space of physical states is larger in SU(N) gauge theory. This is an immediate consequence of the neutrality requirement leading to (124) and (130) respectively. While in the U(N) theory, Fermi surfaces of right and left handed Fermions have to be aligned (130), gaps in the occupation or simultanous occupations of right and left handed Fermion-states is possible in SU(N) (124). However, the lenghts of the intervals in which none or both of the Fermion-states are occupied have to be the same for all colors. This larger Hilbert space of physical states provides the necessary framework to describe states with different baryon numbers. The neutrality requirement allows for U(N) only mesonic excitations built on the vacuum while in SU(N) states of baryonic (color singlet) matter (or antimatter) exist in addition. Given a choice K_0 for the vacuum, $K - K_0$ is the baryon number – or in operator form (cf. 115a, 117a)

$$\hat{K} = \frac{-1}{N} \left[\sum_{\substack{i=1 \\ n}}^{N} a^+_i(n) a_i(n)\, e^{\lambda \cdot n} + b^+_i(n)\, b_i(n)\, e^{-\lambda n} \right] \tag{131a}$$

with the property

$$\hat{K} \mid N_a, N_b >= \frac{1}{N} \sum_i (N_b^i - N_a^i) \mid N_a N_b >= K \mid N_a N_b > \qquad (131b)$$

and therefore the baryon-number is a parameter which enters the Hamiltonian (126) via the mass term. Determination of the lowest energy state as a function of K, i. e. determination of the value of K in the vacuum is a dynamical problem.

As in electrodynamics, both U(N) and SU(N) gauge theories exhibit symmetries related to the residual gauge invariance. Here the U(N) case is a simple generalization of QED_2. Residual (time independent) gauge transformations which preserve the diagonal Coulomb gauge and the Fermion boundary conditions (93) are (cf. 78)

$$U_{ik} = e^{ig\theta_i(x^-)}\,\delta_{ik} \qquad (132a)$$

$$\theta_i = \frac{2\pi m_i}{g} \cdot \frac{x}{L}, \qquad m_i = \pm 1, \pm 2, \ldots;\ i = 1, \ldots N \qquad (132b)$$

Quantum mechanically, these gauge transformations are generated by (cf. 79) e^{iG} with

$$G = \frac{2\pi}{L} \sum_{l=1}^{N} m_l \left[\frac{1}{ig} \frac{\partial}{\partial A_{-l}} - \int_0^L dx\, x (\phi_l^+(x)\,\phi_l(x) + \psi_l^+(x)\psi_l(x)) \right] \qquad (133)$$

with the property ($N_i^u = N_{a,i}^u = N_{b,i}^u$)

$$G \mid N >= G \mid \{N_i^u, \psi_{Ni}(A_{-i})\} >= \mid \{N_i^u + m_i, \psi_{Ni+mi}(A_{-i})\} >= \mid N + m > (134)$$

In SU(N) the unitary transformations are further restricted by the requirement of vanishing trace , which yields

$$\sum_i m_i = 0 \qquad (135)$$

Thus while in U(N) the residual gauge symmetry is the translational symmetry in the N-dimensional space of states characterized by the Fermi-surfaces N_i^u

$$N_i^u \to N_i^u + m_i \qquad (136a)$$

the residual gauge symmetry of SU(N) is that of transformations of the internal coordinates. The center of mass of the Fermi surfaces

$$\overline{N}_b = \frac{1}{N} \sum_i N_b^i = \frac{1}{N} \sum_i N_a^i + K = const. \qquad (136b)$$

remains unchanged under the transformations (132), (135). Indeed the position of this center of mass $\overline{N}_b$ is determined dynamically. $\overline{N}_b$ is a parameter of the Hamiltonian (126) (cf. 115c)

$$\frac{1}{N} \sum_{l=0}^{N} c_{lb}^0 = -\overline{N}_b - \frac{\varphi - \pi}{2\pi} \qquad (136c)$$

We also note that the SU(N) Hamiltonian (126) still depends as that of free theory (24) on the phase φ appearing in the boundary conditions (93), while this is not the case for the U(N) Hamiltonian (129). The additional U(1) symmetry allows to absorb φ into the definition of the gauge variables.

3.5 Excitations in the chiral limit

The Hamiltonian (126) describes a system of self-interacting Fermions of N colors coupled to (N-1) quantum mechanical gauge degrees of freedom, the N-1 neutral zero momentum gluons. In general it is not possible to exactly solve for the stationary states and eigenenergies of this Hamiltonian. The massless case forms a notable exception and we will start our discussion with this important chiral limit. (cf. ref. [31] for an investigations of this limit within bosonized QCD_2).

The Hamiltonian (126) exhibits for m = 0 a local symmetry. We note that the dynamics of the a-modes is exclusively determined by the Coulomb-interaction which in turn depends only via the current (cf. eq. (106c)

$$j_\phi^{ij}(x) = \frac{1}{2}[\phi_j^+(x)\phi_i(x) - \frac{\delta_{ij}}{N}\sum_l \phi_l^+(x)\phi_l(x)] \tag{137}$$

on the right handed Fermions. This current and therefore the Hamiltonian is invariant under local U(1) changes of the phase of the right handed Fermions fields

$$\begin{aligned} \phi_i(x) &\rightarrow \phi_i'(x) = e^{i\vartheta(x)}\phi_i(x) \\ \psi_i(x) &\rightarrow \psi_i'(x) = \psi_i(x) \end{aligned} \tag{138}$$

Infinitesimal phase changes are generated by the U(1) current of right handed Fermions

$$j_\phi(x) = \frac{1}{N}\sum_i \phi_i^+(x)\phi_i(x) \tag{139}$$

The U(1) current commutes with the SU(N) currents (137) and therefore with the Hamiltonian. The Fouriercomponents (cf. 107) $j_\phi^+(n)$ applied to the vacuum $\mid 0 >$ therefore generate zero energy excitations of momentum $2\pi n/L$

$$Hj_\phi^+(n)\mid 0 >= [H, j_\phi^+(n)]\mid 0 >= 0 \tag{140}$$

i. e.QCD_2 in the chiral limit contains massless bosons (cf. [32]). This is not the case for the U(N) theory; here the classical Hamiltonian possesses the same symmetry, however as in QED_2 due to the anomalous current commutation relation (cf. eq. 63 a)

$$[j_\phi^+(n), j_\phi(m)] = -n\delta_{n,m} \tag{141}$$

we have

$$[H_{U(N)}, j_\phi^+(m)] \neq 0$$

Finite and time independent phase transformations $\vartheta(x)$ (138) are again restricted by the boundary conditions (93)

$$\vartheta_m(L) - \vartheta_m(0) = 2\pi m \qquad m = 0, \pm 1, \pm 2, \ldots \tag{142}$$

Quantum mechanically these phase transformations are generated by $e^{iG(m)}$ with (cf. 79a)

$$G(m) = -\int_0^L dx\, \vartheta_m(x) \sum_i \phi_i^+(x)\phi_i(x) \qquad (143a)$$

The physical interpretation of the winding number m becomes clear, if we decompose

$$\vartheta_m(x) = 2\pi m \frac{x}{L} + \vartheta_0(x) \qquad (143b)$$

and consider as in (79 a)

$$\tilde{G}(m) = -2\pi m \int_0^L dx \frac{x}{L} \sum_i \phi_i^+(x)\phi_i(x) \qquad (143c)$$

and obtain (formally as in eq. 81) for any state with quantum numbers $\mid N_a, N_b >$

$$e^{i\tilde{G}(m)} \mid N_a, N_b >= \mid N_a + m, N_b > \qquad (143d)$$

i. e. the operator $e^{i\tilde{G}(m)}$ connects states which differ in the baryon number K (131 b) by m. The winding number m is the baryon number K. In particular this allows to generate by applying $e^{i\tilde{G}(m)}$ to the vacuum, states with non-zero baryon number which are degenerate with the vacuum. The momentum of such a state can be arbitrarily changed by applying $e^{i\tilde{G}(0)}$ with suitably chosen $\vartheta_0(x)$. Thus for any baryon number a massless state exists.

Thus, the common origin of massless mesons and baryons in the chiral limit of the SU(N) QCD_2 is the invariance of the Hamiltonian under phase changes of the right-handed fermions. With the choice (5, 19 b) of the coordinates, no spatial derivative of the right-handed Fermion spinors appears and this guarantees the invariance under local U(1) phase changes. Massless baryons and mesons are both generated by such phase transformation, they are however distinguished by the winding number of the change in phase.

3.6 Mesons in the large N limit

In this last paragraph we outline the derivation of the meson-equation in the limit of large number N of colors which has been derived within perturbation theory by 't Hooft [1]. We restrict our discussion to the massless case.

As in QED_2, the parameter η' controls the relative importance of the various contributions to the Hamiltonian (126). We assume for the following

$$N\eta' = N\epsilon g^2 L << 1 \qquad (144)$$

In this limit, the equation of the gauge degrees of freedom becomes classical i. e. their kinetic energy term can be dropped. The equilibrium position is determined by the harmonic oscillator potential in (119) and the centrifugal- like potential of the Coulomb interaction. The Coulomb interaction is repulsive and its effect is weakest for states satisfying

$$j^{ij}(0) \mid \tilde{\varphi} >= 0 \qquad (145a)$$

i. e. for singlet-Fermion states. In this case, at the equilibrium positions

$$\tilde{c}_\alpha = 0 \qquad (145b)$$

the Hamiltonian is minimal and therefore the gauge degrees of freedom are decoupled from the Fermion degrees of freedom. The b-modes can be eliminated as in QED and the effective a-mode Hamiltonian is obtained

$$H_a = \frac{g^2 L}{4\pi^2} \sum_{\substack{n\neq 0\\ ij}} \frac{j_\phi^{ij}(n) j_\phi^{ji}(-n)}{n^2} \tag{146}$$

where without loss of generality we have assumed $N_a^i = 0$ (cf. 125 a). To solve for the meson states we now start to apply the following sequence of large N approximations. We first neglect in the definition of the right-handed current j_ϕ the U(1) subtraction

$$j_\phi^{kl}(n) \approx \frac{1}{2} \sum_r a_l^+(r)\, a_k(r+n) \tag{147}$$

The basis for the following is the commutator

$$\begin{aligned} \left[\textstyle\sum_i a_i^+(p) a_i(q), H_a\right] &= \tfrac{g^2 L}{8\pi^2} \textstyle\sum_{\substack{n\neq 0\\ slk}} \tfrac{1}{n^2} \big(a_l^+(p-n) a_l(s-n)\, a_k^+(s)\, a_k(q) \\ &\quad - a_l^+(p)\, a_l(s-n)\, a_k^+(s) a_k\, (q+n)\big) \end{aligned} \tag{148}$$

which will be linearized by the large N approximation. To this end, we decompose Fermion bilinears summed over the color into a leading c– number piece and a fluctuating term. The leading c–number is the corresponding vacuum expectation value, the fluctuating term represents a meson operator

$$\sum_l a_l^+(p) a_l(q) = \delta_{p,q} \varrho(p) \cdot N + M(p-q, \frac{p+q}{2}) N^{1/2} \tag{149}$$

The density ϱ is assumed to be given by the perturbative light cone vacuum expectation value

$$\varrho(p) = \theta(-p) \tag{150}$$

The Ansatz for the N dependence of c-number and operator piece in (149) is chosen such that the commutator of the meson operator is to leading order independent of N. Straightforward calculation yields

$$[M(P,r), M^+(P',r')] = \delta_{pp'}\, \delta_{rr'} \epsilon(-P)\, \theta(\frac{P^2}{4} - r^2) \tag{151}$$

With the Ansatz (149) the commutator (148) is approximatively calculated by retaining only linear terms in the large N suppressed fluctuations and reads

$$\begin{aligned} \left[H_a, \textstyle\sum_i a_i^+(p) a_i(q)\right] &= \tfrac{N g^2 L}{8\pi^2} \textstyle\sum_{n\neq 0} \tfrac{1}{n^2} \big\{(\varrho(p) - \varrho(q))\, M(p-q, \tfrac{p+q}{2} + n) \\ &\quad + \ (\varrho(q-n) - \varrho(p-n))\, M\, (p-q, \tfrac{p+q}{2})\big\}\, N^{1/2} \end{aligned} \tag{152a}$$

In order to obtain a well defined large N limit, the coupling constand g has to be adjusted such that

$$N g^2 = const. \tag{152b}$$

Using again the Ansatz (149), eq. (152a) represents the linearized equation of motion for the meson operators in the large N limit

$$-i\partial_+ M(P,r) = \frac{Ng^2L}{8\pi^2} \sum_{n\neq 0} \frac{1}{n^2} \epsilon(-P) \Big\{\theta(\tfrac{P^2}{4} - r^2)M(P,r+n) - \theta(\tfrac{P^2}{4} - (r+n)^2)M(P,r)\Big\} \tag{153}$$

Note that this equation is diagonal in the total meson momentum P. With the Ansatz

$$M(P,r) = \theta(\frac{P^2}{4} - r^2)\, m\,(P,r) \tag{154}$$

the equation is rewritten as

$$\begin{aligned} -i\partial_+ m(P,r) &= \frac{Ng^2L}{8\pi^2}\, \epsilon(-P) \sum_{n\neq 0} \frac{1}{n^2}\, \theta(\tfrac{P^2}{4} - (r+n)^2)\cdot \\ &\quad \cdot(m(P,r+n) - m(P,r)) \qquad (155a) \\ &= \epsilon(-P) \sum_{r'} K(r,r')m(P,r') \end{aligned}$$

We have introduced the kernel K of the algebraic system to determine the stationary states

$$K(r,r') = \frac{Ng^2L}{8\pi^2} \left[\frac{1-\delta_{r,r'}}{(r-r')^2} - \delta_{r,r'} \sum_{n\neq 0} \frac{1}{n^2}\, \theta\,(\tfrac{P^2}{4} - (r+n))\right]\cdot \theta\,(\tfrac{P^2}{4} - r^2)\, \theta(\tfrac{P^2}{4} - r^{2\prime}) \tag{155b}$$

The eigenfunctions f_i and energies E_i of K determine the meson wave functions and masses

$$K(r,r') = \sum_i E_i(P) f_i(P,r) f_i(P,r') \tag{156}$$

and therefore the meson operator $m(P,r)$ can be written as

$$m(P,r) = \sum_i A_i(P)\, f_i(P,r) \tag{157}$$

where $A_i(P)$ satisfy standard Boson commutation relations

$$[A_i(P),\, A_j^+(P')] = \delta_{ij}\, \delta_{P,P'} \tag{158}$$

In contrast to electrodymancis, QCD_2 in the large N-limit contains in general a large and in the continuum limit infinite number of elementary excitations. At the level of the finite interval, these excitations cannot be identified as simply as in QED_2 with particles. The number of excitations (characterized by the index i in (156)) is determined by the dimension of the matrix K which in turn is given by the "meson momentum"; furthermore for finite length L, the dispersion relation between energy $E_i(P)$ and momentum P is not that of free particles (cf. (21b), (A16b)). In QCD_2 interpretation of these excitations as mesons is possible only in the continuum limit which is obtained from (155a) by assuming $P >> 1$ and replacing summation by integration. The requirement of finite energy in the continuum limit specifies the treatment of the singularity in the integration and yields the boundary condition for the meson wave function. More subtle in this context is the continuum limit of certain ground state quantities, in particular the quark condensate. As can be directly verified, eq. (155a) possesses a zero energy solution with constant wave function :

$$f_0(P,r) = \frac{1}{\sqrt{P}} \qquad E_0(P) = 0 \tag{159}$$

i. e. also in the large N limit the zero mass solution appears which we have found above on general grounds for the exact Hamiltonian. The existence of this massless meson in the large N limit is expected to imply a non vanishing quark condensate [8], [21], which however is not present for finite L. Again it can be shown that the finite interval ground states do not determine in a simple limiting procedure an appropriate continuum ground state.

While the zero mass mesons can be obtained in the large N limit, the construction of the massless baryons is problematic in this limit and we start with the effective a-mode Hamiltonian (146) with the correct SU(N) currents j_ϕ (eq. 112). It is straightforward to show that the baryon operator

$$B = \prod_{i=1}^{N} a_i^+ (0) \tag{160}$$

satisfies

$$[B, j_\phi^{kl}(n) j_\phi^{lk}(-n)] \mid 0 >= 0 \tag{161}$$

i. e. $B \mid 0 >$ is a massless baryon state of zero momentum. With the help of the right handed $U(1)$ current $j_\phi(m)$ (eq. 139, 140) baryon states with arbitrary momentum can be constructed, since

$$[j_\phi(m) B, H_a] \mid 0 >= 0 \tag{162}$$

(this construction is analogous to the decomposition (143b) of the phase changes). The final remark concerns the possibility of defining single quark states in the context of the effective theory described by H_a (146). As in the derivation of the meson operator, we calculate the commutator

$$\begin{aligned}\left[H_a,\, a_i^+(p)\right] &= \tfrac{g^2 L}{16\pi^2} \textstyle\sum_{n\neq 0} \tfrac{1}{n^2}(a_i^+(r) a_l(r+n) a_l^+(p+n) \\ &\quad + a_l^+(p-n) a_i^+(r) a_l(r-n))\end{aligned} \tag{163}$$

and replace the color sum over fermion bilinears by the leading c-number piece (eq. 149) to obtain

$$\begin{aligned}&[H_a, a_i^+(p)] \approx \tfrac{Nq^2 L}{16\pi^2} \textstyle\sum_{n\neq 0} (1 - 2\varrho(p+n)) a_i^+(p) \tfrac{1}{n^2} \\ &= \tfrac{Ng^2 L}{16\pi^2} \left\{\tfrac{\pi^2}{3} - 2 \textstyle\sum_{n-\infty}^{-p} \tfrac{1}{n^2}\right\} a_i^+(p)\end{aligned} \tag{164}$$

Thus the single particle energy is

$$\epsilon(p) = \frac{Nq^2 L}{16\pi^2} \left\{\frac{\pi^2}{3} - 2 \sum_{n=-\infty}^{-p} \frac{1}{n^2}\right\} \overset{p>>1}{\longrightarrow} \frac{Ng^2 L}{48} - \frac{Ng^2}{4\pi} \frac{1}{q} \tag{165}$$

where

$$q = \frac{2\pi p}{L} \tag{166}$$

is the Fermion momentum. The single particle energy of an isolated quark diverges in the continuum limit (cf. the infinite constant in 't Hoofts choice of the Fermion propagator [1]).

4 Summary

I have reported here on studies of two-dimensional gauge theories. The aim of these investigations has been to formulate QED_2 and QCD_2 within the canonical formalism on the light-cone and to study in a first application the well known properties of QED_2 and the less well understood vacuum structure and elementary excitations of QCD_2.

In the attempt to formulate a canonical light-cone theory of QCD_2, two major difficulties are encountered. Conceptual problems irrespective of the particular dynamics arise in the canonical quantization on the light-cone. The ultimate reason is that events at equal light-cone time are light-like separated unlike equal time events in ordinary coordinates. This makes the definition of singular operators like the Hamiltonian very subtle. Separation in the (light-cone) space components of bilinear operators is of no help for regularization of such operators on the light-cone; on the other hand, separation in the time components requires knowledge of the time evolution which in turn is given by the not yet properly defined Hamiltonian. These difficulties are avoided in the finite interval formulation of light cone dynamics, provided the fields related by boundary conditions are dynamically independent. This is guaranteed if the space-time points in the boundary condition are space-like separated.

On this basis, the canonical formalism has been developed and the naive light-cone formulation could be interpreted as an effective theory with well defined properties and equally well defined limitations. In this formalism, the continuum limit requires special care. On the light-cone, the long wave-length excitations important for defining a proper continuum limit are the high-energy excitations which therefore are also affected by the ultraviolet regularization. As a result of this connection between infrared and ultraviolet properties, the vacuum of the continuum theory is not obtained by a straightforward limit of finite interval vacua. On the other hand finite momentum excitations are not affected by these subtleties of the continuum limit. In this way, the success of naive light-cone approaches in reproducing the spectrum of elementary excitations and their failure in correctly describing the vacuum properties becomes understandable. I have illustrated this in the discussion of QED_2 with its non-perturbative vacuum structure.

The second class of problems to be resolved on the way to a canonical formalism of QCD_2 has the origin in the non-trivial infrared properties of two-dimensional gauge theories which in turn are responsible for the confinement of the elementary Fermions. Introduction of a finite interval either in ordinary space or in light-cone space variables resolves the difficulties and an unambiguous definition of these gauge theories is obtained. The dynamical degrees of freedom of the finite inverval QCD_2 (with N colors) are the right and left handed Fermion fields and N-1 quantum mechanical gauge degrees of freedom, neutral zero-momentum gluons. Identification of the independent degrees of freedom is a prerequisite for understanding the structure of the Hilbert space of physical states, which is of special relevance for theories with confined elementary degrees of freedom. The Hilbert spaces of both U(N) and SU(N) gauge theories have been explicitly constructed based on the canonical formalism in the Coulomb-gauge. Despite the many similarities, there are far reaching differences in the Hilbert space structure. Clearly such differences are expected, if baryons are to exist in SU(N) QCD but do not correspond to physicsal states in U(N) QCD. The light-cone formulation makes explicit the additional symmetries present for vanishing Fermion masses. The existence of massless mesons and baryons in SU(N) QCD_2 can be read off almost directly from the Hamiltonian and has its origin in the appearence of a local U(1) symmetry, the invariance under phase changes of the right-handed Fermions. Baryons and mesons can be related to the phase transformations with finite and zero winding number respectively.

A Hamiltonian light-cone formulation of QCD_2 has been achieved and the Hilbert space of physical states has been explicitly constructed. In this way, a well defined theoretical framework has been established for further investigations of the strong interaction. An immediate extension concerns the study of massive theories. As is known from quantization of QED_2 in ordinary coordinates (33, 34), the background field does not anymore decouple from the Fermion degrees of freedom. It will be important for a deeper understanding of light-cone dynamics in general and for the color dynamics in particular to investigate these dynamically more complex theories. Furthermore I envisage studies of hadronic matter; in this context the Hamiltonian formalism as described here should be particularly useful for developing analytical approximations as well as numerical approaches.

The manuscript has been typeset by B. Kreisel. I thank M. Thies for a critical reading of the manuscript.

References

[1] G. 't Hooft, Nucl. Phys. B 75 (1974) 461

[2] C. Callan et al., Phys. Rev. D13 (1976) 1649

[3] M. B. Einhorn, Phys. Rev. D14 (1976) 3451

[4] M. B. Einhorn et al., Phys. Rev. D15 (1977) 2282

[5] M. Durgut, Nucl. Phys. B116 (1976) 233

[6] R. C. Brower et al., Nucl. Phys. B128 (1977) 131, 175

[7] I. Bars and M. B. Green, Phys. Rev. D17 (1978) 537

[8] M. Li, Phys. Rev. D34 (1986) 3888

[9] M. Li et al., J. Phys. G Nucl. Phys. 13 (1987) 915

[10] P. J. Steinhardt, Nucl. Phys. B176 (1980) 100

[11] R. E. Gamboa Saravi, Phys. Rev. D30 (1984) 1353

[12] C. J. Hamer, Nucl. Phys. B121 (1977) 159, B195 (1982) 503

[13] D. B. Carpenter, Nucl. Phys. B228 (1983) 365

[14] S. Huang et al., Nucl. Phys. B307 (1988) 661

[15] Y. Frishman et al., Phys. Rev. D15 (1977) 2275

[16] M. H. Partovi, Phys. Lett. 80B (1979) 377

[17] M. K. Pak and H. C. Tze, Phys. Rev. D14 (1976) 3472

[18] T. T. Wu, Phys. Rept. 49 (1979) 245

[19] N. S. Manton, Ann. Phys. 159 (1985) 220

[20] S. Iso and H. Murayama, preprint

[21] A. R. Zhitnitsky, Phys. Lett. 165B (1985) 405; Sov. J. Nucl. Phys. 43 (1986) 999; 44 (1986) 139

[22] T. Eller, H. C. Pauli and S. J. Brodsky, Phys. Rev. D35 (1987) 1493

[23] C. R. Hagen, Nucl. Phys. B95 (1975) 477

[24] F. Lenz, S. Levit, M. Thies and K. Yazaki to be published

[25] H. Leutwyler, J. R. Klauder and L. Streit, N. C. LXVIA (1970) 536

[26] J. Kogut and L. Susskind, Phys. Rept. 8 (1973) 75

[27] H. C. Pauli and S. J. Brodsky, Phys. Rev. D32 (1985) 1993

[28] S. Coleman, Phys. Rev. D11 (1975) 2088

[29] R. Jackiw, in Luctures in Current Algebra and its Applications, Princeton University Press 1972

[30] K. D. Rothe et al., Ann. Phys. 105 (1977) 63

[31] I. Affleck, Nucl. Phys. B265 (1986) 448

[32] W. Buchmüller et al., Phys. Lett. 108B (1982) 426

[33] S. Coleman, R. Jackiw and L. Susskind, Ann. Phys. 93 (1975) 267

[34] S. Coleman, Ann. Phys. 101 (1976) 239

Appendix I

Free Bosons on a light cone interval

Here we outline the finite interval light cone description for a free Boson theory. As in Chapter I we restrict the space variables to the interval [0, L]. The variable transformations (5) and (9 b) amount to introduce the variables

$$x^+ = \frac{1}{\sqrt{2}}((1+\frac{\epsilon}{L})x^0 + (1-\frac{\epsilon}{L})x^1)\,,\; x^- = \frac{1}{\sqrt{2}}(x^0 - x^1) \tag{A1a}$$

and the corresponding derivatives

$$\partial_0 = \frac{1}{\sqrt{2}}((1+\frac{\epsilon}{L})\partial_+ + \partial_-),\; \partial_1 = \frac{1}{\sqrt{2}}((1-\frac{\epsilon}{L})\partial_+ - \partial_-) \tag{A1b}$$

The choice (A1a) guarantees in general space-like separation of equal light cone time events

$$ds^2 = x_0^2 - x_1^2 = 2x^+x^- - 2\frac{\epsilon}{L}(x^-)^2 \tag{A2}$$

We also note that the d'Alembert operator

$$\square = \partial_0^2 - \partial_1^2 = 2\partial_+(\partial_- + \frac{\epsilon}{L}\partial_+) \tag{A3}$$

contains a second order derivative with respect the (light cone) time variable but only a first order derivative with respect to the space variable. Therefore standard treatment of the free theory is possible and we compare explicitly in the following the standard and light cone formulations.

Lagrangean

$$\mathcal{L} = \tfrac{1}{2}((\partial_0\varphi)^2 - (\partial_1\varphi)^2 - m^2\varphi^2) \qquad \mathcal{L} = \partial_+\varphi(\partial_-\varphi + \tfrac{\epsilon}{L}\partial_+\varphi) - \tfrac{m^2}{2}\varphi^2 \tag{A4}$$

Euler Lagrange equations

$$\partial_0^2\varphi - \partial_1^2\varphi + m^2\varphi = 0 \qquad \partial_+\partial_-\varphi + \tfrac{\epsilon}{L}\partial_+^2\varphi + \tfrac{m^2}{2}\varphi = 0 \tag{A5}$$

Canonical Formalism
Conjugate momenta

$$\pi = \tfrac{\delta\mathcal{L}}{\delta\partial_t\varphi} = \partial_0\varphi \qquad \pi = \tfrac{\delta\mathcal{L}}{\delta\partial_+\varphi} = \partial_-\varphi + \tfrac{2\epsilon}{L}\partial_+\varphi \tag{A6}$$

Hamiltonian

$$\mathcal{H} = \pi\,\partial_0\,\varphi - \mathcal{L}$$

$$\mathcal{H} = \tfrac{1}{2}(\pi^2 + (\partial_1\varphi)^2 + m^2\varphi^2) \qquad \mathcal{H} = \tfrac{L}{4\epsilon}(\pi - \partial_-\varphi)^2 + \tfrac{m^2}{2}\,\varphi^2 \tag{A7}$$

Normal mode expansion
Bondary conditions

$$\varphi(L, x_0) = \varphi(0, x_0) \qquad \varphi(L, x_+) = \varphi(0, x_+) \tag{A8}$$

$$p_n = \tfrac{2n\pi}{L} \qquad p_n = \tfrac{2n\pi}{L} \tag{A9}$$

$$\varphi_n(x^0, x^1) = e^{-ip_0^n x^0 + ip_n x^1} \qquad \varphi_n(x^+, x^-) = e^{-ip_+^n x^+ - ip_n x^-} \tag{A10}$$

Eigenmode energies

$$-(p_0^n)^2 + p_n^2 + m^2 = 0 \qquad -p_+^n p_n - \tfrac{\epsilon}{L}(p_+^n)^2 + \tfrac{m^2}{2} = 0 \tag{A11}$$

$$p_0^n = \pm\omega_n = \pm\sqrt{(\tfrac{2\pi n}{L})^2 + m^2} \qquad \begin{aligned} p_+^n &= \tfrac{1}{\epsilon}(-\pi n \pm \sqrt{(\pi n)^2 + m^2\epsilon L/2}) \\ &= \tfrac{1}{\epsilon}(-\pi n \pm \omega_n) \end{aligned} \tag{A12}$$

Normal mode expansion of field operator and conjugate momenta (in the Schrödinger representation) are almost identical for the two cases. We have in ordinary coordinates

$$\varphi(x) = \sum_n \frac{1}{(2\omega_n)^{1/2}}(A(n)e^{ip_n x} + A^+(n)e^{-ip_n x}) \tag{A13a}$$

$$\pi(x) = \sum_n \frac{-i\omega_n^{1/2}}{L \cdot 2^{1/2}}(A(n)e^{ip_n x} - A^+(n)e^{-ip_n x}) \tag{A13b}$$

and on the light cone

$$\varphi(x) = \sum_n \frac{1}{(2\omega_n)^{1/2}}(A(n)e^{-ip_n x^-} + A^+(n)e^{ip_n x^-}) \tag{A14a}$$

$$\pi(x) = \sum_n \frac{-i\omega_n^{1/2}}{L \cdot 2^{1/2}}(A(n)e^{-ip_n x^-} - A^+(n)e^{ip_n x^-}) \tag{A14b}$$

with ω_n given by their respective definitions (A12). The A(n) satisfy standard Boson commutation relations

$$[A(n), A^+(m)] = \delta_{nm} \tag{A15}$$

The Hamiltonian in ordinary coordinate quantization is

$$H = \sum_n \omega_n A^+(n)A(n) \tag{A16a}$$

and on the light cone

$$\begin{aligned} H &= \tfrac{1}{\epsilon}\textstyle\sum_{n>0}[(\pi^2n^2 + \epsilon Lm^2/2)^{1/2} - \pi n]A^+(n)A(n) + \sqrt{Lm^2/2\epsilon}\cdot \\ &\cdot A^+(0)A(0) + \tfrac{1}{\epsilon}\textstyle\sum_{n>0}[(\pi^2n^2 + \epsilon Lm^2/2)^{1/2} + \pi n]A^+(-n)A(-n) \\ &\xrightarrow[m^2\epsilon L \to 0]{} \textstyle\sum_{n>0} \frac{m^2}{2(2\pi n/L)} A^+(n)A(n) \end{aligned} \tag{A16b}$$

So far we have not specified the value of the parameter η (eq. 25) which determines the approach to the light-cone. Actually the present development allows to connect states corresponding to different ϵ via unitary transformations with each other. The method is similar to the one employed in ref. [28] in the definition of normal ordering with respect to different masses. Starting point is the equal light-cone time canonical commutation relation

$$[\pi(x^-), \varphi(y^-)] = \frac{1}{i}\delta(x^- - y^-) \tag{A17}$$

which asserts that π and φ are independent of the choice of ϵ. As a consequence (eq. A14) A(n) must depend on ϵ

$$A(n) = A(n, \epsilon) \tag{A18}$$

and therefore, by comparison of the normal mode expansion (A14) corresponding to two different choices ϵ, ϵ' we have

$$A(n,\epsilon') = \frac{1}{2x_n}((x_n^2+1)A(n,\epsilon) + (x_n^2-1)A^+(-n,\epsilon)) \tag{A19a}$$

$$x_n^2 = \frac{\omega(n,\epsilon')}{\omega(n,\epsilon)} \tag{A19b}$$

This change in the operators is described by the following unitary transformation

$$A(n,\epsilon') = e^{-C_n}A(n,\epsilon)e^{C_n} = U(\epsilon',\epsilon)A(n,\epsilon)U^+(\epsilon',\epsilon) \tag{A20a}$$

$$C_n = -\alpha_n(A(n)A(-n) - A^+(n)A^+(-n)) \tag{A20b}$$

$$x_n = e^{\alpha_n} = \left(\frac{\pi^2 n^2 + m^2\epsilon' L/2}{\pi^2 n^2 + m^2 \epsilon L/2}\right)^{1/4} \tag{A20c}$$

and as a consequence the two respective vacua defined by

$$A(n,\epsilon) \mid 0_\epsilon >= 0 \qquad A(n,\epsilon') \mid 0_{\epsilon'} >= 0 \tag{A21a}$$

are related by

$$\mid 0_{\epsilon'} >= U(\epsilon',\epsilon) \mid 0_\epsilon > \tag{A21b}$$

Appendix II

Gauge Invariance and regularization

In this appendix the gauge invariance of the regularization procedure described in Chapter II is discussed. Operators involving Fermion bilinears at the same space time point have to be regularized since in general they receive an infinite contribution from the filled Dirac sea. A simple regularization procedure is point splitting (cf. ref. [29]) in which the two Fermion operators are evaluated at slightly different space time points. To compensate for the violation of local gauge invariance as a result of the "non-locality" the Fermions at the two space time points are connected by a string of gauge fields.

As an example we consider the regularization of the "b-charge" and define the regularized operator

$$\begin{aligned} Q_b &= \int dx^- \psi^+(x^+,x^-)\psi(x^+,x^-) =: \\ &=: \int dx^- \psi^+(x^+,x^-)\, e^{ig\int_{x^--\delta}^{x^-} dz A_-(x^+,z)}\, \psi(x^+,x^- - \delta) \end{aligned} \tag{A22}$$

This regularization by splitting only the spatial components of the argument of the Fermion-operators does not work exactly on the light cone where an equal time separation is lightlike and consequently $\psi^+(x^+,x^-)\psi(x^+,x^- - \delta)$ is divergent for finite δ.

We now consider the transformation properties of Q_b under the gauge transformations

$$A_\pm \rightarrow A'_\pm = A_\pm + \partial_\pm \theta \tag{A23}$$

$$\rho'_b = \left(\psi^+(x)e^{ig\int_{x-\delta}^{x^-} dz A_-(x^+,z)}\psi(x^+,x^- - \delta)\right)'$$

$$= \psi^+(x)e^{-ig\theta(x^-)}\, e^{ig\int_{x-\delta}^{x^-} dz(A_-(x^+,z)+\partial_-\theta(x^-))} e^{ig\theta(x^- -\delta)}\, \psi(x^+, x^- - \delta) \tag{A24}$$

$$= \psi^+(x)\, e^{ig\int_{x-\delta}^{x^-} dz A_-(x^+,z)}\, \psi(x^+, x^- - \delta) = \rho_b$$

i. e. ϱ_b and therefore Q_b are gauge invariant. We now establish the connection of point splitting and heat kernel regularization for the b-charge defined on the finite interval

$$Q_b = \int_0^L dx^- \psi^+(x^+, x^-)\, e^{ig\int_{x^- -\delta}^{x^-} dz A_-(x^+,z)}\, \psi(x^+, x^- - \delta) =$$

$$= \frac{1}{L}\sum_{m,n} b_m^+ b_n \int_0^L dx e^{i(2\pi m+\varphi)\frac{x}{L}}\, e^{ig\delta A_-(x^+,x)}\, e^{-i(2\pi n+\varphi)\frac{x-\delta}{L}} \tag{A25a}$$

$$\partial_- A_- = 0$$

$$Q_b = \sum_n b_n^+ b_n\, e^{i\delta(2\pi n+\varphi+gLA_-)/L} = \sum_n b_n^+ b_n\, e^{-\lambda(2\pi n+\varphi+gLA_-)} \tag{A25b}$$

For a state with b modes occupied at $+\infty$, regularization by the oscillating exponential can be replaced by the damped exponential in identifying

$$\frac{\delta}{L} = i\lambda \tag{A25c}$$

Appendix III

Anomalous current commutator

In this appendix it is shown by explicit calculation that due to contributions from the Dirac sea the current operators defined by

$$j_\psi(n) = \sum_r b_{r+n}^+ b_r e^{-\lambda(r+c)}; \quad c = (\varphi + gLA_-)/2\pi; \quad n > 0 \tag{A26}$$

have non vanishing commutators. In the regularization we have again used explicitly the fact that infinities may arise only from the occupied states $r \to \infty$. The commutator

$$[j_\psi^+(m), j_\psi(n)] = \sum_{rs} e^{-\lambda(r+s+2c)}[b_s^+ b_{s+m}, b_{r+n}^+ b_r]$$

$$= e^{-2\lambda c}(\sum_s e^{-\lambda(2s+m-n)} b_s^+ b_{s+m-n} - \sum_r e^{-2\lambda r} b_{r+n}^+ b_{r+m}) \tag{A27}$$

$$= e^{-2\lambda c}\sum_r e^{-2\lambda r} b_{r+n}^+ b_{r+m}(e^{-\lambda(m+n)} - 1) = C_{mn}$$

vanishes if naively $\lambda = 0$ in this expression.

Applying C_{mn} to a state which differs from $\mid N_b >$ defined in Chapter 2 by a finite number of particle hole excitations i. e. a state $\mid \varphi >$ satifsying

$$\begin{aligned} n \geq N_1 :&\quad b_n^+ \mid \varphi >= 0 \\ n < N_2 :&\quad b_n \mid \varphi >= 0 \end{aligned} \tag{A28}$$

with appropriately chosen N_1, N_2. For $n \neq m$ the sum in $C_{mn} \mid \varphi >$ contains only a finite number of non vanishing terms and therefore (in the limit $\lambda \to 0$)

$$C_{mn} \mid \varphi >= 0 \tag{A29a}$$

while for $m = n$ we have

$$C_{nn} \mid \varphi > \approx e^{-2\lambda c} \sum_{r=N_1}^{\infty} e^{-2\lambda r}(e^{-2\lambda n} - 1) \approx \frac{e^{-2\lambda c}(e^{-2\lambda n}-1)}{1-e^{-2\lambda}}$$

$$= -n \tag{A29b}$$

An analogous calculation for the current j_ϕ can be carried through and therefore the two anomalous commutators are obtained

$$[j_\psi^+(m), j_\psi(n)] = -n\delta_{n,m} \qquad [j_\phi^+(m), j_\phi(n)] = -n\delta_{n,m} \tag{A30}$$

As a consequence of the commutation relations we have the following identity

$$[-\sum_n n : b_n^+ b_n :, j_\psi(m)] = [-\sum_n n : b_n^+ b_n :, \sum_r b_{r+m}^+ b_r]$$

$$= -m \sum_r b_{r+m}^+ b_r = -m j_\psi(m) = [\sum_p j_\psi^+(p) j_\psi(p), j_\psi(m)] \tag{A31}$$

Appendix IV

Evaluation of the fermion condensate of QED_2

Starting point is the definition of the mixed current

$$g(n) = \sum_m a_{n+m}^+ \, b_m \tag{A32}$$

and its commutation relations with the currents of right and left handed Fermions (eq. 54)

$$[g(n), j_\phi(m)] = -g(n-m); \qquad [g(n), j_\psi(m)] = g(n+m) \tag{A33}$$

We consider the unitarily transformed operator

$$g(n, \alpha) = e^{\alpha B_m} g(n) e^{-\alpha B_m} \tag{A34a}$$

with

$$B_m = j_\phi^+(m) \, j_\psi^+(m) - j_\phi(m) \, j_\psi(m) \tag{A34b}$$

$g(n, \alpha)$ satisfies the differential equation

$$\frac{dg(n, \alpha)}{d\alpha} = e^{\alpha B_m} [B_m, g(n)] e^{-\alpha B_m} = D_1 - D_2 \tag{A35}$$

With the notation

$$c = cosh \, m\alpha \tag{A36a}$$

$$s = sinh \, m\alpha \tag{A36b}$$

The quantities $D_{1,2}$ are easily calculated

$$D_1 = e^{\alpha B_m}(g(n+m) j_\psi^+(m) + j_\phi(m) g(n+m)) e^{-\alpha B_m}$$

$$= \frac{1}{c+s} e^{\alpha B_m} \{(s j_\phi(m) + c j_\psi^+(m)) g(n+m) + g(n+m) \cdot$$

$$\cdot(cj_\phi(m)+sj^+_\psi(m))+c\left[j_\phi(m)-j^+_\psi(m),\,g(n+m)\right]\Big\}\,e^{-\alpha B_m} \quad \text{(A37a)}$$

$$= \tfrac{1}{c+s}\left\{j^+_\psi(m)g(n+m,\alpha)+g(n+m,\alpha)j_\phi(m)+2cg(n,\alpha)\right\}$$

in the last step we have used the transformation properties of the currents (54)

$$e^{\alpha B_m}j_{\phi,\psi}(m)c^{-\alpha B_m}=cj_{\phi,\psi}(m)-sj^+_{\psi,\phi}(m) \quad \text{(A38)}$$

D_2 is calculated in the same way

$$D_2 \;=\; e^{\alpha B_m}(j^+_\phi(m)g(n-m)+g(n-m)j_\psi(m))e^{-\alpha B_m}$$

$$= \tfrac{1}{c+s}\left\{j^+_\phi(m)g(n-m,\alpha)+g(n-m,\alpha)j_\psi(m)+2sg(n,\alpha)\right\} \quad \text{(A37b)}$$

Taking matrixelements in vacuum states $\mid\Omega>,\mid\Omega'>$

$$j_{\phi,\psi}(n)\mid\Omega^{(')}>=0 \quad \text{(A39)}$$

the differential equation (A35) simplifies to

$$\tfrac{d}{d\alpha}<\Omega'\mid g(n,\alpha)\mid\Omega> \;=\; 2\tfrac{c-s}{c+s}<\Omega'\mid g(n,\alpha)\mid\Omega>$$

$$= \; 2\,e^{-2m\alpha}<\Omega'\mid g(n,\alpha)\mid\Omega> \quad \text{(A40a)}$$

which, together with

$$g(n,0)=g(n) \quad \text{(A40b)}$$

and the definitions (A 36), is integrated with the result

$$<\Omega'\mid g(n,\alpha)\mid\Omega>=<\Omega'\mid g(n)\mid\Omega>e^{-\frac{1}{m}(e^{-2m\alpha}-1)} \quad \text{(A41)}$$

Appendix V

SU(N) identities

Elementary properties of λ matrices:

$$[\lambda^\alpha,\lambda^\beta]=2i\,f^{\alpha\beta\gamma}\lambda^\gamma \qquad sp\lambda^\alpha\lambda^\beta=2\delta_{\alpha\beta} \quad \text{(A42)}$$

$$\sum_\alpha\lambda^\alpha_{ij}\lambda^\alpha_{ke}=2\,[\delta_{il}\delta_{jk}-\frac{1}{N}\delta_{ij}\delta_{kl}] \quad \text{(A43)}$$

(ij) versus α representation

$$F_{ij}=\sum_\alpha F^\alpha\lambda^\alpha_{ij}/2 \qquad F^\alpha=\sum_{ij}F_{ij}\,\lambda^\alpha_{ji} \quad \text{(A44)}$$

$$\sum_{ij}F_{ij}\,G_{ji}=\frac{1}{4}\sum_{\substack{\alpha\beta\\ ij}}F^\alpha G^\beta\lambda^\alpha_{ij}\,\lambda^\beta_{ji}=\frac{1}{2}\sum_\alpha F^\alpha\,G^\alpha \quad \text{(A45)}$$

Define the bilinear E_{ij}:

$$E_{ij}=\textstyle\sum_k F_{ik}G_{kj}-F_{kj}G_{ik}=[F,G]_{ij}$$

$$= \frac{1}{4} \sum_{\alpha\beta k} F^\alpha G^\beta (\lambda^\alpha_{ik} \lambda^\beta_{kj} - \lambda^\beta_{ik} \lambda^\alpha_{kj}) = \frac{i}{2} \sum_{\alpha\beta} F^\alpha G^\beta f^{\alpha\beta\gamma} \lambda^\gamma_{ij} \tag{A46}$$

and therefore the following the correspondence is obtained

$$E_{ij} = [F, G]_{ij} \qquad E^\alpha = i f^{\alpha\beta\gamma} F^\beta G^\gamma \tag{A47}$$

For the field strength tensor defined by

$$F_{kl} = \partial_+ A^{kl}_- - \partial_- A^{kl}_+ - ig[A_+, A_-]^{kl} \tag{A48a}$$

this correspondence yields

$$F^\alpha = \partial_+ A^\alpha_- - \partial_- A^\alpha_+ + g\, f^{\alpha\beta\gamma}\, A^\beta_+ A^\gamma_- \tag{A48b}$$

Quark-Gluon Plasma and Space-Time Picture of Ultra-relativistic Nuclear Collisions

L. McLerran
Theoretical Physics Institute
University of Minnesota
Minneapolis, MN 55455

1 Introduction

In these lectures, I will try to review what is known of the properties of matter at very high energy density, and how it might be produced in ultra-relativistic nuclear collisions. These lectures are aimed primarily at graduate students or young postdocs with an interest in learning a little something about these subjects, but who have not been actively involved in research on related problems.

In the first two lectures, I review the properties of hadronic matter at very high energy density. This review attempts to give some qualitative understanding concerning the properties of high energy density matter. It is not intended as a technical review of the very interesting recent results coming from either lattice gauge theory, strong and weak coupling expansions. In these two lectures, I also review various speculations which have been made concerning the possible appearance of quark matter in astrophysics and in cosmic rays. Much of this discussion involves the properties of stable strange quark matter.

In the third lecture, I present the space time picture of ultra-relativistic nuclear collisions. This description is somewhat idealized, and certainly is not applicable for the current AGS or CERN experiments both due to limitations in energy and in the size of the colliding nuclei. If such a description is ever applicable, it is probably only for the case of RHIC, or LHC. I also have not attempted to describe proposed signals for the existence of a quark-gluon plasma as it might be produced in such collisions. This has been the subject of much work, and a thorough review would take me far beyond the scope of what I wish to present here.

In order to not overtax the reader with a bibliography which is too long be very

Hadrons and Hadronic Matter
Edited by D. Vautherin *et al.*
Plenum Press, New York, 1990

useful, I have only referred to review papers in these lectures. References to the original literature may be found there.

2 Lecture 1: Properties of Matter at Very High Energy Density

2.1 Gross Considerations

In this section I shall discuss the properties of hadronic matter at high energy density. The word high implies a scale for the measurement of the energy density. Such a scale may be provided by a variety of estimates, all of which agree on the order of magnitude of a typical density scale for hadronic matter. The first is the energy density of nuclear matter. With m the proton mass, R_A the nuclear radius, and A the nuclear baryon number, the density of nuclear matter is

$$\rho_A \sim \frac{Am}{\frac{4}{3}\pi R_A^3} \sim .14\ Gev/Fm^3 \tag{1}$$

We can also use Eqn. 1 to estimate the energy density inside a proton. If we use a proton radius of .8 Fm, Eqn. 1 gives

$$\rho_p \sim .5\ Gev/Fm^3 \tag{2}$$

There is a good deal of uncertainty in this estimate of ρ_p. We might have instead used the MIT bag radius, or a proton hard core radius, corresponding to an order of magnitude uncertainty in Eqn. 2. Finally, another estimate comes from dimensional grounds using the value of the QCD Λ parameter, suitably defined as Λ_{ms} or Λ_{mom}, as the dimensional scale factor. Using the Λ parameter, we find

$$\rho_{QCD} \sim \Lambda^4 \sim .2\ Gev/Fm^3 \tag{3}$$

Again there is an order of magnitude uncertainty both due to the lack of precise experimental knowledge of Λ, and differences induced by using alternative sensible definitions of Λ.

In all of the above energy density estimates, the typical scale was in the range of several hundreds of Mev/Fm^3 to several Gev/Fm^3. At energy densities low compared to this scale, we presumably have a low density gas of the ordinary constituents of hadronic matter, that is, mesons and nucleons. At such low densities, the typical particle separation is large compared to the range of the nuclear force, and the gas is therefore ideal. At densities very high compared to this scale, we expect an asymptotically free gas of quarks and gluons, that is, because the typical separation between quarks and gluons is small, their interaction energy is small compared to kinetic energies, and the gas is therefore ideal. At intermediate energy densities, we expect that the properties of matter will interpolate between these dramatically different phases of matter. There may or may not be true phase changes at some intermediate densities.

2.2 The Transition to the Quark Gluon Plasma

To understand how such a transition might come about, consider the example of QCD in the limit of a large number of colors, N_C. Recall that extensive quantities such as the energy density, ϵ, or entropy density, σ, measure the number of degrees of freedom of a system. The dimensionless quantities ϵ/T^4 or σ/T^3 should be of the order of the number of degrees of freedom. For hadronic matter, the number of degrees of freedom relevant at low density are the number of low mass hadrons. Since matter is confined at low density, the number of such degrees of freedom is $N_{dof} \sim 1$ in terms of the number of colors. At high energy density, the relevant number of degrees of freedom are those of unconfined quarks and gluons. The gluons dominate and give $N_{dof} \sim N_C^2$. Therefore in the large N limit, the number of degrees of freedom change by an infinite amount.

Assuming that the transition occurs at finite temperature in the large N_C limit, as is verified by Monte-Carlo simulation, this result can be interpreted in two ways. From the vantage point of a high density world of gluons, the asymptotic energy density is finite, but at low energy density at some finite temperature the energy density goes to zero. The energy density itself is therefore an order parameter for a phase transition, and there is a limiting lowest temperature. Viewed from the low density hadronic world, there is some limiting temperature where the energy density and entropy density become infinite. Here there is a Hagedorn limiting temperature.

For $N_C = 3$, the above statements are only approximate. The number of degrees of freedom of low mass mesons is

$$N_{dof} \sim N_F^2 \sim 4 \tag{4}$$

where we have taken the number of low mass quarks to be $N_F \sim 2$ for the up and down quarks. The number of degrees of freedom of a quark-gluon plasma is on the other hand

$$N_{dof} \sim 40 \tag{5}$$

The number of degrees of freedom might change in a narrow temperature range, or there might be a true phase transition where the degrees of freedom change by an order of magnitude, if our speculations concerning the large N_C limit are applicable.

Results of a Monte-Carlo simulation of the energy density are shown in Fig. 1.[1] These results are typical of the qualitative results arising from lattice Monte-Carlo simulation. The precise values of the energy density are difficult to estimate as is the

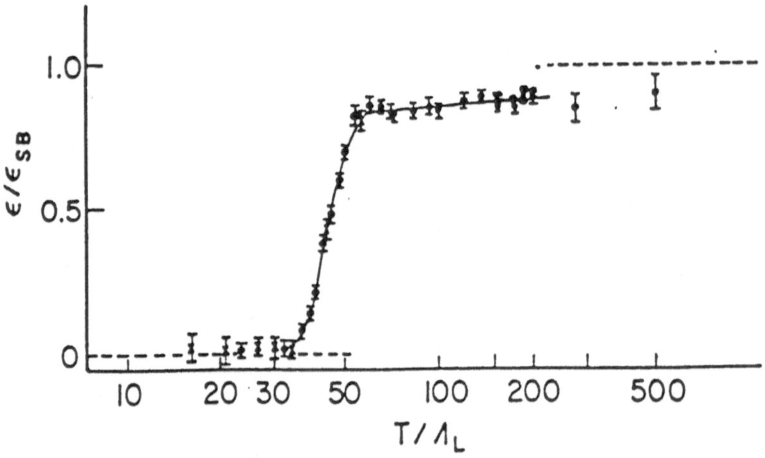

Figure 1

Energy density scaled by T^4 as a function of T.

Figure 2

Back to back lepton bears.

scale for the temperature. The figure does make clear the essential point, on which all Monte-Carlo simulations agree, that the number of degrees of freedom of hadronic matter changes by an order of magnitude in a narrowly defined range of temperature. There is apparently a first order phase transition for SU(3) Yang-Mills theory in the absence of fermions, and a rapid transition which may or may not be a first order transition for SU(3) Yang-Mills theory with two or three flavors of massless quarks.

2.3 Order Parameter for Confinement

For Yang-Mills theory in the absence of dynamical quarks, there is a local order parameter which probes the confinement or deconfinement of a system. This order parameter measures the exponential of the free energy difference between the thermal system with and without the presence of a single static test quark inserted as a probe,

$$< L > = e^{-\beta F_q} \tag{6}$$

As originally proposed by Polyakov and Susskind, and developed in Monte-Carlo studies, the Polyakov loop is a Wilson loop at the position of the quark which evolves only in time and is closed by virtue of the thermal boundary conditions which make the system have a finite extent in Euclidean time. The two phases of the theory are the confined and unconfined phases where

$$e^{-\beta F_q} \sim \; finite\, if\ confined, \quad or\ 0\, if\ deconfined \tag{7}$$

This quantity is an order parameter for a confinement-deconfinement in theories without fermions or in the large N_C limit in theories with fermions (in the fundamental representation of the gauge group). If there are fermions in the fundamental representation, in the 'confined phase' dynamical fermions may form a bound state with a heavy test quark, so the free energy is finite in what would be the confined phase. Since it is already finite in the deconfined phase, the free energy of a static test quark does not provide an order parameter.

Although $< L >$ is not an order parameter, Monte-Carlo simulations with dynamical fermions show that $< L >$ changes very rapidly in a narrow range of temperatures. For SU(3) lattice gauge theory without dynamical quarks, when $< L >$ is a true order parameter, there is a noticeable discontinuous change. It is not entirely clear whether there is a discontinuous change corresponding to a true phase change for the theory with fermions.

2.4 The Chiral Symmetry Order Parameter

In the limit of large dynamical quark mass the quarks are no longer important at any finite temperature and decouple. In this limit the confinement-deconfinement phase transitions is a well defined concept with an order parameter which measures a phase change. At zero quark masses there is another phase transition which may be carefully defined, that is, the chiral symmetry restoration phase transition. Chiral symmetry is a continuous global symmetry of the QCD lagrangian in the limit of zero quark mass. Its realization would require that all non-zero mass baryons have partners of degenerate mass and opposite parity. Since this is far from true for the spectrum of baryons observed in nature, chiral symmetry must be broken. Breaking the continuous global symmetry generates a massless Goldstone boson, which we identify with the light mass pion. As a consequence of the breaking of chiral symmetry, the quarks acquire dynamical masses, which may be seen by computing $< \overline{\Psi}\Psi >$. For the chiral symmetric phase, $< \overline{\Psi}\Psi > = 0$, and is non-zero in the broken phase.

For not unreasonable values of the quark masses, there appears to be a rapid change in $< \overline{\Psi}\Psi >$ at about the same place where the order parameter $< L >$ changes rapidly. We conclude therefore that chiral symmetry is approximately restored at the same temperature where quarks stop being approximately confined. The word approximately is important here since absolute confinement or absolute chiral symmetry is impossible for finite mass dynamical quarks.

2.5 Phase Diagrams

We can now conjecture on the phase diagram in the temperature mass plane. It is important to realize that we may physically vary the temperature, but not the masses of quarks. Theoretically in a Monte-Carlo simulation, these masses may be changed, but they cannot be changed in nature. It is also important to realize that the mass-temperature diagram represents an over simplification to the case of equal mass quarks. With different mass quarks, the diagram has more variables and is more complicated.

To plot this diagram, we first discuss the limiting case $m = \infty$. Here there should be a first order confinement-deconfinement phase transition along the T axis. Since a discontinuous change will not be removed by a large but finite quark mass, this first order phase change must be a line of transitions in the m-T.Along the $m = 0$ axis there is a chiral symmetry restoration transition. By the arguments of Pisarski and Wilczek, this transition is first order, and therefore must generate a line of transitions which extends into the m-T plane.

Of course, we do not know what happens with these two lines of transitions, whether they join or never meet, or pass through one another etc. There may be no true phase transition at the values of masses which are physically relevant, or there may be one or two which are the continuation of the chiral transition from zero mass and the confinement-deconfinement transition from infinite mass. The weight of the evidence from Monte-Carlo numerical simulation suggests a very large transition in the properties of matter in a very narrow temperature range, and not much more than that can be said at present. There are a variety of conflicting claims as to whether or not there is a true first order transition at physically relevant masses.[2]

There have been serious attempts to obtain reliable quantitative measures of the properties of matter from Monte-Carlo simulation. The only truly reliable numbers have been extracted for the unphysical case of $N_F = 0$, that is, no dynamical fermions. It has been shown that the critical temperature of the confinement-deconfinement transition is

$$T_C \; = \; 220 \; \pm \; 50 \; Mev \tag{8}$$

by fitting the potential computed in these theories and comparing it with the potential which fits charmonium. This corresponds to an energy density of $1 - 2 \; Gev/Fm^3$ required to make a quark-gluon plasma. These results now appear to be valid for the continuum limit, and seem to be fairly good.

The numerical situation for QCD with $N_F = 2 - 3$ is not nearly so good. The qualitative results have been summarized above, but it is premature to draw any firm conclusions about numbers.

3 The Quark-Gluon Plasma in Astrophysics and Cosmic Rays

3.1 Quark Matter and Cores of Neutron Stars

A possible place where quark matter might occur naturally is in the cores of neutron stars. Typical neutron star radii are in the range of 1-2 solar masses. The size of a neutron star is about 10-20 km in radius. The density of matter in the core of a neutron star might be as high as 10-20 times that of nuclear matter. Temperatures of neutron stars are however low, $T \sim 10 \; KeV$, compared to typical fermi momenta $k_f \sim .1 - 1 \; GeV$ The matter in neutron stars is therefore cold and dense and may to a good approximation be described as a degenerate fermi gas.

As one proceeds from the surface of a neutron star inward, the density increases monotonically. Near the surface there is a solid crust which rapidly turns into a solid composed of neutron rich nuclei. A little further in, the nuclear matter becomes compressed into a liquid composed primarily of neutrons with some small contamination of protons, electrons, and perhaps heavier unstable baryons. In the core of the neutron star, where the density reaches its maximum, it might be possible that the matter is in the quark phase.

It is of course very hard to estimate the density of transition from quark matter to nuclear matter, as we will soon see. There is also no Monte-Carlo data on the properties of zero temperature high density matter (for as yet insurmountable technical reasons). It is in my opinion impossible to know whether there are quark cores or neutron stars. Even if there are such cores, a variety of computations have shown that all bulk properties of the star are more or less the same except possibly for the cooling rate.

3.2 Strange Quark Matter

I will now turn to a discussion of strange quark matter[1].[4] I will begin the talk with a brief description of our current understanding of the equation of state of nuclear matter at densities near and above that of nuclear matter. I will then turn to a discussion of the properties of zero temperature quark matter, and argue, following Witten, that the ground state of ordinary matter may be strange quark matter. I then describe what the properties of such strange quark matter might be. I then turn to a discussion of some of the phenomenological consequences of the existence of strange quark matter as the ground state of matter. I describe strange quark matter in cosmology, and strange quark stars. The proposals that strange quark matter might explain the Centauro cosmic ray events and the mysterious properties of high energy radiation emanating from Hercules X-1 and Cygnus X-3 are also critically addressed.

3.3 Do We Understand the Equation of State?

In Figure 3, I plot the energy per baryon for ordinary nuclear matter. Since I am not by training a nuclear physicist, I must apologize since I have not drawn a very exact figure. The general shape of the figure is clear, however. At nuclear matter density, marked as ρ_o on the figure, there is a local minimum in the energy density, corresponding to a binding energy per nucleon of about 16 Mev. Beyond that, little is known. At several times nuclear matter energy density, we expect that the typical scale of the energy per baryon is of the order of hundreds of Mev.

The basic point I want to make about our uncertainty concerning the properties of nuclear matter is made in Fig. 4. On this figure, I have drawn the energy per baryon of nuclear matter for nuclear matter computations, and the corresponding quantity for quark matter computations. The way these curves are drawn looks a little silly for a nuclear physicist. I have included the rest mass of the nucleon in the energy per baryon. For those of you who do nuclear matter computations, this may seem a little perverse, since everyone knows the mass of the nucleon, and it never enters directly into the computation of the energy per nucleon except trivially in this way and setting the scale for kinetic energies. In the quark matter computations

[1]There will be no discussion of lepton bears in this talk. See Fig. 2.

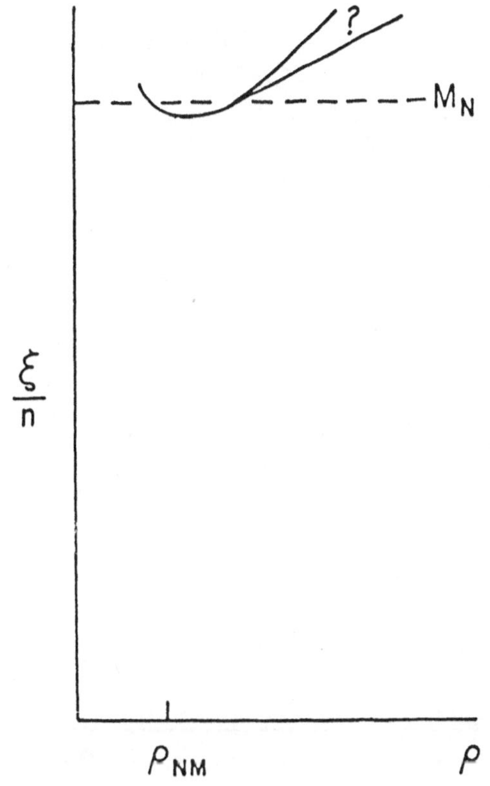

Figure 3

The energy per baryon for ordinary nuclear matter.

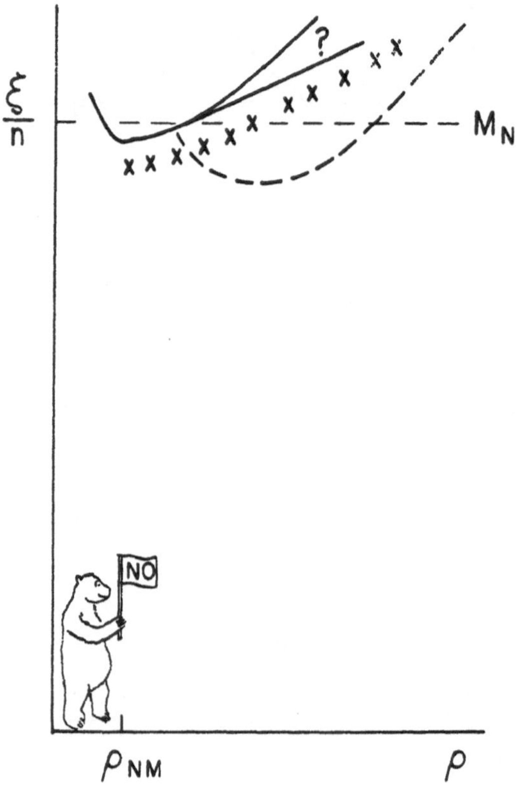

Figure 4

Hypothetical energy per baryon for quark matter superimposed on the equation of state for nuclear matter. Notice the change of scale.

however this is an output, not an input. The quarks are almost massless at the level of the QCD equations of motion, and of course the nucleon mass must be computed dynamically.

There are two essential features of Figure 4. The first is that if there are merely ten percent uncertainties in the computation of the properties of quark matter, the uncertainty is as large as the computed effects of nucleon interactions and nucleon kinetic energy. In fact nobody in their right mind would claim that they could compute the properties of quark matter to an accuracy of ten percent at near nuclear matter energies. The second feature to note is that even if there was a reliable computation of the properties of quark matter, determining the transition density from nuclear matter to quark matter would be extremely difficult since we are comparing two flat curves. A small uncertainty in the computation of the energy per baryon corresponds to enormous uncertainty as to what baryon number density the curves cross. It is therefore very difficult to determine at what density quark matter is the stable ground state of ordinary nuclear matter. Given the uncertainties mentioned above, it is clear that such a baryon number density could be close to that of ordinary nuclear matter.

It is also not clear that the comparison of such curves makes any sense at all if there is anything other than a first order phase transition between nuclear matter and quark matter. If there is no transition or a 2nd or higher order transition, the curves should be computed in one theory for all densities. The fact that this cannot be done reflects the breakdown of various approximation schemes used to compute the curves. The region where the approximations are of course at the point of maximum ignorance, that is, where the curves cross, if they cross. It may turn out that the within the approximations used, the curves never cross, in which case nothing is lost or gained.

Since we are ignorant of the properties of nuclear matter, almost anything can happen. However, unless there is a phase of matter which is not continuously joined to ordinary matter by a smooth variation of the parameters of ordinary nuclear matter, then an economist's knowledge of nuclear matter is probably sufficient near nuclear matter density. Economists predict the future of the economy based on a projection from what is happening today. They use a set of variables, and knowing what happened today and yesterday, can extrapolate into the future. This is what is done in nuclear matter when one knows the binding energy of nuclear matter and the compressibility (the second derivative of the energy per baryon at the minimum of the energy per baryon as a function of density). One can then extrapolate near nuclear matter density.

Of course as we all know, predicting what happens in the economy is difficult to do. For example, the economy may be doing well, and then out of the blue, the stock market takes a dive. In this case, the connection between one day and the next is in some sense not continuous. We shall see that for nuclear matter, there may in fact be a state of matter of lower energy per baryon than ordinary nuclear matter, and this is the stable ground state of nuclear matter. It is not continuously connected to ordinary nuclear matter.

3.4 Properties of Ordinary $T = 0$ Quark Matter

Before discussing the properties of a possible new absolutely stable form of matter, it is first useful to recall a few facts about the properties of ordinary zero temperature quark matter. We should mention that little is known about the properties of quark matter at finite density from non-perturbative computations. For technical reasons which are beyond the scope of these lectures, it has not been possible to perform Monte-Carlo simulations of zero temperature finite density quark matter. The only things which are known are from perturbative studies. These perturbative analysis are in a little better shape than is the case for finite temperature quark-gluon plasma. This is because unlike the case of a plasma, perturbation theory has not been shown to break down at some order due to infrared divergences.

Perturbative analysis always suffers from the disease that it is a weak coupling expansion. All the interesting dynamics however happens when the coupling constant is big, and this is where we cannot perform computations. In particular, if we talk about the properties of quark matter at nuclear matter densities, then surely we are in the non-perturbative region.

Some of the qualitative features extracted from a perturbative analysis must surely have some correspondence in the full non-perturbative theory. For example, the strange quark mass in the QCD Lagrangian is small $M_s \sim 250\ MeV$. At nuclear matter density, typical fermi momenta are large, of order $K_f \geq m_s$. Since we expect that the up and down quarks have a small mass, this fermi momenta is also of the order of the fermi energy.

If the fermi energies are larger than the strange quark mass, then surely the lowest energy configuration of quark matter has converted some of its quarks to strange quarks. This is just because if all the quarks were up and down, they would have a higher energy per quark than would be the case if some of the up and down quarks were converted to strange quarks.

Explicit computations of the perturbative properties of quark matter have been done. At all densities, the strangeness to baryon number density is of order one, even when extrapolated to densities as low as nuclear matter energy density. There is much more strangeness than one might expect from Pauli exclusion principle type arguments. The qualitative conclusions result of detailed and extensive computation of the strangeness to baryon number charge as a function of density is shown in Fig. 5.

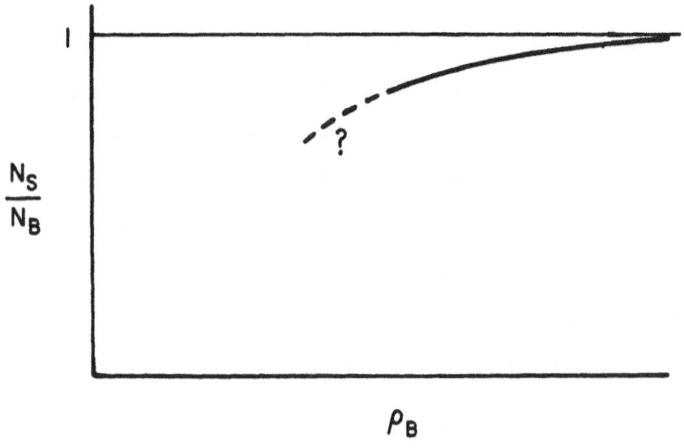

Figure 5

The strangeness to baryon number ratio as a function of baryon number density.

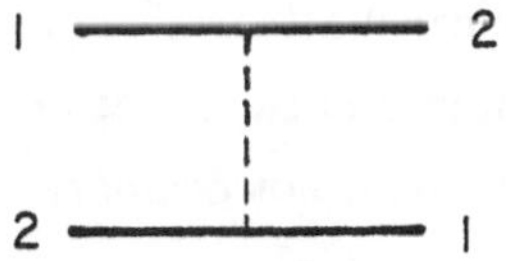

Figure 6

The Feynman diagram which generates the exchange energy.

Although the computation is not reliable at low densities, why is it that there is such a large amount of strangeness? This is a consequence of the exchange energy. The exchange energy is generated by two identical species of quarks exchanging a gluon and changing places in the Fermi sea, as shown in Figure 6. At low densities, the coupling constant becomes big, and this diagram becomes more important. (Of course, we are ignoring contributions from higher orders in perturbation theory which might change this result.)

The basic effect of the exchange energy is that when the Fermi energy is large compared to the mass of a quark, it generates a repulsive contribution to the energy per baryon. This will be the case for the light mass up and down quarks. When the Fermi energy is of the order of the quark mass, the interaction changes sign and is attractive. Even up to ten or so times nuclear matter density, the exchange energy computed in perturbation theory is attractive if the strange quark mass has an energy of $m_s \sim 250\ MeV$.

It is therefore energetically favorable for the up and down quarks with repulsive interactions to convert into strange quarks with attractive interactions. (See Fig. 7)

3.5 Properties of Strangelets

The possibility that strange quark matter might be absolutely stable relative to that of ordinary nuclear matter is originally due to Witten. There were some early arguments that strange quark matter might perhaps be metastable, and in particular be responsible for Centauro cosmic ray events. We will discuss Centauro cosmic ray events in a later section.

The basic idea is that at the minimum energy per baryon, E/N for quark matter is less than that for ordinary nuclear matter. The ground state of baryonic matter is therefore quark matter, not ordinary nuclear matter. At first sight this appears to be a ridiculous proposition. Why would not ordinary nuclear matter spontaneously decay into chunks of strange quark matter?

To see whether this is a real problem or not, we must compute the time it takes for nuclei to decay into strange quark droplets. If we did not have to worry about the weak interactions, then surely the rate would be very rapid, and we could immediately rule out the hypothesis. Also, if for example the deuteron was unstable relative to a baryon number two strange quark drop, the H particle of Jaffe, the decay for this would also be rapid. If however for finite size effects, the ordinary nuclei are absolutely stable up to some minimum baryon number A_{min}, then the only way the matter can convert is by making A_{min} simultaneous weak decays. The probability of this occurring is proportional to $(E^4 G_F^2)^{A_{min}}$. For E a typical nuclear energy scale, and G_F the Fermi constant, we see that the lifetime of a nucleus is stable for $t \geq 10^{60}\ yr$ if $A_{min} \geq 5$.

Is it to be expected that this might be the case? In fact the MIT bag model successfully fits the properties of low energy hadrons, and therefore for small A, it must be true that ordinary nuclei are stable. One can compute properties of strange quark matter droplets in this model, and for parameters which can fit the spectroscopy of hadrons, one finds that it is possible that strange quark droplets may be absolutely stable. In the bag model computations, the value of $A_{min} \sim 10 - 1000$ depending upon the parameters used.

Of course the bag model computations are no doubt unreliable. We are dealing with non-perturbative physics, and estimating small differences in binding energies. Whether or not strangelets exists is an issue which is probably outside the predictive ability of the bag model. It is nevertheless comforting that they do exist in the bag model for some range of parameters. Now that we have allowed for the existence of stable strange quark droplets, we must determine some of their properties. First consider the charge to mass ratio Q/A. For ordinary nuclei, the Fermi exclusion principle requires that the number of protons be approximately that of the number of neutrons. (There are deviations of this for large A due to the Coulomb energy.) Therefore, the charge to mass ratio of ordinary nuclei is $Q/A \sim 1/2$.

Strangelets on the other hand have roughly equal numbers of up, down and strange quarks. The baryon number therefore

$$A = \frac{1}{3}N_u + \frac{1}{3}N_d + \frac{1}{3}N_s \sim N_u \tag{9}$$

and the electromagnetic charge is

$$Q = \frac{2}{3}N_u - \frac{1}{3}N_d - \frac{1}{3}N_s \sim 0 \tag{10}$$

The charge to mass ratio is small and of order zero.

We can now show that strangelets can become as large as a bear, as shown in Fig. 8. To see this, we must first understand why ordinary nuclei never become as big as bears.

When ordinary nuclei become large, their charge grows. Remember that the Coulomb energy grows roughly as

$$E_{coul} \sim Q^2/R \sim A^{5/3} \tag{11}$$

where I have assumed that the number of protons and neutrons is roughly equal. For some sufficiently large A, the Coulomb repulsion makes nuclei unstable with respect to fission. If the nucleus falls apart into two equal halves, the net Coulomb energy is reduced. This is the reason that there is a limiting size for nuclei.

Strangelets on the other hand have a small charge for a given A compared to what would be the case for nuclei. Strangelets become very large with little accumulation of net charge. If the strangelet becomes so large that its size is bigger than the Bohr

Figure 8

Strangelet as big as a bear.

Figure 7

A strange bear arising from the cold Fermi sea.

orbit of a bound electron before it has a fission instability, then it never fissions (Fig. 9). This is so because the electrons are now part of the strangelet, and the system neutralizes itself. The strangelet can grow in size forever. This is precisely what happens in the MIT bag model computations.

We therefore have shown that we can make objects of macroscopic size, with a density of order that of nuclear matter. A clear experimental signature for such objects is that they have an anomolously low charge to mass ratio.

Suppose there are strangelets, and that some small number of them exist in nature. If there was one of them in the sun, why wouldn't it eat up the sun? To understand this, we first must know that there is a non-zero chemical potential for electromagnetic charge in the strangelet. This follows from minimizing the charge of the strangelet. This implies that there will be an attractive electromagnetic interaction as an electron is brought in from infinity. This in turn corresponds to a repulsive electromagnetic barrier for positive charge. Estimates of this chemical potential suggest a Coulomb barrier of order of tens of MeV's.

The dynamics which produces this Coulomb barrier is complicated. What actually happen is that at the surface of the strangelet, the electrons are not well localized within the surface of the strangelet. Recall that the surface thickness of the strangelet for quarks is of order a Fermi. The electrons on the other hand cannot be localized to less than their Compton wavelength. A dipole charge layer therefore develops, and the magnitude of the potential energy change is precisely what is given by the chemical potential.

At low temperatures, $T \ll 10 MeV$, there is little probability that the strangelet will eat charged matter. This is the reason that strangelets do not eat up the sun. Strangelets do not believe in cold fusion (Fig. 10).

Another consequence of the repulsive Coulomb barrier is that the strangelet will electrostatically support matter on its surface. This is potentially important in astrophysics where one might support crusts of ordinary nuclear matter on the surface of a strange quark star.

3.6 Strangelets and Cosmology

One of the original motivations for introducing strangelets was an attempt to solve the dark matter problem. In a nutshell, the dark matter problem is that it appears that there is too little luminous matter in our universe either to account for experimental observation of gravitational effects in clusters of galaxies or the theoretical predispositions of those who either make their living from nucleosynthesis computations or inflationary models of cosmology. Strangelet matter is dark since it is very massive, and at the same time its atomic electrons are bound inside the strangelet. It does not radiate like a hydrogen atom.

Figure 9
Bear fission.

Figure 10
A strangelet trying not to believe in cold fusion.

Witten originally had a scenario for strangelet genesis. To understand how this works, assume that there is a first order chiral symmetry restoration phase transition. By chiral symmetry restoration transition, we mean that in the high temperature quark-gluon plasma phase, the up and down quarks have small mass. In the low temperature hadron phase, the baryon number is bound into nucleons which have a large masses, and typical meson masses such as the ρ are also substantial. (In the low temperature hadron phase the pion has a small mass since it is a Goldstone boson of a dynamically broken chiral symmetry.)

Now as the universe expands, it cools down to the phase transition temperature. At this temperature, the hadron gas and the quark gluon plasma may co-exist together. The quark-gluon plasma has many more degrees of freedom in it than does hadronic matter, and is therefore much more dense. Therefore as the universe expands, the temperature remains fixed, but a larger and larger fraction of the matter becomes associated with the hadronic gas. At some time, the density is so low that all of the matter which can be converted into hadron gas has done so, and then the expansion resume in only the hadron gas phase with the temperature decreasing as a function of time.

It is useful to follow the evolution of the expansion through the region of phase coexistence. We will see that large scale density inhomogeneities result from this expansion. In Fig. 11a, the plasma begins by nucleating isolated centers of hadronic matter. As time evolves, these nucleation centers collide and become larger, making large scale density fluctuations. To get an idea of the time scales involved, the characteristic expansion time at a Gev is about 10^{-5} *sec.* A typical particle collision time is 10^{-23} *sec.* In an ordinary liquid, the collision time is about 10^{-13}*sec* The dynamics of the expansion of the universe at the time is therfore analogous to watching a liquid expand over a time span of about a month. We would expect that the size scale of density fluctuations would indeed be large.

Detailed estimates of the size of the density fluctuations do indeed confirm this expectation. Density fluctuations with a size as large as perhaps 10^{-5} of the size of the universe have been predicted.

As the volume of the system increases, at some point there is more hadron gas than plasma. At some energy density, it is more useful to think about droplets of plasma embedded in a hadron gas, as shown in Fig. 10b. Of course as the energy density is lowered even further, then the hadron gas evaporates away.

Is there any trace left over of these large scale density inhomogeneities? In fact, there is in the net baryon number. Recall that in cosmology there is only a small excess of baryon compared to the entropy of the universe. Most of the entropy of the universe is stored in 3° black body photons. The ratio of the net number of baryons to the total entropy is of order $N_B/S \sim 10^{-10}$.

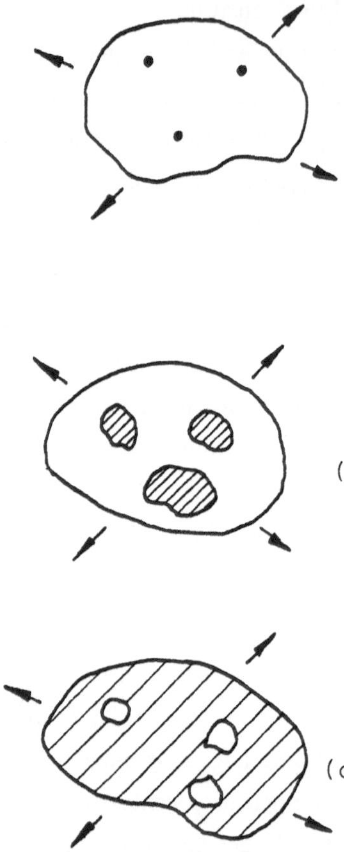

Figure 11

a)Quark matter when nucleation centers of ordinary nuclear matter first form.

b)Quark matter with large islands of ordinary nuclear matter embedded in it.

c) Ordinary nuclear matter with large islands of quark matter in it.

Figure 12

A strangelet which did not survive the big bang.

A difference between a chirally symmetric plasma and a chiral symmetric hadron gas is that the baryons are more massive in the hadron gas than in the plasma. In the plasma, to have excess baryon number only costs a small quark mass, and is weakly suppressed by the Boltzman factor. In the hadron gas, the baryons have a mass of about 1 Gev, and since the transition is expected to occur at a low temperature $T \sim 100 - 200\ MeV$, baryon number will much more preferentially reside in the quark-gluon plasma. For various assumptions about the transition temperature, the baryon number density contrast,

$$N_B^{plasma}/N_B^{hadron} \sim e^{M_B/T} \sim 10 - 100 \tag{12}$$

The phase transition therefore acts like a vacuum cleaner, and sweeps up the excess baryon number into localized regions of space.

It of course takes a good deal of time for the baryon number to diffuse back to a uniform distribution. Recent speculations have suggested that this density inhomogeneity might even persist until as low a temperature as that at which nucleosynthesis occurs, $T \sim 1MeV$, and affect results of computations of element abundances.[4].

Our interest here is whether or not one might be able to use these density fluctuations to generate strangelets, as originally suggested by Witten. The problem is that strangelets have a higher energy density than the surrounding matter, and will tend to evaporate away by neutron emission. Detailed simulations have been performed of the emission process, and the result is that the strangelets evaporate.[4] It does take quite a long time for the strangelets to evaporate, however, and the result is not at all obvious. Strangelets are therefore not produced in the big bang, and it does not seem possible that they can explain the dark matter problem. (Fig. 12)

3.7 Strange Matter and Strange Stars

An interesting possibility is that neutron stars might be strange quark matter stars. This might happen if in the core of a neutron star, or in the supernova explosion which produced the neutron star, the core became unstable with respect to the formation of ordinary quark matter. In this case, the ordinary quark matter could decay into the energetically preferred strange quark matter by single beta decays, and would happen a an unsuppressed rate. The strange quark matter could then expand outwards, eating the neutrons in the neutron matter. Near the surface of the star, the strange quark matter encounters ordinary nuclear matter, and this nuclear matter contains charged nuclei. If the burning was proceeding rapidly through the neutron rich matter, then the heat produced by the burning front will evaporate neutrons from the nuclei, more heat will be generated, more neutrons evaporated etc, and the burning will continue to the surface.

It is possible to estimate the burning rate, and conclude that it is very rapid.[4]. Consider the burning front shown in Fig. 13. On the far right is ordinary nuclear matter. This is absorbed into the leading edge of the burning front. It is transferred to the trailing edge of the burning front, where strangeness is in equilibrium. In the transfer, charged nuclear matter is converted into strange quark matter by weak decays. We shall assume that there is always sufficient heat being generated so that there are enough neutrons being generated to feed the burning front.

To compute the speed of the burning front involves a few simple estimates. If we call the width of the region between the leading and trailing edges of the burning front X, then

$$X = V\tau \tag{13}$$

where V is the speed of the burning front and τ is the time to establish strangeness equilibrium. On the other hand, the equation of motion $F = Ma$ further constrains the burning. Here, the force on a particle is the drop in potential across the burning front divided by the width of the burning front. Drift velocity arguments suggest the Ma is the average momenta the particles have between collisions MV divided by the typical time between collisions λ / V_q, where λ is the scattering mean free path, and v_q is the typical particle velocity.

We can now solve for V, with the result

$$V = \left(\frac{\Delta\phi}{P_f}\right)^{1/2} \left(\frac{\lambda}{\tau}\right)^{1/2} \sim \frac{P_f}{T}\ cm/sec \sim 1\ Meter/sec \tag{14}$$

Here we have used $\Delta\phi \sim 10 Mev$, $P_f \sim 300 MeV$, τ a typical weak interaction time scale, and λ the Pauli blocked collision cross sections. It is remarkable that the velocity in units of Meters/sec is of order one! We therefore conclude that the burning time is very fast, and for a neutron star might take place in a few hours.

We should ask whether the surfaces of quark stars can support a crust. Certainly in a static situation this is true, or if the burning is not sufficiently rapid to go all the way to the surface of the star. The situation in accretion is not at all clear, in the authors opinion. The problem is that the infall of matter in accretion can be quite energetic. The typical gravitational potential energy is after all of order 100 MeV. There can also be fluctuations, and regions of very energetic accretion at isolated places on the star. If the infalling matter is sufficiently energetic so than locally many neutrons are released, then a strange quark matter surface may be able to always keep its surface clean by eating the neutrons, and burning away the remaining charged matter. In any case, the surfaces and accretion are not simple, and it is not clear what will happen.

A distinguishing feature of strange quark stars is that they can have low masses. This is in contrast to ordinary neutron stars, and as well the relation between mass and radius is quit different, as is seen in Fig. 14.[4] The problem is that it may be impossible to dynamically produce small mass neutron stars in supernova explosions.

Another possible signal for strange-quark stars is the cooling rate. In Fig. 15, cooling rates for strange quark matter, ordinary neutron stars, neutron stars with quark matter cores, and with pion condensates are shown[4]. The strange quark matter calculations are very uncertain, and there is therefore much more flexibility in these models. A better and more precise cooling rate computation is required to draw conclusions.

3.8 Are Centauro Cosmics Ray Events Due to Strangelets?

Droplets of strange quark matter in a metastable state were originally proposed as an explanation of Centauro cosmic ray event. They may also be understood as arising from strangelets.

Centauro events are seen in mountain top emulsion chamber experiments. The typical energy of Centauro events is of order $E \sim 10^3\ TeV$. A typical particle multiplicity is 50-100 particles. The typical transverse momentum, which is poorly determined, seems to be larger than that for a typical event at the same energy. Estimates give about 1 Gev per particle.

The striking feature of these events is that there seem to be no photons produced in the primary interaction which makes the Centauro. This is unusual because in high energy collisions, π^o mesons are always produced, and they decay into photons. Typically about 1/3 of the total energy is released in photons.

The Centauro event is much like a nuclear fragmentation. If a nucleus were to fragment, then there would be many nucleons, and if the interaction which produced the fragmentation was sufficiently peripheral, there would be few π's. This hypothesis is ruled out because the typical p_t is so large, and more important because a nucleus would not survive to such a great depth in the atmosphere.

A strangelet seems to explain many of the features however. As has been discussed by C. Greiner et. al. it might be possible at high energy to make metastable strangelets in nuclear collisions with reasonable size cross section. Moreover, the typical ionization energy for a strangelet in a hadronic interaction is large, of order 70 MeV. This is because although relative to ordinary nuclear matter, the strangelet is bound (or if metastable unbound) by 10's of MeV's, in a hadronic interaction, the strangelet ionizes into nucleons and lambdas. The lambda has a much large mass than that of nucleon.

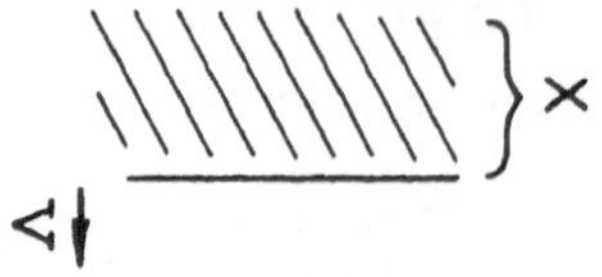

Figure 13

The burning front between ordinary nuclear matter and strange quark matter. The ordinary nuclear matter is on the left, and the strange quark matter is on the left. In the region from the beginning of the burning front, marked with the solid line, to the end of the burning front, marked with the dashed line, strangeness is coming into equilibrium.

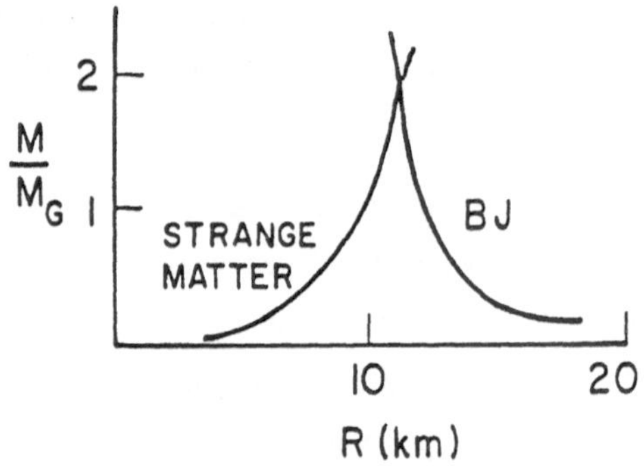

Figure 14

Masses of stars verses radius for strange quark stars, and in the Bethe-Johnson nuclear matter model.

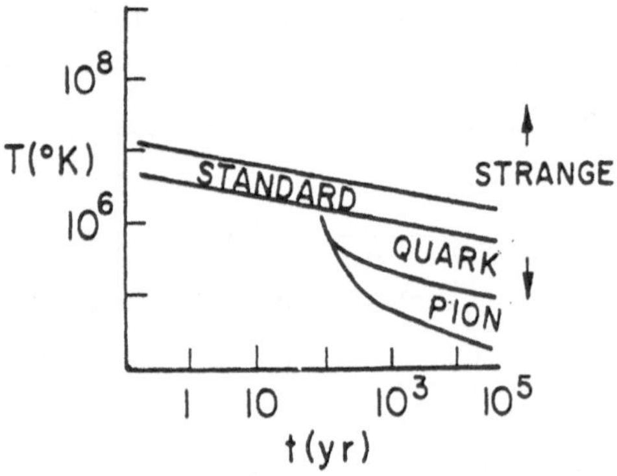

Figure 15

The cooling rates of strange quark stars and neutron stars in various models.

The strangelet is therefore much more tightly held together than an ordinary nucleus. It is conceivable therefore that a strangelet incident upon the top of the atmosphere or produced at the top of the atmosphere could survive to mountain altitude. It may have lost a significant amount of baryon number before getting to this depth however. A peripheral interaction might be sufficient to unbind it, since it certainly will not be so tightly bound with reduced baryon number.

The problem with this explanation is that it does not explain the high transverse momenta. At such large $p_t \sim 1\ GeV$, we would surely expect final state interactions to generate some pions, and therefore an electromagnetic component.

3.9 Cygnus X-3 and Hercules X-1

There have been some recent and baffling observations of high energy gamma ray sources. The typical measurements range in energy from $E \sim 10^3\ TeV$ for air shower experiments on Hercules X-1. For Cygnus X-3, there are underground experiments at about $10\ TeV$, air shower experiments at about $10^3\ TeV$, and observation by the Flys eye at $10^5\ TeV$. There is strong evidence in the air shower experiments, and evidence in the underground experiments that the showers coming from Cygnus X-3 and Hercules X-1 have a muon content typical of cosmic rays.

This is surprising result! Typical cosmic rays at this energy are protons. To arrive from the direction of Cygnus X-1 or Hercules X-1, the particle must be neutral. To survive the trip, it must be long lived. The only known particle which can quantitatively satisfy this constraint is the photon. The neutron does not have a long enough lifetime.

Photon air showers have only a small muon component. Photons shower by producing electron pairs. The probability of pair producing or photo-producing muons is very small $\sim m_e^2/m_\mu^2$. In proton events, the muon content is high because charged pions decay into muons.

When only the underground data was known, it was proposed that the source of the radiation was H particles, the neutral strangeness 2 di-baryon proposed by Jaffe. To make the H long lived enough, it is necessary to make the H have a mass below single weak decay stability. This is however probably not sufficient to give it a long enough lifetime. To generate a large enough flux of H particles, the source is assumed to be a strange quark star.

Moreover, studies of Hercules X-1 rule out this hypothesis. Studies of the correlation in arrival time with the known period of Hercules X-1 give an upper limit of the particle mass of $m \leq 100\ MeV$. The source of radiation must either be either due to anomalous interactions of photons or neutrinos, or from some exotic as yet undiscovered light mass, almost stable particle. (Figure 16).

4 Space-Time Picture of Ultra-relativistic Nuclear Collisions

Ultra-relativistic nuclear collisions provide the possibility for making a quark-gluon plasma in a controlled laboratory environment.[3] If such a plasma is made in these collisions, there remains the enormously difficult problem to measure its properties. In this lecture, I will attempt to describe our current understanding of the space-time evolution of matter produced in such collisions. I will not address the interesting and complicated issue of what are the signals for the production of a quark-gluon plasma.

There are several time scales which characterize the dynamics of ultra-relativistic nuclear collisions:

- t_{coh}: Time to form incoherent distribution of particles from the highly coherent nuclear wave function
- t_{for}: Time to make an almost adiabatic hydrodynamically expanding quark-gluon plasma
- t_{pl}: Time at which the hadron phase first begins to form from the quark-gluon plasma
- t_h: Time at which all the quark-gluon plasma has converted into hadronic matter
- t_{fo}: Time when the pions in the expanding matter will on the average never in their future re-scatter

In Fig. 17, a plot of these various times is given. The scale on the figure is qualitatively correct. The longest time is by far the time the quark-gluon plasma coexists with the hadron matter. (We are assuming there is a first order phase transition with a change in the number of degrees of freedom by an order of magnitude.) The time after the matter entirely hadronizes until it freezes out is typically short compared to the lifetime of the mixed phase. The times t_{coh} and t_{for} are both small.

The physics which occurs in within these various times is

- $0 \leq t \leq t_{coh}$ The quantum mechanics of the nuclear wavefunction is important. Pair production of quarks and gluons is taking place rapidly.
- $t_{coh} \leq t \leq t_{for}$ The matter thermalizes according to transport theory. Space-time gradients in the system are over a size scale large compared to a particle Compton wavelength. For time scales larger than t_{coh} and less than t_{fo} the multiplicity density per unit rapidity dN/dy is approximately conserved. (Numerical simulations indicate there is little change in dN/dy at freeze out also.)

Figure 16

Strangelets headed in the direction of Mount Chacultaya where Centauro events are seen. Mu bears pointing in the direction of Cygnus X-3

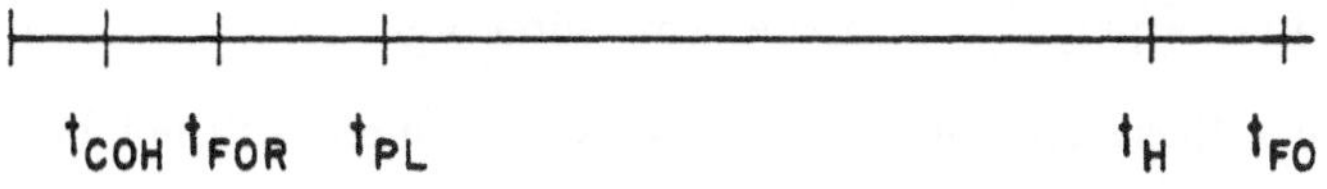

Figure 17

The various time scales for ultra-relativistic nuclear collisions.

- $t_{for} \leq t \leq t_{pl}$ To a good approximation the matter is an adiabatically expanding quark-gluon plasma.

- $t_{pl} \leq t \leq t_h$ The matter expands as a mixed phase of hadron gas and quark-gluon plasma.

- $t_h \leq t \leq t_{fo}$ The matter expands adiabatically as a hadron gas.

- $t_{fo} \leq t$ The matter has frozen out and is expanding as a free distribution of pions.

In this idealized description of time scales, we have assumed that there are well defined times scales for the above regions. Of course in any real collisions, these time scales are spread out and fuzzy.

The time before the nuclear wavefunction yields an incoherent distribution of quarks and gluons, we shall treat as a black box. The dynamics before the coherence time is complicated, and not understood. Therefore at the coherence time, we shall assume that we are given some initial distribution of quarks and gluons. The distribution can be parameterized as

$$\frac{dN}{d^3p/2E} = F(x)g(p_t) \tag{15}$$

where x is the Feynman x variable

$$x = E/E_{CM} \tag{16}$$

which may be expressed in terms of the rapidity

$$y = \frac{1}{2}ln\left(\frac{E+p_z}{E-p_z}\right) \tag{17}$$

and where E_{CM} is the center of mass energy per nucleon in the colliding nuclei, and p_z is the momenta of a particle along the beam direction.

We expect that these distributions of quarks and gluons will not be greatly different in functional from from those directly measure in for example Drell-Yan pair production from nuclei. We expect that

$$g(p_t) \sim e^{-m_t/\lambda} \tag{18}$$

where $\lambda \sim 150\ MeV$, and where the integral over p_t is normalized to 1. The function $F(x)$ is normalized to

$$\int_{x_{min}}^{1} \frac{dx}{x}F(x) = <n> \tag{19}$$

where $<n>$ is the particle multiplicity and $x_{min} = M/E_{CM}$ We should expect that F vanishes as some power of 1-x for x near one and is singular like some power of $ln(x)$ or some small power of x near 0.

Notice that in these variables $y \sim ln(1/x)$, so that in terms of rapidity a finite range corresponds to an infinitesimally small range of x near 0.

What is the value of t_{coh}? At a minimum, we expect that quantum mechanics will be important whenever $\Delta t, \Delta x, \geq 1/p$ where p is a measure of the typical momenta of a particle. Since particles are produced with a more or less uniform distribution in rapidity in high energy collisions, the range of momenta is enormous. To measure time scales, we will therefore do so in a frame which co-moves along the beam axis with the particles of interest. The p which is relevant is the typical scale of momenta fluctuations in this comoving frame, which we shall take to be p_t

In Fig. 18, a picture of two ultra-relativistic nuclei colliding in the center of mass frame is shown. The two nuclei have been shown as Lorentz contracted pancakes. The time after the collision is taken to be about 1 fm/c, which is of the order of magnitude of the coherence time. It might be as much as a factor of 10 smaller, but that will not concern us here. Notice that the nuclei have passed through one another. They have become transparent at high energies.

Transparency here does not mean that they pass through one another and nothing happens however. As the energy increases from very low energies, the total number of produced particles and total amount of energy produced transverse to the beam direction monotonically increases. Transparency means here two things. The first is the trivial fact that at high energy any two objects of finite thickness at sufficiently high energy will pass through one another. The second is that the fast moving particles in the center of mass frame carry the quantum numbers of the fragmenting nuclei (the fragmentation regions). The width of the fragmentation measured in rapidity is several units wide.

Some time after the collision, the collision looks as shown in Fig. 19 At this time, the matter which is slow in the center of mass frame has had time to expand and cool. We can see the fast moving matter has not yet had a time to form. A time which is comoving with the produced particles is the relevant time scale for the time evolution of the matter. This time is Lorentz contracted for the fast moving particles:

$$t \sim \gamma/m_t \sim p_z/m_t^2. \tag{20}$$

where $m_t = \sqrt{m^2 + p_t^2}$ is the transverse mass of a produced particle, and γ is the Lorentz gamma factor for the particles of interest. This picture where the slow particles are produced first and the slow particles last is called the inside-outside cascade, and is a very general feature of field theory.

We can now crudely estimate the size of the fragmentation region. Recall that the projectile nuclei have been Lorentz contracted to a size scale of order $\Delta z \leq 2R/\gamma_{CM}$, where γ_{CM} is the Lorentz gamma factor for the center of mass. Now if the matter is concentrated in such a small region of longitudinal phase space, it must also be

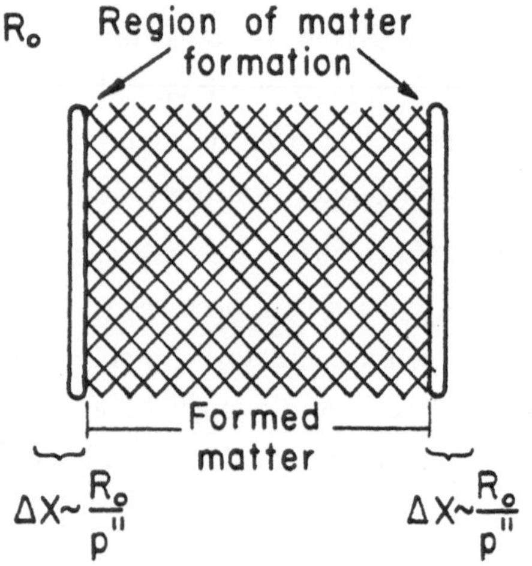

Figure 18

The collision of two ultra-relativistic nuclei in the center of mass frame. The time is about 1 fm/c after the collisions.

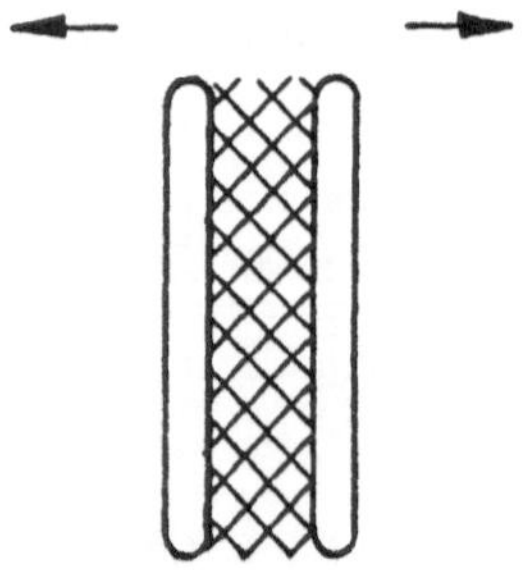

Figure 19

The collision between two nuclei in the center of mass frame at a time much larger than 1 fm/c.

true that the typical spread in longitudinal momenta is $p_z \sim 1/\Delta z$ Therefore, the nuclear fragmentation region is concentrated where $p_z \geq \gamma_{CM}/2R$. The width of the nuclear fragmentation region is therefore of order $\Delta Y_{frag} \sim ln(2Rm_t)$. Using realistic numbers shows that in order to have a well defined central region in high energy collisions requires a center of mass energy per beam of order $E_{CM} \sim 10 - 100\ GeV$.

To analyze ultra-relativistic nuclear collisions, it is useful to introduce variables which have simple transformation properties under Lorentz transformations along the beam direction. We introduce a proper time τ and a space-time rapidity η:

$$\tau = \sqrt{t^2 - z^2} \tag{21}$$

and

$$\eta = \frac{1}{2} ln \left(\frac{t+z}{t-z} \right) \tag{22}$$

Under Lorentz transformations along the beam direction, we find $\tau \rightarrow \tau$ and $\eta \rightarrow \eta + c$. Notice that for free streaming particles, space-time and momentum space rapidities are equal:

$$\begin{aligned}
y &= \frac{1}{2} ln \left(\frac{E+p_z}{E-p_z} \right) && (23) \\
&= \frac{1}{2} ln \left(\frac{1+p_z/E}{1-p_z/E} \right) && (24) \\
&= \frac{1}{2} ln \left(\frac{1+v_z}{1-v_z} \right) && (25) \\
&= \frac{1}{2} ln \left(\frac{1+z/t}{1-z/t} \right) && (26) \\
&= \frac{1}{2} ln \left(\frac{t+z}{t-z} \right) && (27) \\
&= \eta && (28)
\end{aligned}$$

In these variables, we also obviously have that the surfaces of equal τ correspond to equal times as measured in comoving frames.

A space-time picture of ultra-relativistic nuclear collisions is shown in Fig. 20. Here the physics changes on equal τ surfaces, corresponding to hyperbola in the t-z plane.

4.1 Energy Density in the Central Region

We can now use our crude space-time picture to estimate energy densities which might be achieved in nuclear collisions. First let us understand the central region. We will take t to be the time after the collision. The time t can at smallest be taken as a coherence time before the dynamics is complicated by the quantum mechanics of the nuclear wave function, $t_{coh} \sim 1/p_t$ where p_t is the typical transverse momenta of particles at the coherence time. We expect that the average transverse momentum per particle will decrease as the system expands and cools, so that the $p_t \geq p_t^{final} \sim 500\ MeV$. Therefore the coherence time is of the order of $t_{coh} \sim .5\ Fm/c$ which is close to the other early time scales.

The energy density in the region between the two nuclei is the number of particles per unit length, divided by the area of the nuclei, times the typical energy per particle which we take to be p_t. We therefore have

$$\epsilon = \frac{1}{\pi R^2}\, p_t\, \frac{dN}{dz} \tag{29}$$

We can now convert z into space time rapidity since

$$d\eta/dz = 1/t \tag{30}$$

If we further assume that space-time rapidity equals momentum-space rapidity

$$\epsilon = \frac{1}{\pi R^2} \frac{p_t}{t} \frac{dN}{dy} \tag{31}$$

which at the coherence time is

$$\epsilon_{coh} \sim \frac{1}{\pi R^2} p_t^2 \frac{dN}{dy} \tag{32}$$

The approximation that space-time rapidity equals momentum space rapidity is true for free streaming particles and for hydrodynamically expanding matter, and should be a very good approximation at very high energy.

We can also estimate the energy density at the formation time, the time at which we expect it should be a good approximation to describe the quark-gluon plasma as an adiabatically expanding fluid. If we take this time to be $t_{for} \sim 1\ Fm/c$, and approximate p_t from the experimental data from final state transverse momenta, and take the experimentally measured multiplicities, we find for large A that the energy density is

$$\epsilon_{for} \sim 3\ Gev/Fm^3 \tag{33}$$

If the formation time is earlier, the corresponding energy density is greater.

4.2 Energy Density in the Fragmentation Region

We can also crudely estimate the energy density in the nuclear fragmentation region. Recall that the nuclear fragmentation region corresponds to a rapidity interval of order $\Delta y = \Delta y_{pp} + ln(2A^{1/3})$. Taking the fragmentation region in pp collisions y_{pp} to be of order 2 units, we see that for large nuclei, the fragmentation region can easily be as large as 4 units. At CERN SPS energies therefore, the entire kinematic region

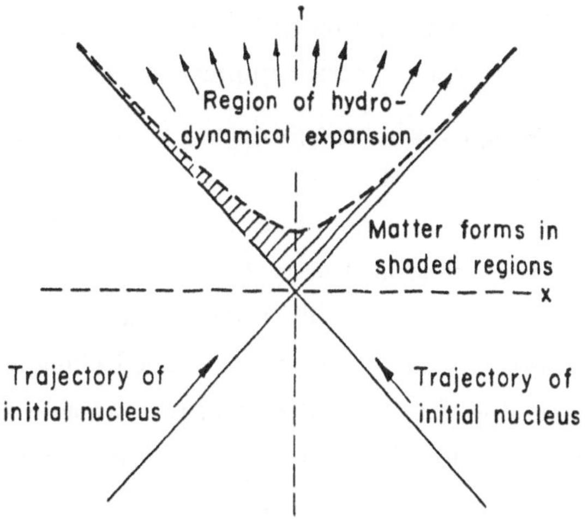

Figure 20

A space-time diagram of ultra-relativistic nuclear collisions.

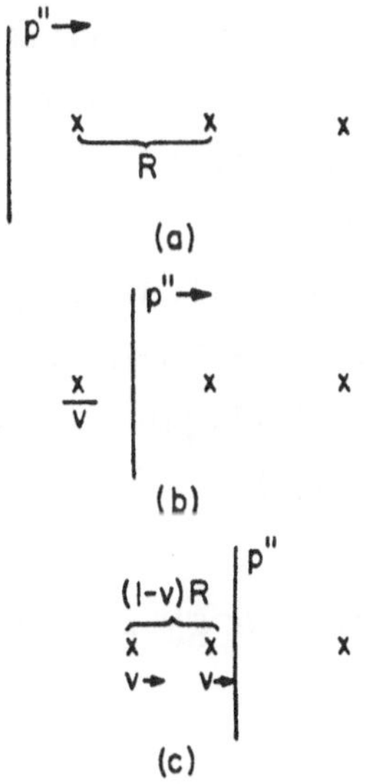

Figure 21

a) A Lorentz contracted nucleus incident on a row of target nucleons b) The projectile nucleus accelerates the first nucleon before colliding with the second. c) The projectile nucleus has passed through two target nucleons.

is in this sense the fragmentation region. (Sometimes the fragmentation region is defined to be the region of Feynman x near one. Our definition includes relatively small values of Feynman x.)

We can estimate the energy density produced in the nuclear fragmentation regions. There are two types of energy density here: compressional energy and heat.

We first compute the compressional energy density Consider a Lorentz contracted projectile nucleus incident on a row of nucleons as is shown in Fig. 21. The projectile nucleus collides with the first nucleon and passes through it imparting a velocity v to it. This happens before the second nucleus is struck, and therfore when the second nucleon is struck, the first nucleon has had its separation form the second reduced to $(1 - v)\Delta x$ Part of this separation is an artifact of not looking at the nucleons in a comoving frame. Correcting for this, the system as a whole is compressed by a factor of γ_{frag}, the Lorentz gamma factor appropriate for the rapidity of the nuclear matter studied. This Lorentz gamma factor can be in the range of 2-10, and the compressional energy associated with the nucleons alone can therefore be in the range of $.3 - 1.5\ GeV/Fm^3$.

Estimating the amount of heat produced is more complicated, and can only be properly done in a hydrodynamical model. Modern estimates of the energy density stored in heat give of the order of a few GeV/Fm^3 as not atypical for most of the fragmentation region. We conclude therefore that the energy density in the central region and fragmentation regions are of comparable magnitude.

The central region at ultra-relativistic energies is probably more simple to understand than the fragmentation region. The kinematics is more complicated in the fragmentation region, since there is no simple correspondence between space-time and momentum space rapidity. Each region in principle probes a different region of the baryon number density temperature plane.

4.3 The Central Region Energy Density for Asymptotically Large Nuclei

For very large nuclei, we expect that the energy density will increase with nuclear size. What is the rate of increase? We shall consider the nuclei at the coherence time. Recall that QCD is approximately scale invariant for high energies. The only scale invariant quantity we can make with the dimensions of an energy density is

$$\epsilon \sim \left(\frac{1}{\pi R^2} \frac{dN}{dy} \right)^2 \tag{34}$$

and the typical particle transverse momenta at the coherence time must be of the order or

$$p_t^2 \sim \frac{1}{\pi R^2} \frac{dN}{dy} \tag{35}$$

If we assume that $dN/dy \sim A - A^{4/3}$, then we find a slow variation of $p_t \sim A^{1/6} - A^{1/3}$. This corresponds to an energy density increase of order $\epsilon \sim A^{2/3}$ which is not insubstantial. The nuclear sizes required for this argument to be true may be far beyond the sizes of nuclei which are stable in nature. It also requires collider energies which are high enough so that the Lorentz contracted size of the nuclei is small compared to the p_t required above, $2R/\gamma_{cm} \leq p_t$

4.4 When is Hydrodynamics Valid?

After the coherence time, the matter produced in a nuclear collision presumably expands to a good approximation according to the equations of transport theory. At some time the scattering time for particles becomes much less than the typical expansion time, $t_{scat} \ll t_{exp}$, and the the system expands adiabatically according to the laws of perfect fluid hydrodynamics.

We shall first estimate the expansion time. Typically for adiabatic expansions, the energy density decreases as $\epsilon \sim t^{-D}$ where D is dimensionality of the expanding system. This is because entropy is conserved in the expansion and if there is no source of energy driving the expansion, each particle may be thought of as on the average moving with a constant velocity. Therfore we expect that characteristic expansion time is

$$t_{exp} \sim \frac{\epsilon}{d\epsilon/dt} \sim t \tag{36}$$

This equation means that the earlier you look at the system, the more rapidly it appears to be expanding.

What is the scattering time? This is given by the mean free scattering time

$$t_{scat} \sim 1/(\sigma\rho) \tag{37}$$

where σ is the particle scattering cross section and ρ is the density of particles. If the energy density is large enough so that perturbative QCD can be used, then the cross section for particles with typical energy p_t is

$$\sigma \sim \alpha_s^2/p_t^2 \tag{38}$$

We find therefore that

$$t_{scat} \sim 1/(\sigma\rho) \sim p_t^2 \frac{\pi R^2}{dN/dy} t \sim t\ p_t^4/\epsilon_{coh} \tag{39}$$

Here we have taken the coupling constant as approximately constant. If we ignore the dependence of particle transverse momenta upon time, then $t_{scat} \sim t$, and the particles are either in equilibrium or they are not. Of course particle scattering will reduce p_t somewhat, and even if the system was not in equilibrium to begin with, there is still a chance it can achieve it. Also, the scattering cross section increases at lower energies, also driving the system to equilibrium. Eventually at very low densities, the system goes out of equilibrium when the particles begin to expand three dimensionally, and it is not a good approximation to take $\rho \sim 1/t$ but the expansion is more like $1/t^3$. In this case the scattering time rapidly increases and the particles decouple.

When more or less realistic numbers are used for particle cross sections and energy densities, $t = t_{scat}$ when $t \sim .05-1\ Fm/c$ with most of the range due to uncertainty in the computation of particle cross sections, and some of the range due to the difference in cross sections for quarks and gluons. Gluons tend to thermalize before quarks do.

4.5 The Hydrodynamic Expansion

For the times when the matter is expanding adiabatically, the system may be described by perfect fluid hydrodynamics. For a perfect fluid, the stress energy tensor of the system is

$$T^{\mu\nu} = \{\epsilon + P\}u^\mu u^\nu + Pg^{\mu\nu} \tag{40}$$

where ϵ is the energy density, P is the pressure and u is the fluid four velocity vector,

$$u = (1, \vec{v})/\sqrt{1 - v^2} \tag{41}$$

with $u^2 = -1$. As a consequence of this form of the stress energy tensor and its conservation,

$$\partial_\mu T^{\mu\nu} = 0 \tag{42}$$

the entropy current

$$s^\mu = u^\mu s \tag{43}$$

is conserved, where s is the local entropy density.

As a consequence of the conservation of the stress energy tensor, we have four equations for the five variables u^μ, ϵ and P. If we are given these variables at $t = 0$, and the equation of state

$$P = P(\epsilon) \tag{44}$$

then the solution is known for all time.

For example is we consider Lorentz invariant 1+1 dimensional expansion, we can compute the time evolution of the temperature and entropy density. Lorentz invariant expansion should be valid in the central region of very high energy nuclear collisions if the multiplicity per unit rapidity is approximately independent of rapidity. In this case, particle distributions are independent of Lorentz boosts, since a Lorentz boost simply translates the rapidity.

For Lorentz invariant expansion, the velocity four vector must be of the form

$$u^\mu = x^\mu/\tau \tag{45}$$

The equation for the conservation of entropy current becomes

$$s = s_0 \tau_0 / \tau \tag{46}$$

where the entropy at time τ_0 is s_0. For an ideal relativistic gas therefore, the temperature is

$$T = T_0 (\tau_0 / \tau)^{1/3} \tag{47}$$

As a consequence of the large number of degrees of freedom in the quark-gluon plasma, it takes a long time to convert the entropy into the entropy of a hadron gas. If initial time is about a fermi or so when the plasma exists, it takes a time of about 10-20 Fm/c to convert this to hadronic degrees of freedom. (If entropy is produced, the time it takes is even longer.)

If the system expands through the mixed phase adiabatically and eventually freezes out as pions, we can relate the entropy to the number of pions. In this case the number density of pions and entropy density are proportional. Therefore in some sense adiabatic hydrodynamic expansion preserves multiplicity.

It may also happen that in the mixed phase there is phase separation and globs of quark-gluon plasma are formed. In this case, explicit simulations by Bertsch et. al. show that the pions more or less decouple after being produced by the globs. It again takes a long time for the globs to radiate away their entropy, and the system live for a long time, $t \sim 10 - 20 Fm/c$. In gross aspects, such as generation of transverse momentum by hydrodynamic expansion, the glob and hydrodynamic picture agree.

At the end of hydrodynamic expansion, the problem of freeze out must be addressed. In the glob picture, this is done by assuming the globs radiate and absorb pions as black bodies. The pion-pion interactions after emission are accounted for by a cascade computation. In the hydrodynamic picture, usually an instantaneous decoupling assumption is made. At some temperature, the system assumed to fall apart into free pions. This is not such a bad assumption if the system is cool at decoupling since typical interaction are weak and significant momentum changes for particles take a long time, much longer than the time of decoupling.

5 Acknowledgments

I thank Dominique Vautherin and John Negele for organizing this excellent school. I also thank my wife Alice who in spite of a broken leg and and operation immediately prior to the meeting had the strength to come.

References

[1] For a review see e. g. L. D. McLerran, *Reviews of Modern Physics* **58**, 102 (1986).

[2] B. Svetitsky, *Phys. Rep.* **132**, 1 (1986).

[3] K. Kajantie and L. McLerran, *Ann. Rev. Nucl. Part. Sci.* **132**, 1 (1986).

[4] C. Alcock and A. Olinto, *Ann. Rev. Nucl. Part. Sci.* **38**, 161 (1988).

THE ROLE OF QUARKS IN ASTROPHYSICS

Charles Alcock

Institute of Geophysics and Planetary Physics
Lawrence Livermore National Laboratory
Livermore, CA 94550, USA

ABSTRACT

There are astrophysical consequences of quark degrees of freedom only at very high density or at very high temperatures. Sufficiently high densities may occur in the cores of neutron stars that a transition (of unknown order) to quark matter might occur; the presence of the quark matter has impact on the maximum possible mass for the star (which is lower because of the quark phase) and on the cooling rate of the neutron star (which is enhanced by the quark phase). It is possible that three-flavor quark matter is absolutely stable at zero pressure; if this is the case then all neutron stars would (given enough time) convert to "strange stars". These stars have some very interesting properties, including the possible existence of low mass objects which have nuclear density; one possible example of such an object has been reported in the discovery of a pulsar in Supernova 1987A.

Arbitrarily high temperatures were achieved during the birth of the universe. Temperatures in excess of 100 MeV occurred at times earlier than ~30μSec. At these high temperatures, the true degrees of freedom were (almost certainly) quarks and gluons, not hadrons. The subsequent appearance of the hadron plasma presumably came about via a phase transition. If this were a first order phase transition, then there are presently observable consequences of the phase transition in the abundances of light elements. Second or higher order transitions do not have such consequences.

I. INTRODUCTION

The most convenient way to introduce this subject is to present elementary models of the physical systems. Since the thermodynamics of QCD is not well understood (why else would we have this Summer School?), there is no unique prescription for the construction of these models. The following guiding principles will be adopted here: (i) the models should be simple to describe; (ii) the physics content of the models should be manifest; (iii) the relationships between assumption and consequence should be obvious. In general these principles can be satisfied by assuming an ideal gas of hadrons for the hadronic phase, and an ideal gas of quarks and gluons plus a "bag term" for the quark-gluon plasma. This choice does have one immediate, important consequence: there is a first-order phase transition between the two phases, which is described below.

Hadrons and Hadronic Matter
Edited by D. Vautherin *et al.*
Plenum Press, New York, 1990

This first-order phase transition occurs because different model descriptions are used for the two different phases, which almost guarantees that there will be discontinuities in first derivatives of the thermodynamic potentials at the equilibria between the two phases. Whether or not this is a realistic description of nature can only be learned by returning to the underlying theory, QCD, and computing the thermodynamics explicitly, for instance, by using the lattice.

It is also convenient at this stage to summarize the important physical content of these models be constructing equilibrium phase diagrams. Phase diagrams widely used in geophysics, geochemistry, materials science and chemical engineering, where they are constructed from empirical data. The phase diagram in QCD is entirely based on theory.

A theoretical phase diagram is computed by minimizing a thermodynamic potential with respect to all internal degrees of freedom(1). An important degree of freedom is the phase; in the model discussed here the system will either be in the hadron phase or in the quark-gluon phase. Phase equilibria occur where the thermodynamic potentials for the two phases are equal. The Gibbs Potential G(T,P) and the Landau* Potential Ω (T,μ) (where T is temperature, P is pressure, and μ is baryon chemical potential) are the two convenient thermodynamic potentials for this procedure since they depend only on intensive quantities which are continuous across phase boundaries. In contrast the energy density, the Helmholtz free energy density and enthalpy density depend upon quantities (such as baryon number density N_B entropy density s) which are discontinuous at phase boundaries.

Before introducing the models it is useful to outline two representative model phase diagrams. To begin, the "standard" phase diagram† for QCD is shown in Figure 1. The axes are temperature and chemical potential, drawn on logarithmic scales to bring out the extraordinary range in magnitudes of the important quantities.

The region of interest for the study of neutron stars is the high μ, low T region. The existence in this region of a transition to quark matter is not in doubt; however, the locus of this transition is very uncertain. It is not possible to establish, therefore, the existence or non-existence of quark matter in the cores of neutron stars.

The region of interest for cosmology is the high T, low μ portion of the phase diagram. The trajectory of the universe in the phase plane is indicated. The most important feature of this trajectory is that $\mu \ll T$ at the phase transition. This is a consequence of the extremely low baryon number of the universe(2); this fact is usually expressed quantitatively using the baryon to photon ratio, $N_B/N_\gamma < 10^{-8}$. For the purpose of building models of the phase dynamics the non-zero baryon number may be ignored. The possible long term consequences of the phase transition, which may be observed by astronomers today, all turn out to depend on the behavior of this small baryon number density, which may be treated perturbatively.

---

* This potential is conventionally referred to as the Landau Potential, even though it was known to Gibbs. Since $\Omega = -P$, where P is pressure, we will henceforth use $P = P(T,\mu)$ as our thermodynamic potential.

† This diagram does not indicate the complex phase structure that occurs in the crusts of neutron stars, in the region $\mu \approx$ 1GeV, T $\leq$ 10KeV.

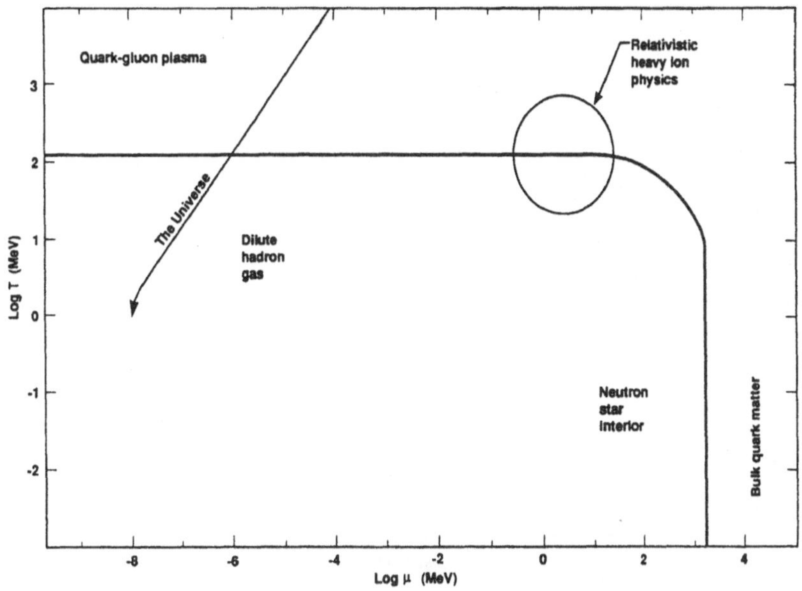

Figure 1. Standard phase diagram in QCD

Also indicated on this phase diagram is the region that may be explored using relativistic heavy ion collisions. There is a real possibility of exploring the thermodynamics of the cosmic quark-hadron phase transition in these experiments, since the central rapidity region of the collision is characterized by low baryon number. This is the possible connection between experimental heavy ion physics and cosmology(3).

One should keep in mind some important differences between the cosmic quark-hadron phase transition and the heavy ion events. The cosmic phase transition lasts about 10^{-5} seconds, whereas the heavy ion event lasts about 10^{-22} seconds. During the cosmic transition the weak interactions have ample time to remain in equilibrium; consequently there is thermodynamic equilibrium among the quark flavors, and there is only one chemical potential which keeps track of the baryon number. In contrast, there are essentially no flavor changing interactions in the heavy ion collision, which means that the numbers of each flavor of quark are separately conserved. There are three chemical potentials (for the up, down and strange quarks; the charm, top and bottom quarks play insignificant roles). For these reasons we have some confidence that elementary models capture the essence of the cosmic quark-hadron phase transition, but probably these models are inadequate for the description of the heavy ion events.

An important new possibility for the phase diagram in QCD was introduced by Witten(4,5). Witten proposed that three flavor quark matter might be absolutely stable. In our language, this means that at zero pressure and zero temperature, the energy per baryon of three flavor quark matter is lower than that of hadronic matter (^{56}Fe in its most bound state (i.e. ^{56}Fe). This profoundly modifies the phase diagram, as shown in Figure 2.

The absolute stability of strange matter now means that much of the phase plane in the low T region is not occupied by the hadron gas, but by a gas of *strangelets*. Strangelets are isolated lumps of strange quark matter. In this region of the phase plane most of space is empty, and the mean density is very low; all of the baryon number is concentrated into isolated, high density lumps.

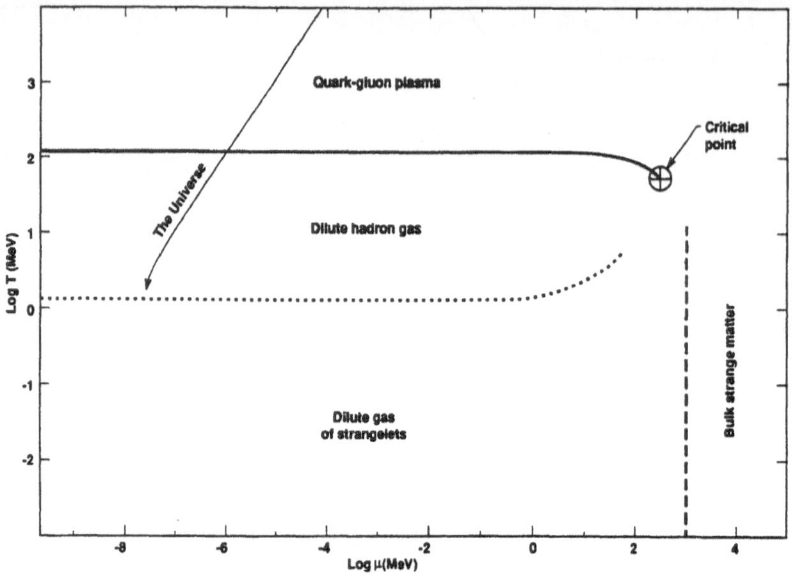

Figure 2. Phase diagram for strange matter

As one moves to high μ in the strangelet portion of the phase plane, a point is reached at $\mu \approx 1$ GeV where the strange matter fills all space. The behavior of the system changes markedly with further increase in μ, since the degenerate gas of quarks supplies the pressure. This change is not a phase transition in the conventional sense: there is no possibility of two phase equilibrium.

If, instead of increasing μ, one leaves the gas of strangelets by increasing the temperature T, the properties of the system change substantially at $T \sim 2$MeV. The strangelets evaporate into a gas of protons and neutrons[6]. This process is analogous to ionization of a gas of atoms, and is not a phase transi-tion. At the higher temperature the gas of strangelets disappears because such a highly condensed state is a configuration of very low entropy. At high temperatures, the free energy of a system of hadrons is reduced by evapora-ting strangelets (a process which costs energy) and producing protons and neutrons (which increases the entropy).

At $T \geq 100$MeV, in the low μ region, there is a hot quark-hadron phase transition that is (at our present level of understanding) identical to that shown in Figure 1. This establishes an important qualitative feature of the phase diagram for QCD: the stability of strange matter is essentially a low temperature phenomenon.

We can now make an interesting qualitative statement about the high μ, high T portion of the phase diagram. There is a first order phase transition in the high T low μ region. There is no phase separation in the high μ low T region. There is no phase transition between the gas of hadrons and the gas of strangelets. Therefore there *must* be a critical point in the diagram (1); most likely it occurs near $\mu \sim T \sim 100$MeV, as marked in Figure 2. There may be interesting consequences of this critical behavior, but this author is not aware of any investigation into this issue.

II. STRANGE STARS

If the strange matter hypothesis is correct, neutron stars are metastable with respect to stars made of strange matter. This in turn means that the objects known to astronomers as neutron stars are probably made of strange matter, not of neutron matter, and should be called "strange stars"(4,7,8). The properties of strange stars are discussed here.

These objects have extremely simple structures, because the zero-temperature equation of state is, to high accuracy, $P = \frac{1}{3}(\rho - 4B)$. This expression is exact in the bag model with massless quarks (either two-flavor or three-flavor). The addition of mass to one of the flavors (the s quarks) causes deviations no greater than 4% from the simple relation because, if the mass is dynamically important, the abundance of the massive quarks becomes small and their contribution to the material insignificant.

This equation of state has the property that as $P \to 0$, $\rho \to 4B$. For $B = (145\ \text{MeV})^4$ this means $\rho = 4 \times 10^{14} \text{g cm}^{-3}$, slightly greater than nuclear density. Thus, there is a sequence of objects with very low internal pressure and uniform density. Their mass (M)-radius (R) relation is $M \text{ a } R^3$.

The pressure at the center of one of these objects is $P_c = (2\pi G/E)\ \rho^2 R^2$, where G is Newton's constant and Newtonian gravity is assumed. For sufficiently large radius R the pressure P_c approaches 4B/e and the density increases toward the center of the object. This effect becomes noticeable at $R \approx 5$ km, $M \approx 0.1\ M_O$, and the mass-radius relation is very different from $M \text{ a } R^3$ for object with $R \approx 10$ km, and $M \approx 1\ M_O$. Relativistic gravity also becomes important in these stars and the Oppenheimer-Volkoff equation for stellar structure must be used to compute the models. The full mass-radius relation is shown in Figure 3. The sequence terminates at the limit of dynamical stability, known as the Chandrasekhar limit. The dynamical stability of relativistic stars is discussed fully in (9).

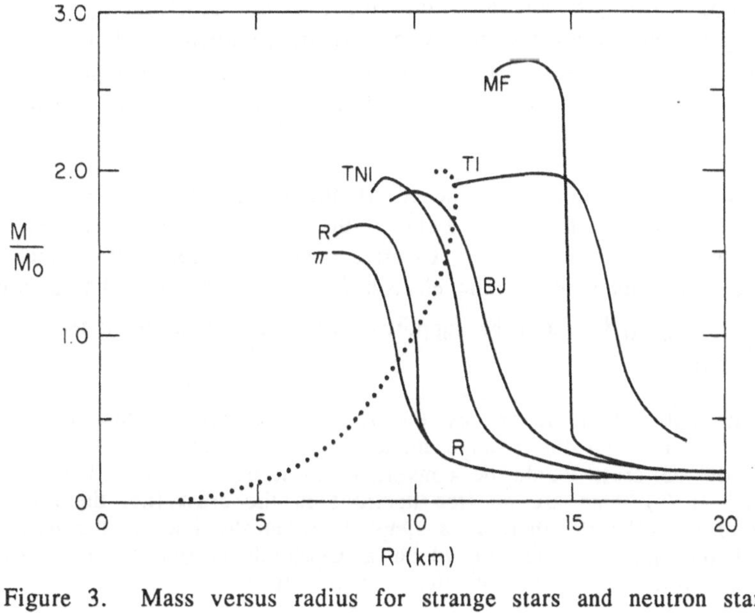

Figure 3. Mass versus radius for strange stars and neutron stars

Figure 3 also shows some well-known mass-radius relations for neutron stars, which are computed for a variety of different nuclear matter equations of state: MF is a mean field theory calculation; "TI is a tensor-interaction model; BJ is a Bethe-Johnson model, which includes hyperons; R is a pure neutron model with a soft-core interaction; π is the R model with pion condensate. These models were reviewed by Baym and Pethick[10,11]. These mass-radius relations are very different from that for strange stars, and the difference arises entirely because, for nuclear matter, $\rho \to 0$ as $P \to 0$. One would hope that such a large, qualitative difference could be exploited to discover the truth regarding the strange matter hypothesis.

Nature has not been kind here. All neutron/strange stars for which masses have been determined have masses near 1.4 M_O, where the two models of compact stars have very similar radii. Should a very low-mass compact star be discovered, the two pictures would be distinguishable.

The fact that strange matter is absolutely stable raises the possibility that strange stars are made exclusively of strange matter, and that the surface of the star is exposed quark matter. Early discussions of strange stars presumed that this would be the case, and some interesting consequences for the appearance of these objects were found. However, as we show below, there is also the strong possibility that the surface of a strange star is made of the same material as the surface of a neutron star.

A bare strange surface has very unusual properties. The thickness of the "quark surface" is ~1 fm; the integrity of this surface is ensured by the strong force. The electrons are held to the quark matter electrostatically, and the thickness of the "electron surface" is several hundred fermis; the electric field in this region is $\sim 5 \times 10^{17}$ V cm^{-1} . since neither component is held in place gravitationally, the traditional "Eddington Limit" to the luminosity that a static surface may emit does not apply, and these objects may (in principle) have photon luminosities much greater than 10^{38} erg s^{-1} .

A strange matter surface will have a low emissivity for x-ray photons[8]. This conclusion is reached by calculating the dispersion relation for photons in strange matter. The result is much like the dispersion relation for photons in an electron plasma, but with characteristic "plasma frequency" $\omega_p = (8\pi\alpha/3)\ N_u{}^2/\rho_u$ (where α is the fine structure constant, N_u the number density of up quarks, ρ_u the energy density of up quarks). For typical parameters, $\omega_p \approx 19$ MeV. This means that the surface of a bare strange star is highly reflective in the x-ray region and has a low emissivity. The emissivity has not yet been calculated.

There is a further consequence of the electrical properties of this surface. The very high electric field in the electron surface will exert a strong outward force on an ion. Clearly, a certain amount of normal ionic material can be supported by this electric field. It turns out that a crust of mass up to $\sim 5 \times 10^{28}$ g may be supported, with density at the inner edge up to 4×10^{11} g cm^{-3} .

This upper limit is set by the requirement that nuclear reactions between the crust and the strange matter must be prevented, or else the ions at the base of the crust would be converted to strange matter. This requirement is satisfied if (a) there are no free neutrons in the crust [i.e. there is no "neutron drip" (12)]; (b) there is a "gap" between the ions at the base of the crust and the quark surface in which a Coulomb barrier prevents direct reactions between the ions and the strange matter.

This thin layer is identical to the "outer crust" of a neutron star. For this reason, a strange star with a crust is not different from a neutron star in regard to its photon emissivity. Furthermore, since the crust is held onto the star by gravitation, this new surface is subject to the Eddington limit.

It seems likely that this latter view of the surface of a strange star is more realistic. The universe is a "dirty" environment, and certainly supernova remnants contain a lot of material that may accrete onto the surface of a newly formed strange star and make a crust. Hence, we are once again driven to conclude that a strange star is very similar to a neutron star in its observable properties.

Radio pulsars are observed to have periods that steadily increase. This is attributed to the loss of angular momentum by magnetic dipole radiation. In some pulsars small "glitches" in this smooth spin-down are occasionally observed. In a glitch, the period abruptly (in less than a day) decreases; over the next 40-80 days most of this decrease is lost as the pulsar appears to "heal" back toward its original spin-down curve.

A model has been developed for this phenomenon involving the behavior of superfluid neutrons in the inner crust of a neutron star (13). There is no equivalent for this model involving strange stars. It is not clear how seriously the lack of a model for glitches should be taken; this may reflect only lack of imagination on our part.

A variety of "routes" from neutron matter to strange matter have been suggested[8,13]. These include conversion via two-flavor quark matter, clustering of lambdas, kaon condensates, direct "burning", and seeding from the outside. The uncertainties in each of these are so large that estimates of conversion rates cannot be made with confidence. It is possible, if unlikely, that neutron stars will not convert to strange stars, even if the strange matter hypothesis is correct.

However, once there is a seed of strange matter inside a neutron star it is possible to calculate the rate of growth[14]. The strange matter front absorbs neutrons, liberating u and d quarks into the strange matter. Weak equilibrium is then reestablished by the diffusion of strange quarks and by the weak interactions. The rate of progress of this front has a strong inverse temperature dependence.

If this conversion happens just after the supernova explosion one expects a neutrino signature of 10^{52} ergs over a period between minutes and hours. Neutrino astronomy will be able to detect neutrinos from nearby supernova and this signature can be tested.

This conversion can happen in later stages of neutron star evolution. If it happens in an active pulsar, a macroglitch will be observed because of the change in moment of inertia. An old defunct pulsar will convert even faster, and a gamma-ray burst will be its signature.

III. A BRIEF REVIEW OF ELEMENTARY COSMOLOGY

For the purposes of discussing the cosmic quark-hadron phase transition it will be necessary to review some elements of physical cosmology. This is most easily accomplished by combining some general considerations with a Newtonian analysis[15]. The Newtonian analysis yields correct, exact expressions which describe the rate of expansion of the universe. The full analysis based on the General Theory of Relativity, which is not discussed here, is essential in order to devise meaningful global expressions which describe the universe.

The universe today is observed to be expanding. The the extent that observers can determine, the universe satisfies Hubble's Law of uniform expansion:

$$v = \dot{r} = H_0 r , \tag{1}$$

where r is the proper vector distance between the observer and a distant galaxy, and v is the measured recession velocity of the galaxy. Note that if one shifts one's "center" from $r = 0$ to $r = r_0$, the law is preserved; this is the property of *homogeneity*. Furthermore, the expansion is *isotropic*.

In addition to the expansion, the universe is filled with thermal radiation at $T \approx 2 \cdot 8K$. This radiation is isotropic to a high degree.

Modern physical cosmology is largely based on the assumption that the universe is homogeneous and isotropic, and that Einstein's General Theory of Relativity applies. We shall now see what consequences that has for a small, spherical "patch" of the universe.

This "patch" has proper radius $r = r_p(t)$, where t is the time; particles within this patch have proper coordinates $r = r(t)$. It is convenient to label each particle within this patch by the coordinate position it has at some time $t = t_0$, i.e. $x \equiv r(t_0)$. It is conventional to choose t_0 to refer to the present epoch, so that t_0 is the age of the universe. The coordinate x is time-independent, and is known as the "comoving coordinate". Uniform expansion can be expressed using an expansion factor a(t),

$$r(t) = a(t)x , \tag{2}$$

where $a(t = t_0) = 1$.

The recession velocity is

$$v = \dot{a}(t)x = \left(\frac{\dot{a}}{a}\right)r = Hr . \tag{3}$$

The expansion coefficient, when evaluated at $t = t_0$, is referred to as the "Hubble Constant" H_0.

Observational cosmologists measure the *redshift* z of photons which are emitted by atomic transitions at some time $t < t_0$ (thus fixing the emission wavelength λ_ε) and received at the telescope at t_0 with wavelength λ, with definition $1 + z \equiv \lambda/\lambda_\varepsilon$. It is easy to show in this model universe that:

$$1 + z = a(t_0)/a(t) = a(t)^{-1} \tag{4}$$

The rate of change of the density of some conserved object can readily be expressed using H. Examples of conserved quantities include baryon number and entropy; for reasons that will become clear in section IV we look at the entropy density, s, which satisfies the equation of continuity

$$\frac{\partial s}{\partial t} + \nabla \cdot (sv) = 0 \qquad (5)$$

Since the universe is homogeneous, $(v \cdot \nabla) = 0$. Equation (3) shows that $\nabla \cdot v = 3H$, and hence we have

$$\frac{\partial s}{\partial t} = -3Hs\,. \qquad (6)$$

We can now describe the evolution of the universe by obtaining a dynamical equation for a(t). Consider the motion of a test particle at the edge of the local patch. It feels a gravitational acceleration due to the material inside the patch, but no acceleration due to the remainder of the matter in the universe (this is a consequence of spherical symmetry, and is rigorously true in General Relativity). Thus, we have:

$$\ddot{r}_p = -\frac{4\pi}{3} G\rho\, r_p, \qquad (7)$$

where ρ is the density of mass and energy inside the patch, and G is Newton's constant. Using equation (3) this becomes:

$$\ddot{a} = -\frac{4\pi}{3} G\rho\, a\,. \qquad (8)$$

In order to make further progress we must have some knowledge of ρ, and how it behaves with expansion. At the present epoch the matter density greatly exceeds the energy density, so we know that $\rho = \rho_o a^{-3}$, where ρ_o is the mean density of the universe at $t = t_0$. The universe is said to be matter dominated at the present epoch. Inserting the expression of ρ into equation (8) and integrating once yields:

$$\left(\frac{\dot{a}}{a}\right)^2 = \frac{8\pi G\rho}{3a^3} - \left(\frac{1}{aR}\right)^2, \qquad (9)$$

where R^{-2} is an integration constant. The General Relativistic analysis reveals that $(aR)^{-1}$ is the curvature of the space-like hypersurface at fixed cosmic time(16).

Precise measurement of ρ_o and of H_o can, in principle, determine R^{-2}, which in turn fixes the geometry of the universe. It is conventional to use the ratio:

$$\Omega_0 \equiv \frac{8\pi G\rho_o}{3Ho^2}\,. \qquad (10)$$

If $\Omega_o > 1$, then $R^{-2} > 0$ i.e. the spatial hypersurfaces are closed. Further integration of equation (9) reveals that the universe will reach a maximum expansion, after which it will collapse back onto itself. If $\Omega_o < 1$, then $R^{-2} < 0$ i.e. the spatial hypersurfaces are open, and the universe will expand forever, with $\dot{a} \to$ constant as $t \to \infty$. The special case $\Omega_o = 1$ has $R^{-2} = 0$, i.e. the spatial hypersurfaces are flat. The universe expands forever, but with $a \propto t^{2/3}$.

The observational situation with regard to Ω_0 is unclear, but analyses of data lead astronomers to conclude that $0 \cdot 3 \leq \Omega_0 \leq 2$. For a review of this complicated subject see (17).

Theories based on the inflationary paradigm(18) predict that $R^{-2} = 0$, and hence that $\Omega_0 = 1$. For this reason the geometrically flat models are taken very seriously. The composition of the matter that makes up the density ρ_0 is a sub-ject of considerable controversy, as we shall see below.

Note that the geometric term R^{-2} plays a significant role in equation (9) only at large a, which means at late times. At early times in the universe this term may be neglected i.e. all models resemble the flat model at sufficiently early times. Furthermore, the energy density of the photons (2·8K background) and neutrinos (which we believe to be present for reasons discussed below) scale as $\rho \propto a^{-4}$; this means that at sufficiently early times (when $a \leq 10^{-4}$) the contribution of the matter to ρ may be neglected. Equation (9) becomes

$$\left(\frac{\dot{a}}{a}\right)^2 = \frac{8\pi G\rho}{3a^3}, \tag{11}$$

where ρ is the energy density. When most of the energy density is in the form of massless particles such as photons and neutrinos, $\rho \propto T^4$, where the constant of proportionality may readily be calculated, and involves the numbers of massless fermion and boson species.

What is important about this result is that we can obtain a unique relationship between temperature T and time t. For a thorough, quantitative summary the reader is referred to (16).

IV. THE COSMIC QUARK HADRON PHASE TRANSITION

In the spirit of this analysis, we describe the quark-gluon plasma as an ideal gas of quarks and gluons, and write an expression that includes only lowest order non-vanishing terms in μ/T. The pressure P_Q of this phase (i.e. the negative of the Landau potential) is:

$$P_Q = \frac{7\pi^2}{180} N_c N_f T^4 \left[1 + \frac{10}{21\pi^2}\left(\frac{\mu}{T}\right)^2\right] + \frac{\pi^2}{45} N_g T^4 - B, \tag{12}$$

where N_c is the number of color degrees of freedom ($N_c = 2$ if we include only up and down quarks, for instance), $N_f = 3$ is the number of flavor degrees of freedom, $N_g = 8$ is the number of gluons, and $B \sim (145\text{MeV})^4$ is the MIT "bag term".

The expression for the ideal gas of hadrons is complicated by the dynamically significant masses of the baryons and mesons, and by the large number of hadronic states. The full mathematical expression describing this model is given in (19); the expression given here for the hadron pressure P_H has been simplified in order to bring out the most important physics:

$$P_H = (2\pi)^{-2} T^{5/2}\left[1 + \left(\frac{\mu}{T}\right)^2\right] \sum_{\text{baryon}} g_b m_b^{3/2} \exp\left(-\frac{m_b}{T}\right) + \frac{\pi^2}{9} T^4. \tag{13}$$

The summation in the equation is over all baryon states, where g_b is the statistical weight of the state and m_b the mass of the state. The second term, proportional to T^4, is the pressure due to pions (all massive mesons have been ignored in this equation).

The pion term is the dominant contributor to P_H. This is because the pion is the only hadronic state with mass less than (or, of order) the temperature. The pion gas also contributes most of the entropy. However, the baryon terms carry all information about baryon number in the system.

We now have a complete prescription for the phase equilibrium. Equilibrium occurs when $P_Q(\mu,T) = P_H(\mu,T)$. Since the chemical potential in this system is so low, it may be neglected when seeking the equilibrium temperature. We define the critical temperature T_c:

$$P_Q(T_c, 0) = P_H(T_c, 0)\,. \tag{14}$$

The latent heat of the transition is

$$L = T_c\left\{\frac{\partial P_Q}{\partial T} - \frac{\partial P_H}{\partial T}\right\}, \tag{15}$$

where the derivatives are evaluated at T_c.

Even though the baryon number is so small that terms containing μ are insignificant in computing pressure and latent heat, it is the behavior of the baryon number than will lead to long term consequences. The baryon number density in either phase is given by:

$$N_B = \frac{\partial P}{\partial \mu} = \left(\frac{\partial^2 P}{\partial \mu^2}\right)_{\mu=0} \mu\,, \tag{16}$$

where the facts that $\mu << T$ and that $(\partial P/\partial \mu) = 0$ when $\mu = 0$ were exploited in deriving the second derivative expression. When equilibrium between the two phases obtains, the chemical potential μ is the same in both phases. The baryon number N_B will not, in general, be the same in both phases. We define the equilibrium ratio $R = N_B\,(\text{quark})/N_B\,(\text{hadron})$, which is given by:

$$R = \frac{\left(\partial^2 P_Q/\partial \mu^2\right)}{\left(\partial^2 P_H/\partial \mu^2\right)}\,, \tag{17}$$

In the ideal gas model described by equations (12) and (13) this ratio gen-erally turns out to be $R \sim 100$(19,20). This means that baryon number is 100x more soluble in the quark-gluon plasma than in the hadron plasma. It is easy to understand this important result. A high penalty in free energy is paid when one unit of baryon number is added to the hadron gas, because all baryons have mass $\geq$ 1GeV. This is reflected in the Boltzmann factor in equation (13). In contrast, when the unit of baryon number is added to the quark-gluon plasma it is carried by massless particles. Furthermore, since three quarks will carry the unit, the entropy of the system is increased, further reducing free energy.

The description of the cosmic quark-hadron phase transition is complicated by a variety of non-equilibrium effects. These include initial supercooling of the quark-gluon plasma below T_c followed by reheating to T_c, and the development of eddy currents in the quark phase. These departures from equilibrium are not expected to be large because the expansion of the universe is so very slow; recall from the introduction that the duration of this period is about 30 μsec.

The initial supercooling arose because, as the universe cooled to temperatures just below T_c, bubbles of the new phase did not appear immediately. This is because there is a positive free energy σ associated with the surface of the bubbles; this free energy is called the "surface tension". At $T = T_c$ the bulk thermodynamic potentials are equal ($P_H = P_Q$), so there is no thermodynamic gain to creating a small bubble of hadron gas since the bubble has a positive surface contribution to the free energy. Only after some supercooling of the system to $T \leq T_c$ does the bulk thermodynamic advantage of the hadron phase compensate for the positive free energy of the bubble surface.

The supercooling has been analyzed using classical nucleation theory(20,21). The degree of supercooling is found to be small, typically $\Delta T \sim 0{\cdot}01\ T_c$, because the very slow rate of expansion (compared to the QCD timescale) permits the random formation of rare, large bubbles for which the surface term plays a smaller role. This same analysis yields an estimate of another physical quan-tity which is of great importance later, during nucleosynthesis. This quantity is the mean separation between nucleating bubbles at the end of the super-cooling phase,

$$\ell \approx 0 \cdot 3 \frac{\sigma^{3/2}\, t}{T_c^{1/2}\, L} . \tag{18}$$

The dependence of ℓ on uncertain quantities, in particular on the surface tension σ, means that a good quantitative estimates of ℓ impossible to obtain. It is possible that good estimates of ℓ may be obtained using lattice calculations, but at this stage we can compute ℓ quantitatively. In the discussion below of nucleosynthesis, ℓ will be treated as an unknown parameter.

Once the small bubbles of hadron gas have nucleated, they grow rapidly, releasing latent heat, and reheating the remainder of the universe. This continues until the universe has heated back up to T_c, at which point the two phases are in equilibrium. Subsequent passage through the phase transition occurs through a sequence of equilibrium states--the process is (thermodynamically) reversible. The reheating process is complex and not fully understood; the first steps toward a good theory of this process are described in (22).

The period described above--supercooling followed by reheating--is brief, with duration ~ 0·5 μ second. The ensuing period of phase equilibrium has some very interesting properties. The phase separation induces very interesting *peculiar* motions in the system, for two very different reasons that we now discuss. *Peculiar velocities* are velocities superimposed on top of the Hubble expansion described by equation (3).

During the ensuing period of near phase equilibrium, the overall expansion of the universe was manifested in a peculiar velocity field during the phase transition. Examining conditions in the "small patch" described above, there must be a velocity field $\mathbf{v}(\mathbf{x})$ which, on a large scale, exhibits Hubble expansion. However, within each phase the energy density is constant, since it is fixed by the thermodynamics. The velocity field must satisfy the equation of continuity,equation (5), whence we find $\nabla \cdot \mathbf{v} = 0$. This appears to contradict the assertion that there is large scale expansion. The apparent paradox is reconciled by noting that $\nabla \cdot \mathbf{v}$ is singular at the phase boundary. All of the expansion of the universe occurs at the phase boundary. This is clearly very different from the normal Hubble Law described above.

The system is further complicated by gravitational instability. There is a substantial difference in energy density between the two phases, $\Delta\rho = \rho_Q - \rho_H \sim L$. Gravitational forces between different regions of the different phases are not in equilibrium. These forces give rise to accelerations and additional peculiar velocities. A crude estimate yields $\Delta v \sim H\ell$, which is of order the mean recession velocity between neighboring distinct regions of like phase.

This response of the system to the gravitational instability creates a very complicated velocity field, as large bubbles of one phase "plow through" the other phase. The new peculiar velocity has non-zero vorticity, and perhaps is best described as a system of "eddy currents", or even slow convection. The Reynolds number for the flow is ~ 100, too small for fully developed turbulence to arise. The significance of these eddy currents will be made clear in the next section.

V. BARYON TRANSPORT DURING THE PHASE TRANSITION

The eddy currents described above lead to the mixing of baryon number within each phase. We describe this mixing in terms of an effective diffusion coefficient $D \sim \Delta v\,\ell \sim H\ell^2$. Working again in the "small patch" we can construct a diffusion equation for the baryon number density n_B . In this patch the mean velocity of the baryons at any point x is:

$$v_B = H\,r - n_B^{-1}\,D\,\nabla\,n_B, \tag{19}$$

where the first term is the Hubble expansion and the second is drift due to diffusion. Combining with the equation of continuity yields:

$$\frac{\partial n_B}{\partial t} = -\,3Hn_B - H\,(r\cdot\nabla)\,n_B, \tag{20}$$

where the terms on the right hand side represent expansion, Hubble drift and diffusion. This equation describes the evolution of n_B within a particular phase. It must be solved in each phase separately, with the two solutions satisfying the ratio R at the phase boundary. Note that this is a boundary condition on a moving surface.

Equation (10) assumes a much simpler form if rewritten in comoving coordinates, since the expansion is removed. The equation becomes:

$$\frac{\partial n_B}{\partial t} = \nabla^c\cdot D^c\,\nabla^c\,n_B^c, \tag{21}$$

where superscript c denotes the use of comoving coordinates. This, of course, is the standard form of the diffusion equation. The "comoving diffusion coefficient" is related to the physical, or proper diffusion coefficient, by $D^c = a^{-2}\,D$.

A complete three dimensional treatment of this mixing is impracticable at present, and unwarranted until clearer understanding of the eddy currents emerge. An "equivalent sphere" model is shown in Figure 4. The sphere is centered at a point where a bubble of quark-gluon plasma will disappear. The full volume of the sphere represents a patch of volume $\sim \ell^3$. The phase boundary is always a spherical shell, which starts at the edge of the sphere and moves toward the center. The proper diffusion coefficient in the quark phase is, in this example, $D = 4 \times 10^9\ cm^2\ s^{-1}$.

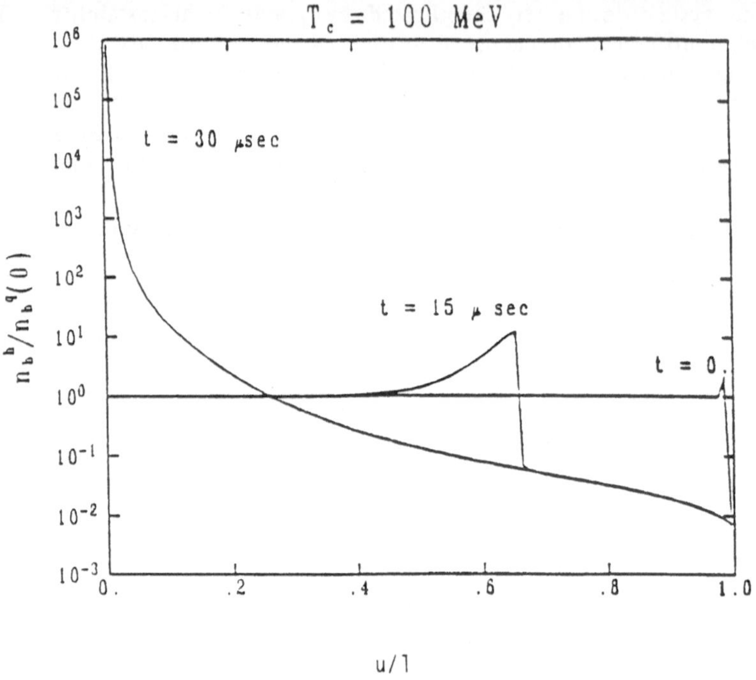

Figure 4. Distribution of baryon number

The inhomogeneity develops as the phase boundary moves inward. Chemical equilibrium sweeps baryon number ahead of the boundary. Diffusion sweeps the excess baryon number deep into the quark-gluon plasma. As the boundary moves inwards, the amplitude of the inhomogeneity grows rapidly. The high central peak in this figure probably would not occur in a more realistic calculation, because in the inner core the phase boundary moves faster than the speed of sound, which means that quasi-equilibrium cannot be maintained.

This model shows that the solubility ratio at the phase boundary, together with mixing, leads to a very substantial inhomogeneity in the distribution of the baryon number.

VI. LIGHT ELEMENT SYNTHESIS

The inhomogeneities described above remain frozen in comoving coordinates until $t \sim 1s$, $T \sim 1MeV$. At this point there is a very interesting development. Neutrons and protons are now the only baryons in the system, and for $T > 1MeV$ the neutrons and protons are strongly coupled by weak interactions with the thermal neutrinos. The ratio of number densities of neutrons and protons is thermal. Below $T \sim 1MeV$, the weak interactions become too slow to affect the number densities, and free decay of neutrons becomes the only reaction of importance.

The neutron has a long scattering mean free path in the ambient plasma of e^+, e^- and γ, because it is uncharged. Accordingly, the neutron has a much larger diffusion coefficient than the proton. For this reason, neutrons begin to diffuse rapidly out of the clumps of baryons. This process can be so efficient that, by the onset of nucleosynthesis, the neutrons are largely separated from the protons(23).

The first stage of nucleosynthesis is the formation of deuterons. Clearly this important reaction is greatly affected by the separation of most of the neutrons from most of the protons. A further complication is that, as most of the neutrons are consumed in the proton rich regions, neutrons begin to diffuse back into the these cores and sustain further nucleosynthesis.

A full description of these processes involves solving the coupled, non-linear differential equations that describe the nuclear reaction rates of importance and include spatial diffusion of neutrons. These processes have been simulated by differencing the equations on an eight-zone representation of the initial baryon number distribution[24]. The evolution of neutron number density in these eight zones is shown in Figure 5, for the case ℓ = 40m . Weak reaction decoupling occurs after about 1 sec when the temperature is ≈ 1 sec, when the neutrons are no longer rapidly converted into protons, the outer zones are able to approach equilibrium. At t ~ 200 sec, the onset of nucleosynthesis causes a factor of 2 drop in the neutron density as the available protons are consumed. Afterwards, the neutron density decreases more gradually, and is given by the timescale for neutron decay into the protons which are required for further nucleosynthesis.

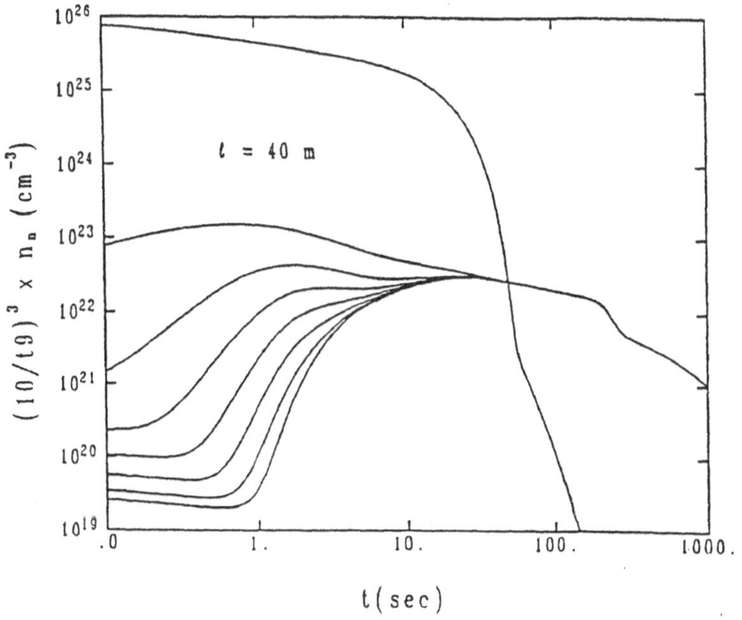

Figure 5. Comoving neutron density versus time for each of 8 zones

The innermost zone appears to evolve almost independently from the outer zones. The density in this zone remains high until t ~ 10 sec at which point the nucleosynthesis begins and the neutron density is quickly diminished as baryons are converted into ^{4}He. The reason that this zone does not communicate efficiently with the outer zones is its high baryon density. The resulting high neutron-proton scattering rate drastically shortens the baryon diffusion length.

Given our calculated fluctuation shapes and diffusion constants, the amount of neutron flow is determined by the average separation between nucleation sites. Figure 6 shows an example of our calculated nucleosynthetic yields as a function of the separation between fluctuations, for a model universe with $\Omega_B = 1$. The optimum separation distance for producing "acceptable" yields or ^2H, ^3He and ^4He appears to be ~30–100m at the time of the phase transition, which is comparable to our earlier estimates based upon classical nucleation theory. For separations which are too close, ℓ<10m, ^4He and ^7Li are overproduced while ^2H and ^3He are underproduced. This is due to the flow of neutrons back into the high density zone after the onset of nucleosynthesis. For separations ℓ ~ 30 m, ^7Li is "overproduced" by some factor which is presently very controversial.

The problem with determining by observations an estimate of the primordial ^7Li abundance is that it cannot be measured in the interstellar gas (for spectroscopic reasons: the strong absorption lines are in the far ultraviolet). Weak absorption due to ^7Li can be found in old stars. The interpretation of this absorption is extremely unclear, however, because of convincing empirical evidence that stars slowly destroy their ^7Li with time. The primordial ^7Li abundance is not known with any confidence, and should not be used to distinguish between cosmological models(25).

The success of the models described above at producing appropriate amounts of ^2H, ^3He and ^4He is very suggestive. These models contain no exotic dark matter. It may turn out that the material of our universe is the familiar mix of baryons and leptons which is studied in the laboratory!

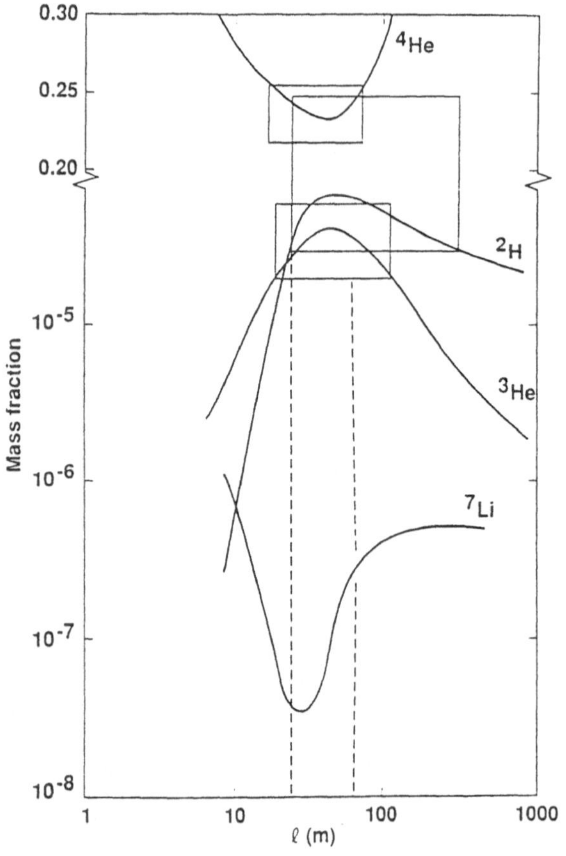

Figure 6. Light element yields (Ω = 1)

VII. CONCLUSION

It is clear that quark phases can have significant astrophysical consequences for compact stars and for primordial nucleosynthesis. Significant consequences occur if zero temperature quark matter is absolutely stable, and if the host quark-hadron phase transition is first order. For the purposes of astrophysics and cosmology, the validity of those two hypotheses are the most interesting problems in QCD.

REFERENCES

1. Alcock, C., and Olinto, A. ,1988, Ann. Rev. Nucl. Part. Sci. 38:161.
2. Boesgaard, A. M., and Steigman, G., 1985, Ann. Rev. Astron. Ap., 23:319.
3. Alcock, C., et al, 1989, Nucl. Phys. A., 498:301.
4. Witten, E., 1984, Phys. Rev., D30:272.
5. Farhi, E., Jaffe, R. L., 1984, Phys. Rev., D30:2379.
6. Alcock, C., Farhi, E., 1985, Phys. Rev., D32:1273.
7. Haensel, P., Zdunik, J. L., Schaeffer, R., 1986, Astron. Astrophys., 160:121.
8. Alcock, C., Farhi, E., Olinto, A., 1986, Astrophys. J., 310:261.
9. Shapiro, S. L., Teukolsky, S. A., 1983, "Black Holes, White Dwarfs and Neutron Stars", Wiley, New York.
10. Baym, G., Pethick, C., 1975, Ann. Rev. Nucl. Part. Sci., 25:27.
11. Baym, G., Pethick, C., 1979, Ann. Rev. Nucl. Part. Sci., 17:415.
12. Baym, G., Pethick, C. J., Sutherland, P. G., 1971, Astrophys. J., 170:415.
13. Pines, D., Alpar, M. A., 1985, Nature, 316:27.
14. Olinto, A. V., 1987, Phys. Lett., 192:71.
15. Peebles, P. J. E., 1980, "The Large Scale Structure of the Universe", Princeton University Press, Princeton.
16. Weinberg, S., 1972, Gravitation and Cosmology", Wiley, New York.
17. Trimble, V., 1987, Ann Rev. Astron.Ap. 25:425.
18. Guth, A. H., 1981) Phys. Rev., D23:347.
19. Alcock, C. R., Fuller, G. M., and Mathews, G. J., 1987, Astrophys. J., 320:439.
20. Fuller, G. M., Mathews, G. J., and Alcock, C., 1988, Phys. Rev., D37:1380.
21. Kajantie, K., and Kurki-Suonio, H., 1986, Phys. Rev., D34:1719.
22. Miller, J. C., and Pantano, O., 1989, Phys. Rev., D40:1789.
23. Applegate, J. H., and Hogan, C., 1985, Phys. Rev., D30:3037.
24. Miller, J. C., and Pantano, O., 1989, Phys. Rev., D40:1789.
25. Mathews, G. J., Alcock, C., and Fuller, G. M., 1990, Ap. J. in press.

STRANGENESS IN THE SKYRME MODEL

Igor Klebanov*
Department of Physics
Joseph Henry Laboratories
Princeton University
Princeton, NJ 08544

ABSTRACT

These lectures review the fundamentals of the Skyrme model of baryons and its relationship with the non-relativistic quark model and the large-N QCD. Special emphasis is placed on the problems the model experiences in perturbative treatment of $SU(3)$ breaking. It is argued that the strange quark mass m_s is large enough to invalidate first-order perturbation theory in m_s. Because of the failure of the perturbation theory, an alternate approach to strangeness is introduced where hyperons emerge as kaon-soliton bound states. Various applications of this approach are discussed, with emphasis on its striking similarity with the quark model.

1. INTRODUCTION

In the late 50's and early 60's Skyrme[1] proposed a unified theory of mesons and baryons, based on a non-linear chiral lagrangian with only meson fields appearing explicitly. Skyrme suggested that the baryons appear in this theory in the form of stable topological solitons. Conservation of the topological winding number then explains the observed baryon number conservation. At the time, separate fermionic fields were introduced to represent baryons, and the idea that lumps of bosonic matter could somehow act as fermions seemed extremely far-fetched. Consequently, Skyrme's theory was ignored for many years.

Several years after Skyrme's papers were published, Gell-Mann and Zweig independently proposed that each baryon consists of 3 quarks, and each meson — of a quark and an antiquark. Deep inelastic scattering revealed that hadrons are, indeed, composed of point-like constituents. In the early 70's, Quantum Chromodynamics (QCD), the theory of interacting colored quarks and gluons, emerged as the fundamental description of hadrons. Although the quark model of Gell-Mann and Zweig was a historical and logical precursor of QCD, it turned out to be difficult to connect to QCD directly. One can argue qualitatively, that the same effects that confine the light quarks to the inside of hadrons,

* Supported in part by DOE grant DE-AC02-76WRO3072

Hadrons and Hadronic Matter
Edited by D. Vautherin *et al.*
Plenum Press, New York, 1990

give rise to rather complicated collective excitations which, at low energies, behave as rather massive ($\sim 350 MeV$) "constituent" quarks. Although such an argument is difficult to make precise, the non-relativistic quark model (NRQM) is remarkably successful for a theory rather loosely connected to QCD.

On the surface, it appears that the view of a baryon as a composite of 3 quarks has nothing to do with Skyrme's idea that it is a stable lump of pionic matter. However, the early 80's witnessed a remarkable revival of the Skyrme model. At that time, it was first realized that the model may be related to QCD in a rather non-trivial way.[2] Some of the particularly compelling arguments were constructed by Witten,[3,4] who was building on 't Hooft's idea[5] of generalizing QCD to a large number of colors. In the 3-flavor Skyrme model, Witten identified the crucial extra term, the Wess-Zumino term,[6] which, as Skyrme envisioned, turns the seemingly bosonic solitons into fermions. The newly discovered connection with QCD has provided the impetus for detailed exploration of the phenomenology of skyrmions. In many applications, it was found that the Skyrme model is not just a good qualitative guide to the properties of baryons, but also provides quantitative information, typically within 30% of the observations.[12–14] However, after some initial success, the model began running into some problems, in particular, in its attempts to describe strange baryons. In the unbroken $SU(3)$ limit, the model predicts the correct baryon multiplets.[7–11] However, if one attempts to describe the $SU(3)$ breaking by first order perturbation theory in the strange quark mass, one finds wrong mass relations among the lightest hyperons.[9–11] In these lectures, I will try to make a case that this perturbative approach breaks down not due to a fundamental failure of the Skyrme model, but because the strange quark is heavy enough to affect the baryon states non-trivially.[15–18] I will develop an approach to strangeness which should be applicable for m_s on the order of the strong interaction scale Λ_{QCD}. In this approach, the starting point is the chiral lagrangian expanded to second order in the kaons fields, and hyperons emerge naturally as kaon-soliton bound states.[19] This description of baryons turns out to have striking similarities with the NRQM. These similarities can be traced to the fact that a kaon bound to a skyrmion acts like a "constituent" strange quark.[19] Further exploration of the phenomenology in this bound state approach to strangeness reveals that essentially all the predictions are qualitatively and quantitatively similar to those in the NRQM. This is remarkable, given that, in the Skyrme model, baryon properties are *derived* from the chiral lagrangian, with only a few reasonable approximations.

Since a number of nice general reviews of the Skyrme model are available, these lectures are not intended to be comprehensive. I will first review the basic results of the rigid rotator treatment of skyrmions, emphasizing its connections with QCD and the NRQM. Subsequently, I will focus on the problems one encounters in the treatment of strangeness, and on their possible resolutions. These lecture notes are organized as follows. In section 2, I review the large-N approach to QCD of 't Hooft and Witten, from which the identification of baryons with solitons follows naturally. Section 3 is devoted to the collective coordinate quantization of solitons in the 2-flavor Skyrme model.[12] In section 4, I review the 3-flavor generalization, and its crucial extra ingredient, the Wess-Zumino term. In section 5, the 3-flavor rigid rotator model is treated in an unconventional way,[16] which allows for a better inclusion of the strange quark mass. In section 6, a further improvement is introduced which departs from the rigid rotator approximation. The resulting bound state approach to strangeness[19] is shown to produce good parameters in the hyperon mass formula.[22] Finally, in section 7, I review further checks on the bound state approach and emphasize its remarkable similarity to the NRQM.

2. BARYONS AND THE LARGE-N QCD

One of the reasons why it is hard to describe the low-energy behaviour of QCD is its asymptotic freedom. Because of renormalization effects, at each energy scale the theory is described by an effective lagrangian with an energy-dependent (running) coupling constant. As the energy scale increases, the coupling constant gets weaker, rendering perturbation theory in the coupling more and more reliable. As a consequence, the high-momentum processes, such as the deep inelastic scattering, are well described by QED-like Feynman graphs with quark and gluon lines – the degrees of freedom explicitly appearing in the QCD lagrangian. On the other hand, in the infrared, where the running coupling becomes strong, conventional perturbative techniques become inapplicable. This is the regime where the qualitatively new phenomena of confinement and chiral symmetry breaking take place. Many years of effort have produced a number of "non-perturbative" approaches to these problems which are, unfortunately, not nearly as reliable as the well-tested diagrammatic expansion in the perturbative domain. Some of these approaches are discussed in other lectures appearing in this volume. Typically, they offer a rather qualitative picture of confinement and chiral symmetry breaking. Some of the most convincing and quantitatively promising results have been obtained with the lattice regularization. Unfortunately, the sheer numerical complexity of the lattice gauge theory is so high that a good quantitative understanding of hadron properties does not appear to be within reach in the near future.

When we deal with a theory which we cannot solve exactly, the best hope is to solve it in perturbation theory. The problem in the low-energy QCD is that the effective coupling constant is large and cannot be used as an expansion parameter. In search of a less obvious expansion parameter, 't Hooft proposed to generalize QCD to gauge groups $SU(N)$.[5] His idea was that the theory could become simpler in the limit of the infinite number of colors. The physically relevant case of $N = 3$ could then be approximated by constructing a systematic expansion in powers of $1/N$. Although it was not *a priori* clear whether $N = 3$ is large enough to make the $1/N$ expansion useful, the development of 't Hooft's idea has produced a number of qualitative, if not quantitative, insights into the structure of hadrons. Here I will review the large-N approach to QCD following Witten's seminal paper, ref. 3.

Let us write down, as a starting point, the lagrangian for QCD generalized to N colors:

$$\mathcal{L} = \bar{q}_{i\alpha}(i\gamma^\mu\partial_\mu - m)q^i_\alpha - ig\bar{q}_{i\alpha}\gamma^\mu A^i_{\mu j}q^j_\alpha + \tfrac{1}{2}\,\mathrm{Tr}(F_{\mu\nu}F^{\mu\nu}), \tag{1}$$

where $A^i_{\mu j}$ are traceless anti-hermitian $N \times N$ matrices and the field strength

$$F_{\mu\nu} = \partial_\mu A_\nu - \partial_\nu A_\mu - g[A_\mu, A_\nu]. \tag{2}$$

The color indices i, j run from 1 to N, and the flavor index α runs from 1 to N_f, the number of flavors. When considering the diagrammatic structure of the large-N theory, the crucial task is to classify various Feynman graphs according to their N-dependence. The simplification that occurs in the large-N limit is that only a special class of diagrams, the planar diagrams, remains non-vanishing.[5] One can also classify the

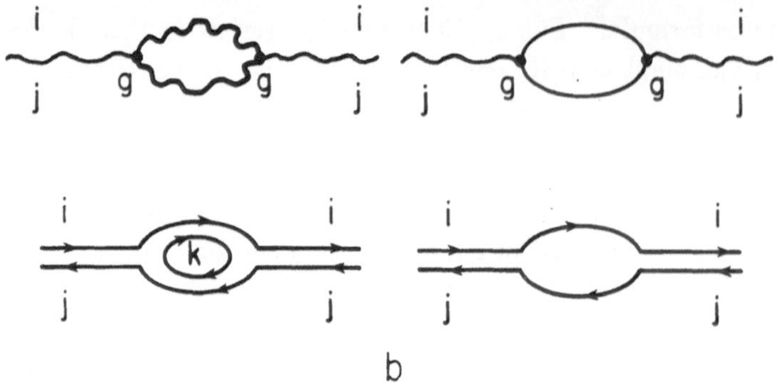

Fig. 1. One-loop corrections to the gluon propagator are shown both in the conventional and in the double-line notation: a) the gluon loop; b) the quark loop.

subleading diagrams according to their N-dependence. The whole classification follows from a simple observation that the quark fields have $O(N)$ components while the gluon fields have $O(N^2)$ components. For the purpose of index-counting, one can think of a gluon matrix A^i_j as a quark-antiquark pair $q^i\bar{q}_j$. Thus, every gluon line in a graph can be replaced by a quark line and an antiquark line. Consider now the one-loop corrections to the gluon propagator (Fig. 1), which are shown both in the conventional and in the double line notation. In Fig. 1a, the color index k of the internal line can assume any value from 1 to N. This multiplies the graph by an extra factor of N, compared to the graph in Fig. 1b. As you see, the double line notation makes the N-counting easy: every closed line contributes a factor of N. We have found that the gluon loop correction is $\mathcal{O}(g^2N)$ while the quark loop correction is only $\mathcal{O}(g^2)$. In general, every extra internal quark loop lowers the value of the graph by a power of N.

In order to achieve a smooth large-N limit, we have to adjust the coupling constant g so that the one-loop correction to the gluon propagator is of the same order in N as the tree-level result. Otherwise, higher loops bring higher powers of N, and the theory diverges uncontrollably. Therefore, we rescale g so that $g^2N \to const$, *i.e.*, $g \sim 1/\sqrt{N}$. As we proceed to the more complicated graphs, we will confirm that this is the only assignment that leads to an interesting theory in the large-N limit.

Now, consider a set of quark bilinears J, such as $\bar{q}q$, $\bar{q}\gamma^\mu q$, $\bar{q}\gamma_5 q$, *etc.* The simplest contribution to the two-point function $\langle J(x)J(y)\rangle$ is shown in Fig. 2a and is $\mathcal{O}(N)$ because the graph has one closed quark line. A gluon correction shown in Fig. 2b is $\sim g^2N^2 \sim N$, which is made obvious by the double-line notation of Fig. 2c. You can convince yourself that every graph that can be simply laid out on a plane, such as in Fig. 2d, is $\mathcal{O}(N)$. In such graphs, for every factor of g^2 coming from the vertices, there is an extra index contraction giving a factor of N. Thus, all planar graphs are of the same order in N, and need to be summed in order to determine the leading behavior of $\langle J(x)J(y)\rangle$ in the large-N limit. The simplest subleading graph is shown in Fig. 3. Because one of the gluon lines has to go underneath another, this graph cannot be laid out on a plane. Going to the double line notation, Fig. 3b, we observe that this graph has only 1 closed line and 4 vertices. Therefore, it is $\sim g^4 N \sim 1/N$, *i.e.*, the simplest nonplanar graphs are suppressed by $\mathcal{O}(1/N^2)$ compared to the planar graphs.

An interesting feature of the graphs in Fig. 2 is that, at every point in time, there is only one quark-antiquark pair present. The sum of such planar graphs, which are of the leading order in the large-N expansion, can be interpreted as creation, propagation, and destruction of mesons. Thus, to the leading order in N, the two-point function of J has only single-meson poles:

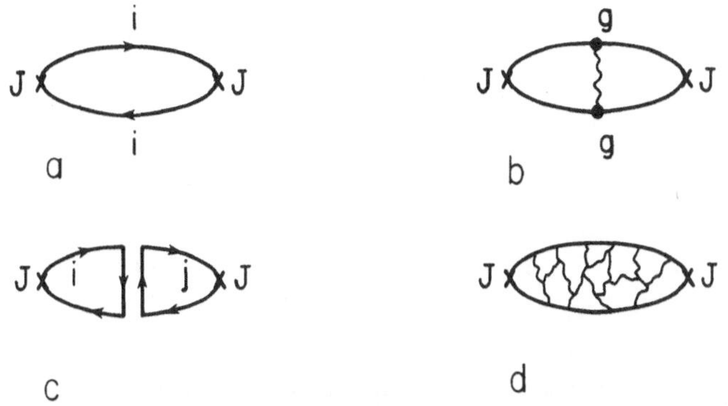

Fig. 2. The 2-point function of quark bilinears J: a) the lowest order graph; b) a 1-gluon correction; c) graph b) in the double-line notation; d) a complicated planar graph with no internal quark loops.

Fig. 3. a) The simplest non-planar graph. b) The same graph in the double-line notation.

$$\langle J(k)J(-k)\rangle = \sum_n |\langle 0|\, J(k)\, |n\rangle\,|^2 \frac{1}{k^2 - m_n^2} \sim N. \qquad (3)$$

Since the meson masses are $\mathcal{O}(1)$, $\langle 0|\, J\, |n\rangle \sim \sqrt{N}$, *i.e.*, the current J acting on the vacuum creates mesons with amplitude $\sqrt{N}$. In other words, meson decay constants f_i (such as f_π) are $\mathcal{O}(\sqrt{N})$.

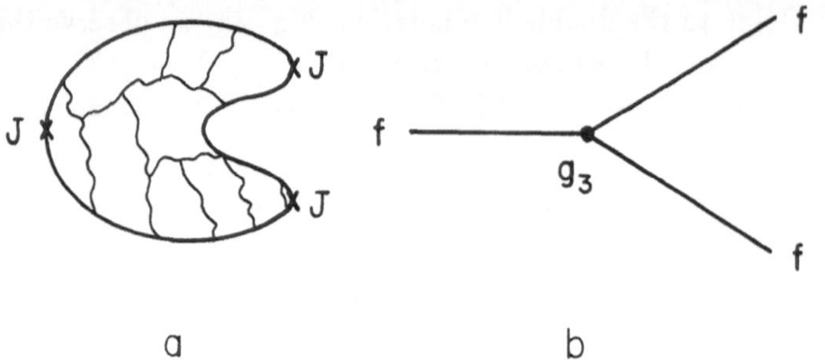

Fig. 4. The 3-point function of J: a) a typical planar contribution; b) a representation in terms of mesons.

We proceed now to the 3-point functions of the operators J. A typical planar contribution is shown in Fig. 4a. If we think of this graph as a sequence of events, then it represents a process where a meson is created, and later decays into 2 other mesons which are subsequently annihilated. In an equivalent field theory for mesons, this process, whose amplitude is $\mathcal{O}(N)$, is shown in Fig. 4b. In terms of the mesonic parameters, the amplitude for this process is $\sim f^3 g_3$, where g_3 is the three-meson coupling. Recalling that $f \sim \sqrt{N}$, we determine $g_3 \sim 1/\sqrt{N}$. Proceeding analogously to the 4-point functions of the J's we find $g_4 \sim 1/N$. In general, one finds that

$$g_{2+n} \sim N^{-n/2}, \tag{4}$$

i.e., all meson interactions vanish in the large-N limit. Since there is an infinite number of quark bilinears, the $N = \infty$ QCD has an infinite tower of stable mesons.

In addition to free mesons, the theory contains free glueballs. The coupling of mesons to glueballs becomes weak in the large-N limit. To see this, we observe that the graph in Fig. 5a is O(1): it is suppressed by a power of N compared to the graph in Fig. 4a. In the meson-glueball language, Fig. 5a can be represented as Fig. 5b: a meson turns into a pure glue state and back into a meson which subsequently decays. Our N counting indicates that the meson-glueball coupling constant g_{mg} is $\mathcal{O}(1/\sqrt{N})$. Another interpretation of this result is that the OZI rule holds exactly in the large-N limit: all processes where a quark and an antiquark annihilate into gluons are exactly forbidden. It seems that this large-N argument is the only simple way to explain why the OZI rule is approximately obeyed in nature.[3]

Another property of the large-N QCD is the suppression of the quark-antiquark (sea) fluctuations. Indeed, the fact that the diagram in Fig. 6a is $1/N$ times the diagram in Fig. 2d means that the processes where virtual $\bar{q}q$ pairs are created and destroyed are completely forbidden in the large-N limit. As N is decreased, in addition to the planar

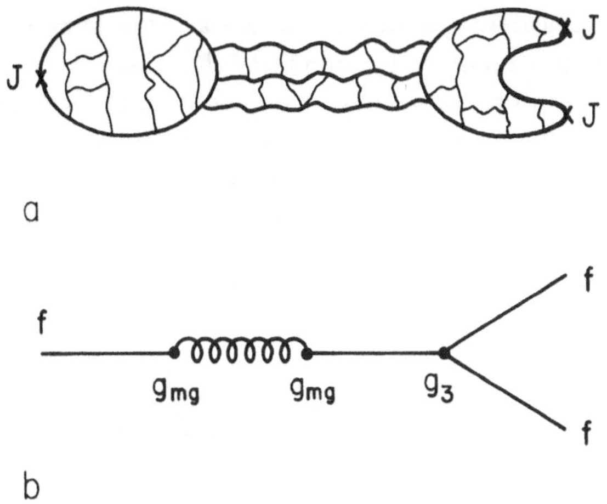

Fig. 5. An OZI-suppressed process: a) a typical graph; b) the representation in the equivalent meson-glueball field theory.

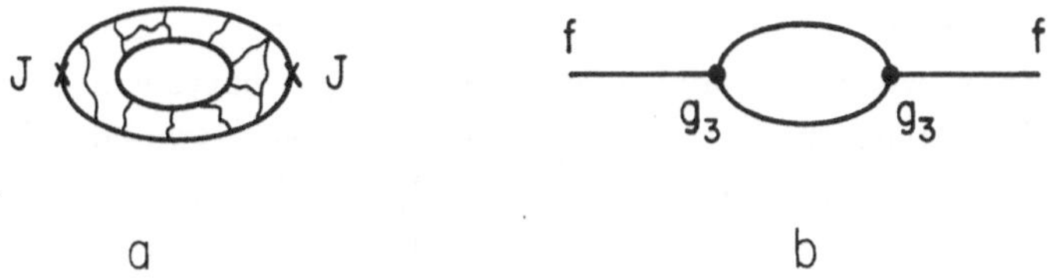

Fig. 6. A typical process with an internal quark loop (a). The sum over such graphs is equal to the sum over 1-loop graphs (b) in the equivalent meson-glueball field theory, with mesons propagating in the loop.

graphs, we have to include graphs of increasing topological complexity. In the limit where gluon lines become dense, the graphs can be viewed as surfaces. These surfaces may contain holes, such as the internal quark loop correction of Fig. 6a, or handles, which are the consequences of non-planarity of gluon lines. Every such class of graphs can be translated into the language of a field theory of interacting mesons and glueballs. For example, the topology of Fig. 6a corresponds, in the meson-glueball theory, to a sum over 1-loop graphs, with mesons propagating in the loop (Fig. 6b). In this sense, QCD with a finite number of colors is a theory of interacting mesons and glueballs. The expansion in powers of $1/N$, which involves Feynman graphs of increasing topological complexity, is similar to the perturbative expansion in string theory. In this analogy, mesons can be thought of as open string states, and glueballs – as closed string states.

If $1/N$ expansion in QCD can be represented by a field theory of interacting mesons and glueballs, the immediate question is: where are the baryons in this picture? In answering this question, we assume for simplicity that there is only one flavor of quarks. In a quark model with three colors, baryons do not seem to be drastically different from mesons. Meson wave functions $q^i\bar{q}_i$ involve a quark and an anti-quark summed over all possible colors; baryon wave functions are constructed out of three quarks having different colors: $q^i q^j q^k \epsilon_{ijk}$. Now, if the number of colors is large, a meson still contains a quark and an anti-quark, but a baryon must have N quarks in order to antisymmetrize N color indices: $B \sim q^{i_1} q^{i_2} \ldots q^{i_N} \epsilon_{i_1 i_2 \ldots i_N}$. Including only effective 2-body forces among quarks, the mass of a baryon can be expressed as

$$M_B \sim N m_q + N T_q + V_{pair} \frac{N(N-1)}{2} \sim N. \tag{5}$$

m_q is the current quark mass which, at first, we take to be large enough to treat each quark non-relativistically. T_q is the quark kinetic energy, and $V_{pair} \sim 1/N$ is the two-body potential, which is found by summing graphs with arbitrary numbers of gluons connecting 2 quark lines. According to Eq. (5), in the large-N limit, all baryon masses are $\mathcal{O}(N)$. This is the reason why baryons do not appear in the $1/N$ expansion. For instance, production of a baryon and an antibaryon by a pair of mesons is a non-perturbative process which cannot be simply described in terms of Feynman diagrams. In a seminal paper,[3] Witten argued that, when the number of colors is large, baryons are best thought of in the Hartree-Fock approximation. Indeed, if only the quark two-body interactions are included, we find that each quark feels a force $\mathcal{O}(1)$ which is composed of the forces due to $N-1$ other quarks, each one $\mathcal{O}(1/N)$. Thus, when N is large, a mean field approach should be a very good approximation. In the non-relativistic Hartree-Fock ground state, all quarks are in the same spatial wave function, which could be determined self-consistently if the interactions were known exactly. To determine the quark interactions one needs to sum infinite classes of planar graphs, which is not feasible. Even without a detailed knowledge of the interactions, Witten used the Hartree-Fock analysis to derive a number of interesting properties of baryons in the large-N limit.[3] In addition to the fact that all baryon masses are $\mathcal{O}(N)$, he found that baryon-baryon interactions are $\mathcal{O}(N)$, meson-baryon interactions are $\mathcal{O}(1)$, baryon sizes are $\mathcal{O}(1)$, and the mass splittings between low-lying baryons are $\mathcal{O}(1/N)$. All these properties are easy to derive in the non-relativistic Hartree-Fock approximation, but Witten also argued that they continue to hold as the current quark mass is dialed to zero. Remembering that the QCD coupling constant g is $\sim 1/\sqrt{N}$, Witten's conclusions can be restated as follows: baryon masses are $\mathcal{O}(1/g^2)$, the lowest mass splittings are $\mathcal{O}(g^2)$, *etc.* These can be recognized as the general characteristics of topological solitons, such as the 't Hooft-Polyakov monopole. In the weak coupling limit, solitons are heavy rigid objects, whose motion can be treated using semiclassical methods. As mentioned earlier, the idea that baryons can be thought of as topological solitons in mesonic theories was developed by T. Skyrme[1] long before the advent of QCD. Skyrme's theory had little influence at the time of its appearance, and was largely ignored for many years until the large-N approach of 't Hooft[5] and Witten[3] showed how it can be connected with QCD. After its revival in the 80's, the Skyrme model has led to quite a number of conceptual insights into the baryon structure. The development of Skyrme's remarkable ideas is the main subject of these lectures, which will be addressed beginning with the next section.

3. THE 2-FLAVOR SKYRME MODEL

According to the results reviewed in section 2, QCD is equivalent to a theory of an infinite number of interacting mesons and glueballs. However, it is extremely difficult and cumbersome to rewrite the QCD lagrangian in terms of these degrees of freedom. In order to test the idea that baryons can be identified with solitons, we will, following Skyrme, truncate the degrees of freedom down to the lightest mesons which would become massless if the light quark masses were identically zero. As a first try, let us consider a world with two light flavors, up and down, and take into account only the pionic degrees of freedom.

Pion interactions for energies up to about 600 MeV are well described[23] by the non-linear chiral lagrangian[1]

$$\mathcal{L}_{Sk} = \frac{F_\pi^2}{16}\,\mathrm{Tr}(\partial_\mu U^\dagger \partial^\mu U) + \frac{1}{32e^2}\,\mathrm{Tr}[\partial_\mu U U^\dagger, \partial_\nu U U^\dagger]^2 i + \frac{m_\pi^2 F_\pi^2}{16}\,\mathrm{Tr}(U + U^\dagger - 2) \tag{6}$$

where $U(\vec{x},t) = \exp(2i\vec{\tau}\cdot\vec{\pi}/F_\pi) \in SU(2)$, $F_\pi = 186 MeV$ and $e \approx 5.6$. Since the pion mass is small, we will ignore it throughout the lectures. The flavor transformation properties of U can be understood through the identification

$$U_\alpha^\beta \sim \left\langle \bar{q}_{Ri}^\beta q_{L\alpha}^i \right\rangle \tag{7}$$

where the right-handed and the left-handed quark fields are defined by

$$q_L = \tfrac{1}{2}(1+\gamma_5)q, \qquad q_R = \tfrac{1}{2}(1-\gamma_5)q. \tag{8}$$

If the quark masses are zero, then the QCD lagrangian is invariant under independent flavor rotations acting on the right-handed and the left-handed quarks. Under $SU(2)_L$,

$$q_{L\alpha} \to A_\alpha^\beta q_{L\beta}, \tag{9}$$

and under $SU(2)_R$,

$$q_{R\alpha} \to B_\alpha^\beta q_{R\beta}. \tag{10}$$

In view of Eq. (7), under the $SU(2)\times SU(2)$ chiral rotations, $U \to AUB^{-1}$. The vacuum value $U=1$ breaks $SU(2)_L \times SU(2)_R$ down to the diagonal ($A = B$) subgroup, the isospin. If we expand U in pion fluctuations near 1,

$$U = 1 + \frac{2i\vec{\tau}\cdot\vec{\pi}}{F_\pi} + \ldots, \tag{11}$$

and assign the N-dependences $F_\pi \sim \sqrt{N}$, $1/e \sim \sqrt{N}$, then, as you can easily check, the multi-pion vertices derived from the chiral lagrangian (6) are consistent with the large-N rule (4) derived in section 2.

Skyrme showed[1] that, in addition to pions, which are small fluctuations about $U=1$, the chiral lagrangian (6) has topologically stable static solutions where U takes values all over $SU(2)$. In order to explain why such solutions exist, let us parametrize the $SU(2)$ field as

$$U = u_0 + i\vec{u}\cdot\vec{\tau}. \tag{12}$$

The unitarity $UU^\dagger = 1$ translates into the constraint $u_o^2 + \vec{u}^2 = 1$, *i.e.*, the $SU(2)$ group manifold has the topology of a 3-sphere S^3. Consider now the static configurations $U(\vec{r})$ such that, as $\vec{r} \to \infty$, $U \to 1$ sufficiently rapidly. In considering these functions, we may think of all the points in space with $\vec{r} \to \infty$ as identified into one point. This replaces the spatial topology R^3 with S^3.* Thus, the interesting static solutions to the non-linear chiral theory are smooth maps from S^3 (space) into S^3 ($SU(2)$). Such maps fall into discrete homotopy classes which are labeled by an integer, the winding number. The simplest situation where winding numbers arise is a lower-dimensional analogue: maps from a circle into another circle. The topologically non-trivial maps can be thought of as follows. The circle, which is the base space of the map, is represented by a rubber band which is cut open, wound N times around the target space circle, and glued together at the ends. No matter how the band is stretched or rearranged, the winding number N cannot be changed, unless the band is torn, *i.e.* the winding number is invariant under any continuous deformations of the map.

In 3 dimensions, the winding number can be expressed as

$$B = \frac{1}{24\pi^2} \int d^3x \epsilon^{ijk} \operatorname{Tr}(U^\dagger \partial_i U U^\dagger \partial_j U U^\dagger \partial_k U). \tag{13}$$

Since (13) is invariant under any smooth deformations of $U(\vec{x})$, it is conserved in any physical process. For that reason, Skyrme proposed to identify the winding number of the solitons with the baryon number. In fact, one can further define the baryon number current

$$B^\mu = \frac{1}{24\pi^2} \epsilon^{\mu\alpha\beta\gamma} \operatorname{Tr}(U^\dagger \partial_\alpha U U^\dagger \partial_\beta U U^\dagger \partial_\gamma U). \tag{14}$$

You can check that $\partial_\mu B^\mu = 0$ identically, without the use of the equations of motion. This is another way to see that B is conserved in any smooth evolution of the field.

The simplest class of non-trivial configurations are the one-to-one maps from space into $SU(2)$ which clearly have the winding number equal to 1. To construct a soliton solution within this class, Skyrme used the hedgehog ansatz

$$U_0(\vec{x}) = \exp(iF(r)\hat{r} \cdot \vec{\tau}) \tag{15}$$

where F satisfies the boundary conditions $F(0) = \pi$, $F(r) \to 0$ as $r \to \infty$. If $F(r)$ is monotonic, then (15) is a one-to-one map, with the origin mapped into the south pole of $SU(2)$, $U_0(0) = -1$, and the point at ∞ mapped into the north pole, $U_0(\infty) = 1$. Indeed, the formula for the winding number yields, after some algebra,

$$B = -\frac{2}{\pi} \int_0^\infty F' \sin^2 F dr = \frac{F(0) - F(\infty)}{\pi} = 1. \tag{16}$$

The special property of the hedgehog ansatz is that, for any profile function $F(r)$, U_0 does not change under the combined spatial and isospin rotations

$$(\vec{I} + \vec{L})\, U_0 = \left[\frac{\vec{\tau}}{2}, U_0\right] - i\vec{r} \times \vec{\nabla} U_0 = 0. \tag{17}$$

The solution within the $B = 1$ sector is found by minimizing the soliton's classical mass with respect to $F(r)$ subject to the boundary conditions above. If there was no

---

* In order to visualize this, think of a 2-dimensional analogue: if the boundary of a large disc is identified into one point, it acquires the topology of a 2-sphere. Recall, for instance, the familiar example from complex analysis: a plane, together with the point at ∞, is in a one-to-one correspondence with the Riemann sphere, obtained through stereographic projection.

4-derivative term in Eq. (6), the hedgehog would shrink to zero size because scaling down would lower its energy. (This is known as Derrick's theorem.) Skyrme introduced the commutator term into the lagrangian as the simplest stabilizer of static solitons. Luckily, this term, which mimicks the ρ-exchange, seems to account for the low-energy pion interactions.[23] Generally, the effective pion lagrangian includes terms up to an infinite order in derivatives which arise after the heavier mesons are integrated out. Discarding these higher-derivative terms is not *a priori* justified because, on the scale of the soliton size, these terms are not suppressed. Thus, the truncation of the lagrangian probably discards some of the important QCD effects. Nevertheless, we will see that much of the essential physics is reflected in the Skyrme lagrangian (6), which is probably a reflection of the ρ-dominance.

By isospin invariance, if U_0 is a solution, so is AU_0A^{-1}. Because of this symmetry, the low-lying excitations are the slow rigid rotations

$$U(\vec{x},t) = A(t)U_0(\vec{x})A^{-1}(t). \tag{18}$$

In the simplest quantum mechanical treatment of the low-lying excitations, one only quantizes the motion of $A(t)$, the collective coordinate.[12] Substituting Eq. (18) into Eq. (6), we find the collective coordinate lagrangian

$$L = -M_{cl} + \Omega\,\mathrm{Tr}(\partial_0 A \partial_0 A^{-1}) \tag{19}$$

where Ω is the soliton moment of inertia. In order to quantize this lagrangian, it is convenient to use the parametrization

$$\begin{aligned} A &= a_0 + i\vec{a}\cdot\vec{\tau}, \\ a_0^2 &+ \vec{a}^2 = 1. \end{aligned} \tag{20}$$

Then Eq. (19) turns into

$$L = -M_{cl} + 2\Omega \sum_{i=0}^{3} (\dot{a}_i)^2. \tag{21}$$

The hamiltonian

$$H = M_{cl} + \frac{1}{8\Omega}\sum_{i=0}^{3} \pi_i^2 \tag{22}$$

is canonically quantized through the replacement $\pi_i \to -i\partial/\partial a_i$:

$$H = M_{cl} + \frac{1}{8\Omega}\sum_{i=0}^{3}\left(-\frac{\partial^2}{\partial a_i^2}\right) = M_{cl} - \frac{1}{8\Omega}\nabla^2 \tag{23}$$

where, because of the constraint (20), ∇^2 is the laplacian on the 3-sphere. One of the purposes of the collective coordinate quantization is to construct baryon wave functions which are eigenstates of the isospin and the angular momentum. The generators of rotations and isospin rotations are found through the Noether procedure to be

$$\begin{aligned} J_k &= -i\,\Omega\,\mathrm{Tr}(\tau_k A^{-1}\dot{A}), \\ I_k &= -i\,\Omega\,\mathrm{Tr}(\tau_k A\dot{A}^{-1}). \end{aligned} \tag{24}$$

Classically, they satisfy $\vec{J}^2 = \vec{I}^2$. Upon canonical quantization, from Eq. (24) we obtain the following operators,

$$J_k = \tfrac{1}{2}i\left(a_k\frac{\partial}{\partial a_0} - a_0\frac{\partial}{\partial a_k} - \epsilon_{klm}a_l\frac{\partial}{\partial a_m}\right), \tag{25}$$

$$I_k = \tfrac{1}{2}i\left(-a_k\frac{\partial}{\partial a_0} + a_0\frac{\partial}{\partial a_k} - \epsilon_{klm}a_l\frac{\partial}{\partial a_m}\right), \tag{26}$$

which satisfy the operator identity

$$\vec{J}^2 = \vec{I}^2 = -\frac{1}{4}\nabla^2. \tag{27}$$

Thus, the wave functions satisfy the $I = J$ rule. Since A and $-A$ give the same matrix U in Eq. (18), there are two consistent ways to quantize the soliton.[24]

1. The wave functions satisfy the constraint $\psi(A) = \psi(-A)$. Since under a 2π rotation

$$A \to A\exp(i\pi\tau_3) = -A, \tag{28}$$

this rule restricts the soliton to bosonic states. For instance, consider the wave functions

$$\chi_{l/2}(A) \sim (a_0 + ia_1)^l \tag{29}$$

You can easily check that they are eigenstates of spin and isospin with $I = J = l/2$. Thus, for an even l, the soliton is indeed a boson.

2. The wave functions satisfy the constraint $\psi(A) = -\psi(-A)$. Now the soliton is a fermion, which is confirmed by the example of wave functions (29) with an odd l.

As we will demonstrate in section 4, in the 3-flavor Skyrme model the solitons are required to be fermions if the number of colors N is odd, and to be bosons if N is even. In the 2-flavor case, we simply adopt the fermionic option. The allowed quantum numbers are $I = J = 1/2, 3/2, \ldots$ The $I = J = 1/2$ states are identified with the nucleons, and the $I = J = 3/2$ states – with the deltas. Their wave functions are given by[12]

$$\begin{gathered} |p\uparrow\rangle = \frac{1}{\pi}(a_1 + ia_2), \qquad |n\uparrow\rangle = \frac{i}{\pi}(a_0 + ia_3), \\ |\Delta^{++},\ J_z = 3/2\rangle = \frac{\sqrt{2}}{\pi}(a_1 + ia_2)^3, \end{gathered} \tag{30}$$

and so forth. In order to find the baryon masses, we use Eq. (27) to write the hamiltonian as

$$H = M_{cl} + \frac{1}{2\Omega} J(J+1). \tag{31}$$

It follows that

$$M_N = M + \frac{3}{8\Omega}, \qquad M_\Delta = M + \frac{15}{8\Omega}, \tag{32}$$

where the parameters are found numerically to be

$$M_{cl} \approx 36.5 \frac{F_\pi}{e}, \qquad \Omega \approx \frac{107}{e^3 F_\pi}. \tag{33}$$

If we fit F_π and e to the nucleon and delta masses, we find $F_\pi \approx 129 MeV$, which is about 30% off the observed value, and $e \approx 5.45$, which is quite close to the value found from $\pi - \pi$ scattering.[23] It is satisfying that the Skyrme model predicts N and Δ, but what are we to do with the states $I = J = 5/2, \ldots$? Actually, with the above fit, all these states have a rotational energy greater than the classical mass of the soliton: they rotate too fast for the rigid rotator approximation to be valid. We expect that, in a treatment where soliton deformations are included, all the higher spin states are destabilized by their rotation.[25] What is the situation in the hypothetical large-N world? By Eq. (33), the parameters of the mass formula M and Ω are both $\mathcal{O}(N)$. Therefore, only for states with $J \sim N$ does the rotational energy in Eq. (31) become comparable to the static soliton energy, and the rigid rotator treatment breaks down. Therefore, the isorotational excitations that appear are $I = J = 1/2, 3/2, \ldots, \mathcal{O}(N)$.

We would like to compare what we found in the Skyrme model with the predictions of the large-N QCD. In the Hartree-Fock approximation, the low-lying baryons have all the quarks in the same spatial orbital. By Fermi statistics, the spin-flavor state must be totally symmetric, which restricts the admissible representations[4] to $I = J = 1/2, 3/2, \ldots, N/2$, in agreement with the Skyrme model. To find what the quark model says about the masses of these baryons, we assume a magnetic moment interaction between every pair of quarks:[26]

$$\delta H = \lambda \sum_{i<k} \vec{j}_i \cdot \vec{j}_k = \frac{\lambda}{2} \left(J(J+1) - \frac{3}{4} N \right). \tag{34}$$

If we identify λ with $1/\Omega$, the spin dependence of the baryon mass is in agreement with the Skyrme model Eq. (31)! It seems almost miraculous that the two models of baryons, which seem so different, lead to such similar predictions. Since the Skyrme model provides a field theoretic description of baryons, based on the action which includes some dynamical information about the low-energy QCD, it is far more complete than the quark model. Thus, in the Skyrme model one can calculate a variety of baryon properties. Adkins, Nappi and Witten[12] carried out such calculations in the context of the 2-flavor model outlined above. They found that such properties of nucleons and deltas as the charge radii, magnetic moments, axial coupling, couplings to pions all come out within 30% of the experimentally observed values. Since the Skyrme action (6) is, undoubtedly, a rather crude approximation to QCD, this agreement is remarkable.

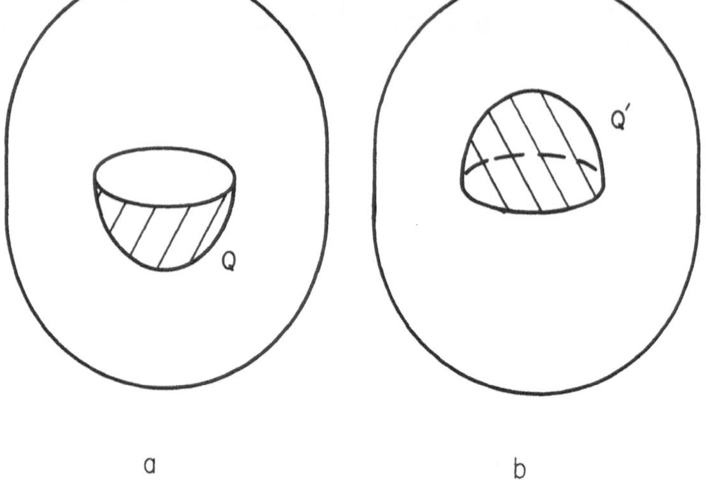

Fig. 7. The image of space-time in $SU(3)$ is shown as a circle. There are 2 topologically distinct discs in $SU(3)$, Q (a) and Q' (b), which have space-time as their boundary.

4. $SU(3)$ COLLECTIVE COORDINATE QUANTIZATION

We have seen that the 2-flavor Skyrme model is rather successful in describing the static properties of nucleons and deltas. The most natural extension of the model is to include the low-lying strange baryons. Since the strange quark is much lighter than the charm, we will first follow the standard route in hadron physics: we will begin with the model where u, d, and s quarks are massless, and find how the low-lying baryons fit into the $SU(3)$ multiplets. Next, we will attempt a perturbation theory in the quark masses, in order to find the mass splittings within the multiplets. To our surprise, we will find that, at this stage, the Skyrme model runs into serious difficulties, which may be due to the fact that the strange quark is much heavier than the up and down quarks. The rest of the lectures will be devoted to the attempts to solve the problems that arise in the treatment of the massive strange quark. First, however, we need to see what goes wrong with the standard, perturbative approach to the quark masses.

As a starting point, we will consider a hypothetical world with 3 massless flavors. The 3-flavor Skyrme model differs from the 2-flavor case in some important ways, which uncover a beautiful new structure. A straightforward generalization would be to adopt the lagrangian (6) with U now an $SU(3)$ field built out of the eight Goldstone bosons (pions, kaons, and the eta) as $U = \exp(2i\lambda_a M^a/F_\pi)$. For $a = 1$ to 3, $M^a = \pi^a$; for $a = 4$ to 7, $M^a = K^a$, and $M^8 = \eta$. Witten[4] has observed, however, that (6) cannot be the correct low-energy description of QCD with 3 flavors: it has a discrete symmetry not realized in QCD. Indeed, (6) is invariant under a "naive parity" operation P_0: $\vec{x} \to -\vec{x}$, $t \to t$; it is also invariant under $U \to U^\dagger$ or, equivalently, $M^a \to -M^a$, which is an operation that counts the number of Goldstone bosons modulo 2. In QCD these two symmetries are separately broken in processes like

$$K^+K^- \to \pi^+\pi^-\pi^0, \tag{35}$$

where an even number of Goldstone bosons produces an odd number. However, all QCD processes conserve the parity, defined by $\vec{x} \to -\vec{x}$, $t \to t$, $U \to U^\dagger$. It turns out that a manifestly $SU(3)$ symmetric action, which incorporates processes like in Eq. (35), is very subtle and cannot be written in a local form. In the cases where the space-time evolution is periodic, which will be of interest to us, the requisite extra term in the chiral action, called the Wess-Zumino term,[6] can be written as a 5-dimensional integral over a disc Q, whose boundary is space-time:[4]

$$S_{WZ} = -\frac{in}{240\pi^2} \int_Q d^5x \epsilon^{\mu\nu\alpha\beta\gamma} \operatorname{Tr}(U^\dagger \partial_\mu U U^\dagger \partial_\nu U U^\dagger \partial_\alpha U U^\dagger \partial_\beta U U^\dagger \partial_\gamma U). \tag{36}$$

where the function $U(\vec{x}, t)$ defined on the boundary has been extended smoothly into the interior of the disc. A fact that turns out to be crucial for the success of the soliton description of baryons is that the coefficient n in the normalization of the Wess-Zumino term must be an integer. To show this, we observe that Eq. (36) does not completely define the value of the action because $U(\vec{x}, t)$ can be extended over a disc in 2 topologically distinct ways: instead of the imbedding of the disc in $SU(3)$ denoted by Q (Fig. 7a), we could have chosen a topologically distinct imbedding denoted by Q' (Fig. 7b), which has the same boundary and therefore corresponds to the same space-time evolution of U. The difference between these two definitions is proportional to an integral over the 5-sphere consisting of the sum of the 5-discs Q and Q':

$$\Delta S_{WZ} = -\frac{in}{240\pi^2} \int_{Q+Q'} d^5x \epsilon^{\mu\nu\alpha\beta\gamma} \operatorname{Tr}(U^\dagger \partial_\mu U U^\dagger \partial_\nu U U^\dagger \partial_\alpha U U^\dagger \partial_\beta U U^\dagger \partial_\gamma U). \tag{37}$$

If n is an integer, Eq. (37) can be shown to be a multiple of 2π.[4] Since quantum mechanical amplitudes only involve the action in the combination $\exp(iS)$, this lack of singlevaluedness is harmless. Thus, action (36) makes quantum mechanical sense only if n is an integer.

The reader may find it strange that the Wess-Zumino action has to be written in a non-local form. In fact, a similar phenomenon occurs in a simpler system: a particle of charge e moving on the surface of a sphere surrounding a magnetic monopole.[4] The gauge potential of a monopole has a well-known Dirac string singularity. Thus, the interaction lagrangian, written in its local form $eA_i\dot{x}^i$, is singular at at least one point on the sphere. For any closed trajectory, the Stokes theorem allows us to write the contribution of this term in a non-singular form $e\int_D \vec{B} \cdot d\vec{A}$. The integral gives the magnetic flux through a disc D whose boundary is the particle trajectory, but this definition is not unique because there are 2 such discs. The difference between the two possible definitions of the action is $e\int_{sphere} \vec{B} \cdot d\vec{A} = 4\pi eg$. Thus, as in the case of the Wess-Zumino term in the chiral action, quantum mechanical consistency imposes a quantization condition on the coefficient of the action which is not single-valued. For the monopole, this is the familiar Dirac condition $4\pi eg = 2\pi n$. In section 6 we will see further analogy between the Wess-Zumino term and particle motion in magnetic field.

In order to find the integer parameter in the fundamental lagrangian of QCD, which corresponds to n in Eq. (36), Witten[4] has coupled the chiral theory to electromagnetism. Then S_{WZ} gives rise to an effective interaction responsible for the decay $\pi^0 \to \gamma\gamma$:

$$\mathcal{L}_{\pi\gamma\gamma} = \frac{ne^2}{48\pi^2 F_\pi} \pi^0 \epsilon^{\mu\nu\alpha\beta} F_{\mu\nu} F_{\alpha\beta} \tag{38}$$

whose normalization agrees with the famous triangle anomaly if n is taken to be equal to the number of colors N. Another interaction found after gauging the Wess-Zumino action is

$$\mathcal{L}_{\pi\pi\pi\gamma} = -\frac{2}{3} ie \frac{n}{\pi^2 F_\pi^3} \epsilon^{\mu\nu\alpha\beta} A_\mu \partial_\nu \pi^+ \partial_\alpha \pi^- \partial_\beta \pi^0 \tag{39}$$

which agrees with the QCD VAAA anomaly if $n = N$. In fact, if $n = N$, the gauged Wess-Zumino term incorporates all the effects of QCD anomalies in low-energy processes with photons and Goldstone bosons. Appropriate gauging of S_{WZ} can also be used to justify the identification of B^μ in Eq. (14) with the baryon number current.

The explicit appearance of the number of colors N in the chiral lagrangian is responsible for the dependence of the statistics of the solitons on N. To show this, Witten calculated the adiabatic phase produced by the Wess-Zumino term as the soliton is rotated through the angle 2π:

$$U(\vec{r}, t) = \mathcal{A}(t) \exp\left(iF(r) \sum_{i=1}^{3} \hat{r}_i \lambda_i \right) \mathcal{A}^{-1}(t) \tag{40}$$

where

$$\mathcal{A}(t) = \exp(it\lambda_3/2), \qquad 0 \le t < 2\pi, \tag{41}$$

and λ are the Gell-Mann matrices. Since time t is a periodic variable similar to an angle, space-time has the topology of S^1(a circle) $\times S^3$. In order to calculate the value of the Wess-Zumino action, Witten found a smooth map from D^2(a 2 − dimensional disc) $\times S^3$ into $SU(3)$ which on the boundary of the disc reduces to the 2π rotation of Eq. (41). For such a function, he found[4] $S_{WZ} = \pi N$ *i.e.*, as a soliton rotates through 2π, its wave function picks up the phase $(-1)^N$. Thus, the soliton is a fermion for an odd N, and it is a boson for an even N, in agreement with QCD! The effect of the Wess-Zumino term goes beyond this simple rule. In the 3-flavor Skyrme model with massless quarks, it imposes powerful constraints on the $SU(3)_{flavor} \times SU(2)_{spin}$ quantum numbers of the baryon multiplets.[7–11] To show this, we will carry out the collective coordinate quantization in the 3-flavor case, mostly following ref. 21.

After substituting the form (40) into the full action, the effective lagrangian for the $SU(3)$ collective coordinate $\mathcal{A}(t)$ is[21]

$$\begin{aligned} L(m_s = 0) = &-\frac{\Omega}{2} \sum_{j=1}^{3} \left(\mathrm{Tr}\, \lambda_j \mathcal{A}^\dagger \dot{\mathcal{A}} \right)^2 - \frac{\Phi}{2} \sum_{a=4}^{7} \left(\mathrm{Tr}\, \lambda_a \mathcal{A}^\dagger \dot{\mathcal{A}} \right)^2 \\ &- i \frac{N}{2\sqrt{3}} \left(\mathrm{Tr}\, \lambda_8 \mathcal{A}^\dagger \dot{\mathcal{A}} \right) , \end{aligned} \tag{42}$$

where Ω and Φ are two moments of inertia $\mathcal{O}(N)$ which are functionals of the hedgehog profile $F(r)$. It is convenient to rewrite Eq. (42) in terms of the local coordinates a_i on the SU(3) manifold, such that $\mathcal{A}^{-1}\dot{\mathcal{A}} = \frac{i}{2}\sum_{i=1}^{8}\lambda_i \dot{a}_i$

$$L(m_s = 0) = \frac{\Omega}{2}\sum_{j=1}^{3}(\dot{a}_j)^2 + \frac{\Phi}{2}\sum_{a=4}^{7}(\dot{a}_a)^2 + \frac{N}{2\sqrt{3}}\dot{a}_8. \tag{43}$$

Upon canonical quantization, we find (in terms of momenta π_i conjugate to a_i) the hamiltonian

$$H = \frac{1}{2\Omega}\sum_{j=1}^{3}(\pi_j)^2 + \frac{1}{2\Phi}\sum_{a=4}^{7}(\pi_a)^2, \tag{44}$$

together with a constraint on the wave functions originating from the Wess-Zumino term

$$\pi_8\psi(\mathcal{A}) = \frac{N}{2\sqrt{3}}\psi(\mathcal{A}). \tag{45}$$

Since π_i are generators of the right rotations on $\mathcal{A}$, the constraint implies

$$\psi(\mathcal{A}Y) = \frac{N}{3}\psi(\mathcal{A}) \tag{46}$$

where $Y = \lambda_8/\sqrt{3}$ is the conventionally normalized hypercharge generator. In order to find the wave functions which satisfy this constraint, we need to recall a few basic facts about the irreducible representations of $SU(3)$. All the irreducible representations are labeled by a pair of integers p and q. The elements of a (p,q) representation of $SU(3)$ form a traceless tensor with p upper indices and q lower indices, which is symmetric separately in the upper and the lower indices. Every $SU(3)$ group element $\mathcal{A}$ can be represented by a $N_{(p,q)} \times N_{(p,q)}$ matrix $D^{(p,q)}(\mathcal{A})$, which acts on the states of the (p,q) representation, where $N_{(p,q)}$ is the dimension of the representation. Now, following ref. 21, we will consider the complete set of functions on the $SU(3)$ manifold given by the components of the group element in the (p,q) representation:

$$\langle I, I_3, Y|\, D^{(p,q)}(\mathcal{A})\,|I', I_3', Y'\rangle\,, \tag{47}$$

where the states are conventionally labeled by their isospin I, third component of isospin I_3, and hypercharge Y. Because of the representation property

$$D^{(p,q)}(AB) = D^{(p,q)}(\mathcal{A})D^{(p,q)}(B), \tag{48}$$

the constraint (46) implies that the allowed wave functions must have the right hypercharge index Y' set to $N/3$:

$$\psi(\mathcal{A}) = \langle I, I_3, Y|\, D^{(p,q)}(\mathcal{A})\,|I', I_3', N/3\rangle\,. \tag{49}$$

For this expression to make sense, the representation (p,q) must contain a state with hypercharge $N/3$. Let us determine the $SU(3)_{flavor} \times SU(2)_{spin}$ quantum numbers of the wave functions (49).

Under $SU(3)_f$, $U \to FUF^{-1}$. For the rigid rotator ansatz (40), this transformation reduces to $\mathcal{A} \to F\mathcal{A}$, under which the wave functions (49) transform as

$$\begin{aligned} &\langle I, I_3, Y | D^{(p,q)}(\mathcal{A}) | I', I_3', N/3 \rangle \to \\ &\langle I, I_3, Y | D^{(p,q)}(F) | I'', I_3'', Y'' \rangle \langle I'', I_3'', Y'' | D^{(p,q)}(\mathcal{A}) | I', I_3', N/3 \rangle . \end{aligned} \tag{50}$$

Therefore, the left set of indices (I, I_3, Y) are the flavor indices of a wave function which transform in the (p, q) representation of $SU(3)$.

Under rotations, the hedgehog $U_0 \to R^{-1}U_0R$ where R is an $SU(2)_{spin}$ matrix. For the collective coordinate $\mathcal{A}$, this reduces to $\mathcal{A} \to \mathcal{A}R^{-1}$. Substituting this into Eq. (49), we find that the right indices (I', I_3') transform under rotations. The angular momentum carried by a wave function is $(J, J_3) = (I', -I_3')$.

From these transformation properties it follows that, in each allowed representation of $SU(3)_{flavor} \times SU(2)_{spin}$, there are wavefunctions with $Y = N/3$ and $I = J$, which are obtained by identifying the left indices with the right indices in Eq. (49). This rule[9] is helpful in determining the allowed baryon multiplets in the Skyrme model.

An important new feature of the $SU(3)$ Skyrme model is that the structure of the low-lying baryon multiplets explicitly depends on the number of colors. First, consider the case of three colors. There, the lowest allowed $SU(3) \times SU(2)$ representations are (1, 1), $J = 1/2$ (the octet); (3, 0), $J = 3/2$ (the decuplet). Both of these multiplets contain the requisite $Y = 1$, $I = J$ states: the nucleons in the octet and the deltas in the decuplet. Thus, remarkably, the quantum numbers of the low-lying baryons are correctly predicted by the Skyrme model. In addition to the observed multiplets, the collective coordinate treatment predicts an infinite tower of exotic multiplets, starting with (0, 3), $J = 1/2$; (2,2), $J = 1/2$ and (2,2), $J = 3/2$. In the rigid rotator approximation, these baryons are $\mathcal{O}(1)$ higher in mass than the octet and the decuplet. We will see later that the rigid rotator methods are not trustworthy in describing these exotic baryons. In a more complete treatment, some of them can be explicitly shown to be unstable. Presumably, all these states are artifacts of the rigid rotator approximation.

The next task is to determine the mass splittings separating the lowest multiplets. When acting on the allowed wave functions (49), the hamiltonian of Eq. (44) can be manipulated into[7–11,21]

$$M^{(p,q)} = \frac{1}{2\Omega} J(J+1) + \frac{1}{2\Phi} \left(C^{(p,q)} - J(J+1) - N^2/12 \right) \tag{51}$$

where we have used Eq. (45), the fact that, for $j = 1$ to 3, $\pi_j = J_j$, and

$$\sum_{i=1}^{8} \pi_i^2 \psi^{(p,q)}(\mathcal{A}) = C^{(p,q)} \psi^{(p,q)}(\mathcal{A}). \tag{52}$$

For the octet and the decuplet, the quadratic Casimir invariants are $C^{(1,1)} = 3$, $C^{(3,0)} = 6$. Thus, if the quark masses are neglected, the two lowest multiplets have masses

$$\begin{aligned} M^{(1,1)} &= M_{cl} + \frac{3}{4\Phi} + \frac{3}{8\Omega}, \\ M^{(3,0)} &= M_{cl} + \frac{3}{4\Phi} + \frac{15}{8\Omega}. \end{aligned} \tag{53}$$

Just as in the 2-flavor case, the splitting between the $J = 3/2$ and the $J = 1/2$ states is $\frac{3}{2\Omega} \sim 1/N$.

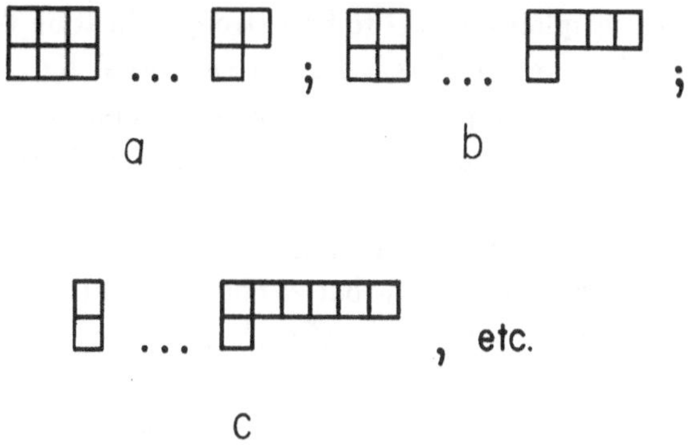

Fig. 8. The low-lying spin-flavor baryon representations are characterized by identical Young tableux for the spin and the flavor parts of the wave function. The total number of boxes is N, the number of colors.

We have found that, if $N = 3$, the Skyrme model predicts the correct quantum numbers of baryons. It is helpful, though, to consider arbitrary odd N for the sake of comparison with large-N QCD. First, using Witten's Hartree-Fock arguments, let us derive the predictions of QCD. Since the low-lying baryons have all the quarks in the same spatial orbital, Fermi statistics demands that the spin-flavor wave function be fully symmetric. Therefore, the $SU(2)$ representation and the flavor representation must be characterized by identical Young tableaux with N boxes,[27] shown in Fig. 8. For $N = 2n + 1$, the $(1, n)$ representation of $SU(3)$ is thus associated with the $J = 1/2$ representation of $SU(2)$ (Fig. 8a); $(3, n-1)$ - with $J = 3/2$ (Fig. 8b); $(5, n-2)$ - with $J = 5/2$ (Fig. 8c); *etc.* The discussion of the splittings between these multiplets proceeds in the same way as in the 2-flavor case: the mass formula of Eq. (34) applies. Comparing these results with the large-N Skyrme model, we immediately find that all the quark model representations described above are allowed. Using Eq. (51) and the formula

$$C^{(p,q)} = \frac{1}{3}\left[p^2 + q^2 + 3(p+q) + pq\right], \tag{54}$$

we find that the masses of these representations can be expressed as[27]

$$M = M_{cl} + \frac{N}{4\Phi} + \frac{1}{2\Omega}J(J+1). \tag{55}$$

Thus, the dependence of mass on spin agrees with the quark model result (34).

In summary, it appears that, if the masses of the 3 lightest quarks were exactly zero, the Skyrme model would explain all the essential features of the baryon spectrum. In reality, the up and down quark masses can be safely treated as negligible, but the situation with the strange quark is much less clear since its mass is comparable to Λ_{QCD}, the scale of strong interactions. The fact that the lightest baryons can be approximately organized into $SU(3)$ multiplets appears to be a strong indication that the strange quark is effectively light. This point of view is further supported by the Gell-Mann–Oakes–Renner (GMOR)[28] treatment of quark masses, based on first-order perturbation theory. In their approach, the smallness of the strange quark mass is the only natural reason why the breaking of $SU(3)$ is successfully modeled without inclusion of operators which transform as a 27. The Gell-Mann–Okubo relations between baryon masses, which hold to a high accuracy, follow from this perturbative approach to the quark masses. Since the Skyrme model comes complete with the hamiltonian and the baryon wave functions, the dependence of the baryon masses on the quark masses can be found explicitly. Thus, we can check for consistency with the GMOR approach. Ignoring m_u and m_d, the simplest mass term in the chiral lagrangian is

$$\delta\mathcal{L}_m = \frac{m_K^2 F_\pi^2}{24}\,\mathrm{Tr}(1-\sqrt{3}\lambda_8)(U+U^\dagger-2). \tag{56}$$

In the effective lagrangian for the collective coordinate, this translates into

$$\delta L_m = \frac{1}{6} m_K^2 \Gamma\left[\mathrm{Tr}(\lambda_8 A \lambda_8 A^{-1}) - 2\right] \tag{57}$$

where

$$\Gamma = F_\pi^2 \int d^3x \sin^2(F/2) \approx 107 F_\pi^{-1} e^{-3}. \tag{58}$$

Evaluating the expectation values of Eq. (57) in the octet wave functions, we find, to first order in m_K^2,[9]

$$\begin{aligned} m_N &= m - \frac{3}{10}x, & m_\Lambda &= m - \frac{1}{10}x, \\ m_\Sigma &= m + \frac{1}{10}x, & m_\Xi &= m + \frac{2}{10}x, \end{aligned} \tag{59}$$

where $x = m_K^2\Gamma/3$, $m = M^{(1,1)} + x$. It is interesting to compare the ratios of mass splittings, rather than the splittings themselves which depend on the details of the chiral lagrangian. In fact, these ratios are in bad disagreement with nature. For instance, the following mass "relations" are found:

$$m_\Lambda - m_N\ \ (176MeV) = m_\Sigma - m_\Lambda\ \ (78MeV) = 2(m_\Xi - m_\Sigma)\ \ (250MeV), \tag{60}$$

where in parentheses we have stated the empirical values. Clearly, first-order perturbation theory in the strange quark mass fails badly in the Skyrme model. This is one of the major surprises in the skyrmion physics. Perhaps, this failure is an indication that something is wrong with the Skyrme model, and that we should give up its study. After all the remarkable results reviewed above, we are reluctant to take that point of view. Instead, we will interpret the difficulties with the quark masses as a clue about what might be happening in the real world. Namely, the strange quark may be heavy enough to invalidate the first order perturbation theory, and to necessitate a more complete inclusion of its mass. We will attack this problem in the spirit of the $1/N$ expansion: instead of concentrating on $N = 3$, we will regard N as a parameter. We can then make some general observations about flavor symmetry breaking.

As shown in section 3, in the 2-flavor case, the same quantum numbers $I = J = 1/2, 3/2, \ldots$ arise for all $N = 2n+1$, *i.e.*, we can pass to a large N without altering the quantum numbers of the low-lying baryons. The situation is more complicated in the 3-flavor case where the size of the baryon multiplets grows rapidly with an increasing N. Indeed, the smallest representation $J = 1/2$, $(1, n)$ contains $2(n+1)(n+3)$ states, with strangeness ranging from 0 to $-n-1$. Since the strange quark mass m_s is, by definition, $\mathcal{O}(1)$, it induces a mass difference $\mathcal{O}(N)$ between the highest and the lowest strangeness members of a multiplet. On the other hand, according to Eq. (55), in the $SU(3)$ symmetric case the splittings between low-lying multiplets are only $\mathcal{O}(1/N)$. Thus, $SU(3)$ is badly broken in the large-N limit. In passing to a large N, instead of looking at the entire multiplets, we will focus on states with strangeness $\mathcal{O}(1)$, which are the natural large-N analogues of $N, \Delta, \Lambda, \Sigma, etc.$ They are the low strangeness members of the representations $J = 1/2$, $(1, n)$ and $J = 3/2$, $(3, n-1)$. Since these multiplets contain baryons with up to $\sim N$ strange quarks, the wave functions of the baryons with fixed strangeness deviate only an amount $\sim 1/N$ into the strange directions of the collective coordinate space. Perturbation theory in these deviations allows us to construct $1/N$ expansions for various baryon observables, with the dependence on m_s calculable analytically.

5. RIGID ROTATOR AND THE $1/N$ EXPANSION

Consider a hedhehog which rotates in the $SU(2)$ subgroup in a way which is employed in the 2-flavor quantization, and also makes small rigid oscillations into the strange directions. In order to describe this, we decompose the $SU(3)$ collective coordinate $\mathcal{A}$ as[16]

$$\mathcal{A}(t) = A(t)S(t)\,, \qquad \text{with}$$
$$A(t) \in SU(2)\,, \qquad S(t) = \exp\left[i\sum_{a=4}^{7} k_a(t)\lambda_a\right]. \tag{61}$$

Since $\dot{A} \sim 1/N$, we can ignore the $SU(2)$ rotation in the treatment of baryons to $\mathcal{O}(1)$. Expanding the lagrangian of Eq. (42) to the second order in the $SU(3)/SU(2)\times U(1)$ coordinates k_a,

$$S(t) \simeq 1 + i(k\cdot\lambda) - \tfrac{1}{2}(k\cdot\lambda)^2 + \ldots, \tag{62}$$

and adding the mass term from Eq. (57), we find $L_K = L_{45} + L_{67}$.

$$L_{45} = 2\Phi(\dot{k}_4^2 + \dot{k}_5^2) + \frac{N}{2}(k_4\dot{k}_5 - k_5\dot{k}_4) - \tfrac{1}{2}m_K^2\Gamma(k_4^2 + k_5^2) \tag{63}$$

describes a particle of mass 4Φ on a plane (k_4, k_5), with normal magnetic field N and a quadratic potential symmetric about the origin. An analogous lagrangian L_{67} describes the motion in the (k_6, k_7) plane. The magnetic field originating from the Wess-Zumino term profoundly affects the motion: it makes the particle move in circles even when the potential is turned off. In section 4 we remarked that the quantization of the coefficient of the Wess-Zumino term is analogous to the quantization of the magnetic charge of a monopole located inside a compact surface. This analogy can be taken further here. By expanding to second order in k_a, we have approximated the surface $SU(3)/SU(2)\times U(1)$ by the surface which is tangent to it at $k_a = 0$, *i.e.*, by the 2 flat planes with a normal magnetic field which approximates the field of the monopole. If the magnetic field is strong, there is a large number of states whose wave functions are confined to the vicinity of $k_a = 0$, which are well described by the approximation of Eq. (63).

The frequencies of the classical circular motions are found by substituting $k_4 - ik_5 \sim \exp(-i\omega t)$ into Eq. (63). The resulting quadratic equation,

$$4\Phi\omega^2 + N\omega = m_K^2\Gamma, \tag{64}$$

has 2 roots, ω_- and $-\omega_+$, which correspond to 2 different senses of rotation and are given by

$$\begin{aligned} \omega_\pm &= \frac{N}{8\Phi}\left(\sqrt{1+\left(\frac{m_K}{M_0}\right)^2} \pm 1\right), \\ M_0^2 &= \frac{N^2}{16\Phi\Gamma} \sim \mathcal{O}(1). \end{aligned} \tag{65}$$

Note that M_0 has the dimensions of mass. In the limit $m_K \to 0$, ω_+ approaches the cyclotron frequency $\frac{N}{4\Phi}$ while $\omega_- \to 0$, which is a sign of degeneracy of the Landau levels. If we define a doublet

$$K = \frac{1}{\sqrt{2}}\begin{pmatrix} k_4 - ik_5 \\ k_6 - ik_7 \end{pmatrix}, \tag{66}$$

the lagrangian for the strange degrees of freedom is conveniently written as

$$L_K = 4\Phi\dot{K}^\dagger\dot{K} + i\frac{N}{2}\left(K^\dagger\dot{K} - \dot{K}^\dagger K\right) - \Gamma m_K^2 K^\dagger K. \tag{67}$$

The Hamiltonian derived from (67) is

$$H = \frac{1}{4\Phi}\Pi^\dagger\Pi + i\frac{N}{8\Phi}(\Pi^\dagger K - K^\dagger\Pi) + (\frac{N^2}{16\Phi} + \Gamma m_K^2)K^\dagger K \tag{68}$$

where $\Pi^\dagger$ is the canonical conjugate of K. In order to diagonalize the Hamiltonian, it is convenient to transform to the basis of creation and annihilation operators:

$$\begin{aligned} K &= \frac{1}{\sqrt{N}}\left(1+\left(\frac{m_K}{M_0}\right)^2\right)^{-1/4}(a+b^\dagger), \\ \Pi^\dagger &= \frac{i}{2}\sqrt{N}\left(1+\left(\frac{m_K}{M_0}\right)^2\right)^{1/4}(a^\dagger - b), \end{aligned} \tag{69}$$

where a and b are doublets, and the parameter M_0 is defined in Eq. (65). This transformation brings the Hamiltonian to the diagonal form

$$H = \frac{1}{2}\omega_-\{a, a^\dagger\} + \frac{1}{2}\omega_+\{b, b^\dagger\}. \tag{70}$$

The eigenstates form a Fock space

$$|n_s,\ n_{\bar{s}}\rangle = (a^\dagger)^{n_s}(b^\dagger)^{n_{\bar{s}}}|0\rangle \tag{71}$$

with $n_s, n_{\bar{s}}$ integers. Since the strangeness operator can be shown to be

$$S = b^\dagger b - a^\dagger a, \tag{72}$$

the state $|n_s,\ n_{\bar{s}}\rangle$ looks as if it contains n_s s-quarks and $n_{\bar{s}}$ $\bar{s}$-quarks. This identification will be made more precise later. The states with $n_{\bar{s}} \neq 0$ correspond to the exotic states found in the Skyrme model, which are usually presumed to be artifacts of the rigid rotator approximation. In the bound state approach to strangeness reviewed in section 6, which improves on the rigid rotator approximation, these exotic states do not appear. Within our rigid rotator approximation, even in the $SU(3)$ limit $m_s \to 0$, it costs energy $\frac{N}{4\Phi}$ to produce a state with an $\bar{s}$, but not to replace a u or a d with an s-quark.

Here is a good place to check the consistency of our quadratic approximation in K, *i.e.*, to confirm that the deviations into the strange directions are small. Using Eq. (69), we find that, in any finitely excited state,

$$\left\langle K^{\dagger}K\right\rangle \sim \frac{1}{N}\frac{1}{\sqrt{1+(m_K/M_0)^2}}. \tag{73}$$

Thus, soliton's mean squared deviation into strange directions is $\mathcal{O}(1/N)$, and our approximations are good in the large-N limit. In our approach, the expansion in powers of K implements the $1/N$ expansion. Further, if N is fixed and m_K is dialed up, the wave functions become more confined to the vicinity of the non-strange $SU(2)$. For a large enough m_K, this effect on the wave functions becomes significant and invalidates first-order perturbation theory in m_K^2. According to Eq. (73), this happens if $m_K \sim M_0$. With the standard fit of the Skyrme model parameters F_π and e, we find $M_0 \approx 245 MeV$, which indeed suggests that the baryon wave functions for $m_K = 495 MeV$ are significantly different from the $SU(3)$ symmetric wave functions ($m_K = 0$). Our simple exercise appears to shed some light on why the first-order perturbation theory in the quark masses fails in the Skyrme model. As we have seen, if the model is treated consistently, the non-linear effects emerge. The same conclusion was reached, using different techniques, in refs. 17, 29.

In the general chiral theory for the pseudoscalar mesons, the symmetry breaking effects due to the current masses of the quarks are parametrized by m_K^2/Λ^2, where Λ is the mass scale $\simeq 4\pi f_\pi \approx 1$ GeV.[30,31] Thus, $SU(3)$ violating effects in the pseudoscalar meson sector are $\lesssim 25\%$, and the terms with the higher powers of m_K^2 can be ignored. We have found that in the baryon sector, in addition to the $SU(3)$ breaking effects parametrized by m_K^2/Λ^2, there are also deviations from $SU(3)$ that are analytic functions of m_K^2/M_0^2. Thus, a new mass scale $M_0 \ll 4\pi f_\pi$ has emerged in baryon physics. It is tempting to identify the new dimensionless parameter $(m_K/M_0)^2$ with the expected parameter m_s/Λ_{QCD}, up to a factor of order 1. Our results suggest that the baryon wave functions exhibit significant departures from the $SU(3)$ symmetry. But can such wave functions be consistent with the $SU(3)$ structure undoubtedly present in baryon physics? The answer to this question probably depends on the specific model under consideration. In order to address this question here, we will carry out $SU(2)$ collective coordinate quantization and study the baryon spectrum to $\mathcal{O}(1/N)$.[16]

The baryon states of Eq. (71) are eigenstates of strangeness but carry no definite isospin or angular momentum. In order to identify the eigenstates of I and J, it is necessary to excite the $SU(2)$ collective coordinate $A(t)$. For each value of strangeness, we will be able to construct a tower of narrowly split states of increasing spin and isospin.

The lagrangian (42) depends on $A(t)$ only through the angular velocity $\dot{\alpha}_i$ defined by

$$A^{\dagger}\dot{A} = \frac{i}{2}\dot{\alpha}_i\tau_i, \tag{74}$$

where τ_i are the Pauli matrices. Since the soliton moment of inertia Ω is $\mathcal{O}(N)$, the angular velocities of the low-lying rotational excitations are $\mathcal{O}(1/N)$. Thus, in addition to rotation in $K^\dagger K$ space with angular velocity of order 1, the soliton undergoes a slow rigid rotation in isospace. The $\mathcal{O}(1/N)$ correction to the lagrangian L_K of Eq. (67) is

$$\delta L_K = \frac{1}{2}\Omega(\dot{\alpha}_i)^2 + i\dot{\alpha}_i(2\Phi - \Omega)(\dot{K}^\dagger \tau_i K - K^\dagger \tau_i \dot{K}) - \frac{1}{2}N\dot{\alpha}_i K^\dagger \tau_i K \tag{75}$$

The Hamiltonian obtained from $L_K + \delta L_K$ through the standard canonical procedure can be expanded in powers of $1/N$ as

$$H = H_0 + H_1 + \mathcal{O}(1/N^2). \tag{76}$$

H_0 is the $\mathcal{O}(1)$ hamiltonian of Eq. (68). As shown in the previous section, this Hamiltonian is diagonal in the basis of Fock states of Eq. (71). Since the full Hamiltonian H cannot be solved exactly, our goal is to determine the energy levels correct to $\mathcal{O}(1/N)$. To this end, we will use first order perturbation theory in

$$H_1 = \frac{1}{2\Omega}\left\{\vec{J}_{ud} - \frac{i}{2}(\Pi^\dagger \vec{\tau} K - K^\dagger \vec{\tau}\Pi)\left(1 - \frac{\Omega}{2\Phi}\right) + \frac{N\Omega}{4\Phi}\, K^\dagger \vec{\tau} K\right\}^2, \tag{77}$$

where $\vec{J}_{ud}$ is the momentum conjugate to $\vec{\alpha}$. Our notation is intended to remind the reader that, in the quark language, $\vec{J}_{ud}$ is the net angular momentum of the u and d quarks. In the soliton model $\vec{J}_{ud}$ is the generator of the right rotations on the collective coordinate A. It acts on $\chi_I(A)$, the collective coordinate component of the wave function, and its explicit form is given in Eq. (25). The wave functions can be written as sums of products of the form

$$|n_s,\ n_{\bar{s}}\rangle\, \chi_I(A), \tag{78}$$

where $\chi_I(A)$ are the wave functions defined on the $SU(2)$ group manifold which were discussed in section 3. The index I labels the spin-isospin representation. In order to identify the wave functions which correspond to various excitations of the total spin and isospin, we need to construct the $\vec{I}$ and $\vec{J}$ operators.

First, let us consider the transformation properties of K and A. Under spatial rotations, $U_0 \to R^{-1}U_0 R$, where R is an $SU(2)$ matrix. This transformation acts on K and A as $K \to RK$ and $A \to AR^{-1}$. Using Noether's theorem, we find the total angular momentum to be

$$\vec{J} = \vec{J}_{ud} + \vec{J}_s, \tag{79}$$

where, restoring the doublet indices,

$$\vec{J}_s = \frac{1}{2}(a^\dagger_\alpha \vec{\tau}_{\alpha\beta} a_\beta - b_\alpha \vec{\tau}_{\alpha\beta} b^\dagger_\beta). \tag{80}$$

Thus, $\vec{J}_s$ is the contribution of the strange excitations to the angular momentum. Eq. (80) shows that each unit of strangeness carries half a unit of angular momentum and further supports our identification of the a-quanta with the s-quarks and the b-quanta with the $\bar{s}$-quarks. This fermionization of the seemingly bosonic degrees of freedom is what makes the Skyrme model so remarkable. The Fock states $|n_s,\ n_{\bar{s}}\rangle$ can be decomposed into irreducible representations of $\vec{J}_s$. Then the irreducible representations of $\vec{J}$ can be constructed as sums of products of the form (78) using the familiar addition of angular momenta.

Further restriction on the baryon wave functions comes from the fact that they must carry definite total isospin. This restriction is particularly easy to implement. Since K does not transform under isorotations, isospin only acts on $\chi_I(A)$: $\vec{I} = \vec{I}_{ud}$, whose explicit operator form is given in Eq. (26). Thus, the subscript I of the collective coordinate component of the wave function *is* the total isospin. It follows that each unit of strangeness behaves as an object with no isospin and half a unit of spin. These are the familiar spin-isospin quantum numbers of a strange quark.

A further constraint on the quantum numbers follows from the fact that all $SU(2)$ rotator wave functions $\chi_I(A)$ satisfy the $I = J_{ud}$ rule, with I taking on either integral or half-integral values. The Wess-Zumino term requires that, if N is odd, the baryon must have half-integral spin. In our approach, this leads to the following rule: states with even strangeness carry half-integral isospin, while states with odd strangeness carry integral isospin. This rule leads to the observed baryon quantum numbers.

For example, labeling the states as $|n_s, n_{\bar{s}}\rangle\, |I_{ud}, J_{ud}\rangle$, the wave functions of the $S = -1$ baryons are

$$\begin{aligned} &|N\rangle = |0,0\rangle\, |\tfrac{1}{2},\tfrac{1}{2}\rangle\ , && |\Delta\rangle = |0,0\rangle\, |\tfrac{3}{2},\tfrac{3}{2}\rangle\ , \\ &|\Lambda\rangle = |1,0\rangle\, |0,0\rangle\ , && |\Sigma\rangle = \left[|1,0\rangle\, |1,1\rangle\right]_{J=\frac{1}{2}}\ , \\ &|\Sigma^*\rangle = \left[|1,0\rangle\, |1,1\rangle\right]_{J=\frac{3}{2}}\ , && \end{aligned} \tag{81}$$

and so forth. We can also construct the wave functions for baryons with $S = -2, -3$, *etc.*. There, Bose statistics of $a^\dagger_\alpha$ requires J_s to assume its maximum possible value $J_s = -S/2$. With this rule, we find

$$\begin{aligned} &|\Xi\rangle = \left[|2,0\rangle\, |\tfrac{1}{2},\tfrac{1}{2}\rangle\right]_{J=\frac{1}{2}}\ , && |\Xi^*\rangle = \left[|2,0\rangle\, |\tfrac{1}{2},\tfrac{1}{2}\rangle\right]_{J=\frac{3}{2}}\ , \\ &|\Omega\rangle = |3,0\rangle\, |0,0\rangle\ . && \end{aligned} \tag{82}$$

Of course, for any $N > 3$, all these states belong to the $J = 1/2$ and $J = 3/2$ multiplets, which contain other baryons. The approximation quadratic in K actually contains states with arbitrarily large $|S|$. This is correct in the $N \to \infty$ limit where the higher powers of K become negligible. For any finite N, our approximation contains $|S| > N$ states which are spurious. Presumably, as we take into account the higher powers of K, they will induce repulsive interactions between the strange quanta, which will prevent the formation of high-strangeness baryons.* In the 2-flavor case, the high-spin collective coordinate states are unstable in a more complete treatment.[25] Similarly, in our approach to hyperons, we expect the spurious states to be eliminated dynamically.

The next step is to determine how the baryon mass splittings found in our approach compare with the mass formulæ of the non-relativistic quark model (NRQM). A nonzero strange quark mass breaks $SU(3)$ down to $SU(2) \times U(1)$. Thus, the most general form for an effective hamiltonian describing the masses of the N, Λ, Σ and Ξ fields may be expressed as

$$M = a_1 + a_2 Y + a_3\left[I(I+1) - \tfrac{1}{4}Y^2\right] + a_4 Y^2. \tag{83}$$

Empirically, $a_1 = 1116 MeV$, $a_2 = -190 MeV$, $a_3 = 38 MeV$, $a_4 = -11 MeV$. The

⋆ In the rigid rotator model, as we sum the whole series in K, we expect the spurious states to be pushed to infinite energy.

Gell-Mann–Okubo (GMO) mass relation in the octet,

$$2(m_N + m_\Xi) = 3m_\Lambda + m_\Sigma \tag{84}$$

holds because a_4 is negligible. We will see that the leading N-dependences of the coefficients in Eq. (83) are $a_1 \sim N$, $a_2 \sim 1$, $a_3 \sim a_4 \sim 1/N$. Thus, large-N reasoning can explain why $|a_3| \ll |a_2|$ and $|a_4| \ll |a_2|$, but it does not by itself explain why $|a_4| \ll |a_3|$. The usual extra ingredient is that, in first order perturbation theory in m_s, one finds that a_4 is zero, and the GMO formula results. If first order perturbation theory is a poor approximation to nature, then the GMO relations may simply be due to a numerical coincidence. Our large-N argument shows that this coincidence may not be as drastic as one might have expected.

A mass formula similar to (83) follows in the NRQM with magnetic moment interactions between quarks. There the Hamiltonian is taken to be

$$M = m_0 + m_1|S| + m_2 \sum_{i<k} \vec{j}_i \cdot \vec{j}_k + m_2 c \sum_{i,I} \vec{j}_i \cdot \vec{j}_I + m_2 \bar{c} \sum_{I<K}{}' \vec{j}_I \cdot \vec{j}_K \tag{85}$$

where the small indices refer to the light quarks and the capital indices – to the strange quarks. m_0 is $\mathcal{O}(N)$ because a baryon contains N valence quarks. m_1 is the difference between the constituent masses of the strange and the light quarks, which is $\mathcal{O}(1)$. m_2, the strength of the phenomenological magnetic moment interaction, is $\mathcal{O}(1/N)$. In the NRQM such interactions are suggested by one gluon exchange calculations.[26] There one finds $\bar{c} = c^2$, and c is equal to the ratio of the light and strange constituent quark masses: $c = M_u/M_s$. In Eq. (85), the $SU(3)$ symmetric values of the parameters are $m_1 = 0$ and $c = \bar{c} = 1$. As we will see, there are significant empirical deviations from the $SU(3)$ symmetry. Eq. (85) can be manipulated into

$$H = m_0' + m_1' Y + \frac{m_2}{2}\left\{ cJ(J+1) + (1-c)\left[I(I+1) - \tfrac{1}{4}Y^2\right] + \frac{1+\bar{c}-2c}{4}\, Y^2 \right\}. \tag{86}$$

From the point of view of Eq. (86), the GMO mass relations are successful if

$$1 + \bar{c} - 2c \approx 0\,. \tag{87}$$

Of course, there are other, tighter constraints on the values of these parameters which follow from further empirical data. In fact, Eq. (86) provides a good fit to the masses of all the octet and decuplet baryons with[32]

$$\begin{gathered} m_0' \approx 1062 MeV, \qquad m_1' \approx -192 MeV, \qquad m_2 \approx 213 MeV, \\ c \approx .67, \qquad \bar{c} \approx .27. \end{gathered} \tag{88}$$

This fit shows how far c and $\bar{c}$ deviate from their $SU(3)$ symmetric values. To some extent, the fit makes the conventional explanation of the GMO relations suspect: if first order perturbation in m_s changes $\bar{c}$ by 70%, then why are we allowed to ignore higher order perturbations? In fact, in the quark model,

$$\begin{aligned} c &= \frac{M_u}{M_s} \simeq \frac{M_u}{m_s + M_u} = \left(1 + \frac{m_s}{M_u}\right)^{-1}, \\ \bar{c} &= \left(\frac{M_u}{M_s}\right)^2 \simeq \left(1 + \frac{m_s}{M_u}\right)^{-2}. \end{aligned} \tag{89}$$

Since $m_s/M_u \simeq .5$, the higher-order terms in the expansion of c and $\bar{c}$ in powers of m_s play an important role. Perhaps, there exists a class of models with appreciable nonlinearities in m_s, which manage to yield output parameters close to the empirical. How well does our rigid rotator treatment do in this respect?

A mass formula of the form (86) follows in our approach after we find the expectation values of H_1 from Eq. (77) in the baryon states constructed in Eqs. (81) and (82). For negative strangeness baryons,

$$
\begin{aligned}
M =&M_{cl} - S\omega_- + \frac{1}{2\Omega}(\vec{J}_{ud} + c\vec{J}_s)^2 = \\
M_{cl} - S\omega_- +& \frac{1}{2\Omega}\left[cJ(J+1) + (1-c)I(I+1) + \frac{c(c-1)}{4}S(S-2)\right]
\end{aligned}
\tag{90}
$$

where

$$
c = 1 - \frac{4\Omega\omega_-}{8\Phi\omega_- + N} \,. \tag{91}
$$

Comparing with Eq. (86), the parameters we find are

$$
\begin{aligned}
m_2 &= \frac{1}{\Omega}, & \bar{c} &= c^2, \\
m_1' &= -\omega_- + \frac{c(1-c)}{2\Omega}, & m_0' &= M_{cl} + \omega_- - \frac{3c(1-c)}{8\Omega}
\end{aligned}
\,. \tag{92}
$$

It is interesting that the NRQM relation $\bar{c} = c^2$ holds in this approach. However, the numerical agreement of these parameters is rather poor. In the model stabilized by the 4-derivative Skyrme term, $\Omega \approx \Gamma \approx 107$ and $\Phi \approx 39$ in units of $F_\pi^{-1}e^{-3}$. With the values $F_\pi = 129MeV$ and $e = 5.45$ from the 2-flavor fit of ref. 12, we find $\omega_- \approx 248MeV$, $m_2 \approx 196MeV$, $c \approx .24$, $\bar{c} \approx 0.06$. The calculated values of c and $\bar{c}$, which offer an important test of the model, grossly violate relation (87), *i.e.*, the GMO relations are not obeyed. Also, their magnitudes are very far from the best fit of Eq. (88). As a result, the model poorly accounts for the observed hyperon splittings. For instance, c being far from its empirical value leads to the incorrect relative splittings between Λ, Σ, and Σ^*. Although our input parameters may be far from the best fit in the 3-flavor case, the drastic discrepancy in the calculated values of c and $\bar{c}$ is another way to see why the rigid rotator treatment breaks down, as reported in the earlier papers.[9–11]⋆ So far, our improved handling of the $SU(3)$ breaking has not produced a successful model. However, further improvements introduced in the next section will be shown to work better.

6. THE BOUND STATE APPROACH TO STRANGENESS

Let me emphasize that the approach to hyperons outlined in section 5 is simply a variant of the $SU(3)$ rigid rotator quantization. The novel feature is that the $SU(3)$ rotations are factored into two parts: the ones that rotate the $SU(2)$ soliton into the "strange directions" ($S(t)$ in Eq. (61)), and the ones that do not ($A(t)$ in Eq. (61)). As explained above, the expansion about $S(t) = 1$ provides us with a simple way of performing the $1/N$ expansion both in the case of unbroken $SU(3)$ and when arbitrary strange quark mass is included. Thus, we have developed an approximation to the large-N analogue of Yabu and Ando's solution[33] of the $SU(3)$ rigid rotator at $N = 3$. Let us

---

⋆ H_1 does not include the full $1/N$ correction to the masses: the terms quartic in K, which we have so far omitted, are also $\mathcal{O}(1/N)$. They incorporate the strange quark – strange quark interactions, and could shift the value of $\bar{c}$, but cannot eliminate the discrepancy in c.

point out that all approaches based on the rigid $SU(3)$ rotator have a common deficiency once the kaon mass term is included: the rotations in the "strange directions" become only approximate collective coordinates. Strictly speaking, as the soliton deviates into strange directions, it has a tendency to deform. For a sufficiently large m_s this effect is significant. Further, it is useful to study strange deformations which are not smoothly connected to the rigid rotations. With these ideas in mind, refs. 19 and 22 developed a treatment of general strange fluctuations about the $SU(2)$ skyrmion based on the parametrization

$$U = \sqrt{U_\pi} U_K \sqrt{U_\pi} \tag{93}$$

where

$$U_\pi = \exp(i\frac{2}{F_\pi}\sum_{j=1}^{3}\lambda_j \pi^j), \qquad U_K = \exp(i\frac{2}{F_\pi}\sum_{a=4}^{7}\lambda_a K^a). \tag{94}$$

λ are the generators of $SU(3)$ normalized to $Tr(\lambda_k\lambda_l) = 2\delta_{kl}$.

We have argued that m_s is large enough to make some aspects of $SU(3)$ breaking significant in the baryon sector. At the same time, symmetry breaking effects in meson physics, parametrized by $(m_K/4\pi f_\pi)^2$, are quite small. Therefore, we substitute ansatz (93) into the effective mesonic action where all terms except for the kaon mass term are $SU(3) \times SU(3)$ symmetric. As in the approach of section 5, an increase in both N and m_K acts to restrict the fluctuations of K_a. This makes the expansion in K_a tantamount to the large-N expansion. The difference from section 5 is that here $K_a(\vec{x}, t)$ are not simple variables, but fields which can take us beyond the rigid rotator approximation.

By the general theory of non-linear realizations of $SU(2) \times SU(2)$,[34] the expansion in kaon fluctuations can be organized in terms of the complex kaon isodoublet

$$K = \frac{1}{\sqrt{2}}\begin{pmatrix} K_4 - iK_5 \\ K_6 - iK_7 \end{pmatrix} = \begin{pmatrix} K^+ \\ K^0 \end{pmatrix}, \tag{95}$$

its "covariant derivative"

$$D_\mu K = \partial_\mu K + \frac{1}{2}(\sqrt{U}^\dagger_\pi \partial_\mu \sqrt{U}_\pi + \sqrt{U}_\pi \partial_\mu \sqrt{U}^\dagger_\pi)K, \tag{96}$$

and

$$A_\mu = \frac{1}{2}(\sqrt{U}^\dagger_\pi \partial_\mu \sqrt{U}_\pi - \sqrt{U}_\pi \partial_\mu \sqrt{U}^\dagger_\pi). \tag{97}$$

After expanding Eq. (6) and Eq. (56) to second order in K, we get

$$\begin{aligned}\mathcal{L} = \mathcal{L}_{Sk}(U_\pi) + (D_\mu K)^\dagger D^\mu K - m_K^2 K^\dagger K + \frac{1}{8}K^\dagger K\left(4\mathrm{Tr}\, A_\mu^2 - \frac{1}{e^2F_\pi^2}\mathrm{Tr}\,[A_\mu, A_\nu]^2\right) \\ + \frac{2}{e^2F_\pi^2}\left((D_\mu K)^\dagger D^\mu K \mathrm{Tr}\, A_\nu^2 - (D_\mu K)^\dagger D_\nu K \mathrm{Tr}\,(A^\mu A^\nu) + 3(D_\mu K)^\dagger [A^\mu, A^\nu] D_\nu K\right).\end{aligned} \tag{98}$$

An additional term is found in the expansion of the Wess-Zumino term (36):

$$\mathcal{L}_{WZ} = \frac{iN}{F_\pi^2} B^\mu (K^\dagger D_\mu K - (D_\mu K)^\dagger K), \tag{99}$$

where B^μ is the baryon number current of the $SU(2)$ soliton configuration. Terms of this kind describe the interaction of a charged field with a vector potential. In our case, the charge is strangeness, and the role of the vector potential is played by the baryon current.† The term in Eq. (99) is not a peculiarity of the Skyrme model: it is necessary in the conventional meson-baryon lagrangians which include baryons as elementary fields. Let us show how this term arises if we fix $N = 3$ and consider the interactions of the baryon octet. If we assemble the baryon fields into a traceless 3×3 matrix B, and the Goldstone bosons — into $U = \exp(2i\lambda_a M^a/F_\pi)$, then the $SU(3) \times SU(3)$ invariant kinetic term for baryons is

$$\begin{aligned} &i\mathrm{Tr}\ \bar{B}\gamma^\mu\partial_\mu B + \mathrm{Tr}\ \bar{B}\gamma^\mu[V_\mu, B], \\ &V_\mu = \frac{1}{2}(\sqrt{U}^\dagger\partial_\mu\sqrt{U} + \sqrt{U}\partial_\mu\sqrt{U}^\dagger). \end{aligned} \tag{100}$$

One of the interaction terms found in the expansion of Eq. (100) is

$$\frac{3i}{F_\pi^2}\bar{\mathcal{N}}\gamma^\mu\mathcal{N}(K^\dagger D_\mu K - (D_\mu K)^\dagger K), \tag{101}$$

where $\mathcal{N}$ is the isodoublet of nucleon fields. This is the precise analogue of the term in Eq. (99), including the normalization. Thus, in both approaches to baryon phenomenology, the normalization of this term is fixed in the $m_K \to 0$ limit by $SU(3) \times SU(3)$ symmetry. The argument above is yet another way to see how naturally the Wess-Zumino term fits into the interpretation of baryons as solitons. In fact, in the bound state approach, the term in Eq. (99) is the only interaction which distinguishes between positive and negative strangeness in a fixed baryon number background and, therefore, plays a crucial role.

In section 5 we found baryon states which are hedgehogs rotating in the $SU(3)$ collective coordinate space in such a way that their departures into strange directions are small. However, it is these departures that allow the soliton to acquire strangeness. In a more complete theory of strange deformations, based on Eqs. (98) and (99), baryon states of this nature should manifest themselves as kaon modes localized near the soliton. In the following, we will show that such bound states indeed exist and occur precisely in the right channels, with roughly the right excitation energies, to allow us to construct sensible hyperon phenomenology.

Let us show how the kaon-soliton bound states arise and how to interpret them as hyperons. Eq. (98) reduces the problem to the motion of kaons in the classical background of the $SU(2)$ soliton. Our goal is to calculate the kaon energy levels in the large N expansion. In the semiclassical approximation, solitons rotate slowly, with velocities of order $1/N$. Therefore, to find the kaon energy levels to $O(1)$, it is sufficient to replace U_π by the static background hedgehog $U_0(\vec{r})$ from Eq. (15). Since the hedgehog is symmetric under combined spatial and isospin rotations $\vec{T} = \vec{I} + \vec{L}$, we can write the kaon eigenmodes as

$$K(\vec{r}, t) = k(r, t) Y_{TLT_z}. \tag{101}$$

After substituting this into $\mathcal{L}_{Sk} + \mathcal{L}_{WZ}$ we find the following effective Lagrangian for the radial field $k(r,t)$ (It is convenient to transform to dimensionless variables with the mass

† This analogy is incomplete because the bound state Lagrangian has no term $\sim B_\mu B^\mu K^\dagger K$.

scale set by eF_π.):

$$L = 4\pi \int drr^2 \left[f(r)\dot{k}^\dagger \dot{k} + i\lambda(r)(k^\dagger \dot{k} - \dot{k}^\dagger k) - h(r)\frac{d}{dr}k^\dagger \frac{d}{dr}k - k^\dagger k \left(m_K^2 + V_{eff}(r;T,L) \right) \right] \tag{102}$$

where

$$f(r) = 1 + \frac{2\sin^2 F}{r^2} + (F')^2, \qquad h(r) = 1 + \frac{2\sin^2 F}{r^2},$$
$$\lambda(r) = -\frac{Ne^2}{2\pi^2 r^2} F' \sin^2 F, \tag{103}$$

and $V_{eff}(r;T,L)$ is a lengthy expression given in terms of $F(r)$ in Eq. (3.1) of ref. 19. The term linear in time derivatives originates in the W-Z term. The Lagrangian (102) is analogous to the interaction of a relativistic charged field with a background static, radial electric field. We find that the negative strangeness particles are attracted to the origin while the positive are repelled. In ref. 19 it was shown that, in the absence of the W-Z term, the lowest bound state has quantum numbers $T = 1/2$, $L = 1$. As we dial the coefficient of the W-Z term up, the $S = -1$ state becomes more tightly bound to the soliton, while the $S = 1$ state gets repelled. In fact, this repulsion is strong enough to prevent the existence of either bound states or resonances with $S = 1$. This is the essential role of the W-Z term in our treatment of strangeness.

The lagrangian (102) is quadratic in fields and at most quadratic in time derivatives. In general, physical systems described by such lagrangians can be treated exactly. The observation crucial for our purposes is that the quantum energy levels of the system are given by the classical eigenfrequencies. We demonstrate how this works in our specific example. The variational equation resulting from (102) is

$$-f(r)\frac{d^2}{dt^2}k + 2i\lambda(r)\frac{d}{dt}k + Ok = 0, \tag{104}$$

where

$$O = \frac{1}{r^2}\frac{d}{dr}h(r)r^2\frac{d}{dr} - m_K^2 - V_{eff}(r;T,L) \tag{105}$$

is a hermitian operator. Let us expand the field k in terms of its eigenmodes:

$$k(r,t) = \sum_{n>0} \left(\tilde{k}_n(r) e^{i\tilde{\omega}_n t} b_n^\dagger + k_n(r) e^{-i\omega_n t} a_n \right) \tag{106}$$

where ω_n and $\tilde{\omega}_n$ are assumed to be positive. By substituting (106) into (104) we find that the eigenmodes satisfy

$$\left(f(r)\omega_n^2 + 2\lambda(r)\omega_n + O \right) k_n = 0 \tag{107}$$

and

$$\left(f(r)\tilde{\omega}_n^2 - 2\lambda(r)\tilde{\omega}_n + O \right) \tilde{k}_n = 0. \tag{108}$$

Using the hermiticity of O we derive the following orthogonality relations:

$$4\pi \int drr^2 k_n^\star k_m \left(f(r)(\omega_n + \omega_m) + 2\lambda(r) \right) = \delta_{nm}, \tag{109}$$

$$4\pi \int drr^2 \tilde{k}_n^\star \tilde{k}_m (f(r)(\tilde{\omega}_n + \tilde{\omega}_m) - 2\lambda(r)) = \delta_{nm}, \tag{110}$$

$$4\pi \int drr^2 k_n^\star \tilde{k}_m (f(r)(\omega_n - \tilde{\omega}_m) + 2\lambda(r)) = 0. \tag{111}$$

Upon carrying out canonical quantization we find that the momentum conjugate to k is

$$\pi^\dagger(r,t) = f(r)\dot{k}^\dagger + i\lambda(r)k^\dagger. \tag{112}$$

Canonical commutation relations between the fields and their conjugate momenta and eqns. (109)-(111) imply that the oscillators have the usual algebra

$$[a_n, a_m^\dagger] = \delta_{nm}, \qquad [b_n, b_m^\dagger] = \delta_{nm}, \tag{113}$$

with the rest of the commutators vanishing. In terms of the creation and annihilation operators the Hamiltonian reduces to

$$H = \sum_{n>0} (\omega_n a_n^\dagger a_n + \tilde{\omega}_n b_n^\dagger b_n). \tag{114}$$

This proves that the quantum energy levels are given by the classical eigenfrequencies. The strangeness charge is

$$S = \sum_{n>0} (b_n^\dagger b_n - a_n^\dagger a_n). \tag{115}$$

It follows that a_n and b_n annihilate the modes of strangeness -1 and 1 respectively. The bound state energy in the $S = -1$ sector can be found by solving for the lowest eigenfrequency of eqn. (107). This can easily be done numerically.

We find that in the lowest partial wave ($T = 1/2$, $L = 1$) there is exactly one bound state with $S = -1$. We have determined its energy to be $0.23eF_\pi$. This is the bound state on the basis of which we construct the Λ, Σ and $\Sigma^\star$ baryons. We find no bound states corresponding to the exotic baryons with $S = 1$. In ref. 35 it is also shown that there are no $S = 1$ resonances. A notable feature of our approach is the presence of a bound state in the lowest negative parity partial wave ($S = -1, T = 1/2, L = 0$).[19,22] This state probably corresponds to the observed $\Lambda(1405)$ which is indeed below the KN threshold. To $O(N^0)$ we find the energy for this state to be 0.50 eF_π.

In collective coordinate quantization, strange baryons acquire definite spin and isospin quantum numbers through a slow rotation of the soliton together with the bound meson:[19]

$$U(\vec{r},t) = A(t)U_0(\vec{r})A^{-1}(t), \qquad \tilde{K}(\vec{r},t) = A(t)K(\vec{r},t). \tag{116}$$

With this definition, K is the kaon field as observed in the "rest frame" of the rotating skyrmion. Under an isospin rotation, these variables transform as

$$A \to FA, \qquad K \to K, \tag{117}$$

which shows that the kaon has lost its isospin. Under a spatial rotation,

$$U_0 \to \exp(i\vec{\alpha}\cdot\vec{L})U_0 \exp(-i\vec{\alpha}\cdot\vec{L}) = \exp(-i\vec{\alpha}\cdot\vec{I})U_0 \exp(i\vec{\alpha}\cdot\vec{I}),$$

$$K \to \exp(i\vec{\alpha}\cdot\vec{L})K = \exp(-i\vec{\alpha}\cdot\vec{I})\exp(i\vec{\alpha}\cdot\vec{T})K, \tag{118}$$

where we have used the invariance of the hedgehog U_0 under $\vec{T} = \vec{I} + \vec{L}$. From Eq. (118)

we deduce that the effect of a spatial rotation on U and $\tilde{K}$ is to transform A and K as

$$A \to A \exp(-i\vec{\alpha} \cdot \vec{I}), \qquad K \to \exp(i\vec{\alpha} \cdot \vec{T})K. \tag{119}$$

While A transforms the same way as in the 2-flavor case, the angular momentum operator for K is $\vec{T}$, the operator used to classify kaon partial waves.

We can now list the $(I,\ J)$ eigenvalues for kaons bound to the skyrmion rotator. The lowest bound state lies in the $T = 1/2, L = 1$ partial wave, and each occupation of it contributes $(I,\ J,\ S)$=$(0,\ 1/2,\ -1)$. Apart from the baryon number assignment, these are precisely the quantum numbers of an s-wave strange quark.* Perhaps by now we should not be very surprised by this transmutation of quantum numbers, because a similar phenomenon took place in the simpler setting of section 5. This transmutation makes the quantum number counting virtually identical to that in NRQM.

The construction of wave functions of low-lying baryons is identical to the construction in Eqs. (81) and (82), with $a^\dagger$ replaced by the bound state creation operator $a_1^\dagger$. Their mass splittings are also described by the mass formula of Eq. (86). After a calculation which is quite analogous to the one in section 5, we find that the parameters in this mass formula are given by Eq. (92), with ω_- replaced by the bound state energy ω_1, and the important parameter c given by

$$c = 1 - \omega_1 \frac{\int dr k_1^\star k_1 \left(\frac{4}{3} f r^2 \cos^2 \frac{F}{2} - 2\left(\frac{d}{dr}(r^2 \frac{dF}{dr} \sin F) - \frac{4}{3} \sin^2 F \cos^2 \frac{F}{2}\right)\right)}{\int dr r^2 k_1^\star k_1 (f\omega_1 + \lambda)} \tag{120}$$

where $k_1(r)$ is the bound state eigenfunction. With $F_\pi = 129 MeV$ and $e = 5.45$, the calculated values of parameters are

$$\begin{gathered} m_2 = 196 MeV, \qquad c = .60, \qquad \bar{c} = .36, \\ m_1' = -138 MeV, \qquad m_0' = 1007 MeV. \end{gathered} \tag{121}$$

The parameters c and $\bar{c}$, which govern the hyperfine splittings of baryons, are much better than in section 5: they are not far off the relation (87), from which the GMO relations follow, although not as close as the values in the best fit of Eq. (88). The calculated values of m_1' and m_0' are somewhat low, compared to Eq. (88). Inclusion of the $\mathcal{O}(K^4)$ terms, which incorporate $\mathcal{O}(1/N)$ kaon-kaon interactions, will shift the values of $\bar{c}, m_1'$ and m_0'. This has not yet been carried out because of computational complexity, but, we hope, will further improve the parameters. Another missing ingredient is the careful determination of the zero-point energy of the kaons, which has been ignored in this calculation. When included, it could improve the value of m_0'. In view of the good prediction of the hyperfine splittings in the bound state approach, it is compelling to test the model in other ways and to explore its possible improvements.

There is an interesting connection between Eq. (120) and the corresponding Eq. (91) in the rigid rotator treatment. In the rigid rotator approximation, the strange

---

⋆ We expect $\mathcal{O}(1/N)$ baryon number to be carried by strange quarks. In our approach, it has not been sorted out precisely how the baryon number is divided between the strange and non-strange fields. The only fact necessary for consistency is that the net baryon number of the bound system is equal to 1, which is insured by the quantization of the winding number.

deformation is obtained by a small rotation of the hedgehog into a strange direction, *i.e.*,

$$U \approx U_0 + [i\epsilon\lambda_4, U_0]. \tag{122}$$

Comparing this with Eq. (93) we find, after a short calculation, that such a deformation corresponds to a kaon in the $T = 1/2, L = 1$ partial wave, with the profile function

$$k_1(r) \sim \sin(F(r)/2). \tag{123}$$

If we substitute this into Eq. (120), and replace ω_1 by the corresponding energy ω_-, we find the rigid rotator formula (91). The advantage of the bound state approach is that it allows the kaon profile to adjust itself away from the $SU(3)$ symmetric value (123) as m_K increases. Is this effect significant? The answer is unambiguous: the important coefficients c and $\bar{c}$ have been changed from 0.24 and 0.06 to 0.6 and 0.36 respectively. It is satisfying that the effect has worked in the right direction to improve agreement with experiment. This seems to confirm our hypothesis that it is important to understand the effects of $SU(3)$ symmetry breaking on baryon wave functions. In section 7 we will review further comparisons of the bound state approach to strangeness with the available experiments.

7. FURTHER TESTS OF THE BOUND STATE APPROACH

Since the Skyrme model is only a rough model of nature, and its predictions in the 2-flavor case are only within 30 % of observations, in the bound state approach to strangeness we can only expect agreement at this level of accuracy. The best way to explore whether the qualitative picture of hyperons as kaon-soliton bound states is approximately valid, is to compare its predictions with nature. In the remainder of these lectures, I will review some of the recent work in this direction.

Since there is extensive data on the magnetic moments of the octet baryons, a calculation of magnetic moments serves as a good test of the bound state approach to strangeness. In the 2-flavor Skyrme model, it was found that the neutron and proton magnetic moments come out too low numerically, but their ratio agrees very well with experiment.[12] A recent calculation of hyperon magnetic moments in the bound state approach reveals that most of them also turn out somewhat low.[36] With the exception of μ_{Ξ^-}, the calculated values are in good agreement with the data, as shown in Table 1.[†] Perhaps more interesting than the calculated numbers is the fact that all magnetic moments can be expressed in terms of 4 quantities a_i which depend on the form of the kaon lagrangian and on the bound state mode.[36] Thus, there is a set of "model-independent" relations among the magnetic moments, which test our method of quantization rather than the detailed form of the lagrangian.

† Inclusion of the pion mass and of a contribution which is $\mathcal{O}(1/N)$ (ref. 37) improves the numerical agreement; most notably, for μ_{Ξ^-}. However, the term $\mathcal{O}(1/N)$ reduces by one the number of model-independent relations among magnetic moments.

Table 1. The magnetic moments of baryons.[36]

particle	μ	μ_{exp}	$\mu(a_1, a_2, a_3, a_4)$
p	1.870	2.793	$\frac{1}{2}a_1 + \frac{2}{3}a_2$
n	-1.313	-1.913	$\frac{1}{2}a_1 - \frac{2}{3}a_2$
$N \to \Delta$	2.251	~ 3	$\frac{2\sqrt{2}}{3}a_2$
Λ	-0.414	-0.613 ± 0.004	$\frac{1}{2}a_4$
Σ^+	2.066	2.42 ± 0.05	$\frac{2}{3}a_1 + \frac{2}{3}a_2 + \frac{2}{3}a_3 - \frac{1}{6}a_4$
Σ^0	0.509	-	$\frac{2}{3}a_1 - \frac{1}{6}a_4$
Σ^-	-1.048	-1.157 ± 0.025	$\frac{2}{3}a_1 - \frac{2}{3}a_2 - \frac{2}{3}a_3 - \frac{1}{6}a_4$
$\Lambda \to \Sigma$	-1.557	-1.61 ± 0.08	$-\frac{2}{3}a_2 - \frac{2}{3}a_3$
Ξ^0	-1.153	-1.250 ± 0.014	$-\frac{1}{6}a_1 - \frac{2}{9}a_2 - \frac{4}{9}a_3 + \frac{2}{3}a_4$
Ξ^-	-0.138	-0.69 ± 0.04	$-\frac{1}{6}a_1 + \frac{2}{9}a_2 + \frac{4}{9}a_3 + \frac{2}{3}a_4$

Derivation of model-independent relations is similar to the quark model approach where all magnetic moments are expressed in terms of 3 constituent quark magnetic moments, which are usually treated as unknown. The result of such a calculation is a set of relations between magnetic moments. In the Skyrme model, one of these relations arises purely in the 2-flavor context:[12]

$$\mu_{N\to\Delta} = (\mu_p - \mu_n)/\sqrt{2}. \tag{124}$$

In the quark model, this relation is replaced by $\mu_{N\to\Delta} = 2\sqrt{2}(\mu_p - \mu_n)/5$, which is in somewhat worse agreement with the data than Eq. (124). In the bound state approach one finds a further set of relations

$$\begin{aligned}
&\mu_{\Sigma^+} + \mu_{\Sigma^-} - 2\mu_{\Sigma^0} = 0, \\
&\mu_{\Sigma^+} + \mu_{\Sigma^-} + \frac{2}{3}\mu_\Lambda - \frac{4}{3}(\mu_p + \mu_n) = 0, \\
&\mu_{\Xi^0} + \mu_{\Xi^-} - \frac{8}{3}\mu_\Lambda + \frac{1}{3}(\mu_p + \mu_n) = 0, \\
&\mu_{\Xi^0} - \mu_{\Xi^-} + \frac{2}{3}(\mu_{\Sigma^+} - \mu_{\Sigma^-}) - \frac{1}{3}(\mu_p - \mu_n) = 0, \\
&\mu_{\Sigma^+} - \mu_{\Sigma^-} + 2\mu_{\Lambda\to\Sigma} = 0.
\end{aligned} \tag{125}$$

In the quark model, the first 4 relations in Eq. (125) hold while, in the last relation, the factor 2 is replaced by $4/\sqrt{3}$, which slightly improves its agreement with the data. The first relation in Eq. (125) constitutes a prediction for the magnetic moment of Σ^0; the other relations are obeyed by the data with good accuracy. Generally, the model-independent structure in the bound state approach is surprisingly similar to the one in the quark model. The relative merit of the former is that everything is calculable from the starting chiral lagrangian. The fact that all the model-independent relations are good gives us faith that changes in the chiral lagrangian can improve agreement with the data in Table 1.

One of the attractive features of the bound state approach is that it incorporates higher kaon partial waves. Baryon states based on these modes are not accessible to the rigid rotator analysis. For instance, as we mentioned before, there is a rather weakly bound state with $T = 1/2$, $L = 0$, $S = -1$, which acquires the quantum numbers of a p-wave strange quark. From the collective coordinate excitations of a single kaon bound to the soliton in this partial wave, we obtain p-wave baryons Λ, $\Sigma_{J=1/2}$ and $\Sigma_{J=3/2}$, which have negative parity (opposite to the parity of the lightest hyperons). The hyperfine splittings among these negative parity hyperons were determined in ref. 38. The masses were approximately predicted to be 1360 MeV for the Λ, 1380 MeV for the $\Sigma_{1/2}$, and 1670 MeV for the $\Sigma_{3/2}$. The Λ state undoubtedly corresponds to the well-established $\Lambda(1405)$, which is a major non-trivial prediction of the bound state approach. The well-established D_{13} Σ resonance at 1670 MeV has $L = 2$ and therefore cannot be identified with the heaviest of the predicted states.‡ However, there are 1-star and 2-star Σ resonances at 1480 and 1670 MeV, whose quantum numbers have not yet been experimentally determined, which may turn out to be the states predicted above. Further, the model contains doubly strange baryons where 1 or 2 kaons are excited to the $L = 0$ bound state. These are the analogues of the quark model states where 1 or 2 strange quarks are excited to the p-orbital. One of these states, with $J^P = (3/2)^-$, was identified in ref. 38 with $\Xi(1820)$. It would be interesting to see if other such states can be observed.

Beyond the baryons based on kaons bound to the soliton, we would like to explore the states which manifest themselves as kaon resonances. The corresponding problem of $\pi - N$ scattering in the 2-flavor Skyrme model has by now been thoroughly explored, with the conclusion that the Skyrme model can account for many of the features of the baryon spectrum up to $\sim 3GeV$.[39] Kaon-nucleon scattering was explored in the $SU(3)$ symmetric formalism in ref. 40.§ Recently, Scoccola has begun the study of kaon resonances in a model based on the bound state approach to strangeness.[35] The model he used explicitly includes vector mesons to stabilize the hedgehog. In fact, the commutator term in Eq. (6) results when the ρ meson is integrated out from such a model. The lagrangian of ref. 35, when used in the soliton sector, has consistently produced predictions almost indistinguishable from the predictions based on the Skyrme lagrangian.[41]

In order to look for kaon resonances, one needs to solve Eqs. (107) and (108) for the eigenmodes in the background of a static soliton. First, consider the $S = -1$ modes where the Wess-Zumino term is attractive. There, pronounced resonances are found for $L = 2, 3, 4, 5$, with $T = L - 1/2$. These resonances get higher in energy and broader with an increasing L. The splittings between them are $\mathcal{O}(1)$. Also, there is an $\mathcal{O}(1)$ splitting between the $L = 2, T = 3/2$ resonance and the $L = 0$, $T = 1/2$ bound state, which corresponds to the spin-orbit splitting found in the quark model between the $J = 3/2$ and $J = 1/2$ states of a p-wave strange quark. In the $L = 1, T = 1/2$ channel there is some attraction (after all, this channel contains the bound state), but it is not sufficient to form a resonance. In all the $T = L + 1/2$ partial waves V_{eff} is more repulsive than in

‡ As we will mention later, in the bound state approach this state does appear as a resonance in the $L = 2$ partial wave.

§ The major discrepancies between refs. 35 and 40 seem to be due in part to the factor of e^2 omitted from the normalization of the Wess-Zumino term in ref. 40. This rendered the Wess-Zumino term negligible and resulted in KN scattering being almost identical to $\bar{K}N$ scattering. In fact, when properly normalized, the Wess-Zumino term plays the crucial role of distinguishing between positive and negative strangeness.

the corresponding $T = L - 1/2$ partial waves, and no resonances are found. In the quark model, this effect would be attributed to spin-orbit interactions.

In all the $S = 1$ channels, the repulsion due to the Wess-Zumino term turns out to be the dominant effect which prevents resonance formation for any L and T. This is encouraging because no $S = 1$ baryons have been experimentally confirmed. In order to interpret a resonant mode about an unrotated hedgehog in terms of $\bar{K}N$ scattering, it is necessary to excite the $SU(2)$ rotational degree of freedom A. In both the $|in\rangle$ and the $|out\rangle$ states, when the kaon is far from the soliton, the isorotation of the soliton is described by an appropriate nucleon wave function $f_{1/2}(A)$ while the kaon is unaffected by A. Since the soliton angular velocity is $\mathcal{O}(1/N)$ and the kaon velocity is $\mathcal{O}(1)$, the method commonly used is the adiabatic approximation where the entire scattering event is assumed to take place at a fixed soliton orientation AU_0A^{-1}. The physical scattering amplitude is then found by taking the matrix element of the scattering amplitude at a fixed soliton orientation A between the appropriate initial and final nucleon wave functions. Clearly, this approximation breaks down whenever a long-lived metastable intermediate state forms. In such a state, the kaon is localized near the soliton for a long time, rotating together with it in isospace similarly to a bound state (see Eq. (116)) The adiabatic approximation is also invalid in processes where the kaon moves slowly, such as the low-energy scattering in the low partial waves. In general, although the adiabatic approximation is valid in the large-N limit, it may not be very good at $N = 3$. However, the adiabatic approximation should improve as the resonances get broader. Also, it is the simplest way to study scattering in the Skyrme model, and it seems to give good results in $\pi - N$ scattering for $L > 2$.[39]

In the adiabatic approximation, each resonance about the fixed hedgehog in the $T = L - 1/2$ channel gives rise to 3 separate $\bar{K}N$ resonances with the [isospin, spin, orbital angular momentum] quantum numbers given by $[I, J, L] = [0, L - 1/2, L]$, $[1, L - 1/2, L]$, $[1, L + 1/2, L]$.[35] Therefore, this approach predicts that some of the resonances should come in triplets. In nature, some of the Λ and Σ resonances can indeed be grouped into triplets. In the traditional $L_{I,2J}$ notation there are 2 complete triplets: $D_{0,3}(1520)$, $D_{1,3}(1670)$, $D_{1,5}(1775)$ and $F_{0,5}(1820)$, $F_{1,5}(1915)$, $F_{1,7}(2030)$. For $L = 4$ and 5, only the $I = 0$ components of the triplets, $G_{0,7}(2100)$ and $H_{0,9}(2350)$, have been reported. It is tempting to conjecture the existence of their Σ partners. The $\mathcal{O}(1)$ approximation of ref. 35 predicts roughly correct splittings between the centers of the triplets. In order to find the splittings within the triplets, $\mathcal{O}(1/N)$ corrections need to be taken into account.

Although these results are encouraging, not all the observed resonances are readily found in the model: some $L = 0$ resonances are missing, along with $D_{0,3}(1690)$, $D_{0,5}(1830)$, $F_{0,5}(2110)$. All of these, except $\Lambda(1830)$ are not the first resonances in their partial waves. Perhaps, some small changes in the chiral lagrangian, which has only been treated rather crudely, could increase the number of resonances in these attractive channels. Another possibility is that, when pion fluctuations are taken into account, some of the missing states will be identified more easily in the processes involving pions in the initial and/or final states. Comparing with the Skyrme model results on pion-nucleon scattering, we find that there too some of the low partial waves posed a problem. This could be due in part to the failure of the adiabatic approximation. In spite of some discrepancies, which could be eliminated in a more precise treatment, kaon-nucleon scattering in the bound state approach undoubtedly exhibits some important effects observed in nature, such as the absence of $S = 1$ states and the presence of the triplets of $S = -1$

resonances. This gives us hope that this approach to the Skyrme model will eventually provide a good description of the hyperon spectrum up to $\sim 3GeV$.

I cannot cover every application of the bound state approach to hyperons in the span of these lecture notes. Some of the most recent investigations involve, for instance, the electric and magnetic mean square radii,[42] meson-hyperon coupling constants,[43] and the strange dibaryons.[44] I would like to finish these lectures with a discussion of one application of the model — to dense baryonic matter — which seems quite interesting.

In the Skyrme model, dense nuclear matter can be described by solitons arranged on a periodic lattice. In the first investigation of such a crystal,[45] it was found that, if appropriate relative orientations between the nearest neighbor solitons are chosen, the interactions are attractive at low densities. Since the high densities are dominated by the hard core repulsion, it is clear that, if the baryon kinetic energy is ignored, then there exists a stable high-density state of skyrmion matter. Upon a closer look[46] at the soliton matter, it was found that, as the lattice constant is decreased in the low-density phase with widely separated skyrmions, a phase transition eventually takes place. On the low-density side of the phase transition, the crystal consists of deformed and compressed hedgehogs; on the high-density side the hedgehogs entirely lose their identity. In the high-density phase, which is characterized by the vanishing of the order parameter[47]

$$\int_{unitcell} d^3x \mathrm{Tr}\,(U + U^\dagger - 2), \tag{126}$$

baryon number and energy are very smoothly distributed over the unit cell. In ref. 48 the order parameter (126) was identified with $\langle \bar{q}q \rangle$, and the phase transition was interpreted as the chiral symmetry restoration. It is fascinating that, in the Skyrme model, chiral symmetry restoration can be observed without an explicit mention of quarks. This phase transition appears to be generic: it occurs not only in the cubic crystal of ref. 45, but also in the more complicated crystalline arrangements of refs. 49 and 50 which are somewhat preferred energetically. In all these crystals, the minimum of energy per baryon number occurs in the high-density phase.

Since extensive numerical work is needed to treat the skyrmion crystals, Manton invented a much simpler model where the phase transition can be studied analytically.[51] In his toy model of dense matter, space is taken to be explicitly compact, with the metric of the 3-sphere. If the radius R of the spatial 3-sphere is taken to be much larger than the skyrmion scale $(eF_\pi)^{-1}$, then the soliton is localized in a small region of space, much like if space is flat. However, for R below some critical value, the lowest energy configuration has energy and baryon number uniformly distributed over space.

All the above studies of dense matter were done in the 2-flavor context. However, various general arguments suggest that, at some high density, matter becomes unstable with respect to the formation of strangeness. An interesting problem is to estimate this critical density. In the Skyrme model, stability of nucleonic matter with respect to kaon condensation can be investigated using the techniques developed in the bound state approach.[52] The motion of kaons in the multi-soliton backgrounds characteristic of nucleonic matter of varied density can be studied using the chiral lagrangian expanded to second order in K, as in Eqs. (98) and (99).

The critical density is where the lowest kaon energy eigenvalue becomes zero, which we expect to occur for an $S = -1$ mode. Beyond this density, it is energetically favorable to populate this mode, and the $\bar{K}$ condensate begins to develop. In the quadratic approximation for kaons, the energy is not bounded from below beyond the critical density. However, we expect that the positive $\mathcal{O}(K^4)$ terms will stabilize the dense matter at some finite strangeness per baryon number. Since the calculation in the background of a crystal is formidable, Forkel et. al.[53] recently used Manton's toy model of dense matter[51] to study kaon condensation. They found that the bound state energy does drift down with increasing density and kaon condensation does occur, but only at densities comparable to the density of the chiral symmetry restoration. Thus, it seems that strangeness appears at densities so high that nuclear matter has become quark matter. In order to test the results found in ref. 53, it is necessary to implement the full calculation in a 3-dimensional crystalline background. This was first attempted in ref. 54, where it was also found that kaon condensation occurs at an extremely high density. However, in ref. 54 the soliton crystal was treated in a crude Wigner-Seitz approximation where each skyrmion is packed into a hard sphere. In this approximation, chiral symmetry restoration does not occur, and it is impossible to compare the densities of the two phase transitions, kaon condensation and chiral symmetry restoration. Although the evidence accumulated so far suggests that the two phase transitions occur at comparable densities, it would be nice to confirm this in a full calculation, with the crystalline backgrounds of refs. 45, 49 and 50.

ACKNOWLEDGEMENTS

I am indebted to C. G. Callan, my thesis adviser and collaborator on the bound state approach, for getting me started on skyrmions and for sharing his ideas. I thank K. Hornbostel and D. Kaplan for fruitful collaborations. Finally, I thank the organizers of the Summer Institute on Hadrons and Hadronic Matter, D. Vautherin, J. Negele and F. Lenz, for giving me the opportunity to present these lectures in the enjoyable setting of Cargese.

REFERENCES

1. T. Skyrme, *Nucl. Phys.* **31** (1962), 550, 556, *Proc. Roy. Soc.* **A260** (1961), 127.
2. A. P. Balachandran, V. P. Nair, S. G. Rajeev and A. Stern, *Phys. Rev. Lett.* **49** (1982), 1124, *Phys. Rev.* **D27** (1983), 1153.
3. E. Witten, *Nucl. Phys.* **B160** (1979), 57.
4. E. Witten, *Nucl. Phys.* **B223** (1983), 422, 433.
5. G. 't Hooft, *Nucl. Phys.* **B72** (1974), 461, *Nucl. Phys.* **B75** (1974), 461
6. J. Wess and B. Zumino, *Phys. Lett.* **37B** (1971), 95.
7. E. Guadagnini, *Nucl. Phys.* **B236** (1984), 35.
8. P. Mazur, M. Nowak and M. Praszalowicz, *Phys. Lett.* **147B** (1984), 137.
9. A. Manohar, *Nucl. Phys.* **B248** (1984), 19.
10. M. Praszalowicz, *Phys. Lett.* **158B** (1985), 264.
11. M. Chemtob, *Nucl. Phys.* **B256** (1985), 600.
12. G. Adkins, C. Nappi and E. Witten, *Nucl. Phys.* **B228** (1983), 552.
13. M. Rho, A. Goldhaber and G. Brown, *Phys. Rev. Lett.* **51** (1983), 747.

14. A. D. Jackson and M. Rho, *Phys. Rev. Lett.* **51** (1983), 751.

15. R. L. Jaffe, *Phys. Rev.* **D21** (1980), 3215.

16. D. Kaplan and I. Klebanov, preprint SLAC-PUB-4964, UCSD/PTH 89-02, to appear in *Nucl. Phys. B.*

17. N. Park, J. Schechter and H. Weigel, *Phys. Lett.* **B224** (1989), 171.

18. H. Yabu, *Phys. Lett.* **218 B** (1989), 124.

19. C. Callan and I. Klebanov, *Nucl. Phys.* **B262** (1985), 365.

20. I. Zahed and G. Brown, *Phys. Rep.* **142** (1986), 1.

21. A. P. Balachandran, Lectures at TASI '85, M. Bowick and F. Gürsey, eds., World Scientific (1986).

22. C. Callan, K. Hornbostel and I. Klebanov, *Phys. Lett.* **202B** (1988), 269.

23. J. Donoghue, C. Ramirez and G. Valencia, *Phys. Rev.* **D38** (1988), 2195, *Phys. Rev.* **D39** (1989), 1947.

24. D. Finkelstein and J. Rubinstein, *J. Math. Phys.* **9** (1968), 1762.

25. J. P. Blaizot and G. Ripka, Saclay preprint SphT/87-181 (1988).

26. A. De Rujula, H. Georgi and S. L. Glashow, *Phys. Rev.* **D12** (1975), 147.

27. V. Kaplunovsky, unpublished.

28. M. Gell-Mann, R. Oakes and B. Renner, *Phys. Rev.* **175** (1968), 2195.

29. H. Weigel, J. Schechter, N. W. Park, U. G. Meissner, Syracuse preprint SU-4228-416 (1989).

30. S. Weinberg, *Physica* **96A** (1979), 327.

31. H. Georgi and A. Manohar, *Nucl. Phys.* **B234** (1984), 189.

32. A. Th. M. Aerts, P. J. G. Mulders and J. J. de Swart, Phys. Rev. D17 (1978) 260.

33. H. Yabu and K. Ando, *Nucl. Phys.* **B301** (1988), 601.

34. C. Callan, C. Coleman, J. Wess and B. Zumino, *Phys. Rev.* **177** (1969), 2247.

35. N. Scoccola, University of Regensburg preprint TPR-89-35.

36. J. Kunz and P. J. Mulders, NIKHEF preprint P-10 (1989).

37. E. Nyman and D. O. Riska, Helsinki preprint HU-TFT-89-50.

38. K. Dannbom, E. Nyman and D. Riska, *Phys. Lett.* **B227** (1989), 291.

39. For a review, see B. Schwesinger, H. Weigel, G. Holzwarth and A. Hayashi, *Phys. Rep.* **173** (1989), 173.

40. M. Karliner and M. Mattis, *Phys. Rev. Lett.* **56** (1986), 449, *Phys. Rev.* **D34** (1986), 1991.

41. N. N. Scoccola, H. Nadeau, M. A. Nowak and M. Rho, *Phys. Lett.* **B201** (1988), 425; J. P. Blaizot, M. Rho, and N. N. Scoccola, *Phys. Lett.* **B209** (1988), 27; N. N. Scoccola, H. Nadeau, D. P. Min and M. Rho, Nucl. Phys. A, in press.

42. J. Kunz and P. J. Mulders, NIKHEF preprint P-11a (1989).

43. E. Nyman, *Phys. Lett.* **B224** (1989), 21.

44. J. Kunz and P. J. Mulders, *Phys. Lett.* **B215** (1988), 449.

45. I. Klebanov, *Nucl. Phys.* **B262** (1985), 133.

46. G. Brown, A. Jackson and E. Wüst, *Nucl. Phys.* **A468** (1987), 137.

47. A. D. Jackson and J. Verbaarschot, *Nucl. Phys.* **A484** (1988), 419.

48. H. Forkel, A. D. Jackson, M. Rho, C. Weiss, A. Wirzba and H. Bang, Nordita preprint 89/16 N (1989).

49. M. Kugler and S. Shtrikman, *Phys. Lett.* **208B** (1988), 491.

50. L. Castillejo, P. Jones, A. D. Jackson, J. Verbaarschot and A. Jackson, preprint Print-89-0195 (1989).

51. N. Manton, *Comm. Math. Phys.* **111** (1987), 469.

52. I. Klebanov, in Proceedings of the Workshop on Nuclear Chromodynamics, S. Brodsky and E. Moniz, eds. World Scientific (1986).

53. H. Forkel, A. D. Jackson, M. Rho and N. Scoccola, preprint (1989).

54. J. Yeh, B. S. thesis (1987), Princeton University.

PROBING THE QUARK STRUCTURE OF MATTER

Anthony W. Thomas

Department of Physics and Mathematical Physics
The University of Adelaide
P.O. Box 498, Adelaide, SA 5001, AUSTRALIA

INTRODUCTION

Perhaps the most profound question in nuclear physics is the extent to which the internal structure of the nucleon matters in microscopic calculations of phenomena such as the N-N force, nuclear binding energies and densities and the response of nuclei to weak and electromagnetic probes. In answering it we would forge a long needed link between high energy and nuclear physics. My own interest in the problem is over a decade old, beginning with naive considerations of size based on the MIT bag model and its chiral extensions, notably the cloudy bag model. From such a beginning it is possible to take two divergent paths. The first is to ask what constraints more elementary, phenomenological considerations can put on the structure of the nucleon and therefore on the existing QCD-motivated models. The second is to ask how the consequences of different models might vary for the phenomena mentioned.

In these lectures we shall briefly address both of these questions. We begin with a discussion of deep-inelastic scattering (DIS) of leptons from nucleons. In particular we explain the most recent approach to calculating the quark/parton distributions corresponding to a particular quark model. Since DIS probes yield unambiguous information about valence and sea quarks separately, this promises to be a powerful tool in constraining models of nucleon structure.

Our discussion will include a clear quantitative explanation of the connection between such apparently diverse phenomena as the N-Δ mass splitting due to one-gluon-exchange, the ratio of the down- to up-quark distributions and the polarisation asymmetry of the proton and the neutron. We also discuss the recent surprise from the European Muon

Hadrons and Hadronic Matter
Edited by D. Vautherin *et al.*
Plenum Press, New York, 1990

Collaboration (EMC) concerning the proton spin sum-rule (the missing spin of the proton) within the same context.

Turning to the second approach we review a recent phenomenological model of nuclear matter due to Guichon. It seems to indicate that it would be inconsistent to ignore the internal structure of the nucleon within a relativistic mean-field theory. On the other hand, once this structure is included using a bag model, results of the admittedly toy model are surprisingly good. Furthermore, having a quark level description of nuclear matter we are able to use the approach described earlier to calculate the nuclear structure functions in a consistent, microscopic way. These calculations indicate that the usual impulse approximation, in which the off-mass-shell dependence of the nucleon structure function is ignored, may be quite unreliable. We also present some promising initial results for the ratio of the structure functions of Fe and D - the so-called EMC effect.

In the final section we turn to a recent surprise consequence of our improved knowledge of nucleon structure for the N-N force. In particular, we shall see that the "good old" one pion exchange force is not so well understood after all.

DEEP INELASTIC SCATTERING

The elementary ideas of deep-inelastic scattering (DIS) can be found in many text books and reviews.[1-8] In introducing this section we simply define our notation and collect together the essential formulae.

Consider the inclusive scattering of a high energy lepton (initial energy E, final energy E′) from a hadronic target (mass m and initial four-momentum p) with the transfer of spacelike four-momentum q. For an unpolarised target the laboratory differential cross-section for electromagnetic scattering is:

$$\frac{d^2\sigma}{dE'\,d\Omega} = \frac{4\alpha^2(E')^2}{q^4}\left[\cos^2 {}^{\theta}\!/_{2}\,\frac{F_2}{\nu} + 2\sin^2 {}^{\theta}\!/_{2}\,F_1/_{m}\right]. \quad (1)$$

All of the information concerning the structure of the target is contained in the structure functions F_1 and F_2 which can depend on at most two variables. It is most usual to choose those to be the Lorentz invariant quantities $Q^2(= -q^2>0)$ and Bjorken $x(= Q^2/2m\nu$, with ν the photon energy in the laboratory frame).

For ν ($\bar{\nu}$) scattering from an unpolarised target we find a third structure function, F_3, associated with parity violation:

$$\frac{d^2\sigma^{\nu(\bar\nu)}}{d\Omega dE'} = \frac{G_F{}^2(E')^2}{2\pi^2}\left[\frac{M_W{}^2}{M_W{}^2+Q^2}\right]^2\left[\cos^2{}^{\theta}/_2\,\frac{F_2{}^{\nu(\bar\nu)}}{\nu} + 2\sin^2{}^{\theta}/_2\,\frac{F_1{}^{\nu(\bar\nu)}}{m}\right.$$

$$\left.\pm\frac{(E+E')}{m\nu}\sin^2{}^{\theta}/_2\,F_3{}^{\nu(\bar\nu)}\right] . \qquad (2)$$

Finally, for scattering of a polarised electron (or muon) from a polarised, spin-½ target there are (at least in principle) two more structure functions (g_1 and g_2) which can be measured. Denoting beam and target helicity with arrows top and bottom respectively we find:

$$\frac{d^2\sigma}{dE'\,d\Omega}\left(\begin{smallmatrix}\rightarrow\\ \rightarrow\end{smallmatrix} - \begin{smallmatrix}\rightarrow\\ \leftarrow\end{smallmatrix}\right) = \frac{4\ \alpha^2}{m\nu Q^2}\left(\frac{E'}{E}\right)\left[(E + E'\cos\theta)g_1 - 2mx\ g_2\right] . \qquad (3)$$

(We note in passing some recent theoretical arguments[9] which imply a relationship between g_1 and g_2, but we cannot discuss it further here.) In the deep-inelastic regime Q^2 and ν are both very large ($Q^2 > 2$ GeV^2, $\nu > 1$ GeV) but $x \in (0,1)$. Clearly the second term in equ.(3) will be negligible if g_1 and g_2 are of the same order. For this reason only g_1 has been measured so far. To determine g_2 one would need to work with a longitudinally polarised beam and a transversely polarised target.

$$\frac{d^2\sigma(\rightarrow\uparrow\ -\ \rightarrow\downarrow)}{dE'\,d\Omega} = \frac{4\alpha^2\ (E')^2}{m\nu E\ Q^2}\left(g_1 + \frac{2Eg_2}{E+E'}\right)\sin\theta . \qquad (4)$$

In the late 60's tremendous excitement was generated by the discovery at SLAC that the structure functions were almost independent of Q^2 over a very wide range. That is, they were functions of the single variable Bjorken x. It is very easy to see that this is what one would expect if the nucleon contained a collection of elementary constituents (initially called partons by Feynman but later identified with quarks) with low mass, which do not interact strongly during the DIS collision.

For simplicity it is usual to consider this problem in a so-called infinite momentum frame - e.g. one where the nucleon has momentum $P \gg m$ in the z-direction so that its 4-momentum is $p = (P,00P)$. Suppose a constituent with "momentum fraction y", that is 4-momentum $(yP,00yP) = yp$ absorbs the photon. Its final invariant mass squared will be $(yp+q)^2$, and:

$$(yp+q)^2 = y^2p^2 - Q^2 + 2yp.q . \qquad (5)$$

But $p^2 \ll Q^2$ and p.q, and the invariant mass squared of the parton must be small (≈ 0) by assumption. Then we find $y = Q^2/2p.q$ which was called

Bjorken x above. Thus we see that under the assumptions of the parton model, only a parton with fraction x of the momentum of the nucleon can absorb the exchange boson. In the case that the impulse approximation is valid, DIS structure functions then measure the number density of partons in the nucleon with momentum fraction x.

It is usual to define distributions $f_i^{\uparrow\downarrow}(x)$ which give the number density of quarks in the target with helicity parallel or anti-parallel to that of the target. For example, u(x)x dx gives the fraction of momentum of u quarks in the proton with momentum between xP and (x+dx)P in the infinite momentum frame (and with either helicity). By charge symmetry u also gives the distribution of down quarks in the neutron.

The structure functions mentioned earlier are directly related to these distribution functions. For an electromagnetic probe one finds:

$$F_1 = \tfrac{1}{2} \sum_i e_i^2 f_i \tag{6}$$

$$F_2 = 2 x F_1 \tag{7}$$

$$g_1 = \tfrac{1}{2} \sum_i e_i^2 (f_i^{\uparrow} - f_i^{\downarrow}) \tag{8}$$

with e_i the charge in units of e. Equation (7) is the Callan-Gross relation and relies on the partons having spin - ½ and no transverse momentum (in the infinite momentum frame). In general we have

$$F_2 = 2 x F_1 \frac{(1 + R)}{(1 + 2mx/\nu)} \tag{9}$$

where R is the ratio of cross sections for absorbing a longitudinal to that for a transverse photon. Experimentally R is small [10] (<0.1) for all x, for $Q^2 \gtrsim 5$ GeV^2.

For $\nu(\bar{\nu})$ scattering from an isoscalar target one finds

$$F_2^{(\nu)} = (u + \bar{u} + d + \bar{d} + s + \bar{s}), \tag{10}$$

which measures the total quark content of the proton. Even more important, by combining ν and $\bar{\nu}$ data one can measure the combination

$$F_3^{(\nu,\bar{\nu})} = (u - \bar{u} + d - \bar{d}), \tag{11}$$

which isolates the excess of quarks over anti-quarks - i.e. the valence quark distribution of the nucleon. Clearly we would expect the sum-rule (due to Gross and Llewellyn Smith)

$$\int_0^1 dx \quad F_3^{(\nu,\bar{\nu})}(x) = 3, \tag{12}$$

to be obeyed. It will also be useful to define the n'th moment of a structure function like xF_3, F_2 or xF_1 as, e.g.

$$M_{3n} = \int_0^1 x^{n-2} [xF_3(x)] \, dx. \tag{13}$$

Initially the major experimental activity in this field was at SLAC, but for the past 10-15 years, the emphasis has shifted to the muon and neutrino beams of CERN and Fermilab. For a thorough summary of the experiments at these laboratories we refer to the recent review by Morfin.[11] While most of data has been accumulated for Q^2 between (5,20) GeV^2 and x between 0.1 and 0.65, Q^2 as high as 200 GeV^2 and x as low as 0.02 have been obtained. (The latter is particularly relevant for certain sum- rules as we shall see.) Figure 1 illustrates the results typically obtained - "this experiment" is the CDHS neutrino experiment.[12] We note that the anti-quarks, which form half of the sea of virtual $q-\bar{q}$ pairs in the nucleon, are concentrated at low $x(x \lesssim 0.3)$. The valence quarks (cf. equ.(11)) dominate the large x region. There is impressive agreement between the weak and electromagnetic experiments once one allows for the appropriate charges. Actually the situation is a little worse than Fig.1 might suggest. Because of systematic errors, different muon data sets on the same target may differ by as much as (10-20) percent. These differences are typically within the quoted systematic errors. Again such differences can be important whenever an absolute measurement is required - e.g. in the spin sum-rule.

It is clear from the analysis of the experimental data that even in the Bjorken region the structure functions have a weak Q^2 - dependence, and therefore so do the distribution functions which we write as $F_i(x,Q^2)$. If one sticks to any one data set in order to (partially), avoid systematic errors, this variation of the structure functions (scaling violation) is essentially logarithmic. In order to understand it one must go beyond the naive parton model to QCD.

Scaling Violations

Figure 2 shows the DIS process as we have described it. By assumption the wave function of the target has no high momentum components (i.e. $\vec{p}_T^{\,2} \ll Q^2$). Thus any Q^2 - dependence can only come from the lepton-quark scattering process. Scaling results if the quark is treated as point-like and the trivial Q^2 dependence of the Mott cross-section is factored out. On the other hand, in an interacting field theory, the lepton-quark scattering amplitude will involve radiative corrections, some of which add coherently (e.g. wave function and vertex renormalisation) while others are incoherent (e.g. bremmstrahlung). It is well known that such radiative processes lead to corrections which vary logarithmically with the appropriate cut-off scale - in this case Q^2.

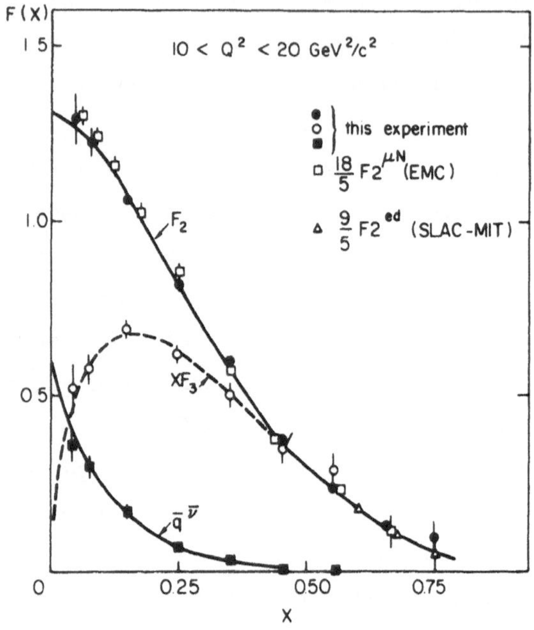

Figure 1. Typical results for the structure functions extracted from neutrino experiments ("this experiment" is CDHS)[12] as well as electron and muon scattering.[56,57] Notice that the sea quarks ($\bar{q}^{\bar{\nu}}$) are concentrated at small x, while the valence distribution (xF_3 - c.f. equ.(12)) dominates at large x.

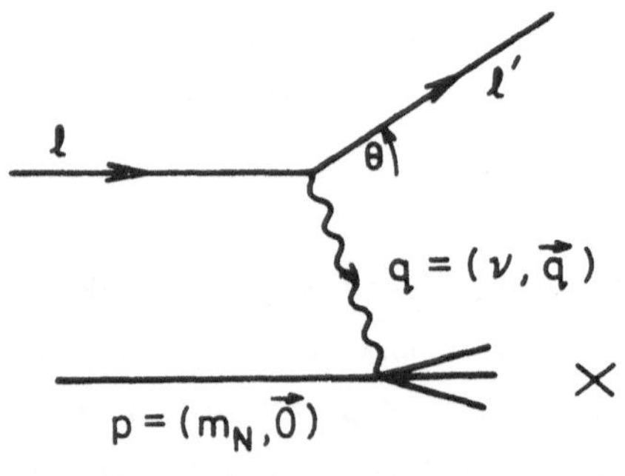

Figure 2. The DIS cross section involves a sum over all final states X. It is therefore proportional to the imaginary part of the forward scattering amplitude of the exchanged boson at momentum q.

For explicit calculations of these radiative corrections, we refer to some excellent texts.[1-8] It is particularly important from the point of view of application to other systems, that one can develop a very physical interpretation of this Q^2 variation. This is perhaps best expressed through Close's "onion skin" picture, whereby every time we increase Q^2, we increase the resolution at which we observe the structure of the target - hence revealing more and more of its previously virtual quarks and gluons. The mathematical description of the variation of the structure function with Q^2 is given by the Altarelli-Parisi equations. If one sticks to any one of the data sets mentioned above (in order to avoid systematic errors), the Q^2 variation of the structure functions is well described by these equations. However there are discrepancies between data sets and difficulties have also been encountered when trying to make a consistent fit to the EMC data on different nuclear targets.

The most rigorous approach to the calculation of structure functions and the description of their Q^2 variation, comes through the operator product expansion and the renormalisation group. As these are also discussed in many texts, we highlight only those features needed for our main consideration, namely the the prediction of structure functions from particular quark models.

The Operator Product Expansion

As electromagnetic DIS involves a total cross section for lepton scattering with a single photon exchanged, the structure functions are proportional (through the optical theorem) to the imaginary part of the forward Compton amplitude for a photon of 4-momentum q. The Compton amplitude is written:

$$T_{\mu\nu} = \int d^4z \; e^{iq\cdot z} \langle N|T(j_\mu(z) \; j_\nu(0)|N\rangle. \qquad (14)$$

The essential idea of the operator product expansion (OPE), is that one can expand the time-ordered product of the currents in what is essentially a generalisation of familiar Taylor series. That is one writes it as an infinite series, whose terms each involve a (possibly singular) function of z^2 times a local operator, in general involving $\bar{\psi}(0)$ and derivatives of $\psi(0)$, contracted with products of z^λ. It is crucial that this is an expansion of the operator which is therefore target independent.

Dropping Lorentz indices, for purposes of illustration, one has schematically:

$$T(j(z) \; j(0)) = \sum_{i,n} \tilde{C}_{i,n}(z^2) \; z_{\mu_1} \ldots z_{\mu_n} \; O^{\mu_1 \ldots \mu_n}_{i,n} \qquad (15)$$

where for example one might have an operator

$O_{i,n}{}^{\mu_1 \cdots \mu_n}$ of the form:

$$\bar{\psi}(0)\ \gamma^{\mu_1}\ D^{\mu_2}\ \ldots\ D^{\mu_n}\ \psi(0), \tag{16}$$

$$\bar{\psi}(0)\ \gamma^{\mu_1}\ D^{\mu_2}\ \psi(0)\ \bar{\psi}(0)\ \gamma^{\mu_3} \ldots\ D^{\mu_n}\ \psi(0), \tag{17}$$

and so on. Now ignoring renormalisation ψ has dimensions of $[\text{energy}]^{3/2}$, which is usually called simply dimension-3/2. Therefore the l.h s. of equ. (15) has dimension-6. Suppose the operator $O_{i,n}$ has dimension N_i, then $\tilde{C}_{i,n}$ must have dimension $6 + n - N_i$. After Fourier transformation it will therefore behave like $(Q^2)^{6+n-N_i}$. Since $Q^2 \to \infty$ in the Bjorken limit, the dominant operators will be those with the largest exponent, or the smallest value of $(N_i - n)$ - which is usually called "twist". One can easily check that the operator in equ.(16) is twist-2, while that in equ. (17) is twist-4. With a little thought one can see that twist-2 is the best that can be done, and therefore <u>DIS in the Bjorken limit will be entirely determined by twist-2 operators</u>.

The operator given in equ.(16) is called a singlet, twist-2 operator because it involves a trace over flavours. The only other twist-2, singlet operator involves the gluons:

$$g_{\alpha\beta}\ F^{\mu_1 \alpha}\ D^{\mu_2}\ \ldots\ D^{\mu_{n-1}}\ F^{\mu_n\ \beta}\ . \tag{18}$$

One can also write down a set of non-singlet operators

$$\bar{\psi}(0)\ \ \gamma^{\mu_1} D^{\mu_2}\ \ldots\ D^{\mu_n}\ \lambda_i \psi(0), \tag{19}$$

where λ_i are the flavour - SU(3) matrices. For simplicity we shall concentrate on the non-singlet operators from now on.

In general the matrix elements of these operators must be of the form

$$\langle N| O_i^{\mu_1 \cdots \mu_n} |N\rangle = p^{\mu_1} \ldots\ p^{\mu_n}\ \langle N| O_{i,n} |N\rangle. \tag{20}$$

Returning to equ.(14) we realize that the Fourier transform of z_μ is essentially q_μ/Q^2, which contracted with p^μ gives $1/x$. Thus we find that in the large - Q^2 limit, schematically (i.e. corresponding to equ.(15))

$$T \sim \sum_n\ C_n(Q^2)\ x^{-n}\ \ \langle N| O_n |N\rangle, \tag{21}$$

and we have incorporated an appropriate number of factors of Q^{-2} into the (n'th derivative of the) Fourier transform of $\tilde{C}_n$ to give C_n. The latter is easily seen to be dimensionless.

Finally, for the full electromagnetic case one finds

$$i\int d^4z\, e^{iq.z}\, T(j_\mu(z)\, j_\nu(0))$$

$$= \sum_{i,n} \Big[(g_{\mu\mu_1} g_{\nu\mu_2} Q^2 + g_{\mu_1\mu} q_\nu q_{\mu_2} + g_{\nu\mu_2} q_\mu q_{\mu_1} - g_{\mu\nu} q_{\mu_1} q_{\mu_2})\, C_{i,n}(Q^2)$$

$$+ \left(g_{\mu\nu} - \frac{q_\mu q_\nu}{q^2}\right) q_{\mu_1} q_{\mu_2}\, C^L_{i,n}(Q^2) - i\,\epsilon_{\mu\nu\alpha\beta} g_{\alpha\mu_1} q_\beta q_{\mu_2} C^3_{i,n}(Q^2)\Big]$$

$$q_{\mu_3} \cdots q_{\mu_n} \cdot \left(\frac{2}{Q^2}\right)^n O_i^{\mu_1 \cdots \mu_n} . \tag{22}$$

Here we have kept only the twist-2 operators, $C_{i,n}$ and $C^L{}_{i,n}$ correspond to the parity conserving transverse and longitudinal contributions respectively while $C^3{}_{i,n}$ is the parity violating term leading to F_3 in $\nu(\bar{\nu})$ scattering - c.f. equ.(2).

The Renormalisation Group

The arguments just presented must be modified in a field theory like QCD. The matrix element of the currents on the l.h.s. of equ.(22) must be renormalised, as must the matrix elements of the operators in the OPE on the r.h.s. This procedure introduces a new mass scale (or re-normalisation scale), μ^2, upon which no physical results can depend. It is very important that the μ^2 dependence of the coefficient functions $C_{i,n}$ $(Q^2,\mu^2,g^2(\mu^2))$ be chosen such that equ.(22) is true after renormalisation. From the practical point of view it is crucial that the OPE is an operator relationship which holds independent of the target. One can therefore calculate the μ^2-dependence for a simple target, such as a free quark. As this too is found in many texts[3-8,13,14] we just review it briefly.

Assuming that the operator O_n is multiplicatively renormalised by $Z_{O_n}{}^{-1}$, we define

$$\beta(g) = \frac{\partial g(\mu^2)}{\partial \ell n\, \mu^2} \tag{23}$$

and

$$\gamma_{O_n} = \frac{\partial \ell n\, Z_{O_n}}{\partial \ell n \mu^2} \tag{24}$$

(Note that we have dropped the label i for convenience.)

We shall show below that the product C_n (Q^2,μ^2,g^2) $< N|O_n|N>$ is measurable and therefore cannot depend on μ^2. (It is proportional to the n'th moment of the DIS structure function defined in equ.(13)). For the present we write

$$\mu \frac{d}{d\mu} C_n \langle N|O_n|N\rangle = 0, \tag{25}$$

and hence (in an appropriate gauge)

$$\left[\frac{\partial}{\partial \ell n\mu^2} + \frac{\partial g}{\partial \ell n\mu^2} \frac{\partial}{\partial g} - \gamma_{0_n} \right] C_n(Q^2,\mu^2,g^2) = 0. \tag{26}$$

However we saw above that for twist-two C_n is dimensionless and therefore can only depend on Q^2 as Q^2/μ^2. Therefore we can replace $\partial/\partial \ell n\mu^2$ by $-\partial/\partial \ell nQ^2$ and equ.(26) becomes

$$\left(\frac{\partial}{\partial \ell nQ^2} - \beta(g(Q^2)) \frac{\partial}{\partial g(Q^2)} + \gamma_{0_n} \right) C_n = 0. \tag{27}$$

The first two terms are easily identified as $-\beta(g(Q^2))d/dg(Q^2)$ so that equ.(27) implies

$$\frac{d\ \ell n\ C_n}{d\ g(Q^2)} = \frac{-\gamma_{0_n}}{\beta} . \tag{28}$$

Finally we obtain:

$$C_n(Q^2,\mu^2,g^2(\mu^2)) = C_n(Q^2,Q^2,g^2(Q^2))e^{-\int_{g(\mu^2)}^{g(Q^2)} \frac{\gamma}{\beta} dg'} \tag{29}$$

In practice one has a series expansion for $\gamma(g')$ and $\beta(g')$ to only a few terms. In the so-called leading order we have

$$\beta(g) = -\frac{\beta_o\, g^3}{16\pi^2} \quad ; \quad \gamma(g) = \gamma_o{}^n \frac{g^2}{16\pi^2} , \tag{30}$$

and

$$C_n\ (Q^2,Q^2,g^2(Q^2)) = 1 . \tag{31}$$

Then the integral in equ.(29) is easily performed and we find

$$C_n(Q^2,\mu^2,g^2) \overset{L.O.}{=} \left[\frac{\alpha(Q^2)}{\alpha(\mu^2)} \right]^{d_o{}^n} , \tag{32}$$

and using the calculated value of β_o (for N_f quark flavours) and $\gamma_o{}^n$ the anomalous dimension, $d_o{}^n$, is

$$d_o{}^n = \frac{\gamma_o{}^n}{2\beta_o} = \frac{4}{33-2N_f} \left(1 - \frac{2}{n(n+1)} + 4 \sum_{m=2}^{n} m^{-1} \right) . \tag{33}$$

<u>The Moments of the Structure Functions</u>

As we hinted above, there is a direct connection between the moments of the structure functions and the Compton amplitude which we have calculated so far. In fact as a function of ν for fixed Q^2, $T_{\mu\nu}$ has two cuts, (ν_{th},∞) corresponding to the physical region and $(-\infty,-\nu_{th})$ corresponding to crossed processes. In terms of x, again for fixed Q^2, the corresponding cuts run from (0,1) and (-1,0). Thus the dispersion relation for $T_{\mu\nu}$ at fixed Q^2 has the form

$$T_{\mu\nu}(x,Q^2) = \frac{1}{2\pi i} \int_{-1}^{+1} \frac{d\,\nu' \; (\text{disc})\; T_{\mu\nu}(x',Q^2)}{\nu' - \nu} \quad . \tag{34}$$

Replacing ν' by x' and using the optical theorem to replace Im $T_{\mu\nu}$ by the total cross-section, which is by definition the structure function, we find

$$T_{\mu\nu}(x,Q^2) = \sum_n x^{-n} \int_0^1 dx' \; (x')^{n-1} \; W_{\mu\nu}(x',Q^2) \; . \tag{35}$$

Note that for the various terms in $T_{\mu\nu}$ (see equ.(22)) the sum on n is restricted to even or odd values depending on the crossing properties of the corresponding piece of $W_{\mu\nu}$.

Comparing equ.(35) with equ.(21) we see that the product of the coefficient function C_n and the operator matrix element is measurable. Indeed it is equal to the moment of the appropriate structure function as defined in equ.(13). Using the result of the above analysis based on the OPE and the renormalisation group, we see that the Q^2 variation of the moments of the structure functions is given by perturbative QCD. To leading order one finds:

$$M_n(Q^2) = M_n(Q_0{}^2) \left(\frac{\alpha(Q^2)}{\alpha(Q_0{}^2)} \right)^{d_o{}^n} \quad . \tag{36}$$

For fixed $Q_0{}^2$ it is then easily shown that

$$\ell n \; M_n(Q^2) = \frac{d_0{}^n}{d_0{}^m} \quad \ell n \; M_m(Q^2) \quad + \text{constant}, \tag{37}$$

and therefore a log-log plot of any two moments should be a straight line whose slope is predicted by QCD.

Singlet Distributions

All of the above discussion of Q^2 evolution involves non-singlet operators. The Q^2 evolution of the operators given in equs.(16) and (18) is more complicated because they mix under renormalisation. While the corresponding analysis is not much more difficult (it involves a 2x2 matrix), it would divert us too much to explain it here. Instead we refer to the appropriate texts[3-8,13,14] – for example there is a concise summary in Table 2 of the review by Altarelli.[7]

The Inverse Mellin Transform

Given an analytic continuation of a set of moments, $M_n(Q^2)$, there is a standard method for reconstructing the corresponding function - this is the Inverse Mellin Transform (IMT):

$$xF_3\ (x,Q^2) = \frac{1}{2\pi i} \int_{C-i\infty}^{C+i\infty} dn\ x^{1-n}\ M_{3n}(Q^2)\quad . \tag{38}$$

(Here C is chosen so that the integral exists.) If the moments can be written as a product, as in equ.(21) or (22):

$$M_{3n}(Q^2) = C_n(Q^2,\mu^2)\quad \langle N|O_n(\mu^2)|N\rangle \tag{39}$$

then the IMT xF_3 is just a convolution of the IMT of C_n and $\langle N|O_n|N\rangle$, viz:

$$xF_3\ (x,Q^2) = \int_x^1 \frac{dy}{y}\ \mathfrak{C}_3\left(\frac{x}{y},Q^2,\mu^2\right) \left(y\mathfrak{F}_3(y,\mu^2)\right)\ . \tag{40}$$

This is an extremely important result. In particular $\mathfrak{C}_3$ is totally independent of the structure of the target - a property known as factorisation. Clearly if we can evaluate the structure function of the target at any renormalisation scale μ^2, equ.(40) allows us to calculate it at all higher values of Q^2. Higher order QCD corrections do not alter this result in principle, they just make $\mathfrak{C}_3$ harder to compute. For this reason μ^2 cannot be too low.

The Relation to Quark Models

At last we have collected all the results necessary to understand how to relate quark models to QCD. (Of course, if one could use non-perturbative QCD (e.g. on the lattice) to calculate $\langle N|O_n(\mu^2)|N\rangle$ this would be unnecessary. However this is not feasible for more than a few moments at the present time - see Negele[15] and Sachrajda[16].) Since the models are only "QCD motivated", the connection cannot be rigorous. On the other hand we know of no sensible alternative to what is proposed here.

Apart from lattice technology, the only known technique for solving a bound state problem in field theory is to make quantum corrections about a classical solution. That is one calculates radiative corrections at some renormalisation scale using perturbation theory and then solves (non-perturbative) classical equations of motion. One can then systematically add quantum corrections to the classical solutions. We assume that whatever quark model we are using represents just such a solution, renormalised at some scale μ^2. Although physically measurable quantites cannot depend on μ^2, the classical approximation may be better for some value. We treat the value of μ^2 appropriate to a given model, which we shall call $Q_0{}^2$, as a free parameter. If one can evaluate the twist-2 target matrix elements $\langle N|O_n(Q_0{}^2)|N\rangle$ within the model, then through equ.(22) and (29) (or equivalently (39) and (40)), one can calculate the twist-2 structure function at all Q^2. Even though the twist-2 contribution may not be dominant at $Q_0{}^2$, the general considerations presented earlier (see the discussion of higher twist below equ.(17)) ensure that at high enough Q^2 it will eventually dominate.

The most convenient practical method for evaluation the twist-2 moments of the structure function of some target follows from important work by Jaffe.[17] However, his article which was entitled "parton distribution functions" involved no discussion of renormalisation group corrections, and the calculations in it were only made in the Bjorken limit. It would therefore not be surprising if the student were confused as to the connection between his parton distributions (e.g. calculated for some model) and experimental data. We shall make that connection quite clear.

Following ref.(17) we define a function $H(\alpha)$ (in the $A^+ = 0$ gauge):

$$H(\alpha) = \frac{m}{2\pi} \int_{-\infty}^{\infty} e^{-im\alpha z} \langle N|T(\psi_+{}^\dagger(\xi^-)\ \psi_+(0))|N\rangle_c \ . \tag{41}$$

Here we understand a sum over the spins of the target ($|N\rangle$, mass m), c denotes a connected matrix element and $\psi_+\ (\xi^-)$ is an abbreviated notation for $\frac{(1+\alpha^3)}{2}\ \psi(z;00-z)$.

To understand why the second field operator is evaluated at a point on the light-cone with respect to the first we recall equ.(14). In the target rest frame we can choose the photon four momentum q to be $(\nu;00-\nu-mx)$ with $\nu\to\infty$. (Clearly Q^2 is $2m\nu x$.) The argument of the exponential in equ.(14) is i q.z which becomes $i[\nu(z^0+z^3)/2 + mxz^3]$. The rapid oscillations as $\nu\to\infty$ drive $(z^0 + z^3)$ to zero in the Bjorken limit and hence the process is light-cone dominated. (Causality implies that $z_T{}^2$ must be zero if we are to obtain a non-zero contribution to the connected matrix element.)

As a further matter of some practical importance, Jaffe has argued that the time-ordering in equ.(41) can be dropped. In particular, it was shown in ref.(17) that for a connected matrix element involving QCD field operators separated on the light-cone one can write either

$$<N|T(\psi_+^\dagger(\xi^-)\psi_+(0))|N>_c = <N|\psi_+^\dagger(\xi^-)\psi_+(0)|N>_c \qquad (42)$$

or equivalently

$$<N|T(\psi_+^\dagger(\xi^-)\psi_+(0))|N>_c = -<N|\psi_+(0)\psi_+^\dagger(\xi^-)|N>_c \ . \qquad (43)$$

The only reason for preferring one form to the other is calculational simplicity, because equ.(42) has no semi-disconnected contributions for $\alpha>0$ while the second form has none for $\alpha<0$. In general one can show that $H(\alpha)=0$ for $|\alpha|>1$, or alternatively that $H(\alpha)$ has support on $(-1,+1)$.

The next step in establishing the significance of $H(\alpha)$ is to show that its n'th moment, A_n:

$$A_n = \int_{-1}^{+1} d\alpha\ \alpha^{n-1}\ H(\alpha) \qquad (44)$$

is just $<N|O_{2,n}|N>$ (see sect. 3.2 of ref.(17)) where $O_{2,n}$ is the local twist-two operator of order n associated with the structure function $F_2^{(\nu)}$. Finally one has

$$F_2^{(\nu)} = x(H(x) - H(-x)) \ , \qquad (45)$$

while

$$F_3^{(\nu,\bar{\nu})} = (H(x) + H(-x)) \ , \qquad (46)$$

Comparing with the parton model formulas in equs.(10) and (11) we identify $q(x) = H(x)$ and $\bar{q}(x) = -H(-x)$ - for $x > 0$.

Since, as explained earlier, we view the quark model which we use to evaluate $H(\alpha)$ as an approximate solution of the QCD equations at a renormalisation scale Q_0^2, we add the Q_0^2 label to q and $\bar{q}$. Using the relations (42) and (43) to simplify the calculation, we find therefore:

$$q(x,Q_0^2) = \frac{m}{2\pi} \int_{-\infty}^{\infty} dz\ e^{-imxz}\ <N|\psi_+^\dagger(\xi^-)\psi_+(0)|N>_c, \qquad (47)$$

and

$$\bar{q}(x,Q_0^2) = \frac{m}{2\pi} \int_{-\infty}^{\infty} dz\ e^{-imxz}\ <N|\psi_+(\xi^-)\psi_+^\dagger(0)|N>_c. \qquad (48)$$

(In the last equation we used the translational invariance of the field operators to shift the argument ξ^- in $\psi_+^\dagger$ to $-\xi^-$ in ψ_+. We then changed the integration variable from z to -z.)

Before describing some numerical results obtained from equs.(47) and (48), some remarks must be made. Whereas we followed ref.(17) in taking the Bjorken limit ($Q^2\to\infty$) in order to obtain these parton distributions, the operator matrix elements needed in equ.(14) in order to reproduce data at some finite Q^2 should be evaluated at that Q^2. Although there is no rigorous proof yet, we believe that the difference between the exact results and the equations we use should be of order $1/Q^2$. (This is the case for the simple models which we have considered.) Effectively it amounts to a smearing of $\langle N|\psi_+^\dagger(z^0;00z^3)\psi_+(0)|N\rangle$ about the point $z^0=-z^3$ by an amount of order $1/\nu$ (and therefore of $O(1/Q^2)$).

Since we only intend to compare our twist-2 predictions with data at high Q^2, it is consistent to ignore this correction. Indeed, from this point of view, equs.(47) and (48) are the twist-2 quark/parton distributions whose evolution is governed by perturbative QCD. In order to establish this result it is crucial that the renormalisation scale is not a momentum cut-off (c.f. ref.(18)). It is of course also crucial that the OPE has allowed us to factorise the target dependence from the Q^2 dependence of the quark-probe interaction.

Some Numerical Results for the MIT Bag Model

The procedure presented here was first applied by Signal and Thomas[19] to the calculation of the structure function of the 1 dimensional bag. Here we restrict the discussion to the 3D case which is clearly of greater phenomenological interest. To evaluate the equs.(47) and (48), Signal and Thomas suggested inserting a complete set of eigenstates of the total energy and momentum operators. These are written $|n\vec{p}\rangle$, with n labelling the internal quantum numbers of the state of mass m_n and $\vec{p}$ its 3-momentum. Using translational invariance and performing the z integration we find easily

$$q(x,Q_0^2) = \frac{m}{(2\pi)^3} \sum_n \int d\vec{p}\ \delta[m(1-x)-p_n^+]\ |\langle n\vec{p}|\psi_+(0)|N\rangle|^2\ , \tag{49}$$

with

$$p_n^+ = (m_n^2 + \vec{p}^2)^{\frac{1}{2}} + p_z . \tag{50}$$

Because $p_n^+>0$ the δ-function can only be satisfied for $x\leq 1$ and $q(x,Q_0^2)$ is zero for $x\geq 1$. Similarly one finds that $\bar{q}(x,Q_0^2)$ is given by

$$\bar{q}(x,Q_0^2) = \frac{m}{(2\pi)^3} \sum_n \int d\vec{p}\ \delta(m(1-x)-p_n^+)|\langle n\vec{p}|\psi_+^\dagger(0)|N\rangle|^2\ . \tag{51}$$

The secret to ensuring the correct support of these quark/parton distributions is to introduce no approximations before equs.(49) and (51). At that stage one has ensured translation -invariance in space <u>and</u> time regardless of the approximation used for $|N\rangle$ and $|n\vec{p}\rangle$. If, as was typical before, one approximates the field operations in equs.(47) and

(48) it is very easy to destroy this invariance. Of course, in working with our version there is a major approximation. The states $|n\vec{p}>$ are colored, but we choose to use a finite mass, m_n, ignoring their color field. This is clearly in the spirit of the parton model and the MIT bag model. The physical idea is that the intermediate, colored, spectator state only exists while the struck quark travels a fermi or so at the velocity of light. Therefore it does not have time to develop the infinite flux tube necessary for it to have infinite energy. (In fact the true intermediate state is $|n\vec{p}>$ plus a fast quark, which is overall color neutral.)

For the harmonic oscillator (naive non-relativistic) quark model it is straightforward to write down translationally invariant states. In the case of the bag model we use the Peierls-Yoccoz approximation:[20]

$$|n\vec{p}> = \phi_n^{-1}(\vec{p}) \int d\vec{x}\, e^{i\vec{p}.\vec{x}}\, |n;\vec{x}>, \tag{52}$$

where $|n;\vec{x}>$ is a bag centred at $\vec{x}$ with internal state n. The wave-packet $\phi_n(p)$ is fixed by requiring δ-function normalisation for $|n\vec{p}>$. This method is essentially a non-relativistic one which is best for matrix elements between states of similar momenta. It should not be taken too seriously for large values of $|\vec{p}|$ in equs.(49) and (51). Solving the δ-function for p_z at fixed $p_T{}^2$ we find:

$$p_z = \frac{m^2\,(1-x)^2 - m_n{}^2 - p_T{}^2}{2m\,(1-x)} \quad . \tag{53}$$

Clearly large, negative p_z corresponds to $x \to 1$, while large positive p_z corresponds to negative x where we do not need it. The p_z integral can be performed leaving a Jacobian and an integral over transverse momentum. This integral converges rapidly because of the behaviour of the bound-state wave functions.

We now turn to the question of which intermediate states contribute to $q(x,Q_0{}^2)$. Supposing that $|N>$ contains just three quarks in the 1s level the obvious term, n=2, is one where the intermediate state has two quarks in the 1s state. This would be expected to have a mass order 3/4m, or 700 MeV, in the bag model. (For the present, subtle effects like gluon exchange are ignored). There is an additional (n=4) term corresponding to the insertion of an anti-quark into any bag energy level. In fact, the term where the $\bar{q}$ is put into the 1s state will dominate in the physical region. (Note that for wavefunctions peaked at small $|\vec{p}|$ the δ-function implies that q will be a maximum at $x = (m-m_n)/m$. This is negative for n=4 and rapidly becomes more negative as the excitation energy of the state into which the $\bar{q}$ is inserted rises).

Turning next to $\bar{q}(x,Q_0^2)$ the same line of argument suggests that the dominant term will be that where a fourth quark is inserted into the 1s-state in the bag (n=4 again). Clearly there are three contributions to $q(x,Q_0^2)$ for n=2 and twelve (three colours, two flavours, two spins) for n=4. For $\bar{q}$ we find nine n=4 terms, because three quarks are already in the 1s-state. These nine go with nine of the twelve n=4 terms in q to give a finite, intrinsic "sea". The remaining three n=4 terms in q go with the naive n=2 terms to give the valence distribution ($F_3 = q - \bar{q}$ in leading order).

For the bag in 1-space and 1-time dimensions Signal and Thomas found the integral of $F_3(x)$ to be 96% of the naive parton result of three. Of this 91% was from n=2 and 5% from n=4. Past work has almost invariably calculated the n=2 term only, and fixed the normalisation to 100% by hand. In three space dimensions the normalisation varies between 75% and 90% depending on the bag radius. We also find that each valence quark carries about 27% of the momentum of the nucleon (in both 1D and 3D). This is consistent with a scale (Q_0^2) somewhat less than 1 GeV^2.

In order to compare with a measured structure function we need to convolute the inverse Mellin transform of the coefficient functions $C_n(Q^2,Q_0^2,g^2)$ with q and $\bar{q}$ - c.f. equ.(40). Of course in leading order the C_n are all unity at $Q^2 = Q_0^2$. Thus if one takes experimental data at high enough Q^2 and evolves down to Q_0^2 using leading order QCD the resulting "data" can be compared directly with $q(x,Q_0^2)$ and $\bar{q}(x,Q_0^2)$. Below we shall compare with such "data" for the valence quark distribution obtained by Bickerstaff and Thomas.[21] However we must first discuss a rather subtle but very important correction.

Consider the n=2 term which dominates at large x. If we remove the d-quark two u-quarks remain which necessarily have spin-1. If we hit a u-quark the spectator u-d pair will be equally likely to be found with spin-0 or spin-1. However we know that the chromomagnetic interaction is repulsive for spin-1 and attractive for spin-0. With α_s chosen to fit the N-Δ mass difference this mass splitting would be 200 MeV. This is something of an overestimate because of the neglect of pionic corrections,[22,23] nevertheless we shall use it for our first estimates.

Figure 3 shows the calculated valence distributions when the n=2 term uses a spin-0 pair of mass 650 MeV and a spin-1 pair of mass 850 MeV with equal weight.[24] The results are shown for three bag radii in comparison with the "data" obtained in the manner described above. Clearly the overall agreement is rather good up to $x \approx 0.7$. This corresponds to a struck quark of momentum about 1 GeV. It is not surprising that the bag, which is essentially a mean field theory, does

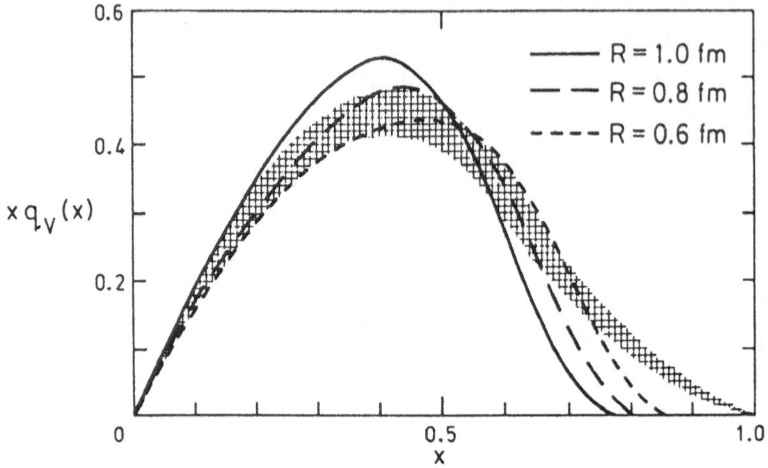

Figure 3. The curves show the valence quark distributions calculated for several bag radii including the effect of the chromomagnetic interaction - from Close and Thomas.[24] They are compared with "data" obtained by evolving various fits at high-Q^2 to a scale $Q_0{}^2$ where each valence quark has about 25% of the nucleon's momentum (hatched area).[21]

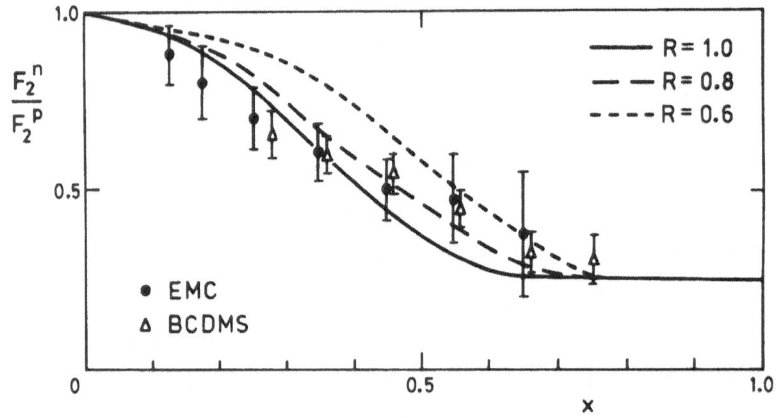

Figure 4. Comparison of the calculated ratio of neutron to proton structure functions (including the chromomagnetic correction described in the text) for several bag radii, with recent data - from ref.(24).

not fit the data at higher momenta. This would be improved by including configuration mixing.[25] Furthermore, the corrections to the leading order result ($C_n = 1$) are bigger[2,3] for the high moments and therefore at large x. Until these corrections are properly accounted for we would not like to draw conclusions about the preferred bag size.

Our discussion on the peak position for q(x) should make it clear that u(x) will dominate over d(x) at large x once the chromomagnetic splitting is included. Indeed we find that $d(x)/u(x) \rightarrow 0$ as $x \rightarrow 1$. It is well known[26] that this implies $F_2^{en}/F_2^{ep} \rightarrow 1/4$ as $x \rightarrow 1$. Our calculations of this ratio do agree rather well with recent data - see Fig.4.

Finally we note that if the proton and neutron are in the 56 of SU(6) there are distinct correlations between the spin and flavour of a quark and the spin of the spectator pair. For example only a u-quark with a spin parallel to that of the proton is found opposite a spin-0 pair. From this information we can calculate the spin polarisation asymmetry A^p and A^n ($A^T = 2xg^T_1(x)/F_2^{eT}(x)$), measured in the scattering of polarised electrons from a polarised target. The agreement between those calculations and the EMC data on the proton is quite impressive,[24] and this leads us to take seriously the predictions for the neutron, which were also give in ref.(24). It would be very nice to have data with which to compare. In particular there is a rapid rise between x of 0.3 and 0.7 which is directly related to the chromomagnetic interaction so familiar from spectroscopy.

Comparison with Earlier Work

Many authors have made contributions to the calculation of structure functions over the past fifteen years. The French group of Le Yaouanc et al.[27] deserves a great deal of credit for emphasising very early that one should use DIS data to constrain nucleon models in the same way that one uses other measured nucleon properties. Since asymptotic freedom was built into the MIT bag model[28] (albeit in a crude way), it was natural to use it too to calculate structure functions. The pioneering work was carried out by Jaffe,[29] and continued by many others.[30-32] The suggestion that one could make a connection to QCD by identifying the calculated quark distributions with data at a low momentum scale was made by Parisi and Petronzio.[33]

A major problem found in all the early work was that of bad support (i.e. $q(x) \neq 0$ for $x \geq 1$). Whether it was the constituent quark model or the static cavity of the MIT bag, the essential reason for the poor support was that translational invariance (in space and/or time) was broken. A more sophisticated method aimed at restoring Lorentz invariance in the bag model, the L_0 approximation,[34] has since been abandoned. It has no physical interpretation and the next term in the series does not seem to be calculable.

Shakin and co-workers[35] obtained the correct support in a di-quark-quark picture of the nucleon by putting the spectator di-quark on mass shell. (This is mathematically the same reason we obtained good support as described above.) No effort was made to restrict the calculation to twist-2, but rather the structure function was calculated as a function of Q^2 and the rapid onset of scaling demonstrated. As there was no discussion of QCD corrections, it is not clear how the calculations should be compared with data.

Finally we mention the work of Benesh and Miller[36] and Bickerstaff and Thomas.[21] These authors used the Peierls-Yoccoz approximation to restore (spatial) translational invariance for the bag model (as well as a particular soliton model). Otherwise the calculations were very similar to that of Jaffe and Ross.[31] Although the problem of bad support was improved in their work, it was not solved.

Remaining Problems

Undoubtedly the major problem experienced in these calculations is that of constructing translationally invariant states. The Peierls-Yoccoz approximation is essentially non-relativistic and also has well known defects. One might expect some improvement from the Peierls-Thouless approach, but there is no easy answer. Boosting a composite state is one of the hardest problems in theoretical physics, and the corresponding uncertainties also enter into the calculation of hadronic moments and form-factors.[37]

We also mentioned the need to include higher-order corrections in the coefficients C_n if $Q_0{}^2$ turned out to be low. At present it seems that a value of $Q_0{}^2$ between 0.8 and 1.0 GeV^2 might provide the best match between the bag model calculations and experimental data. While such a value is higher than the work of Jaffe, Ross and collaborators suggested[31,38] (i.e. $Q_0{}^2 \sim (0.5, 0.65)\ GeV^2$), one would still expect significant next-to-leading-order corrections.

There must also be some concern over the intermediate states $|n\vec{p}\rangle$ which we inserted in equ.(49). In particular one needs a better understanding of the limits of validity of our physically reasonable ansatz concerning the corresponding masses. The bag model also violates chiral symmetry and one must eventually examine systematically the effect of spontaneous chiral symmetry breaking and the pion cloud of the nucleon.[39,40] Finally the promising approach that we have outlined should be used to explore the spin structure functions in more detail.

The Spin of the Proton

Before the discovery of QCD, Ellis and Jaffe showed that under the assumption of SU(3) symmetry, and an unpolarised strange sea, the proton and neutron spin structure functions would obey the sum rule[41]

$$S^{p(n)} = \int_0^1 g_1^{p(n)}\, dx = \frac{1}{12}\left(\frac{g_A}{g_V}\right)\left[\pm 1 + \frac{5}{3}\,\frac{3F-D}{F+D}\right] . \tag{54}$$

Taking the difference one finds the Bjorken sum rule

$$S^p - S^n = \frac{1}{6}\, g_A/g_V , \tag{55}$$

which is actually a much more general result. Leading order QCD modifies both of these results by a factor of $(1 - \alpha_s/\pi)$, which means only a small reduction at even moderate Q^2.

Using reasonable values of F and D one finds $S^p \simeq 0.18$ whereas the EMC group found a value 0.116±0.009±0.019 - a 3 standard deviation effect.[42] The reason this has created so much interest is that what enters the sum-rule are quantities Δq:

$$\Delta q = \int_0^1 (q^{\uparrow}(x) - q^{\downarrow}(x) + \bar{q}^{\uparrow}(x) - \bar{q}^{\downarrow}(x)\ )\ dx, \tag{56}$$

where (modulo the QCD correction)

$$S^p = \Delta q_3/12 + \Delta q_8/36 + \Delta\Sigma/9. \tag{57}$$

The subscripts refer to flavour SU(3) so that $\Delta q_3 = \Delta u - \Delta d$, which is also g_A (= 1.26) and $\Delta q_8 = 3F-D$ (= 0.6). Finally $\Delta\Sigma$ is the flavour singlet combination $(\Delta u + \Delta d + \Delta s)$ which is the total spin carried by all the quarks in the proton. The problem is that the first two terms in equ.(57) reproduce the data, and therefore imply $\Delta\Sigma$=0! Hence the question, where is the spin of the proton?

In fact the naive result $\Delta\Sigma$=1 is a non-relativistic one. In such an approximation g_A would be 5/3. If one instead solves the Dirac equation for a relativistic quark it is well known that g_A is reduced.[23] One finds the same reduction for $\Delta\Sigma$ - e.g. for massless quarks confined in a scalar bag $\Delta\Sigma$=0.65. The physical reason is that the quark wave function has a lower component with L=1, so that some of the quark "spin" is actually orbital angular momentum.

A number of additional corrections associated with the breaking of SU(3) symmetry have been considered. For example, the familiar pion cloud in the cloudy bag model yields a reduction in S^p of between 0.01 and 0.03 depending on the bag radius.[43] Other chiral quark models yield similar results.[44] Admixtures of q-$\bar{q}$ pairs in the nucleon wave function due to one gluon exchange are known to explain certain anomalies in the nucleon octet - e.g. the Ξ^- magnetic moment compared with that of the Λ, and the ratio $\Lambda \to pe\bar{\nu}/\Sigma \to ne\bar{\nu}$ being 2 rather than the SU(6) value of 3.[45] With the same parameters one obtains a reduction in S^p by about 0.02.[46] A d-state admixture in the nucleon wave function would also lower S^p. At least some of these corrections are certainly there and would reduce the difference between theory and experiment to perhaps 2 standard deviations. If one believes that the residual difference is real, something more drastic must be done.

One proposal, which it has been claimed is quite natural in the Skyrme model, is to allow for a negative polarisation of the strange sea. For example the combination $\Delta u \simeq 0.74$, $\Delta d \simeq -0.51$ and $\Delta s \simeq -0.23$, which yields $\Delta\Sigma = 0$, provides a reasonable fit to decays in the nucleon octet. Clearly such a large strange-quark component would be in disagreement with most quark models - hence the speculation of Ellis and others that the quark model may be dead.[47]

Following a proposal by Efremov and Teryaev, Altarelli and Ross[48] (and also Carlitz et al.)[49] have proposed a different explanation which would not require the overthrow of either QCD or the quark model.

It has been known for some years that in next-to-leading order QCD there is an additional contribution to g_1 arising from polarised gluons. That is, at order α_s the gluon turns into a q-$\bar{q}$ pair which is scattered by the incident lepton. Then g_1 has the form

$$g_1(x,Q^2) = \tfrac{1}{2} \sum_i e_i^2 \, \Delta q_i(x,Q^2) - \frac{f}{9} \left(\frac{\alpha_s}{2\pi} \right) \Delta P_{qg} \, x \, \Delta g(x,Q^2), \tag{58}$$

where the x denotes a convolution of a known splitting function (ΔP_{qg}) with $(g^\uparrow - g^\downarrow) \equiv \Delta g$, and $f = 3$ is the number of relevant flavours. The gluon correction might be naively thought small since at 10 GeV^2 $(\alpha_s/2\pi) \sim 1/25$. However, Altarelli and Ross noted that the combination $\alpha_s(Q^2)$ $\Delta G(Q^2)$ (where ΔG is the integral of Δg) is independent of Q^2 in lowest order. Thus if $\alpha_s \Delta G$ is big anywhere, say at $Q^2 = Q_0{}^2$ where a quark model may describe hadron structure, it stays large for all Q^2!

So far as the sum rule is concerned there is one other subtlety. A naive integration of the second term in equ.(58) yields zero (because the integral of ΔP_{qg} vanishes). However, a more sophisticated study shows that there is in fact a non-zero contribution to S^p, namely $- f(\alpha_s/2\pi)\Delta G$, related to the U(1) axial anomaly.[48,49] A very simple summary of all this is that $\Delta\Sigma$ in equ.(57) should be replaced by $\Delta\Sigma - f(\alpha_s/2\pi)\Delta G$. The EMC result can then be fitted by choosing $(\alpha_s/2\pi)\Delta G = -0.23$, that is without any polarised strange quarks in the proton.

One major problem with this approach is that no-one has ever calculated $\Delta g(x,Q^2)$, so we do not know whether the value required is reasonable. Gluck and Reya recently proposed choosing the scale $Q_0{}^2$ at which valence quarks dominate in order that the perturbatively generated glue fitted the EMC value for the sum-rule. One criticism of this work is that the required value of $Q_0{}^2$ (namely 0.06 GeV^2 for $\Lambda_{LO} = 0.2GeV$) is too low to use leading order formulas. However, even allowing this Schreiber et al.[52] have shown that it does not seem possible to fit both the sum-rule <u>and</u> the shape of $g_1\ (x,Q^2)$ using any reasonable starting distribution at $Q_0{}^2$.

In a purely phenomenological study, Altarelli and Stirling have shown that one can fit $\Delta g(x,Q^2)$ without contradicting any other piece of data.[50] However, it is clear that one would like independent experimental confirmation. A number of such tests have been proposed. For example, as one passes through a charm threshold one should see a drop in S^p by $(\alpha_s/2\pi)\Delta G$. Alternatively a semi-inclusive measurement in which $g_1(x,Q^2)$ is measured with various cuts on the transverse momentum of outgoing jets would test the gluon correction. (The anomaly is associated with very high transverse momentum.) Of course if the Ellis-Jaffe sum rule is broken one naturally wonders about the Bjorken sum rule too. This has led to a proposal to measure $g_1^n(x,Q^2)$ at HERA. While we have doubts that this experiment can really help with the sum rule, a first measurement of g_1^n is extremely important for its own sake.[24]

SOME EFFECTS OF NUCLEON INTERNAL STRUCTURE

For much of the history of nuclear physics the structure of the nucleon itself has been considered to be of little relevance. Certainly one needs the proton's charge form factor (and even that of the neutron) to accurately calculate nuclear charge form-factors, but one can measure the latter directly and treat them as phenomenological input. Even Yukawa's insight that the long-range structure of the nucleon (e.g. the negative neutron charge radius) must be associated with a pion cloud has been largely forgotten, while its consequence for the two-nucleon force, namely the one-pion-exchange potential (or OPEP), is used widely.

As a consequence an incredible variety of models of nucleon structure co-exist in the literature without unduly disturbing the majority of nuclear physicists. We would like to give just two examples where nucleon structure may have important consquences. The first is with respect to quantum hadrodynamics (QHD), and the second is the OPEP. Much of this discussion was also presented at the São Paulo meeting.[53]

Quantum Hadrodynamics

Quantum hadrodynamics is built from the idea that one should treat nuclear matter as a system of point-like, relativistic nucleons coupled in a renormalisable way to elementary scalar and vector meson fields.[54] In many ways it has been a very successful idea, and it provides a widely accepted mechanism for nuclear saturation - especially following the success of Dirac phenomenology in intermediate energy nuclear reactions. The essential idea of the model is that in mean-field approximation nuclear matter can be treated as a collection of point-like nucleons interacting via the exchange of scalar and vector mesons (σ and ω respectively). As the density of the system grows the nucleons near the top of the Fermi-sea become relativistic. This reduces the strength of the coupling to the attractive, σ-field (proportional to $\bar{\psi}\psi$), while the repulsive coupling to the ω-field is not density dependent. Solving the problem self-consistently, one finds that eventually the repulsion begins to win and the system saturates. That is, nuclear saturation is a direct effect of the nucleon motion becoming relativistic.

A disturbing feature of this model is that the resulting self-consistent scalar and vector potentials felt by the nucleons are many hundreds of MeV - comparable with the excitation energy of the nucleon itself. The obvious question is, therefore, how would the inclusion of nucleon structure alter one's conclusions? Drastically it turns out! In particular, Guichon has recently constructed a beautiful model incorporating the internal structure.[55] He considered a self-consistently bound collection of MIT bags whose effective interaction was represented by σ and ω exchange between quarks. With two parameters he obtained not just the saturation energy and density of nuclear matter, but also the correct incompressibility. More important, the whole saturation mechanism was altered. It was a consequence of the relativistic motion of the quarks, not the nucleons!

One can be in no doubt after this work that nucleon internal structure must be incorporated in any such model of nuclear matter. Incidentally, we note that from the point of view of those wishing to relate the EMC effect[56,57] to changes in nucleon properties, Guichon's model was somewhat disappointing. Neither the magnetic moment nor the charge radius of the nucleon changed by more than one or two percent. On the other hand, because it involves quarks explicitly, this model, together with the formalism described earlier, enables us to estimate the structure function of a bound nucleon. That is, in equ.(49) we can replace $|N\rangle$ by the quark wave function of a nucleon bound in nuclear matter. For the intermediate states we can use (e.g.) a diquark bag - again interacting with the mean σ and ω fields. Although the calculations have so far been very crude (with the nucleons replaced by nuclear matter of some average density - $\rho_0/3$ for Fe), the results yield the qualitative features of the data.[58] Even more important, the results suggest that the conventional impulse approximation (involving a convolution of a nucleon distribution with a free structure function[59-62]) may be rather inaccurate.

The Long Range N-N Force

It has become increasingly clear over the past decade that the pion-nucleon form-factor needed in conventional nuclear physics, notably in one-boson-exchange (OBE) potentials,[63] is inconsistent with what we know about the free nucleon. In particular, the range of the NNπ form-factor needed in the two cases differs by (50-100)%. As a consequence, our knowledge of the N-N force in the crucial 1-2 fm range cannot be considered satisfactory.

For an isolated nucleon it is unfortunately difficult to obtain direct information on the NNπ form-factor. (For simplicity we shall refer all estimates to a monopole form, $f(\Lambda^2 - m_\pi^2)/(\Lambda^2 + \vec{q}^2)$, where f_c and f_o correspond to charged and neutral pions.) More than a decade ago Primakoff suggested using the same value of Λ obtained for the axial form-factor - that is, about 730 MeV.[64] (The usual fits actually involve a dipole form with mass 1.03 ± .04 GeV.[65]) In an estimate of the σ-commutator Gasser rejected such a value as unreasonably large (or "exotic"), preferring 350 MeV $< \Lambda <$ 730 MeV.[66] His reason was that a bigger value of Λ would give too large a pionic correction to the nucleon self-energy.

In many models of nucleon structure the axial and the pion-nucleon form-factors are closely related. For example, Guichon et al. considered a whole class of chiral bag models in which the pion was excluded from a sphere of arbitrary radius within a bag.[67] In all cases, ranging from the little bag to the cloudy bag the resulting πNN form-factor was softer than the axial form-factor (i.e. $\Lambda < \Lambda_A$). This result was confirmed in later work on the constituent quark model by Beyer and Singh.[68] The only model of which we are aware which gives a different result is that of Weise and collaborators.[69] Working in a Skyrme model supplemented with vector mesons they found the cut-off mass in the axial form factor, Λ_A, to be about 750 MeV while Λ was about 850 MeV - that is, slightly larger. Nevertheless, even in this very different model, the general result that $\Lambda_A \approx \Lambda_F$ still holds.

The most direct evidence concerning this form-factor comes from the observation that the pion cloud of the nucleon constitutes a (non-perturbative) sea of $\bar{q}$-q pairs which breaks SU(3)-flavour symmetry.[40] A recent refinement of this idea by Frankfurt et al., has led to an even more stringent limit, namely $\Lambda < 500$ MeV.[70] Actually there is a little uncertainty in this because the process is sensitive to virtual pion momenta squared of order 2-3 m_π^2. This gives some shape dependence, with the preferred dipole mass being 900 MeV (or Λ (equivalent) ~630 MeV).

We conclude that the cut-off mass Λ, for an isolated nucleon, is close to that measured for the axial form-factor, namely 730 MeV. It may perhaps be as low as 500 MeV or as large as 800 MeV. Such a mass provides a natural explanation of the Goldberger-Treiman relation, as we have noted elsewhere.[67] In particular, the pion coupling constant at $\vec{q}^2=0$ would be of the order (3-8)% smaller than at the nucleon pole.

With this realization we are in a position to understand the recent result of the Nijmegen group concerning a very large apparent difference between the charged and neutral pion-nucleon coupling constants.[71] By a careful analysis of the world data on p-p elastic scattering, they extracted a $pp\pi^0$ coupling constant some (3-4)% below the charged one (obtained from πN dispersion relations). We believe that the coupling constant they extracted was actually $f_0(\vec{q}^2=0)$ rather than f_0 at the nucleon pole.[72] If we allow for this small shift in momentum transfer using the value of Λ just discussed, we find a reduction in $f_0(\vec{q}^2=0)$ compared with f_0 of 3 to 8% with 3½% preferred (for Λ = 730 MeV). This is in good qualitative agreement with the value of $f_0(\vec{q}^2=0)$ found by Bergervoet et al.

Thus, while be believe that Bergervoet et al. have not found a large violation of charge independence, they have nevertheless obtained a very important result. The conventional picture of the N-N force would have it well understood, outside 1 fm, in terms of one and two pion exchange with an $NN\pi$ form-factor of mass 1.3 GeV or higher (plus the tail of heavy boson exchange). As stressed by Ericson, changing Λ from 1.3 GeV to 0.73 GeV has a dramatic effect on the tensor force between 1 and 2 fm.[73] In view of the evidence presented, including this re-interpretation of the Nijmegen work, the conventional description of the N-N force in the region 1-2 fm must be viewed as an effective one. The real physics is more complicated. It must be an urgent priority for our field to develop a more realistic picture.

CONCLUSION

In the latter part of these lectures, we have given two examples which suggest that the issue of nucleon structure can no longer be regarded as purely cultural in nuclear physics. Indeed it may well be vital to an understanding of nuclear saturation and the intermediate range N-N potential. In this context the recent progress that has been made in relating models of hadron structure to deep-inelastic scattering data is quite exciting. This work was our major topic. We tried to highlight not only the successes of our approach but also the problems and hopefully the next few years will see some of those solved. If new experimental facilities like CEBAF and PEGASYS are to fulfil their potential they will have to be solved!

ACKNOWLEDGEMENTS

Above all I would like to express my appreciation to Mrs. A. Shaw, without whose dedication these notes would never have reached the publishers on time. I would also like to thank S. Bass, R.P. Bickerstaff, F.E. Close, P.A.M. Guichon, K. Holinde, C.H. Llewellyn Smith, J.T. Londergan, A. Michels, A.W. Schreiber and A.I. Signal, for their contributions to my understanding of the issues discussed here.

This work was supported by the Australian Research Council.

REFERENCES

1. F.E. Close, "At Introduction to Quarks and Partons", Academic Press, New York (1979).
2. E. Leader and E. Predazzi, "An Introduction to Gauge Theories and the New Physics", Cambridge (1982).
3. T. Muta, "Foundations of Quantum Chromodynamics", World Scientific, Singapore (1987).
4. C.T.C. Sachrajda, in "Gauge Theories in High Energy Physics", North Holland, Amsterdam (1983).
5. F. Yndurain, "Quantum Chromodynamics", Springer, Berlin (1983).
6. J.L. Alonso and R. Tarrach (eds.), "Quantum Chromodynamics", Springer, Berlin (1980).
7. G. Altarelli, Physics Reports, 81:1 (1982).
8. N.S. Craigie et al., Phys. Rep. 99:69 (1983)
9. J.D. Jackson, G.G. Ross and R.G. Roberts, Rutherford Lab. preprint RAL-89-038 (1989).
10. S. Dasu et al., Phys. Rev. Lett. 64:2591 (1988).
11. J.G. Morfin, Fermilab Report - Conf. - 89/20 (1989).
12. H. Abramowitz et al., Zeit. Phys. C15:19 (1982).

13. C. Itzykson and J.-B. Zuber, "Quantum Field Theory", McGraw-Hill, New York (1980).
14. J. Collins, "Renormalization", Cambridge (1987).
15. J. Negele, in these proceedings.
16. C.T. Sachrajda, Southampton preprint, SHEP 88/89-2.
17. R.L. Jaffe, Nucl. Phys. B229:205 (1983).
18. C.H. Llewellyn Smith, Oxford University preprint (1988).
19. A.I. Signal and A.W. Thomas, Phys. Lett. B211:481 (1988); A.I. Signal and A.W. Thomas, ADP-89-118/T65, to appear in Phys. Rev. D (1989).
20. R.E. Peierls and J. Yoccoz, Proc. Phys. Soc. A70:381 (1957).
21. R.P. Bickerstaff and A.W. Thomas, ADP-87-1/T29 (unpublished).
22. S.Théberge, A.W. Thomas and G.A. Miller, Phys. Rev. D22:2838 (1980); D23:2106(E) (1981).
23. A.W. Thomas, Adv. Nucl. Phys. 13:1 (1984); G.A. Miller, Int. Rev. Nucl. Phys. 1:189 (1984).
24. F.E. Close and A.W. Thomas, Phys. Lett. B212:227 (1988).
25. S. Bass and A.W. Thomas, Contributed paper H7 to XII Int. Few Body Conference, Vancouver (1989).
26. F.E. Close, Phys. Lett. B43:422 (1973).
27. A. Le Yaouanc et al., Phys. Rev. D11:2636 (1975).
28. A Chodos et al., Phys. Rev. D10:2599 (1974).
29. R.L. Jaffe, Phys. Rev. D11:1953 (1975).
30. J.S. Bell and A.J.G. Hey, Phys. Lett. B74:77 (1978); J.S. Bell, A.C. Davis and J. Rafelski, Phys. Lett. B78:67 (1978).
31. R.L. Jaffe and G.G. Ross, Phys. Lett. B93:313 (1980).
32. R.J. Hughes, Phys. Rev. D16:622 (1977).
33. R. Parisi and G. Petronzio, Phys. Lett. B62:331 (1976).
34. V. Krapchev, Phys. Rev. D13:329 (1976); R.L. Jaffe, Ann. Phys. 132:32 (1981).
35. L.S. Celenza and C.M. Shakin, Phys. Rev. C27:1561 (1983); (E) C39:2477 (1989).
36. C.J. Benesh and G.A. Miller, Phys. Rev. D36:1344 (1987).
37. L. Wilets, in "Workshop on Nuclear Chromodynamics", eds. S. Brodsky and E. Moniz, World Scientific, Singapore (1986).
38. F.E. Close et al., Phys. Rev. D31:1004 (1985).
39. J.D. Sullivan, Phys. Rev. D5:1732 (1972).
40. A.W. Thomas, Phys. Lett. B126:97 (1983).
41. J. Ellis and R.L. Jaffe, Phys. Rev. D9:1444 (1974).
42. J. Ashman et al., Phys. Lett. B206:364 (1988).
43. A.W. Schreiber and A.W. Thomas, Phys. Lett. B215:141 (1988).

44. H. Høgaasen and F. Myhrer, Phys. Lett. B214:123 (1988).
45. K. Ushio, Zeit Phys. C30:115 (1986).
46. F. Myhrer and A.W. Thomas, Phys. Rev. D38:1633 (1988).
47. S.J. Brodsky, J. Ellis and M. Karliner, Phys. Lett. B206:309 (1988).
48. G. Altarelli and G.G. Ross, Phys. Lett. B212:391 (1988).
49. R.D. Carlitz et al., Phys. Lett. B214:229 (1988).
50. G. Altarelli and W.J. Stirling, CERN preprint TH-5249/88 (1989).
51. M. Glück and E. Reya, Zeit. Phys. C39:569 (1988).
52. A.W. Schreiber, A.W. Thomas and J.T. Londergan, ADP-89-129/T77, to appear in Phys. Lett. (1989).
53. A.W. Thomas, ADP-89-127/T75, to appear in Proc. Int. Nucl. Phys. Conf., São Paulo (August 1989).
54. B.D. Serot and J.D. Walecka, Adv. Nucl. Phys. 16:1 (1986).
55. P.A.M. Guichon, Phys. Lett. B200:235 (1988).
56. J.J. Aubert et al., Phys. Lett. B123:275 (1983); Nucl. Phys. B293:740 (1987).
57. A.C. Benvenuti et al., CERN-EP/89-06; S. Dasu et al., Phys. Rev. Lett. 61:1061 (1988).
58. A.W. Thomas, A. Michels, A.W. Schreiber and P.A.M. Guichon, ADP-89-125/T73 to appear in Phys. Lett. (1989).
59. S.V. Akulinichev et al., Phys. Rev. Lett. 55:2239 (1985).
60. G.V. Dunne and A.W. Thomas, Nucl. Phys. A455:701 (1986).
61. R.P. Bickerstaff and A.W. Thomas, ADP-89-121/T68, to appear in J. Phys. G (1989).
62. C.H. Llewellyn Smith, Phys. Lett. B128:107 (1983).
63. R. Machleidt et al., Phys. Reports 149:1 (1987).
64. H. Primakoff, in "Nucl. and Part. Physics at Int. Energies", ed. J.B. Warren, Plenum, New York (1976).
65. T. Kitagaki et al., Phys. Rev. D28:436 (1983); L.A. Ahrens et al., Phys. Rev. D35:785 (1987).
66. J. Gasser, Ann. Phys. 136:62 (1981).
67. P.A.M. Guichon et al., Phys. Lett. B124:109 (1983).
68. M. Beyer and S.K. Singh, Phys. Lett. B160:26 (1985).
69. U.G. Meissner et al., Phys. Rev. Lett. 57:1676 (1986).
70. L.L. Frankfurt, L. Mankiewicz and M. Strikman, Leningrad preprint (1989).
71. J.R. Bergervoet et al., Phys. Rev. Lett. 59:2255 (1987).
72. A.W. Thomas and K. Holinde, ADP-89-126/T74, to appear in Phys. Rev. Lett. (1989).
73. T.E.O. Ericson, Prog. Part. Nucl. Phys. 11:245 (1982).

MANY-BODY THEORY OF NUCLEI AND NUCLEAR MATTER

V. R. Pandharipande

Department of Physics, University of Illinois at Urbana-Champaign
1110 West Green Street, Urbana, IL 61801 USA

I. INTRODUCTION

Historically nuclei and neutron stars have been considered as bound states of nucleons. Thus the simplest description of these systems is based on a non-relativistic many-body theory in which only the nucleon degrees of freedom are retained, and the effects of all subnucleonic degrees of freedom, such as those associated with mesons or quarks and gluons, are absorbed into phenomenological nuclear forces. A vast amount of nuclear phenomenon can be understood from the resulting many-body theory, and in these lectures we will review some of the recent developments in it.

The failures of this simple theory are probably as important as its successes, for they may provide an insight into the dynamics of the subnucleonic degrees of freedom. With this point of view we will try to obtain accurate predictions of this theory and confront them with the available experimental and observational data on nuclei and neutron stars.

We will assume that low energy nuclear states can be well described by a wave function $\Psi(x_1,x_2...x_A,t)$, where x_i denote the spin, isospin and position of the ith nucleon, and that its evolution is given by the Schrödinger eq:

$$-\frac{i}{\hbar}\frac{\partial}{\partial t}\Psi(x_1,x_2\ ...x_A,t) = H\ \Psi(x_1,x_2...x_A,t) \tag{1.1}$$

$$H = \sum_i -\frac{\hbar^2}{2m}\nabla_i^2 + \sum_{i<j} v_{ij} + \sum_{i<j<k} V_{ijk} + \ ....\ , \tag{1.2}$$

where the ... indicate the possible four- and more-body interactions. In principle the two- and three-nucleon interactions v_{ij} and V_{ijk} can be obtained from a more fundamental theory like QCD or from a deeper

Hadrons and Hadronic Matter
Edited by D. Vautherin *et al.*
Plenum Press, New York, 1990

phenomenology which incorporates the dynamics of hadrons and mesons. Unfortunately this has not yet been possible, and hence we will work with models of nuclear forces constructed with some theoretical insights and the experimental data. The uncertainties in nuclear forces, and the difficulties in solving the many-body equation (1.1) with the required accuracy are the two major problems in this theory.

A third possible problem is that the relativistic corrections to this non-relativistic theory may be large enough to substantially reduce its applicability. Unfortunately the relativistic many-body problem is much more difficult to solve than the non-relativistic. Thus, even though there has been a strong effort in recent years, we do not have reliable estimates of relativistic effects in nuclei or neutron matter.

Realistic models of two-nucleon interaction and electromagnetic current operators are reviewed in sections II and III. A brief overview of the many-body techniques used to study nuclei having three or more nucleons, nuclear matter and neutron stars is given in section IV, models of three- and many-nucleon interactions are discussed in section V, and the optical potential in section VI.

Properties of interest in understanding electron-nucleus scattering experiments are discussed in section VII. These include electromagnetic form factors, pair distribution functions and the Coulomb sum rule, momentum distributions, spectral functions, and the longitudinal and transverse response functions. In sections VI and VII we will attempt to make contact with the available experimental data.

In this review we will not discuss properties of hot nuclear matter such as the liquid-gas phase transition, occurrence of thermal pisobar excitation, density and momentum dependence of nuclear mean field, etc. These properties, of interest in heavy-ion collision physics, have been recently reviewed at the Les Houches winter school.[1] The properties of neutron stars calculated within this approach have been recently compared with the available data by Wiringa, Fiks and Fabrocini.[2] These will also not be reviewed here.

II. THE NUCLEON-NUCLEON INTERACTION

The N-N interaction v_{ij} is the main input in the many-body theory of nuclei, nuclear matter and neutron stars, and most of the properties of these systems can be calculated from it. Unfortunately we do not, as yet, have a complete theoretical understanding of v_{ij}, and have to work with models fitted to the N-N scattering data. This is not too surprising because nucleons are made up of quarks and gluons, and it is difficult to calculate the interaction between composite objects. For example, the electronic structure of helium atoms is well understood, however the calculation of the interaction between two helium atoms starting from the fundamental Coulomb's law is still considered to be a difficult numerical problem. Consequently the many-body theory of helium liquids and solids uses phenomenological potentials[3] whose form is chosen on theoretical basis and parameters are fitted to experimental data.

The N-N interaction has an intricate dependence on the total spin S, isospin T, orbital and total angular momenta L and J; the potential is essentially different in every LSJT partial wave. Empirically it is well known that the interaction has at least fourteen distinct terms, and may be written as:

$$v_{ij} = \sum_{p=1,14} v^p(r_{ij})\, O^p_{ij}. \tag{2.1}$$

The first eight operators $O^{p=1,8}_{ij}$ are rather unambiguously indicated by both the data and theoretical considerations to be:

$$O^{p=1,8}_{ij} = 1,\ \tau_i\cdot\tau_j,\ \sigma_i\cdot\sigma_j,\ \sigma_i\cdot\sigma_j\tau_i\cdot\tau_j,\ S_{ij},\ S_{ij}\tau_i\cdot\tau_j,$$

$$\vec{L}\cdot\vec{S} \text{ and } \vec{L}\cdot\vec{S}\,\tau_i\cdot\tau_j, \tag{2.2}$$

where S_{ij} is the tensor operator:

$$S_{ij} = 3\sigma_i\cdot\hat{r}_{ij}\ \sigma_j\cdot\hat{r}_{ij} - \sigma_i\cdot\sigma_j. \tag{2.3}$$

The form of operators $O^{p=9,14}_{ij}$ is not uniquely constrained by the scattering data. In the Urbana[4] and Argonne[5] models these are taken as:

$$O^{p=9,14}_{ij} = (\vec{L}\cdot\vec{S})^2,\ (\vec{L}\cdot\vec{S})^2\tau_i\cdot\tau_j,$$

$$L^2,\ L^2\tau_i\cdot\tau_j,\ L^2\sigma_i\cdot\sigma_j \text{ and } L^2\sigma_i\cdot\sigma_j\tau_i\cdot\tau_j, \tag{2.4}$$

where as in the Paris[6] model the operator ∇^2_{ij} is used instead of L^2. Theoretically one would prefer to use ∇^2_{ij} terms in the interaction, however the data can be equally well fit with either. Accurate many-body calculations are difficult when the interparticle forces depend upon either L^2 or ∇^2_{ij}. The main reason to choose L^2 over ∇^2_{ij} is that the L^2 terms give relatively small contribution to the energy of nuclear matter, since $L^2 = 0$ in the S-waves. For example the contribution of the L^2 interactions in the Urbana model to the energy of nuclear matter at density 0.28fm^{-3} (equilibrium density of nuclear matter, $\rho_o = 0.16\text{fm}^{-3}$) is ~6 MeV per nucleon as compared to ~40MeV for the ∇^2_{ij} interactions in the Paris model.[4] Day and Wiringa[7] have carried out accurate calculations of nuclear matter binding energies with the Paris and Argonne models, which are quite similar except for the choice of ∇^2_{ij} or L^2. The two interactions give rather similar results at least up to $2\rho_o$ indicating that this choice may not be very critical to the theory.

It is difficult to extract the fourteen functions $v^p(r_{ij})$ by fitting the scattering data. Hence all models of the N-N interaction take some guidance from theory. The long range part of the interaction is assumed to be given by one-pion-exchange potential (OPEP) v^π (fig. 1.1):

$$v^\pi_{ij} = \frac{f^2_\pi}{4\pi}\,\frac{m_\pi}{3}\,(Y_\pi(r)\sigma_i\bullet\sigma_j + T_\pi(r)S_{ij})\tau_i\bullet\tau_j, \tag{2.5}$$

$$Y_\pi(r) = \frac{e^{-\mu r}}{\mu r}\,\xi(r), \tag{2.6}$$

$$T_\pi(r) = \left(1 + \frac{3}{\mu r} + \frac{3}{\mu^2 r^2}\right) Y_\pi(r)\ \xi'(r), \tag{2.7}$$

where $\mu = 0.7\text{fm}^{-1}$ is the pion mass and $\xi(r)$ and $\xi'(r)$ are short range cutoff functions. In Urbana and Argonne models these are taken as:

$$\xi(r) = \xi'(r) = 1 - e^{-cr^2}. \tag{2.8}$$

We note that $T_\pi(r) >> Y_\pi(r)$, particularly in the important 1-2 fm range, and thus the tensor part of OPEP is dominant.

In most potential models the physical pion mass m_π and the πNN coupling constant f_π determined from the observed scattering of charged pions by nucleons are used for OPEP. The differences between the masses of the charged and neutral pions are taken into account in the Paris potential but not in the Urbana and Argonne models. Recently the Nijmegen group[8] has analyzed the p-p scattering data taking f_{π^o} and m_{π^o} as adjustable parameters. Only the exchange of neutral π^o's contributes to the OPEP between two protons. They find that the data is best explained with $m_{\pi^o} = 135\pm2$ MeV and $f^2_{\pi^o} = 0.0725 \pm 0.0006$. The value of m_{π^o} is consistent with the mass 134.96 MeV of real, on-shell π^o, however that of $f^2_{\pi^o}$ is significantly smaller than the $f^2_\pi = 0.079\pm0.001$ obtained from the observed scattering of charged pions by nucleons. The difference is interesting, it could be either due to a charge dependence of the pion-nucleon coupling constant, or due to the difference between the q^2 of real and virtual pions.[9] Real pions have $q^2 = -m^2_\pi$ whereas the virtual pions exchanged in OPEP have $q^2 = |\vec{q}|^2$.

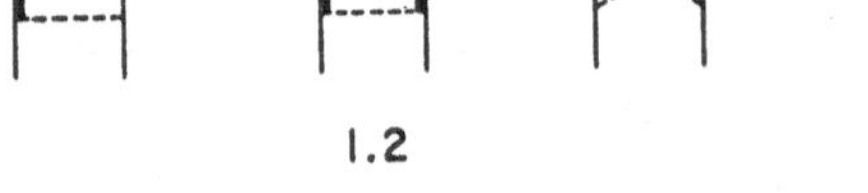

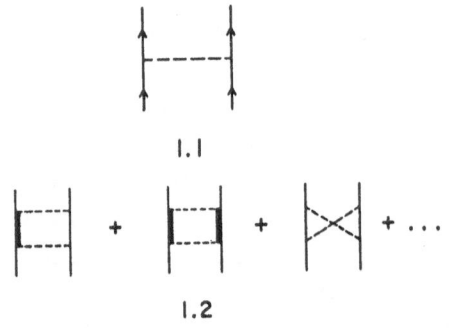

Fig. 1.1 The one-pion exchange interaction.

Fig. 1.2 Examples of two-pion exchange interactions.

The complete N-N interaction can be considered as

$$v_{ij} = v^S_{ij} + v^I_{ij} + v^\pi_{ij}, \tag{2.9}$$

where v^S and v^I denote short and intermediate range parts. The v^I is attributed to various two-pion exchange graphs illustrated in fig. 1.2, and there have been attempts to calculate it.[10] It is often attributed to the exchange of an effective scalar σ-meson with $J^\pi = 0^+$ and mass ~ 550 MeV in phenomenological meson exchange models.[11] In the simplistic Urbana and Argonne models it is taken as:

$$v^I_{ij} = \sum_{p=1,14} I^p \, T^2_\pi(r) \, O^p_{ij}, \tag{2.10}$$

and the strengths I^p are determined by fitting the scattering data. The dominant-term of v^I is the p=1≡c central attraction. (We will often denote the superscripts p = 1,14 by the more informative superscripts, c, τ, σ, $\sigma\tau$, t, tτ, b, bτ, bb, bbτ, q, qτ, qσ and q$\sigma\tau$. Thus $v^{p=1} \equiv v^c$ etc.). In the Argonne model $I^c = -4.8$ MeV, while all other $I^{p\neq c}$ have magnitudes $\lesssim$ 1 MeV.

The v^S is assumed to be given by:

$$v^S = \sum_{p=1,14} S^p \, W(r) \, O^p_{ij}, \tag{2.11}$$

$$W(r) = [1 + \exp\{(r-R)/a\}]^{-1}, \tag{2.12}$$

and it is dominated by the central repulsive core. The S^c ~ 2000 MeV, while the magnitudes of other $S^{p\neq c}$ are $\lesssim$ 500 MeV. In meson exchange models the v^S is attributed to the exchange of the vector mesons ω and ρ having masses of ~750 MeV. In the Urbana model $S^t = S^{t\tau} = 0$ so that the tensor potentials go to zero as $r \to 0$, and the parameters R and a of the W(r) are allowed to be smaller for the L•S and L•S $\tau_i \bullet \tau_j$ potentials than for others as suggested by meson-exchange models. In the Argonne model the W(r) is the same for all p, however S^t and $S^{t\tau}$ are non-zero. The parameters c (of $\xi(r)$), R and a (of W(r)), I^p and S^p are determined by fitting the neutron-proton scattering data at $E_{lab} \lesssim$ 400 MeV where the scattering is dominantly elastic. Both models can quantitatively explain the data as well as the Paris potential which has a much more detailed treatment of the v^I.

The expectation values of the kinetic energy and the interactions in nuclear matter at equilibrium density ρ_0 are given in table I. These expectation values for the Urbana model are calculated from variationally determined ground state wave functions[12], and they could be wrong by ~10%. Nevertheless we note that there are large cancellations; the magnitudes of all the contributions are larger than the total. The OPEP v^π provides ~1/2 of the potential energy of nuclear matter, and $\langle T + v \rangle$ is larger than the experimental energy (-16 MeV) by ~ 3 MeV.

Table I. Contributions to the energy of nuclear matter at equilibrium density ρ_o in MeV per nucleon.

$\langle T \rangle$	+39
$\langle v^{\pi} \rangle$	-27
$\langle v^{I} \rangle$	-149
$\langle v^{S} \rangle$	+124
$\langle T + v \rangle$	-13
Experimental $\langle H \rangle$	-16

The large and cancelling expectation values of v^I and v^S have raised speculations regarding significant relativistic effects in nuclear matter. In relativistic field theories[13] the attractive $\langle v^I \rangle$ is attributed to the coupling of Dirac nucleons to an effective scalar field, while the repulsive $\langle v^S \rangle$ is due to the coupling to the vector fields. The mass of the nucleons in nuclear matter is reduced by the coupling to the scalar field by ~30% which represents a substantial relativistic effect. However the attractive v^I interaction between the nucleons has rather complex origins in the composite nature of the nucleon, and it is not clear that it can be represented by the coupling of nucleons (treated as Dirac particles) to an effective scalar field.

Much of the attractive $\langle v^{\pi} \rangle$ is due to the higher order effects, or equivalently due to the tensor correlations induced by the $v^{t\tau}$ interaction. Both the structure and the equation of state of dense hadronic matter is sensitive to these correlations, and hence the uncertainties of $v^{t\tau}$ have often been discussed in the literature. The $v^{t\tau}(r)$ in Paris, Urbana and Argonne models is compared with the $v^{t\tau}$ in OPEP, calculated with a monopole form-factor having $\Lambda = 7m_{\pi}$ at the π-N vertex, in fig. 2. The $v^{t\tau}$ is poorly determined by the scattering data at $r < 1$fm, however most models have rather similar $v^{t\tau}$ at $r>1.5$ fm. We note that in fitting the parameters of the Urbana potential, attempts were made to keep the $v^{t\tau}$ as small as possible in that model.

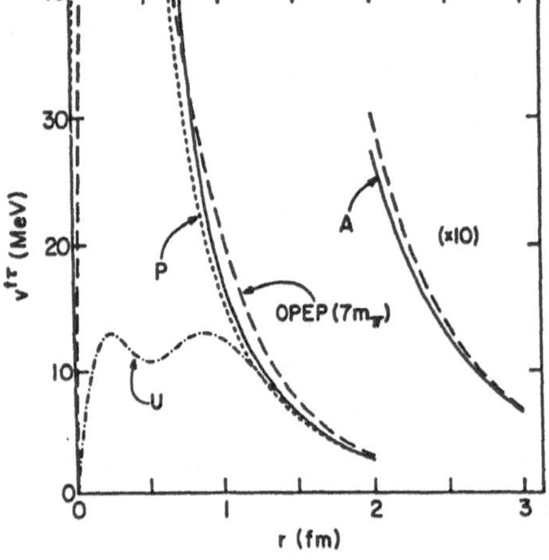

Fig. 2. The $v^{t\tau}$ in Paris (P), Argonne (A) and Urbana (U) models and the OPE $v^{\pi,t\tau}$ with $\Lambda = 7\mu$.

Within the meson-exchange model it is believed that the π- and ρ-mesons generate most of the $v^{t\tau}$ and $v^{\sigma\tau}$ potentials. Following Riska[14] we define

$$v^{\sigma\tau}(k) = \frac{4\pi}{k^2} \int dr\, r^2\, (j_o(kr)-1)\, v^{\sigma\tau}(r), \tag{2.13}$$

$$v^{t\tau}(k) = \frac{4\pi}{k^2} \int dr\, r^2\, j_2(kr)\, v^{t\tau}(r). \tag{2.14}$$

Assuming that the $v^{t\tau}$ and $v^{\sigma\tau}$ potentials are entirely due to π- and ρ-exchange we obtain:

$$\tilde{v}^{\pi}(k) \equiv \frac{1}{3}\left(\frac{f_\pi}{m_\pi}\right)^2 \frac{1}{m_\pi^2 + k^2} [F_{NN\pi}(k)]^2 = \frac{2}{3} v^{t\tau}(k) - \frac{1}{3} v^{\sigma\tau}(k), \tag{2.15}$$

$$\tilde{v}^{\rho}(k) \equiv -\frac{1}{3}\left(\frac{g_\rho}{2m_N}\right)^2 \frac{(1+\kappa)^2}{m_\rho^2 + k^2} \left[F_{NN\rho}(k)\right]^2 = \frac{1}{3} v^{t\tau} + \frac{1}{3} v^{\sigma\tau}(k). \tag{2.16}$$

Thus we can use the $v^{t\tau}$ and $v^{\sigma\tau}$ potentials in the Paris, Urbana and Argonne models to study the effective meson nucleon form-factors $F_{NN\pi}$ and $F_{NN\rho}$ consistent with these models. It is found that the form factors are very hard (effective monopole $\Lambda > 1$ GeV). The OPE $S_{ij}\tau_i{\bullet}\tau_j$ potentials $v^{\pi,t\tau}$ calculated from the $\tilde{v}^{\pi}(k)$ extracted from the Paris, Urbana and Argonne models are shown in fig. 3. They are surprisingly close to the $v^{\pi,t\tau}$ calculated with point pions and nucleons down to ~0.2fm. In this context the main difference between the Urbana and Argonne models is that the former has a stronger ρ-exchange interaction. Both $v^{\pi,t\tau}$ and $v^{\pi,\sigma\tau}$ are positive while $v^{\rho,t\tau}$ is negative and $v^{\rho,\sigma\tau}$ is positive. At $r < 1$fm the Urbana $v^{t\tau}$ is smaller than that of Argonne, while the $v^{\sigma\tau}$ (fig. 4) is larger, thus clearly there is a stronger ρ-exchange interaction in the Urbana model.

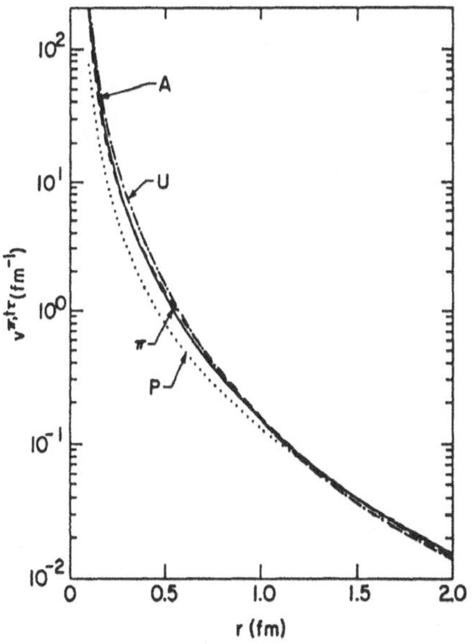

Fig 3. The $v^{\pi,t\tau}$ from Paris (P), Argonne (A) and Urbana (U) models, and in OPEP for point pions and nucleons.

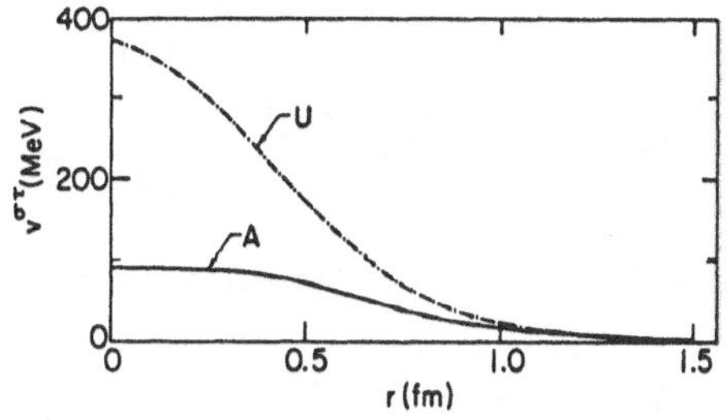

Fig. 4. The $v^{\sigma\tau}$ in Urbana (U) and Argonne (A) models.

It has been suggested[9] that the apparent hardness of the πNN form factor indicated by realistic models of N-N interaction may be in contradiction with the anti-quark distributions in nucleons indicated by deep inelastic neutrino scattering.[15] The virtual pions surrounding a nucleon contribute to the antiquarks in the nucleon, and this contribution increases with the cut off mass Λ of the πNN form factor. As discussed by Thomas[9] at this school, the current calculations of the anti-quark distribution in the nucleon indicate that $\Lambda < 600$ MeV. If these are correct, then mechanisms other than π and ρ exchange, such as exchanges of higher mass $T=1$ $J^{\pi} = 0^-$ mesons, $q\bar{q}$ pairs, or quark-gluon exchanges between nucleons, must be responsible for the $v^{t\tau}$ and $v^{\sigma\tau}$ at small r. For example, if we admit the possible exchange of many $T=1$, $J^{\pi} = 0^-$ pseudoscalar (PS) bosons b, eq. (2.15) then becomes

$$\frac{2}{3} v^{t\tau}(k) - \frac{1}{3} v^{\sigma\tau}(k) \equiv \dot{v}^{PS}(k)$$

$$= \sum_b \frac{1}{3} \left(\frac{f_b}{m_b}\right)^2 \frac{1}{m_b^2 + k^2} [F_{NNb}(k)]^2, \qquad (2.17)$$

where m_b, f_b and F_{NNb} are the mass, coupling constant and form-factor of boson b. In such a scenario the potentials $v^{t\tau}(r)$ and $v^{\sigma\tau}(r)$ have a rather complex origin, but it is often not necessary to get into their origin.

III. THE ELECTROMAGNETIC CURRENT OPERATOR

In order to relate the nuclear many-body theory to the observed electromagnetic processes in nuclei and electron-nucleus scattering, it is necessary to express the nuclear charge and current by operators which act on the nucleon degrees of freedom. These operators have one-body terms corresponding to those of free nucleons, and pair and many-body terms associated with nuclear interactions. In momentum space

$$\vec{j}(\vec{q}) = \sum_i \vec{j}_{1,i}(\vec{q}) + \sum_{i<j} \vec{j}_{2,ij}(\vec{k}_i, \vec{k}_j) + \ldots\ldots, \qquad (3.1)$$

$$\rho_c(\vec{q}) = \sum_i \rho_{c1,i}(\vec{q}) + \sum_{i<j} \rho_{c2,ij}(\vec{k}_i, \vec{k}_j) + \ldots\ldots, \qquad (3.2)$$

where $\vec{k}_i$ and $\vec{k}_j$ are momenta transferred to nucleons i and j, and

$$\vec{q} = \vec{k}_i + \vec{k}_j. \qquad (3.3)$$

The one-body operators $\vec{j}_1$ and ρ_{c1} are obtained by formally starting from the covariant current:[16]

$$j_\mu = \bar{u}(\vec{p}') \left[\gamma_\mu F_1(q^2) + \frac{i\sigma_{\mu\nu} q^\nu}{2m} \kappa F_2(q^2) \right] u(\vec{p}), \tag{3.4}$$

where $\vec{p}$ and $\vec{p}'$ are the initial and final momenta of the nucleon, and $F_1(q^2)$ anf $F_2(q^2)$ are Dirac form factors. The current is expanded in powers of 1/m, and terms up to order $1/m^2$ are retained to obtain:

$$\vec{j}_{1,i}(\vec{q}) = \frac{1}{4m} \left[G_E^S(q^2) + G_E^V(q^2)\, \tau_{z,i} \right] (\vec{p}_i' + \vec{p}_i)$$

$$+ \frac{i}{4m} \left[G_M^S(q^2) + G_M^V(q^2)\, \tau_{z,i} \right] \vec{\sigma}_i \times \vec{q}, \tag{3.5}$$

$$\rho_{c1,i}(\vec{q}) = \left(1 - \frac{q^2}{8m^2}\right) \frac{1}{2} \left[G_E^S(q^2) + G_E^V(q^2)\, \tau_{z,i} \right] - i\, \frac{\vec{\sigma}_i \bullet \vec{q} \times (\vec{p}_i + \vec{p}_i')}{8m^2}$$

$$\times \frac{1}{2} \left\{ \left[G_E^S(q^2) - 2G_M^S(q^2) \right] + \left[G_E^V(q^2) - 2G_M^V(q^2) \right] \tau_{z,i} \right\}. \tag{3.6}$$

The superscripts S and V of the Sachs form-factors G_E and G_M denote isoscalar and isovector combinations. The G_E and G_M are related to the Dirac form-factors in eq. (3.4) by:

$$G_E = F_1 + \frac{\kappa q^2}{4m^2} F_2, \tag{3.7}$$

$$G_M = F_1 + \kappa F_2, \tag{3.8}$$

and they are normalized so that

$$G_E^S\,(q^2 = 0) = G_E^V\,(q^2 = 0) = 1, \tag{3.9}$$

$$G_M^S\,(q^2 = 0) = \mu_p + \mu_n = 0.88\mu\,, \tag{3.10}$$

$$G_M^V\,(q^2 = 0) = \mu_p - \mu_n = 4.71\mu, \tag{3.11}$$

where μ_p and μ_n are the magnetic moments of the proton and neutron. Several authors[17-19] have studied the q^2 dependence of the Sachs form factors by fitting the available electron-nucleon scattering data.

The pair currents $\vec{j}_2$ and ρ_{c2} have been studied by many authors using the meson exchange theory of nuclear interactions. Much of the recent work has been recently reviewed by Mathiot.[20] The currents and charges associated with the exchange of T=1 pion like pseudoscalar (PS) and ρ-like vector (V) bosons give the largest contributions at $|\vec{q}| < 5\ \mathrm{fm}^{-1}$. Contributions of all known exchange currents to the elastic form-factors of ^3H, ^3He and ^4He have been recently estimated by Schiavilla,

Pandharipande and Riska and the details omitted in these lectures may be found in ref. 21 and 22. Here we discuss only the PS-exchange contributions to illustrate the method.

The leading term, in the 1/m expansion, of the contribution of the pair current due to the exchange of T=1 PS bosons is obtained as:

$$\vec{j}^{PS}_{2,ij}(\vec{k}_i,\vec{k}_j) = -3i(\tau_i \times \tau_j)\left\{\tilde{v}^{\pi}(k_j)\,\vec{\sigma}_i\,\vec{\sigma}_j\cdot\vec{k}_j - \tilde{v}^{\pi}(k_i)\,\vec{\sigma}_j\,\vec{\sigma}_i\cdot\vec{k}_i - \frac{\vec{k}_i-\vec{k}_j}{k_i^2-k_j^2}\,\vec{\sigma}_i\cdot\vec{k}_i\,\vec{\sigma}_j\cdot\vec{k}_j\left[\tilde{v}^{\pi}(k_j)-\tilde{v}^{\pi}(k_i)\right]\right\} G_E^V(q^2) \tag{3.12}$$

Several comments on the above expression are in order. If we take

$$\tilde{v}^{\pi}(k) = \frac{1}{3}(f_\pi/m_\pi)^2\,\frac{1}{m_\pi^2+k^2}, \tag{3.13}$$

and set $G_E^V(q^2) = 1$, the above expression gives the pair current due to the exchange of point pions between point nucleons. In practice the $\tilde{v}^{\pi}(k)$ extracted from the N-N interaction (eq. 2.15) is used to take into account the effects of πNN form-factors and the possible contributions from other T=1 PS bosons. The continuity equation

$$\nabla\cdot\vec{j}^{PS}_{2,ij}(\vec{r};\vec{r}_i,\vec{r}_j) + i\,[v^{PS}(\vec{r}_i,\vec{r}_j),\rho_c(\vec{r})] = 0, \tag{3.14}$$

where $v^{PS}(\vec{r}_i,\vec{r}_j)$ is the contribution of all T=1 PS exchanges to the v_{ij}, is satisfied when the Sachs form-factor $G_E^V(q^2)$ is used in eq. (3.12). The current $\vec{j}^{PS}_{2,ij}$ does not contain any new adjustable parameters, and it may be directly obtained from the N-N interaction v_{ij} without invoking the meson-exchange theory.[23]

The total pair current $\vec{j}_{2,ij}$ is divided into contributions, such as those from exchange of T=1 PS and V bosons, L•S interactions, etc., that can be determined from the interaction v_{ij} by Riska's method.[14] The rest of the contributions, such as those due to $\gamma N\Delta$, $\gamma\rho\pi$, $\gamma\omega\pi$ vertices, etc. are not trivially related to v_{ij}. It is difficult to determine these contributions from the experimental data; fortunately they become important only when $|\vec{q}| > 5$ fm^{-1}, so that a meaningful comparison with data at $|\vec{q}| \lesssim 5$ fm^{-1} is possible.[21]

The leading contribution of T=1 PS bosons to $\rho_{c2,ij}$ is obtained as:[22]

$$\rho^{PS}_{c2,ij}(\vec{k}_i,\vec{k}_j) = \frac{3}{2m}\left\{[F_1^S(q^2)\,\tau_i\cdot\tau_j + F_1^V(q^2)\,\tau_{z,j}]\,\tilde{v}^{\pi}(k_j)\,\vec{\sigma}_i\cdot\vec{q}\,\vec{\sigma}_j\cdot\vec{k}_j + [F_1^S(q^2)\,\tau_i\cdot\tau_j + F_1^V(q^2)\,\tau_{z,j}]\,\tilde{v}^{\pi}(k_i)\,\vec{\sigma}_i\cdot\vec{k}_i\,\vec{\sigma}_j\cdot\vec{q}\right\} \tag{3.15}$$

It is determined by the v_{ij} and the electromagnetic form-factors of the nucleons, and has no adjustable parameters. Nevertheless it represents a relativistic correction obtained by assuming a $\gamma_5 \gamma_\mu \partial^\mu$ coupling between PS bosons and Dirac nucleons, and its use with nonrelativistic wave functions has not been rigorously established. One can similarly extract the contribution of T=1 V-bosons to ρ_{c2} from v_{ij}. The other contributions[22] to ρ_{c2} can not be determined from v_{ij}, but they are relatively small for $|\vec{q}| \lesssim 8fm^{-1}$. The ρ_{c1} and ρ_{c2} are often denoted by ρ_1 and ρ_2 for brevity.

IV. MANY-BODY CALCULATIONS

It is relatively simple to solve the two-body Schrödinger equation for the deuteron and N-N scattering. Unfortunately it is much more difficult to solve it for $A \geq 3$ bound and continuum states. The ground states[24] and low-energy continuum[25] states of the trinucleons ^3H and ^3He can be very accurately calculated with Faddeev's method. Recently Carlson[26] has calculated the ground state of the ^4He with semi-realistic interaction models which retain only the terms associated with operators $O^{p=1,8}$ (eq. 2.2), and the Green's function Monte-Carlo (GFMC) method. All other studies of nuclei, nuclear and neutron-star matter use approximate methods based either on the Brueckner-Bethe-Goldstone (BBG) approach or on the variational and correlated-basis theories pioneered by Feenberg.[27] Since most of the results presented here are obtained with these theories we will review them briefly; the BBG theory of nuclear matter has been recently reviewed by Mahaux.[28]

IV.1 VARIATIONAL MONTE CARLO

The deuteron wave function

$$\Psi_d^M = \frac{u(r)}{r} y_{011}^M + \frac{w(r)}{r} y_{211}^M , \tag{4.1}$$

where y_{LSJ}^M are spin-angle parts, and u(r) and w(r) are the L=0 and 2 radial functions, can be written in the form

$$\Psi_d^M = F_{12} \Phi_d^M \tag{4.2}$$

where Φ_d^M is the wave function of an uncorrelated np pair, for example

$$\Phi_d^{M=1} = \frac{1}{\sqrt{2}} \quad |\uparrow p \uparrow n - \uparrow n \uparrow p>, \tag{4.3}$$

and F_{12} the correlation operator:

$$F_{12} = f^c(r) + f^t(r)\ S_{12}, \tag{4.4}$$

$$f_c(r) = u(r)/r, \tag{4.5}$$

$$f^t(r) = \frac{w(r)}{r\sqrt{8}}\ . \tag{4.6}$$

The well known coupled differential equations for u(r) and w(r) can be obtained by using (4.2) as a variational wave function and minimizing the energy with respect to variations in f^c and f^t. Since the exact Ψ_d had the form (4.2) this variational calculation gives the exact result.

The variational wave functions of nuclei having $A \geq 3$ are generally taken to be of a similar form:

$$\Psi_v = \left(S \prod_{i<j} F_{ij}\right) \Phi_v, \tag{4.7}$$

$$F_{ij} = \sum_p\ f^p(r_{ij})\ O^p_{ij}, \tag{4.8}$$

where S is a symmetrizer necessary because the F_{ij} and F_{ik} do not commute, and Φ_v is antisymmetric and uncorrelated. The Φ_v of S-wave nuclei ^2H, ^3H, ^3He and ^4He is purely a spin-isospin state (like 4.3) with no spatial dependence, whereas that of heavier nuclei like ^{16}O is a determinant of single-particle wave functions determined variationally. When A>2 the exact ground state may not have the form of this Ψ_v (eq. 4.7). Hence variational calculations with it will not give the exact results. Nevertheless the method is interesting because this Ψ_v is not too bad an approximation, and also because it provides the input to the more accurate calculations.

It is useful to factor the $\prod f^c(r_{ij})$ from $\prod F_{ij}$ and rewrite Ψ_v as:

$$\Psi_v = \left(S \prod_{i<j} \left\{1 + \sum_{p\geq 2}\ u^p(r_{ij})\ O^p_{ij}\right\}\right) \prod_{i<j} f^c(r_{ij})\ \Phi_v, \tag{4.9}$$

$$u^p(r_{ij}) = f^p(r_{ij})/f^c\ (r_{ij})\ . \tag{4.10}$$

The Jastrow wave function

$$\Psi_J = \prod_{i<j}\ f^c(r_{ij})\ \Phi_v, \tag{4.11}$$

contains only spatial correlations. It is a very poor approximation; nuclei are not bound when Ψ_J is used as a variational wave function.

The spin-isospin correlations between pairs ij and ik interfere. A better Ψ_v, with which many variational Monte Carlo calculations[29] have been carried out, is found to be:

$$\Psi_v = \left[S \prod_{i<j} \left\{ 1 + \sum_{p\geq 2} u^p(r_{ij})\, O^p_{ij} \left[\prod_{k\neq i,j} f^p_{ijk} \right] \right\} \right] \Psi_J \tag{4.12}$$

$$f^p_{ijk} = 1 - t_{1,p} \left(\frac{r_{ij}}{R_{ijk}} \right)^{t_{2,p}} e^{-t_{3,p} R_{ijk}} \tag{4.13}$$

$$R_{ijk} = r_{ij} + r_{jk} + r_{ki}. \tag{4.14}$$

The parameters $t_{1,p}$, $t_{2,p}$ and $t_{3,p}$ are chosen such that when r_{ij} is small $f^p_{ijk} \sim 1$ but f^p_{ikj} and f^p_{jik} are < 1. The best available variational calculations, carried out by Wiringa[30] with realistic models of v_{ij}, for the A=3 and 4 nuclei give energies that are ~ 4% above the exact result. These use terms p = 1,6 in the F_{ij}; spin-orbit terms (p = 7,8) are expected to have little ($\lesssim$ 1%) influence on the energies of A = 2,4 nuclei. Many other calculations[29], which use only f^c, $f^{t\tau}$ and f^σ in the F_{ij} give energies that are above the exact value by ~6%. In first order in u the effects of f^τ, $f^{\sigma\tau}$ and f^t correlations in A=3,4 nuclei can be included in f^σ, f^c and $f^{t\tau}$ respectively.[31]

In VMC calculations the nuclear wave function is represented by a vector function:

$$\Psi = \sum_{n=1,N} \Psi_n(\vec{R})\, |n\rangle, \tag{4.15}$$

where $|n\rangle$ denotes the spin-isospin states and

$$\vec{R} = \vec{r}_1, \vec{r}_2 \ldots \vec{r}_A. \tag{4.16}$$

The various spin-isospin states $|n\rangle$ in ^{3}He are, for example

$$|\uparrow p\uparrow p\uparrow n\rangle, \quad |\uparrow p\uparrow p\downarrow n\rangle, \quad |\uparrow n\downarrow p\uparrow p\rangle, \ \ldots \tag{4.17}$$

In general the number of these states is given by

$$N = 2^A \frac{A!}{(A-Z)!\, Z!}, \tag{4.18}$$

where the factor 2^A comes from allowing the spin of each particle to be $\uparrow$ or $\downarrow$, and the rest from letting any Z of the A nucleons to be protons. N increases very rapidly with A, it is 24, 96, 1280 and 843,448,320 in ^{3}H, ^{4}He, ^{6}Li and ^{16}O respectively. Hence VMC calculations are possible with this method only for the light H, He and Li nuclei.

In this representation the $F_{ij}(R)$ and H are matrix functions of dimension N×N. The energy expectation value is written as:

$$\langle H\rangle = \frac{\int \Psi_v^\dagger(R)\; H\; \Psi_v(R)\; dR}{\int \Psi_v^\dagger(R')\; \Psi_v(R')\; dR'} = \int I(R)\; P(R)\; dR. \tag{4.19}$$

The local energy I(R) and the probability density P(R) are given by:

$$I(R) = \Psi_V^\dagger(R) \; H \; \Psi_V(R) / \Psi_V^\dagger(R) \Psi_V(R), \tag{4.20}$$

$$P(R) = \Psi_V^\dagger(R) \; \Psi_V(R) / \int \Psi_V^\dagger(R') \; \Psi_V(R') dR', \tag{4.21}$$

and the $\langle H \rangle$ is evaluated with standard Monte Carlo methods.[32] The functions $f^p(r_{ij})$, or rather the parameters in the differential equations[29] used to generate the $f^p(r_{ij})$ and the parameters $t_{i,p}$ of f^p_{ijk} are determined by minimizing $\langle H \rangle$.

IV.2 GREEN'S FUNCTION MONTE CARLO

The exact ground state Ψ_o can be projected out of the variational Ψ_v with the GFMC method[26] using evolution in imaginary time τ

$$\Psi(\tau) = e^{-(H-E_o)\tau} \; \Psi_v, \tag{4.22}$$

$$\Psi_o = \Psi(\tau \to \infty) . \tag{4.23}$$

The τ can also be thought of as an inverse temperature. For a small enough time step $\Delta\tau$

$$\Psi(R, \tau + \Delta\tau) = e^{-(H-E_o)\Delta\tau} \; \Psi(R', \tau), \tag{4.24}$$

$$= e^{-(V(R)-E_o)\Delta\tau} \; e^{-T\Delta\tau} \; \Psi(R', \tau), \tag{4.25}$$

$$= e^{-(V(R)-E_o)\Delta\tau} \int dR' \exp\left[\frac{(R-R')^2}{2\Delta\tau\hbar^2/m} \right] \Psi(R', \tau) . \tag{4.26}$$

The V(R) and T in the above equations denote the total potential, i.e. sum of all pair potentials, and the total kinetic energy. Equation (4.26) is multiplied with the importance function $\Psi_v(R)$ to obtain:

$$\Psi^*_{v,n}(R) \; \Psi_n(R, \tau + \Delta\tau) = \sum_{n'} \Psi^*_{v,n} (R) \left[e^{-(V(R)-E_o)\Delta\tau} \right]_{nn'}$$

$$\times \int dR' \exp\left[\frac{(R-R')^2}{2\Delta\tau\hbar^2/m} \right] \frac{1}{\Psi^*_{v,n'}(R')} \; \Psi^*_{v,n'}(R') \Psi_{n'}(R', \tau), \tag{4.27}$$

where the spin-isospin state labels n and n′ are shown explicitly. This equation is solved by Monte Carlo methods.

Using this method Carlson[26,33] has done exact calculations for ^3H and ^4He with a v_8 potential (i.e. a potential containing operators $O^{p=1,8}$) based on that of Reid.[34] He found that the VMC energy of ^4He (-23.1 MeV) obtained with the wave function (4.12) is 6% above the exact energy

(-24.6 MeV) calculated with GFMC. He also calculated various one and two-particle distribution functions using VMC and GFMC and found that they are quite similar.[33]

IV.3 NUCLEAR MATTER

The wave function

$$\Psi_V = \left(S \prod_{i<j} F_{ij} \right) \Phi_{FG} \tag{4.28}$$

where Φ_{FG} is the Fermi-gas ground state, has been used to study the ground states of nuclear and neutron matter at various densities.[2,35] In extended matter $F_{ij} \rightarrow 1$ as $r_{ij} \rightarrow \infty$, and hence $F_{ij} -1$ has a short range. The F_{ij} are written as $1 + (F_{ij} - 1)$ and the $\langle H \rangle$ is expanded in powers of $(F_{ij}-1)$. The terms in this expansion are rearranged to obtain a linked cluster expansion.[36] It is believed that the important terms of this expansion can be summed by using Fermi hypernetted chain[37] and single operator chain[36] summation methods, and the $\langle H \rangle$ can be calculated with an accuracy better than 10%.

The results[7] obtained with the Argonne[5], Paris[6] and Urbana[4] models of v_{ij} are shown in fig. 5. The BBG calculations of Day[7] include two-, three- and parts of four-hole line terms, and give almost identical results with the Argonne and the Paris models. The variational energies obtained with the Argonne v_{ij} are ~2MeV above the BBG energies, while the Urbana v_{ij} gives ~1.5 MeV more binding presumably due to its weaker tensor force (fig. 2). It is likely that further improvements in the variational wave function, such as the inclusion of the f^{p}_{ijk} correlations (eq. 4.12-14), and more accurate determination of the correlations $f^{p}(r_{ij})$ will lower the variational energies by ~2MeV, and thus Day's BBG results[7] and the variational results of Wiringa and coworkers[2,7] seem to be consistent with each other.

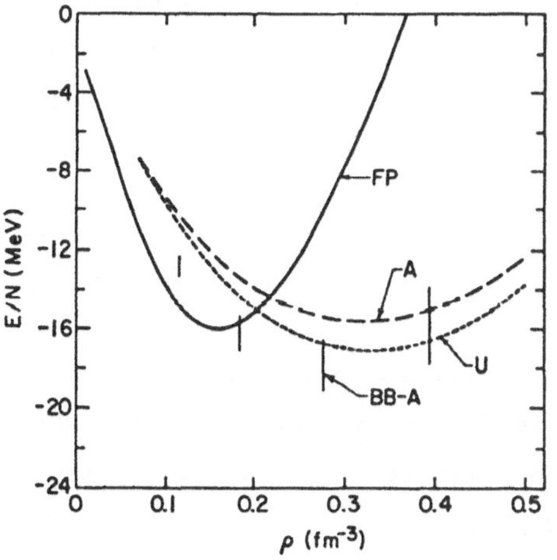

Fig. 5. The dashed lines give results of variational calculations of the energy of nuclear matter with Argonne (A) and Urbana (U) potentials, and the error bars give results of Brueckner-Bethe (BB) calculations with Argonne potential. The curve labeled FP has the correct empirical minimum.

IV.4 NUCLEI

There are rather few calculations of finite nuclei having $A>6$ with accuracies comparable with those of the $A=3$ and 4 nuclei and nuclear matter. The most notable work is that of Kümmel, Lührmann and Zabolitzky[38] who studied the ground states of ^4He, ^{16}O and ^{40}Ca using the BBG theory in the coupled cluster form. All two- and three- and parts of four-body cluster contributions were included in the calculations. The energy obtained for ^4He with the Reid interaction is -23 MeV, quite close to the exact -24.6 MeV. A major problem revealed by these calculations is that with realistic models of v_{ij} one obtains less binding for ^{16}O (~5.1 MeV per nucleon) than for ^4He (~5.8 MeV per nucleon) to be compared with the experimental values of 8.0 and 7.1 MeV per nucleon respectively. Thus the stability of ^{16}O could not be explained.

The present VMC methods can not be directly used to study nuclei like ^{16}O, purely because they have too many spin-isospin states. We note that ground states of ~100 particles interacting with spin-independent central forces can be easily studied with the VMC method.[39] Since central forces do not change the spin of a particle one can work with a single spin-state for the many-body system. Thus the main problem is to develop Monte Carlo or other methods that can treat spin, isospin, tensor and spin-orbit forces.

Recently we[40] have attempted a VMC calculation of ^{16}O using the wave function 4.9. A cluster expansion is obtained by expanding the expectation value in powers of u^p's, and rearranging the terms suitably. The cluster contributions of this expansion are obtained as expectation values with the Ψ_J (eq. 4.12), and those of two-, three- and four-body clusters are evaluated without approximations by the Monte Carlo method. The results indicate that it is necessary to include contribution of five- and six-body clusters to determine the energies accurately. Particularly when three-body forces are considered, the clusters with more than four particles are estimated to give ~8% of the total binding energy of ^{16}O.

IV.5. EXCITED STATES

We will discuss the excited states of nuclear matter. These are simpler, and the ideas can be carried over to finite systems. The Fermi-gas (FG) states are denoted by square bra and kets $|]$ and $[|$. $|0]$ is the ground state, and the excited states are:

$$|\vec{p}] = a^{\dagger}_{\vec{p}} \; |0], \quad |p|>k_F, \tag{4.29}$$

$$|\vec{h}] = a_h \; |0], \quad |h|<k_F, \tag{4.30}$$

$$|\vec{p}_1 \; \vec{p}_2 \; \ldots \; \vec{h}_1 \; \vec{h}_2 \; \ldots] = a^{\dagger}_{\vec{p}_1} \; \ldots \; a_{h_1} \; \ldots \; |0]. \tag{4.31}$$

Let the number of particles in $|0]$ be A, which is assumed to be large. The states $|\vec{p}]$, $|\vec{h}]$, $|\vec{p}\;\vec{h}]$ etc then have A+1, A-1 and A particles respectively. We will assume that the total number of particles and holes is finite, or equivalently the excitation energy is finite. The total energy, being proportional to A is not finite in the limit $A \to \infty$.

The correlated states (CS) are denoted by round bra and kets $|)$ and $(|$. We define,

$$G \equiv S \prod_{i<j} F_{ij}, \tag{4.32}$$

$$|p_1 \; p_2 \; \ldots \; h_1 \; h_2 \; \ldots) = \frac{G|p_1p_2 \; \ldots \; h_1h_2\ldots]}{[p_1p_2 \; \ldots \; h_1h_2\ldots \; |G^{\dagger}G|p_1p_2 \; \ldots \; h_1h_2 \; \ldots]^{1/2}} \tag{4.33}$$

where the correlation functions $f^p(r)$ in F_{ij} are determined by minimizing the ground state energy $(0|H|0)$. The energy expectation values of the CS are not unreasonable, for example

$$(p|H|p) - (0|H|0) = e_v(p), \tag{4.34}$$

$$(0|H|0) - (h|H|h) = e_v(h), \tag{4.35}$$

provide a good approximation to the energies of particles and holes in nuclear matter.[41] The CS however are not orthonormal because the G is not restricted to be a unitary operator in variational calculations. It is possible[42] to orthonormalize the CS such that the orthonormal correlated states (OCS), denoted by the standard bra and kets $|\rangle$ and $\langle|$, have the same energy expectation value as the corresponding CS,

$$\langle p_1p_2\ldots h_1h_2\ldots|H|\; p_1p_2\ldots h_1h_2\ldots\rangle = (p_1p_2\ldots h_1h_2\ldots|H|p_1p_2\ldots h_1h_2 \; \ldots). \tag{4.36}$$

These OCS provide the first approximation to the excited states in correlated basis theories.

Correlated-basis perturbation theory (CBPT) is used to study the excitations, as well as to improve upon the variational ground states. It is obtained from:

$$H = H_o + H_I, \tag{4.37}$$

$$\langle I|H_o|J\rangle = \langle I|H|J\rangle \; \delta_{IJ}, \tag{4.38}$$

$$\langle I|H_I|J\rangle = \langle I|H|J\rangle \; (1-\delta_{IJ}), \tag{4.39}$$

where $|I\rangle$ denote OCS. Since H_I has no diagonal elements, there are no first order terms in CBPT. If the correlated states $|I\rangle$ are close to the eigenstates of H, then the H_I is weak and CBPT has good convergence.

As an example of CBPT we consider the leading second order contribution to the ground-state energy[43] given by:

$$\delta E_o^{(2,2)} = \frac{1}{4} \sum_{p_2p_2h_1h_2} \frac{|\langle p_1p_2h_1h_2|H|0\rangle|^2}{e_v(h_1) + e_v(h_2) - e_v(p_1) - e_v(p_2)}, \quad (4.40)$$

where $e_v(p)$ and $e_v(h)$ are single particle energies given by equations 4.34, 35. In CBPT the total second-order $\delta E_o^{(2)}$ has other contributions,

$$\delta E_o^{(2)} = \sum_{n>2} \delta E_o^{(2,n)}, \quad (4.41)$$

for example, the $\delta E_o^{(2,3)}$ comes from the coupling $\langle p_1p_2p_3h_1h_2h_3|H|0\rangle$. The $\delta E_o^{(2,2)}$ depends upon the correlations $f^p(r)$ in the F_{ij}. The range d_t of the tensor correlations in the F_{ij} is an important parameter of the Ψ_v. In fig. 6 we show the variational ground state energy, per nucleon, E_o^v and the total $E_o^v + \delta E_o^{(2,2)}$ for nuclear matter as a function of d_t/r_o where r_o is the unit radius

$$\frac{4\pi}{3} \rho r_o^3 = 1. \quad (4.42)$$

As expected the perturbative corrections are smallest when the F_{ij} is optimized by minimizing the ground state energy. In many applications only the second order corrections are retained with the hope that the convergence is excellent. Note that these perturbative corrections can explain most of the difference between variational and BBG energies of nuclear matter (fig. 5).

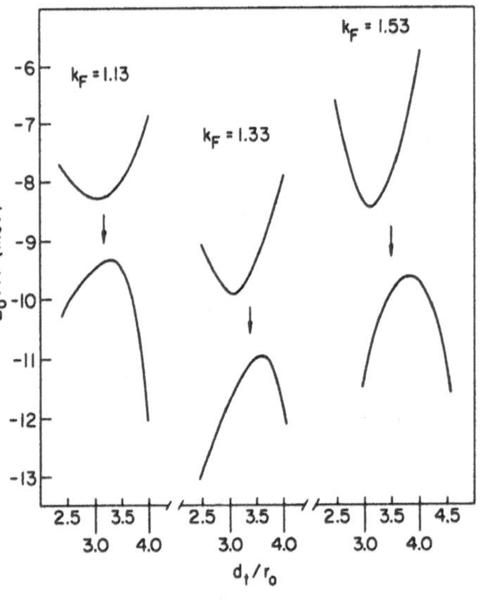

Fig. 6. The upper curves show the variational energy E_o^v as a function of d_t/r_o, and the lower curves show $E_o^v + \delta E_o^{(2,2)}$. The arrows mark values of d_t/r_o where the perturbative correction $\delta E_o^{(2,2)}$ has the smallest magnitude.

The techniques of orthonormalizing the CS were developed quite recently[42]; most of the earlier work, including ref. 43 from which fig. 6 is taken, used perturbation theory with non-orthogonal CS. A review of this theory may be found in ref. 27 and 44. The theory using OCS has many formal advantages, and hopefully a better convergence.

V. THREE NUCLEON INTERACTION

The binding energies of ^3H and ^4He obtained with the Paris, Nijmegen, Argonne and Reid potentials are given in table II. The triton energies are obtained from almost exact Faddeev calculations from refs. 24, 45 and 46, and have errors <0.01 MeV. The ^4He energies are from VMC calculations[26], and the true energies may be ~1.5 MeV below the listed values; that for Reid is known to be -24.6 ± .2 MeV from GFMC.[26] We see that these interactions underbind the light nuclei, ^3H by ~0.8 MeV and ^4He by ~4.MeV.

Table II. Calculated Ground Energies of ^3H and ^4He with several two-body interactions in MeV.

Interaction	^3H	^4He
Reid	-7.59	-23.1
Nijmegen	-7.62	-23.2
Paris	-7.47	
Argonne	-7.67	-23.2
Experiment	-8.48	-28.3

We also see from fig. 5 that the $E(\rho)$ of nuclear matter (NM) obtained from these realistic interactions has less binding at low densities and too much binding at high densities, as compared with the "empirical curve." Moreover, it appears that both in light nuclei and nuclear matter the differences between the results of different potential models are smaller than those between experiment and theory. For some time it was thought that the coordinate space Bonn potential, which gives ^3H energy of -8.29 MeV provides an exception. However this result may not be very significant since the improvement in the ^3H energy probably comes from the unrealistic energy dependence of the $^3S_1$ - $^3D_1$ mixing phase ε_1 of this potential.[47] Calculations with the energy-dependent Bonn-potential[48] give only -6.7 MeV, and thus it too is exceptional.

It is well known that there are three-nucleon forces[49] which we denote by V_{ijk}. Particularly the two-pion exchange $V^{2\pi}_{ijk}$ is theoretically well established for a long time[50], and it can provide the additional attraction necessary to obtain the correct binding energy of ^3H and ^4He.[51] However, if one takes only the attractive $V^{2\pi}_{ijk}$, NM gets overbound and also has too large an equilibrium density. Hence it is necessary to have other terms in V_{ijk} to obtain reasonable NM. Our phenomenological approach[29] has been to consider a model of V_{ijk}:

$$V_{ijk} = V^{2\pi}_{ijk} + V^{R}_{ijk} , \tag{5.1}$$

$$V^{2\pi}_{ijk} = A_{2\pi} \sum_{cy} \left(\{\tau_i \cdot \tau_j, \tau_i \cdot \tau_k\} \{x_{ij}, x_{ij}\} + \frac{1}{4} [\tau_i \cdot \tau_j, \tau_i \cdot \tau_k] [x_{ij}, x_{ij}] \right), \tag{5.2}$$

$$x_{ij} = S_{ij} T_\pi(r) + \sigma_i \cdot \sigma_j Y_\pi(r), \tag{5.3}$$

$$V^{R}_{ijk} = \sum_{cy} U_o T^2_\pi(r_{ij}) T^2_\pi(r_{ik}), \tag{5.4}$$

where $T_\pi(r)$ and $Y_\pi(r)$ are radial functions of the tensor and Yukawa parts of OPEP (eq. 2.6,7) in the two-nucleon interaction. The short range cutoff's of these are assumed to be the same as in v_{ij}, and the strength $A_{2\pi}$ of $V^{2\pi}_{ijk}$ is taken as a parameter. V^{R}_{ijk} is meant to represent the quenching of the isobar-box diagrams (fig. 1.2) by the third nucleon. The strengths $A_{2\pi}$ and U_o are adjusted to obtain the experimental ^3H and ^4He binding energies and reasonable NM properties by variational calculations.

The obvious problems with this approach are that the variational calculations are not exact, and the assumed form of V^{R}_{ijk}, which in principle should contain all the components of V_{ijk} other than $V^{2\pi}_{ijk}$, may be too simplistic. Better calculational methods, such as GFMC, are being developed, and attempts to calculate the energies of more nuclei like ^6Li, ^8He, ^{16}O etc. are underway to address these problems. For example ^8He will be more sensitive to three neutron interaction which is absent in ^3H and ^4He.

The latest model[29] is called Urbana VII, and it has $A_{2\pi} = -0.0333$ and $U_o = 0.0038$ MeV. These parameters appear to be reasonable.[52] In particular the contribution of V^{R}_{ijk} to the energy of the triton is 0.7 MeV, in agreement with the dispersive correction to the two-pion exchange NN interaction with NNΔ intermediate states. The latter has been calculated by Sauer[53] with Faddeev calculations including NNΔ intermediate states.

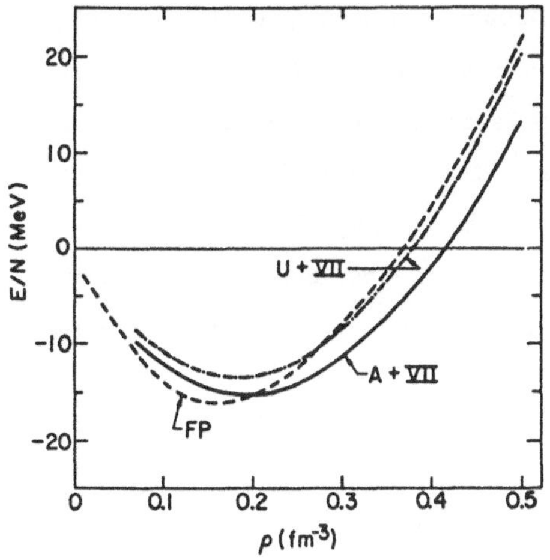

Fig. 7. The E(ρ) of nuclear matter obtained from the Urbana and Argonne potentials and Urbana VII three-nucleon interaction. The curve labeled FP has correct empirical minimum.

This model is meant to be used with the Argonne or Urbana v_{ij}, and the NM results obtained by adding it to these potentials are shown in fig. 7. Note that just by itself the Urbana potential gives lower energies for NM than the Argonne (fig. 5) because of its weaker tensor force. However, the Argonne v_{ij} + V_{ijk} gives lower energies than Urbana v_{ij} + V_{ijk} because the $V_{ijk}^{2\pi}$ can better exploit the stronger tensor correlations induced by Argonne v_{ij}. Even though the V_{ijk} has a rather small effect on the potential energy, it has to be treated non-perturbatively because it enhances the tensor correlations significantly. The results obtained for S-shell nuclei in ref. 29 are given in table III.

Table III. Calculated properties of S-shell nuclei with the Argonne v_{ij} and Urbana VII V_{ijk}, in MEV and fm

	^2H	^3H	^3He	^4He
$\langle T\rangle$	19.17	46.8	46.8	98.7
$\langle v_{ij}\rangle$	-21.40	-53.9	-53.9	-121.5
$\langle v_c\rangle$	0.0	0.0	0.66	0.75
$\langle V_{ijk}^{2\pi}\rangle$	0.0	-1.95	-1.95	-9.84
$\langle V_{ijk}^{R}\rangle$	0.0	0.72	0.72	4.11
E	-2.225	-8.4±.1	-7.7±.1	-27.8±.4
E(expt)	-2.225	-8.48	-7.72	-28.3
R_c(fm)	2.134	1.79	1.92	1.71
R_c(expt)	2.095	1.70	1.95	1.67

Unfortunately neither Argonne or Urbana $v_{ij} + V_{ijk}$ (VII) models give the empirical NM energy (-16MeV) or density ($\rho_0 = 0.16\ fm^{-3}$). One can develop even more phenomenological models that are constructed to explain the equilibrium properties of NM. The model developed by Lagaris and Pandharipande[54] (LP) is based on the Urbana v_{ij} (eq. 2.9). The V^R_{ijk} essentially makes the attractive v^I_{ij} weaker as the density increases, while the $V^{2\pi}_{ijk}$ makes v^{π}_{ij} density dependent. LP assume that the energy of NM (per nucleon) is given by:

$$E(\rho) = E\ [v_{ij}(\rho)] + \gamma_2\ \rho^2\ \exp(-\gamma_3\rho)\ \left(3-2\left(\frac{N-Z}{A}\right)^2\right), \tag{5.5}$$

where $E[v_{ij}(\rho)]$ is the energy obtained with the density dependent interaction:

$$v_{ij}(\rho) = v^{\pi}_{ij} + v^{I}_{ij}\ \exp\ (-\gamma_1\rho) + v^{s}_{ij}, \tag{5.6}$$

and the γ_2 term represents the contribution of all other many-body interactions. The γ_1, γ_2 and γ_3 are varied to obtain $E_o = -16MeV$, $\rho_o = 0.16fm^{-3}$ and K = 240 MeV for NM. the LP interaction is used by Friedman and Pandharipande[55] (FP) to study properties of hot and cold NM and neutron matter. Their results are shown in figs. 5 and 7.

Unfortunately the LP model cannot be used to study the light nuclei. In this respect the models with explicit three-body interactions are superior. The main problems with these are lack of attraction in the $0.1fm^{-3}$ density region (fig. 7), and superluminal sound velocities at very high densities.[2] Thus it appears that the long standing problem of nuclear forces is still not satisfactorily resolved.

The two-pion exchange interaction v^I_{ij} has contributions from NΔ and ΔΔ intermediate states[5], and the V^R_{ijk} was introduced to represent changes in it due to the presence of a third nucleon. However, it has been suggested that relativistic effects[56,57] could also generate a three-body force similar to V^R_{ijk}. In fact some of the recent relativistic Dirac-Brueckner calculations[58,59] give a NM $E(\rho)$ similar to that obtained by FP.

The compressibility K of NM is not well known. The values of K extracted from the observed breathing mode energies of nuclei range from ~220 MeV[60] to ~300 MeV[61]. However, it has been suggested that there may be large uncertainties in these extrapolations.[62]

VI. THE OPTICAL POTENTIAL

The motion of a nucleon with energy e in nuclear matter at density ρ can be described by a complex optical potential $U(\rho,e) - iW_o(\rho,e)$. Its real part gives the dispersion relation for nucleons in nuclear matter:

$$\frac{p^2(e)}{2m} = e-U(\rho,e), \tag{6.1}$$

and its imaginary part gives the damping due to collisions:

$$W_o(\rho,e) \equiv \frac{p(e)}{2m\lambda(\rho,e)}, \tag{6.2}$$

where $\lambda(\rho,e)$ is the mean free path. Both $U(\rho_o,e)$ and $W_o(\rho_o,e)$ can be studied from nucleon-nucleus scattering, and there are compilations of empirical values of $U(\rho_o,e)$ and $W_o(\rho_o,e)$.[63,64]

The velocity of nucleons of energy e in nuclear matter is given by:

$$v(e) = \frac{de}{dp} = \frac{p(e)}{m^*(e)} = \frac{p(e)}{m(1-dU(\rho,e)/de)}, \tag{6.3}$$

and the mean free path

$$\lambda(\rho,e) = v(e)\tau(e) = \frac{p(e)}{m^*(e)}\tau(e), \tag{6.4}$$

where $\tau(e)$ is the lifetime of the single-particle state. The effective mass $m^*(\rho,e)$ of nucleons in nuclear matter is generally less than the bare mass m, and causes the $\lambda(\rho,e)$ to increase, or equivalently the $W_o(\rho,e)$ to decrease.[65,66]

Traditionally the optical potential has been discussed within the framework of Brueckner theory[67]; here we will discuss it using CBPT. The latter is conceptually simple because it uses only low order perturbations. The variational, or 0th order U_v is obtained from:

$$e_v(p) = \frac{p^2}{2m} + U_v(\rho,e) = \langle p|H|p\rangle - \langle 0|H|0\rangle, \tag{6.5}$$

where $|0\rangle$ and $|p\rangle$ are correlated orthonormal states defined in sect. IV.5. The $U_v(\rho_o,e)$ calculated[41] with the LP Hamiltonian provides a fair description of the empirical data as shown in fig. 8. There are no first order corrections in CBPT. In second order the states $|p\rangle$ acquire a lifetime:

$$\frac{1}{\tau(p)} = 2\pi \sum_{p'<p'',h} |\langle p|H|p',p'',h\rangle|^2 \, \delta(e_v(p) + e_v(h) - e_v(p') - e_v(p'')) \tag{6.6}$$

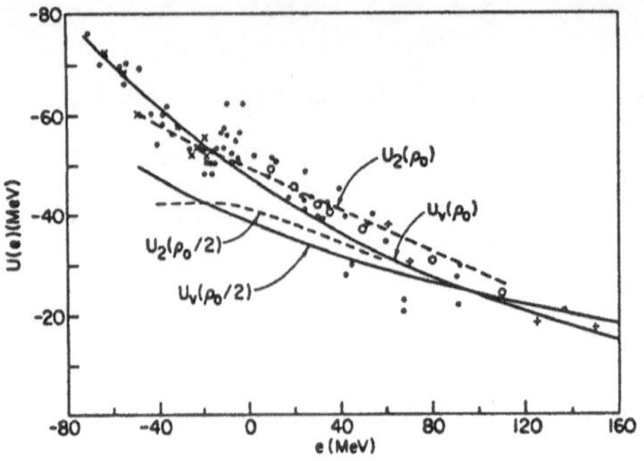

Fig. 8. The real part of the optical potential calculated with variational (full lines) and 2nd order CBPT (dashed lines) methods. The o's give empirical data compiled by Bohr and Mottelson, and the sources of other data are given in ref. 41.

from which the $W_o(\rho_o,e)$ shown in fig. 9 is calculated.[68] The second order terms also give a correction to $U(\rho,e)$[43] which is included in the curves U_2 in fig. 8. The $|U_2-U_v|$ is typically < 5 MeV and hence the CBPT appears to have a good convergence when the correlation operator has the form given by equations 4.8 and 4.32. The empirical values of $U(\rho_o,e)$ and $W_o(\rho_o,e)$ compiled by Bohr and Mottelson[63] are shown in figures 8 and 9 by hollow and filled circles respectively. The theory seems to agree best with these.

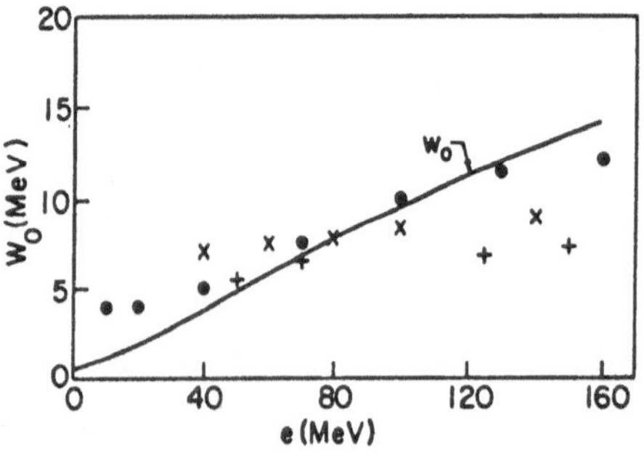

Fig. 9. The imaginary part of the optical potential calculated with 2nd order CBPT. The dots give empirical data compiled by Bohr and Mottelson, and the sources of other data are given in ref. 68.

The $U(\rho,e)$ can as well be expressed as $U(k,\rho)$ where k is the momentum of the particle. Time-dependent mean-field theories of heavy-ion collisions[69] require the $U(k,\rho)$ over a wide range of ρ and k to calculate the motion of particles. Wiringa[70] has calculated the $U_v(k,\rho)$ with the LP, Argonne v_{ij} + V_{ijk}(VII) and Urbana v_{ij} + V_{ijk}(VII) Hamiltonians. His results are shown in fig. 10. Particularly at high densities the $U(k,\rho)$ has a significant model dependence. Calculations of heavy-ion collisions using a $U(k,\rho)$ similar to that given by the LP Hamiltonian have been carried out by Gale et al[71], and they seem to reproduce the observed trends.

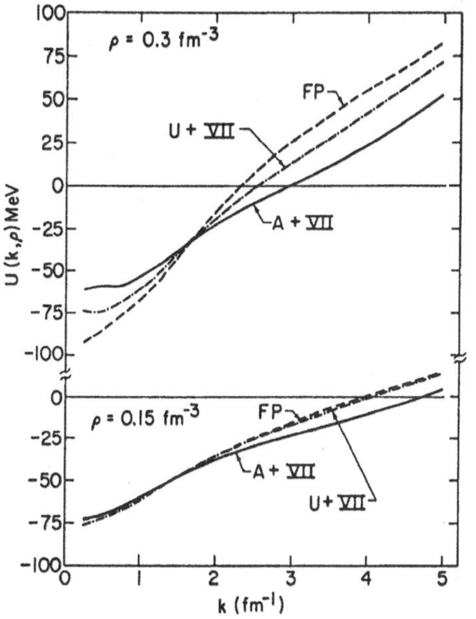

Fig. 10. The real part of the optical potential U(k,ρ) calculated with the variational method using the LP Hamiltonian (FP) and the Urbana and Argonne potentials with model VII of the three-nucleon interaction.

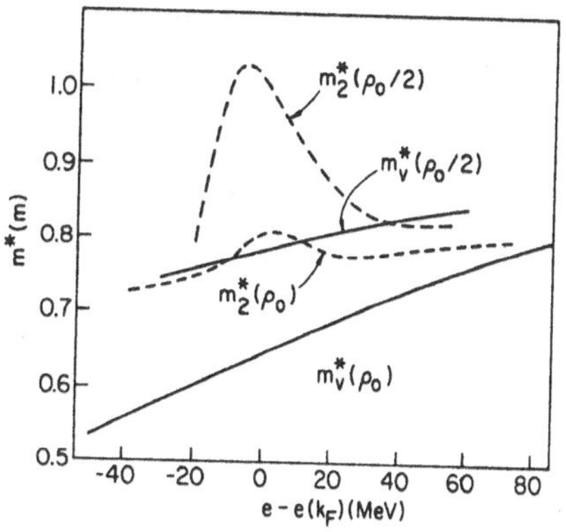

Fig. 11. The effective mass in nuclear matter at densities ρ_o and $\frac{1}{2}\rho_o$. The full and dashed lines give results of variational and 2nd order CBPT calculations.

In relativistic theories[72] using Dirac equation to describe the motion of nucleons in NM, the difference between the m^* and m is attributed to a mean scalar field. Where as in the non-relativistic many-body theories this difference is primarily due to the exchange nature of v_{ij}, i.e. the N-N interaction is observed to be more attractive, on average, in even-ℓ partial waves than in odd-ℓ partial waves. The connection between the two approaches is not very clear.

The second order corrections to $U(\rho,e)$ change sign at the Fermi energy e_F (fig. 8). This implies that the effective mass $m^*(\rho,e)$ has an enhancement over a narrow energy region at e_F (fig. 11). Such an enhancement is observed in nuclei[73] and it is much more pronounced in atomic liquid ^{3}He.[74]

VII. ELECTRON-NUCLEUS SCATTERING

A wealth of nuclear structure information can be extracted from various electron and photon scattering experiments. Here we review a small subset closely related to the many-body theory discussed in sections I-V.

VII.1 ELASTIC SCATTERING FROM FEW-BODY NUCLEI.

Chung et al.[75] find that the relativistic effects, as estimated in Hamiltonian light-front dynamics are quite small in e-d elastic scattering up to $q^2 = 4\text{GeV}^2$, and that the deuteron A and B form factors can be well explained up to large q^2 (fig. 12) with the Argonne potential; the Paris potential does not do too badly either. The main feature is the diffractive minimum in $B(q^2)$ at $q^2 \sim 2\text{GeV}^2$ or $|\vec{q}| \sim 7\text{fm}^{-1}$. The form factors are determined by the Fourier-Bessel transforms of the deuteron wave functions u(r) and w(r) at momenta $|\vec{q}|/2$. In fig. 13 we compare the $u^2(r)$, $w^2(r)$ and $u(r)w(r)$ obtained with the Argonne potential with $j_o(kr)$ for $k=3.5\text{fm}^{-1}$. The $B(q^2)$ at $|\vec{q}| = 7\text{fm}^{-1}$ has a substantial contribution from the r~1fm region where $j_o(3.5r)$ has its first minimum. Thus it appears that the wave functions of nuclear many-body theory have a meaning even when two nucleons are ~1fm apart.

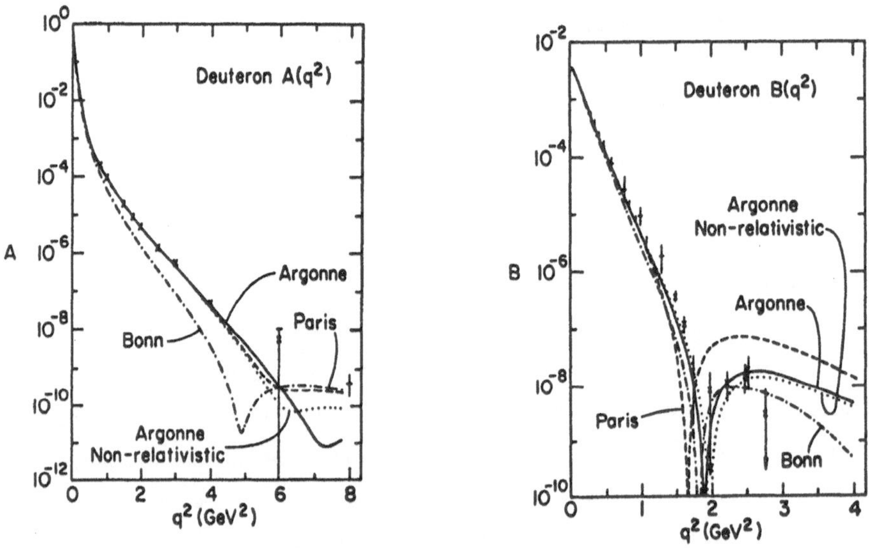

Fig.12. Deuteron A and B form factors.

The pair current and charge operators discussed in section III give significant contributions to the magnetic and charge form factors of ^{3}H and ^{3}He. These contributions are necessary to obtain agreement with the experimental data (fig. 14). A number of terms contribute to the $\vec{j}_2$ and ρ_2 operators[21,22], and their contributions to the isoscalar and isovector form factors are shown in fig. 15. The calculated charge form factor of ^{4}He is compared with the experimental data in fig. 16. Results shown in figures 14-16 use ground state wave functions obtained for the Argonne v_{ij} and V_{ijk} (VII) interactions with the VMC method.

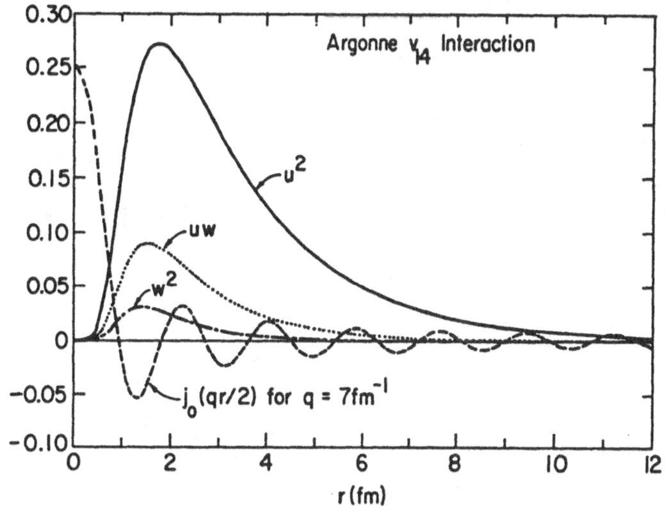

Fig. 13. Deuteron wave functions from Argonne model.

The leading contributions to the isovector magnetic form factor $F_M^V(q)$ of the trinucleons at $q < 6$ fm^{-1}, are from the one-body current $\vec{j}_1$ and pair currents due to the exchange of π-like pseudoscalar (PS) and ρ-like vector (V) mesons. These are uniquely determined from the Argonne v_{ij} by Riska's method, and a reasonable agreement with the data is obtained by just including these contributions. Many of the the other contributions such as those associated with the $\omega\pi\gamma$ and the $\gamma N\Delta$ vertices can not be extracted from the v_{ij}, but they are negligible at $q < 6$fm^{-1}. The $\gamma N\Delta$ contribution becomes important at $q > 6$fm^{-1} where there is no data at present. The observed $F_M^V(q)$ is best reproduced with the Argonne v_{ij} and the IJL[18] or Höler[17] $G_E^V(q^2)$ multiplying the $\vec{j}_2$ (sect. III). The Urbana v_{ij}, which has a much weaker tensor force, fails to reproduce the data at $q > 4$fm^{-1}, while earlier calculations[76] indicate that the data can be explained with the Paris potential if the Dirac form factor $F_1^V(q^2)$ is used in the $\vec{j}_2$ contribution. The G_E and F_1 form factors differ by relativistic terms of order q^2/m^2 (eq. 3.7) and charge conservation requires that G_E^V be used as discussed in section III. Thus one may be tempted to conclude that Argonne v_{ij} provides a more realistic description of the N-N interaction. However it has been shown[77], also using the Paris potential, that the electrodisintegration of the deuteron is better explained when the F_1 rather than G_E is used in the $\vec{j}_2$ contribution. Thus a better understanding of the relativistic q^2/m^2 correction is needed.

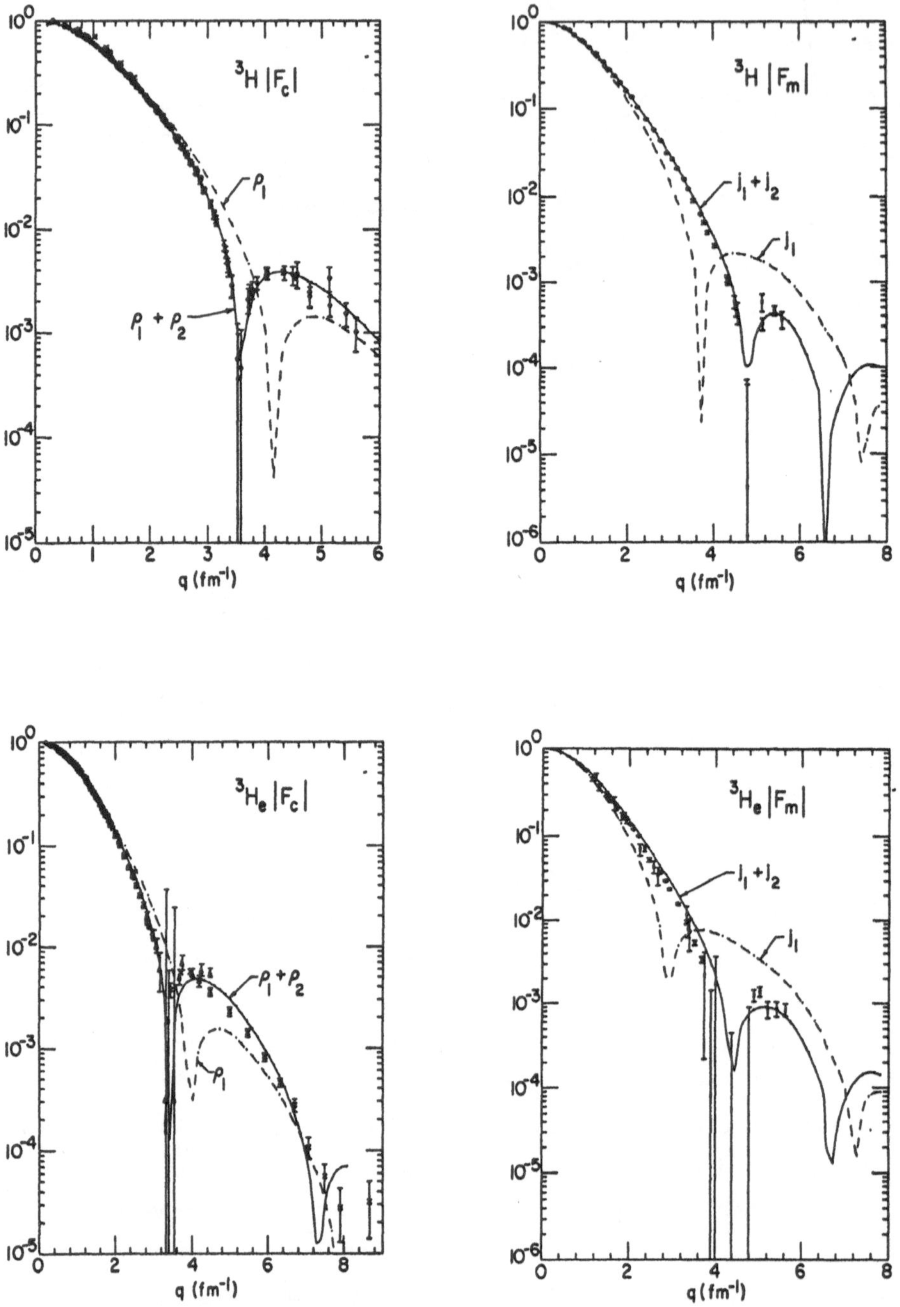

Fig. 14. The charge and magnetic form factors of the trinucleons. The results obtained with one-body and one-plus two-body operators are compared with the experimental data. The wave function and the two-body operators are calculated using the Argonne v_{ij} and model VII V_{ijk}.

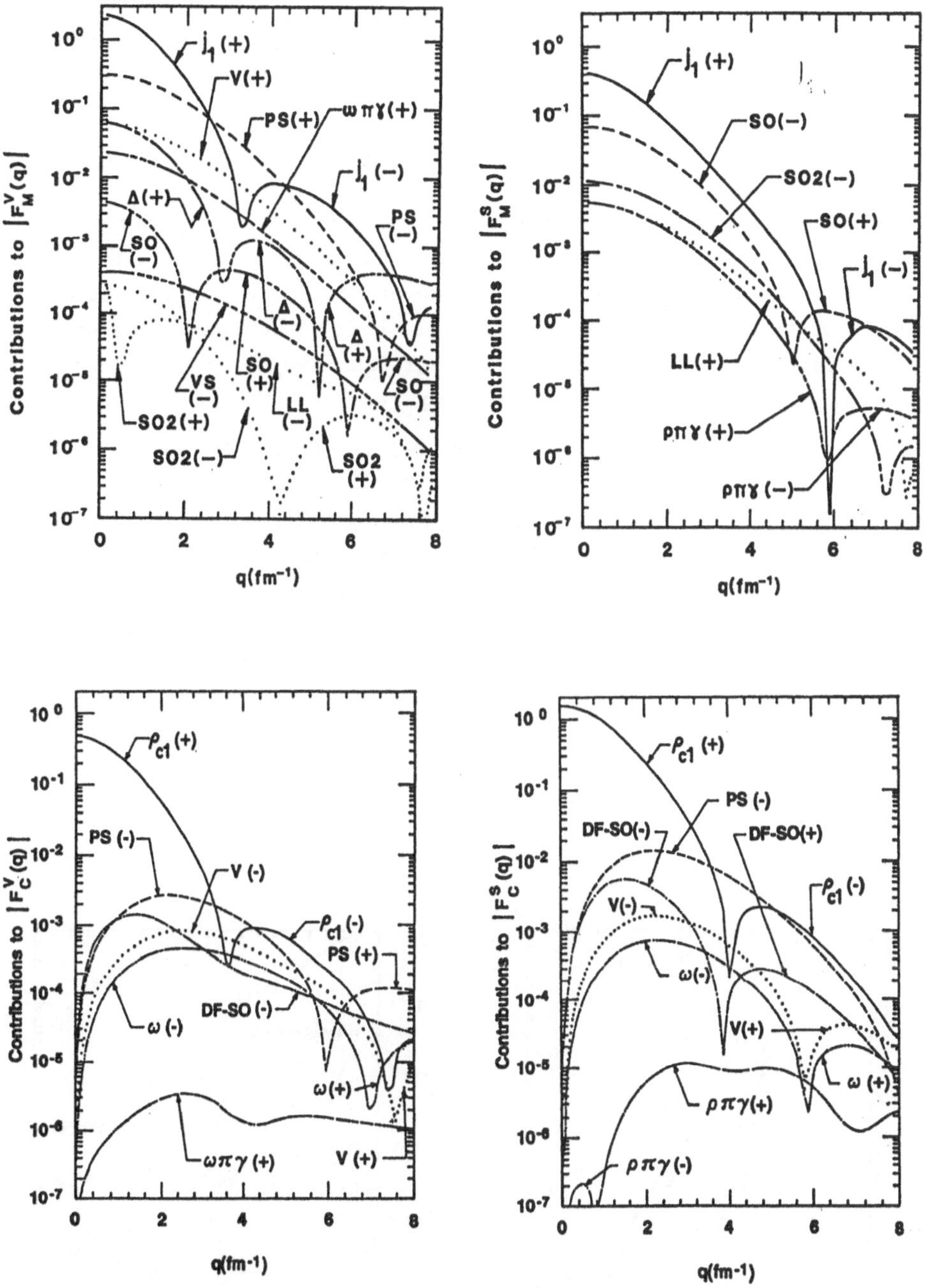

Fig. 15. Contributions to the charge and magnetic form factors of the trinucleons given in fig. 14.

The isoscalar magnetic form factor $F_M^S(q)$ of the trinucleons is dominated by the one-body current $\vec{j}_1$. The pair currents associated with the spin-orbit (SO) and quadratic spin-orbit interactions become important only at $q > 6\text{fm}^{-1}$ (fig. 15). The experimental $F_M^S(q)$ is best explained with $\vec{j}_1$ alone suggesting that the total $\vec{j}_2$ contribution to F_M^S is small. It is possible that the Argonne v_{ij} has too long ranged SO potential. Both Paris and Urbana models have shorter range SO potentials, and the isoscalar $\vec{j}_2$ obtained from them is smaller than that from Argonne v_{ij}. These interactions provide a marginally better description of the observed $F_M^S(q)$.

The isoscalar $\vec{j}_2$ also contributes to the magnetic form factor $B(q^2)$ of the deuteron. Its contribution is not included in the results shown in fig. 12. It is small, but it can account for the difference between Paris $\vec{j}_1$ contribution and the experimental data.[78] Calculations of deuteron form factors and electrodisintegration are being carried out by Schiavilla with the Argonne v_{ij} including $\vec{j}_2$.

The non-relativistic charge density ρ_1, the PS and V meson contributions to ρ_2 and the Darwin-Foldy (DF) and SO corrections to the ρ_1 (eq. 3.6) are the main contributors to the isoscalar and isovector charge form factors F_C^V and F_C^S of the trinucleons (fig. 15). The ρ_2 as well as DF+SO contributions can be regarded as relativistic corrections of order $1/m^2$. In Riska's approach[22] these are determined uniquely by the N-N interaction v_{ij} after making the standard assumptions that $v^{t\tau}$ and $v^{\sigma\tau}$ are due to exchange of isospin T=1 PS and V mesons. Thus there are no free parameters in their calculation, and the experimental data is well reproduced. The isoscalar ρ_2 is more important than the isovector, and it gives large contributions to the F_C of ^{4}He (fig. 16).

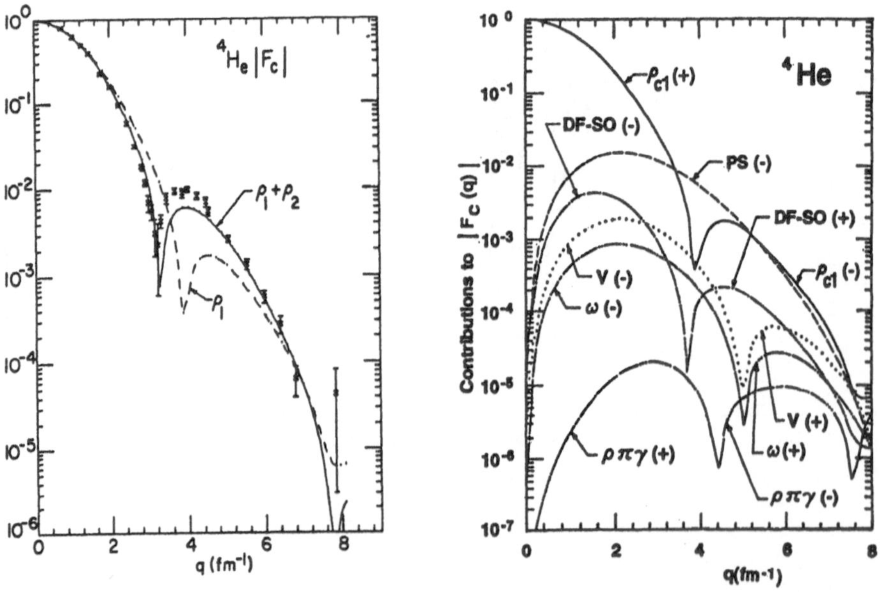

Fig. 16. The charge form factor of ^{4}He is compared with the experimental data in fig. 16a, while its contributions are shown in 16b.

In the one-photon-exchange approximation the cross section for inclusive electron scattering can be written as[79]

$$\frac{d\sigma}{d\Omega dE'} = \sigma_M \left\{ \frac{q^4}{k^4} R_L(\vec{k},\omega) + \left[\tan^2\frac{\theta}{2} - \frac{q^2}{2k^2}\right] R_T(\vec{k},\omega) \right\}, \tag{7.1}$$

$$\sigma_M = \frac{\alpha^2 \cos^2(\theta/2)}{4E^2 \sin^4(\theta/2)}, \tag{7.2}$$

$$q^2 \equiv k^2 - \omega^2. \tag{7.3}$$

In this and the next section we will discuss electron-nucleus scattering at low ($k < 3\text{fm}^{-1}$) momentum transfers. The interaction of the longitudinal and transverse photons with the nucleus is taken as:

$$\rho_L(\vec{k}) = \sum_{i=1,A} e^{i\vec{k}\cdot\vec{r}_i} \frac{1}{2}(1 + \tau_3(i)), \tag{7.4}$$

$$\vec{\rho}_T(\vec{k}) = \frac{1}{m} \sum_{i=1,A} e^{i\vec{k}\cdot\vec{r}_i} \left\{ \frac{1}{2}(1 + \tau_3(i))\, \vec{p}_{i,T} - \frac{i}{2}\left[\mu_p \frac{1}{2}(1 + \tau_3(i)) + \mu_n \frac{1}{2}(1-\tau_3(i))\right] \vec{k} \times \vec{\sigma}_i \right\}, \tag{7.5}$$

ignoring all relativistic and pair contributions. Here $\vec{p}_{i,T}$ is the momentum of nucleon i transverse to $\vec{k}$. In some calculations parts of relativistic corrections of order k^2/m^2 are included in $\rho_L(\vec{k})$, but not very systematically. The response functions $R_x(\vec{k},\omega)$, x = L or T, are given by:

$$R_x(\vec{k},\omega) = \sum_I |\langle I|\rho_x(\vec{k})|O\rangle|^2\, \delta(\omega - E_I + E_o)\, G(k,\omega)^2 \tag{7.6}$$

where $|O\rangle$ and $|I\rangle$ are the ground and excited states of the nucleus at energies E_o and E_I respectively, and $G(k,\omega)$ are the nucleon form factors. The structure functions are obtained from these:

$$S_L(k,\omega) = R_L(k,\omega)/[G(k,\omega)^2 Z], \tag{7.7}$$

$$S_T(k,\omega) = R_T(k,\omega)\Big/\left[G(k,\omega)^2 (Z\mu_p^2 + N\mu_n^2)\frac{k^2}{2m}\right]. \tag{7.8}$$

The covariant dipole approximation[80]

$$G(k,\omega) = (1 + (k^2-\omega^2)/833\ \text{MeV}^2)^{-1} \tag{7.9}$$

is used in most of the results presented here, and the normalization factors in $S_x(k,\omega)$ remove the trivial dependences on the number of neutrons and protons in the target nucleus.

The above equations give only the response due to the nucleon degrees of freedom. At $k < 3\text{fm}^{-1}$ and $\omega < 500$ MeV the longitudinal photons do not excite the nucleons significantly[81], and hence one expects to understand the $S_L(k,\omega)$ in this region from nucleon degrees of freedom alone. The $S_T(k,\omega)$ has a nucleonic contribution $S_T^N(k,\omega)$ that has a peak in the $\omega \sim k^2/2m$ quasi-free region, and other contributions from subnucleonic degrees of freedom, that have a peak at $N \to \Delta$ transition energy.

The sums and energy weighted sums of these responses can be obtained from the ground state wave functions[27]. For example, for the longitudinal response, one has:

$$S_L(k) = \int_{\omega_{el}^+}^{\infty} d\omega \quad S_L(k,\omega)$$
$$= \frac{1}{Z} \quad <0|\ \rho_L^\dagger(k)\ \rho_L(k)\,|0> - (F_L(k))^2, \tag{7.10}$$

$$W_L(k) = \int_{\omega_{el}^+}^{\infty} d\omega \quad \omega\ S_L(k,\omega)$$
$$= \frac{1}{2Z} \quad <0|\,[\rho_L^\dagger(k),\ [H,\rho_L(k)]]\,|0> - \omega_{el}(F_L(k))^2, \tag{7.11}$$

$$W_L^{(2)}(k) = \int_{\omega_{el}^+}^{\infty} d\omega \quad \omega^2\ S_L(k,\omega)$$
$$= \frac{1}{Z} <0|\,[\rho_L^\dagger(k),H]\ [H,\ \rho_L(k)]\ |0> - \omega_{el}^2\ (F_L(k))^2, \tag{7.12}$$

In these expressions $F_L(k)$ is the elastic scattering amplitude normalized to one at $k=0$, and the elastic scattering contribution is omitted from the integrals. The S_L and W_L have been calculated[82,83,84] for ^{2}H, ^{3}H, ^{3}He, ^{4}He and nuclear matter using the variational ground state wave functions, and $W_L^{(2)}$ is also calculated[84] for ^{2}H, ^{3}H, ^{3}He and ^{4}He.

If there are no proton-proton correlations in the nucleus we obtain:

$$S_L^{UC}(k) = 1 - (F_L(k))^2, \tag{7.13}$$

thus the difference $S_L^{UC}(k) - S_L(k)$ provides a measure of the p-p correlations. Similarly when the interactions commute with $\rho_L(k)$,

$$W_L(k) = W_L^K(k) = \frac{k^2}{2m} - \omega_{el}(F_L(k))^2 \tag{7.14}$$

is obtained from the commutator of $\rho_L(k)$ with the kinetic energy. The difference between $W_L(k)$ and $W_L^K(k)$ is mostly due to $\tau_i \cdot \tau_j$ interactions (one-pion-exchange interaction v^π gives a large contribution to $W_L(k)$), and momentum dependent interactions, which do not commute with $\rho_L(k)$. The isospin independent $\vec{L} \cdot \vec{S}$ interaction however does not contribute to

$W_L(k)$. Similar expressions for the sums of the nucleonic part $S_T^N(k,\omega)$ of the transverse response have been given by Schiavilla[85].

Unfortunately, direct comparisons of these sums with experimental data are not possible because $S_L(k,\omega)$ is measured only up to some maximum value ω_{max} and $S_T(k,\omega)$ has contributions due to Δ excitation, etc. We have assumed that

$$S_L(k,\omega>\omega_{max}) = S_L(k,\omega_{max}) \left\{ \lambda e^{-(\omega - \omega_{max})/D_1} + (1-\lambda)\, e^{-(\omega - \omega_{max})/D_2} \right\}, \qquad (7.15)$$

and obtained the parameters λ, D_1 and D_2 from the theoretical values of $W_L(k)$, $W_L^{(2)}(k)$ and the experimental slope of $S_L(k,\omega)$ at ω_{max}. Using the response (7.15) at $\omega>\omega_{max}$, and the experimental $S_L(k,\omega \le \omega_{max})$ the Coulomb sum $S_L(k)$ has been calculated[84] and compared with theory (Fig. 17) for ^2H, ^3H, ^3He and ^4He. The o's in this figure represent the contribution of the observed response at $\omega \le \omega_{max}$, while the •'s include the contribution of the tail (7.15). The tail fortunately does not contribute much to $S_L(k)$ of light nuclei. The data on ^2H, ^3H and ^4He are from Bates[86,87], while those on ^3He are from Saclay[88] and Bates.[86] Since ^2H and ^3H have only one proton their $S_L(k) = S_L^{UC}(k)$, while the $S_L^{UC}(k)$ of ^3He and ^4He is shown by dashed lines. The data is in fair agreement with theory, the $S_L(k)$ of ^3He and ^4He is smaller than $S_L^{UC}(k)$ indicating effects of correlations between the protons in ^3He and ^4He.

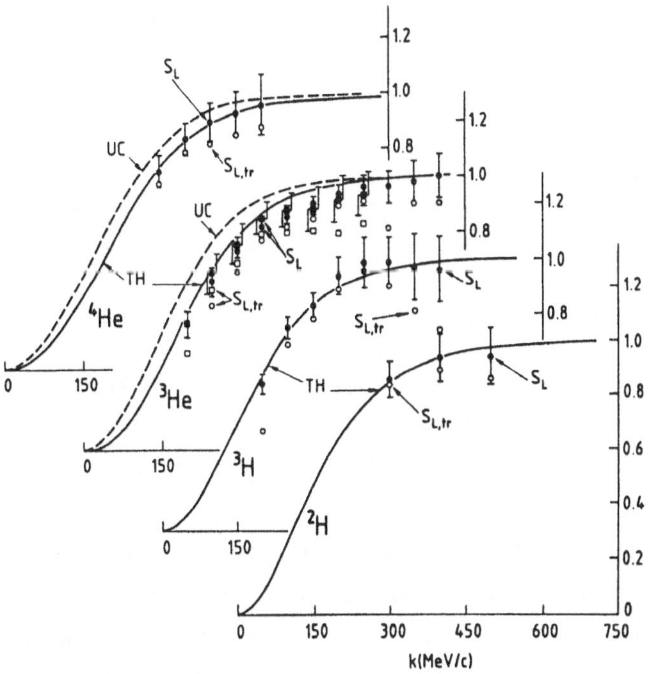

Fig. 17. Coulomb sum rule for light nuclei. The dashed and full lines show results obtained with and without pair correlations. The integrals of the observed response at $\omega<\omega_{max}$ are shown by hollow circles and squares, while the filled circles and squares include the estimated $\omega>\omega_{max}$ tail contribution.

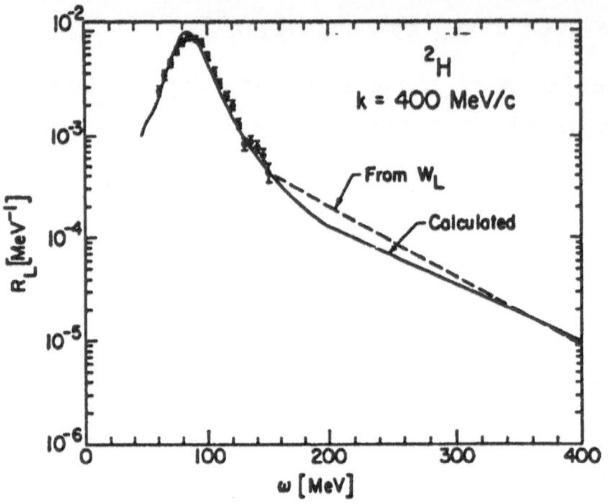

Fig. 18. The experimental (data points) and theoretical (full line) longitudinal response of ^2H. The dashed line shows the tail estimated from the experimental data using the calculated energy weighted sum.

The validity of using $W_L(k)$ to estimate the contribution of the tail has been studied by Arenhovel and Bernstein[89] in the deuteron, where $R_L(k,\omega)$ can be theoretically calculated with good accuracy. They assume for simplicity that $\lambda = 1$ in eq. (7.15), and determine D_1 from $W_L(k)$. The $R_L(k,\omega>\omega_{max})$ obtained from this simple calculation compares well with the accurate theoretical $R_L(k,\omega>\omega_{max})$ as shown in Fig. 18.

Let $\rho_{pp}(r)$ be the probability of finding two protons at a distance r apart in the nucleus. It is called the two-parton distribution function and $\rho_{pp}(k)$ is its Fourier transform normalized to (Z-1) at k = 0. Apart from small corrections of order k^2/m^2 the $\rho_{pp}(k)$ is given by:[82]

$$\rho_{pp}(k) = S_L(k) - 1 - Z|F_L(k)|^2, \tag{7.16}$$

and

$$F_L(k) = F_C(k)/G(k,\omega_{el}). \tag{7.17}$$

Thus the $\rho_{pp}(k)$ can be extracted from the measured $S_L(k)$ and charge form-factor $F_C(k)$. Using the Bates data[86], Beck[90] has extracted the $\rho_{pp}(k)$ in ^3He and compared it with predictions[82] of the Argonne $v_{ij} + V_{ijk}$(VII) model. The experimental $\rho_{pp}(k)$ is smaller by ~ 0.1 in the k = 250 to 550 MeV/c region (Fig. 19). This difference appears also if the Saclay data[88] is used to extract $\rho_{pp}(k)$ and it is related to the fact that the theoretical $S_L(k)$ (Fig. 17) is larger than the experimental $S_L(k)$ by 5-10%. If we assume that the difference between theory and experiment exhibited in Fig. 19 is entirely due to faults in the assumed Argonne v_{ij}, then the analysis indicates that the p-p interaction has a stronger repulsive core than that in Argonne v_{ij}.

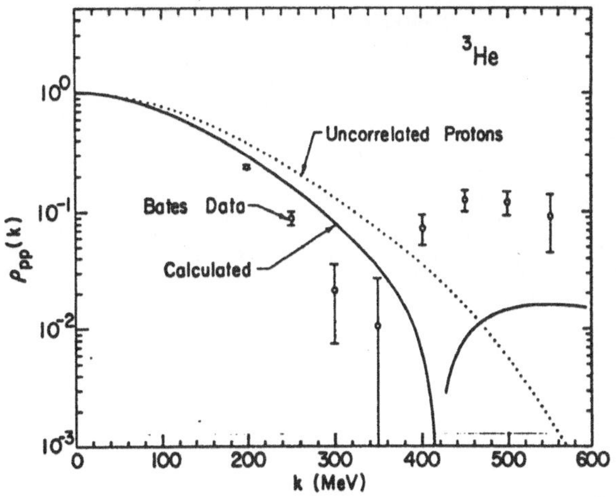

Fig. 19. The calculated two-proton distribution function in ^{3}He is compared with the experimental data and that obtained on neglecting pair correlations.

The $S_L(k)$ and $W_L(k)$ of ^{4}He nucleus and nuclear matter are not too different for k>250MeV/c. The $W_L(k)$ of ^{12}C, ^{40}Ca, ^{48}Ca and ^{56}Fe is estimated from nuclear matter calculations and used to extend the observed response[91,92] of these nuclei beyond ω_{max}. In these nuclei the $\omega > \omega_{max}$ tail of the response gives substantial contribution to $S_L(k)$ (~20%), and hence it is estimated assuming either square or exponential tails[83]. The data are compared with theory in Fig. 20. There appears to be a systematic difference of ~15% at $k > 2fm^{-1}$, particularly for ^{40}Ca. The data have ~15% uncertainties, and Traini[93] has estimated that the observed $S_L(k)$ of ^{40}Ca at k ~ 2.5 fm^{-1} will increase by ~10% on taking into account distortions of electron waves. Thus it is not clear if definite conclusions can be drawn from this comparison; nevertheless this difference between theory and data needs to be investigated further.

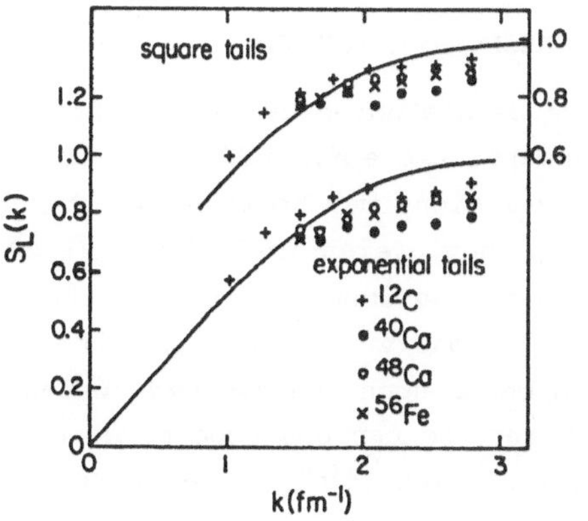

Fig. 20. The Coulomb sum in medium mass nuclei is compared to that of nuclear matter. Two sets of results are obtained by summing the experimental data up to ω_{max} and using either exponential or square tails at $\omega > \omega_{max}$.

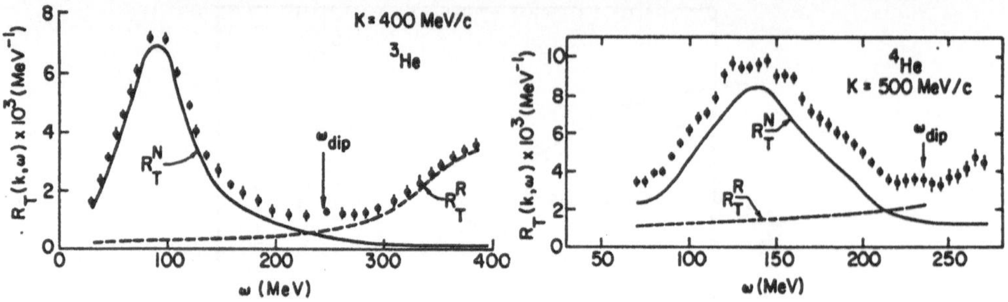

Fig. 21. The transverse response of ^{3}He and ^{4}He separated into nucleonic and sub-nucleonic parts by using sum rules.

As proposed by Nobel[94] and elaborated by Shakin[95], it may be possible to understand the lack of strength in the Coulomb sum $S_L(k)$ if the proton charge radius is changed by its interactions with other nucleons. A 5% increase in the radius increases the experimental $S_L(k)$ by ~ 10% at k_F = 2.5 fm^{-1}; the effect is entirely due to the change in form factor $G(k,\omega)$ relating the $S_L(k,\omega)$ to the observed $R_L(k,\omega)$ (eq. 7.7). However, a y-scaling analysis carried out by Sick[96] suggests that the radii of nucleons in the nucleus are within ~ 3% of those of free nucleons.

It is difficult to extract sums of $S_T^N(k,\omega)$ from the data on $S_T(k,\omega)$ due to the N→Δ peak. Schiavilla[85] has assumed that

$$S_T(k,\omega) = S_T^N(k,\omega) + S_T^R(k,\omega), \tag{7.18}$$

and that $S_T^N(k,\omega > \omega_{dip})$ and $S_T^R(k,\omega < \omega_{dip})$ are given by exponential tails. The tails are determined from the calculated sum $S_T^N(k)$ and energy weighted sum $W_T^N(k)$, and the resulting $R_T^N(k,\omega)$ and $R_T^R(k,\omega)$ are shown in Fig. 21 The contributions of nucleonic and subnucleonic degrees of freedom to R_T seem to be well separated in ^{3}He, but they get mixed in heavier nuclei. Even in ^{4}He there seems to be significant subnucleonic contribution under the quasi-free peak.

VII.3 INCLUSIVE ELECTRON SCATTERING - RESPONSE FUNCTIONS

The response functions $R_X(k,\omega)$ have been studied by many authors using impulse approximation and improvements thereof, and/or effective Hamiltonians. The results of these calculations may be found in the experimental papers.[86,88,91,92] Typically they overestimate the $R_L(k,\omega)$ because the Coulomb sum rule is violated by the impulse approximation. A calculation starting from the bare interactions v_{ij} and V_{ijk} is difficult essentially because $R_X(k,\omega)$ (eq. 7.6) depends upon the excited states $|I\rangle$. Only for the three-body nuclei ^{3}H and ^{3}He can one hope to calculate $R_X(k,\omega)$ exactly by continuum Faddeev calculations.[25,97] Here we review the present status of calculations using correlated basis perturbation theory.

Let us consider for simplicity the response of ^{3}He. In this case there are three types of final states |I>: (i) ^{3}He nucleus with momentum $\vec{k}$ (contributes only to elastic scattering); (ii) d+p states; and (iii) three-body continuum states. A plane wave approximation of these states gives the correct energy E_I, but it is not useful because the wave functions at small distance are wrong, and the three sets of states are not orthogonal to each other. As a matter of fact the last set (iii) includes both (i) and (ii).

It is relatively simple to include short range correlations in the final states[98] via pair correlation operators F_{ij} (eq. 4.8). Parts of the long range correlations in d-p states can also be included via a d-p potential.[99] Such correlated final states have improved behavior at small distances and the correct energies E_I, but they are not necessarily orthogonal. Their correct energies have to be preserved while orthogonalizing them.[98] This can be done by first projecting out the ^{3}He state of the d+p states and then orthogonalizing the d+p states to each other via a Löwdin transformation. Next the ^{3}He and the d+p states are projected out of the p+p+n states and then the p+p+n states are orthogonalized. Use of an orthonormal set |I> ensures that the total strength in the response, such as the Coulomb sum, is correctly treated.

Only the 0th order correlated basis calculations have been carried out for the ^{3}He and ^{3}H response functions[99], and these too contain some approximations in the computational part. Their results are compared with the experimental data in Fig. 22. The d+p final states are most carefully treated; the R_L of ^{3}He is dominated by them and hence well described by the theory. The n+p+p states are relatively crudely treated and they dominate the R_L of ^{3}H. Naturally the R_L of ^{3}H is not well described by the theory. It seems that the general features of the response are reproduced, but a quantitative understanding is not at hand.

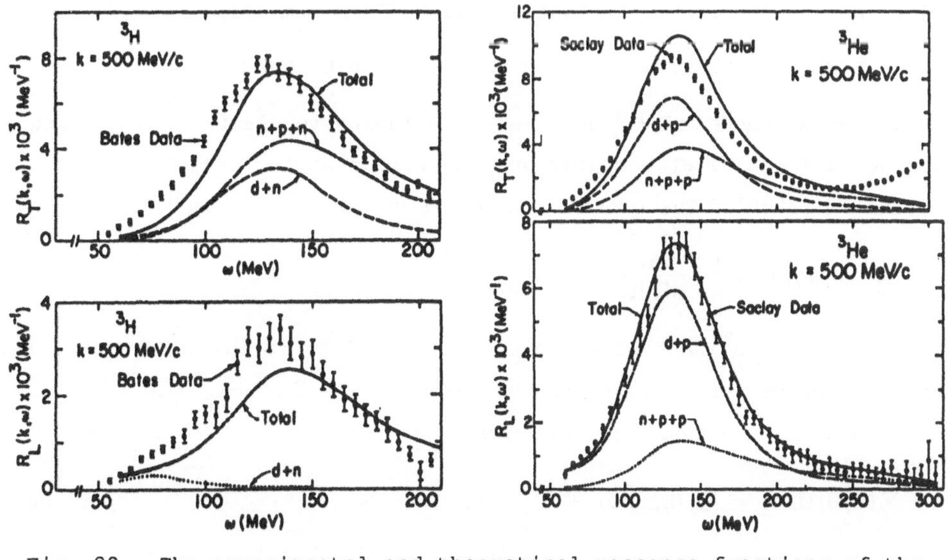

Fig. 22. The experimental and theoretical response functions of the trinucleons. The contributions of two-body (d+p and d+n) and three-body final states are shown separately.

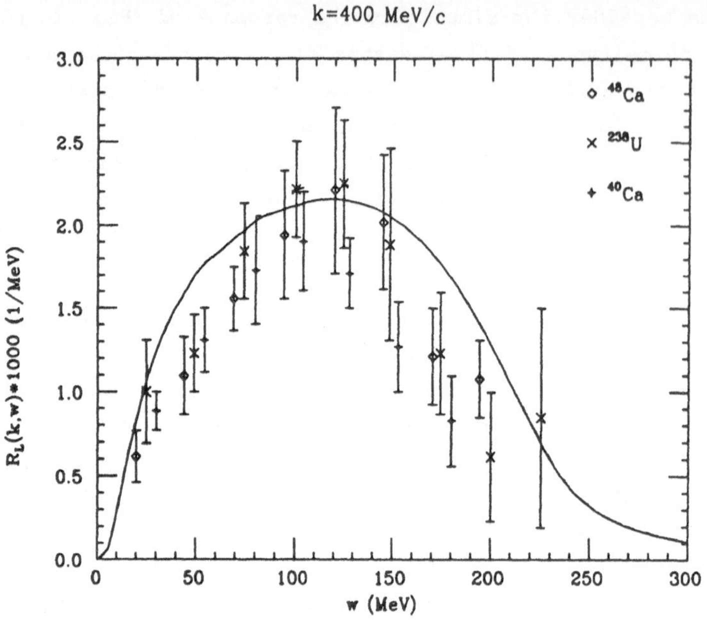

Fig. 23. The longitudinal response of nuclear matter at k = 400 MeV/c is compared with the experimental data.

The longitudinal response of nuclear matter has been studied using correlated orthogonal states[100,101] (Sect. IV.5). The $|ph>$ and $|p_1p_2h_1h_2>$ states are included in the response, and the coupling between $|ph>$ states is treated exactly while that between $|ph>$ and $|p_1p_2h_1h_2>$ states is treated perturbatively. The typical results are compard with the data on ^{40}Ca and ^{48}Ca from ref. 92 and on ^{238}U from ref. 102 in Fig. 23. The measured response is generally smaller than the calculated, the difference is comparable to that in the Coulomb sum (Fig. 20).

VII.4 INCLUSIVE ELECTRON SCATTERING AT HIGHER ENERGIES

Recently Day et al.[103] have measured inclusive electron scattering from several nuclei at energies and scattering angles ranging from 2 to 4 GeV and 15 to 30°. The cross section per nucleon is defined as:

$$\Sigma(E,\theta,\omega,A) \equiv \frac{1}{A}\frac{d\sigma}{d\Omega d\omega}, \qquad (7.19)$$

where E is the incident electron energy, $\omega = E-E'$ is the energy transfer and A the number of nucleons in the target. The Σ of ^{12}C, ^{27}Al, ^{56}Fe and ^{197}Au has a liquid-drop type behavior:

$$\Sigma(E,\theta,\omega,A) = \Sigma_V(E,\theta,\omega) + \Sigma_S(E,\theta,\omega)A^{-1/3} + \ldots \qquad (7.20)$$

The volume term Σ_V can be identified as the response of nuclear matter and Σ_S is the surface correction. These two terms can be well determined from the data; higher terms such as the dependence of Σ on the neutron excess (N-Z)/A cannot yet be extracted.

The momentum transfer $\vec{q}$ is given by:

$$|\vec{q}|^2 = E^2 + E'^2 - 2EE' \cos\theta; \tag{7.21}$$

it is typically 3 to 11 fm^{-1} in these experiments, and $\omega \sim 200$ MeV to 2 GeV. Thus the non-relativistic theory[100,101,104], keeping only the nucleon degrees of freedom, is not adequate to describe them. Here we briefly review an approach taken by Benhar et al.[105] in which the nuclear ground state is calculated with nonrelativistic potentials, but the scattering is treated relativistically including all degrees of freedom.

The spectral function of nuclear matter:

$$P(k,\nu) = \sum_I |\langle I|a_k|0\rangle|^2 \, \delta(E_I+\nu-E_o) \tag{7.22}$$

plays a central role in this approach. Here a_k annihilates a nucleon of momentum k and $|I\rangle$ are states of the system with A-1 nucleons. The $P(k,\nu)$ can be considered as the probability of having a nucleon of energy ν and momentum k in nuclear matter. For example, it is easy to verify that

$$P_{FG}(k,\nu) = \delta(\nu-k^2/2m) \, \theta(k_F-k) \tag{7.23}$$

for ideal nonrelativistic Fermi gas. In nuclear matter, since $E_I-E_o > 16$ MeV the $P(k,\nu)$ is nonzero only for $\omega < -16$ MeV. It has been calculated by Benhar, Fabrocini and Fantoni[106] using correlated basis theory and is shown in Fig. 24 for a couple of values of k.

When $k < k_F$ the $P(k,\nu)$ has a sharp peak at the energy of the one-hole state $|h\rangle$ with h = k. The width of the peak is due to the life-time of $|h\rangle$. The $P(k,\nu)$ also has a tail extending to large negative values of ν due to the overlap of $a_k|0\rangle$ with states like $|h_1p_2p_1\rangle$ etc.

When $k > k_F$ the $P(k,\nu)$ has a broad peak at an energy of $\sim -k^2/2m$. It arises due to ground state pair correlations. If we assume that the total momentum of the pair is zero, then the particles of the pair can be in states with momenta $-\vec{k}$ and $+\vec{k}$ with $k > k_F$ due to correlations. On annihilating the particle with momentum $+\vec{k}$ one is left with a $|h_1h_2p_1\rangle$ state in which $\vec{p}_1 \sim -\vec{k}$, and its energy is $\sim k^2/2m$.

The momentum distribution of nucleons in nuclear matter is given by:

$$n(k) = \int_{-\infty}^{-16} P(k,\nu)\,d\nu, \tag{7.24}$$

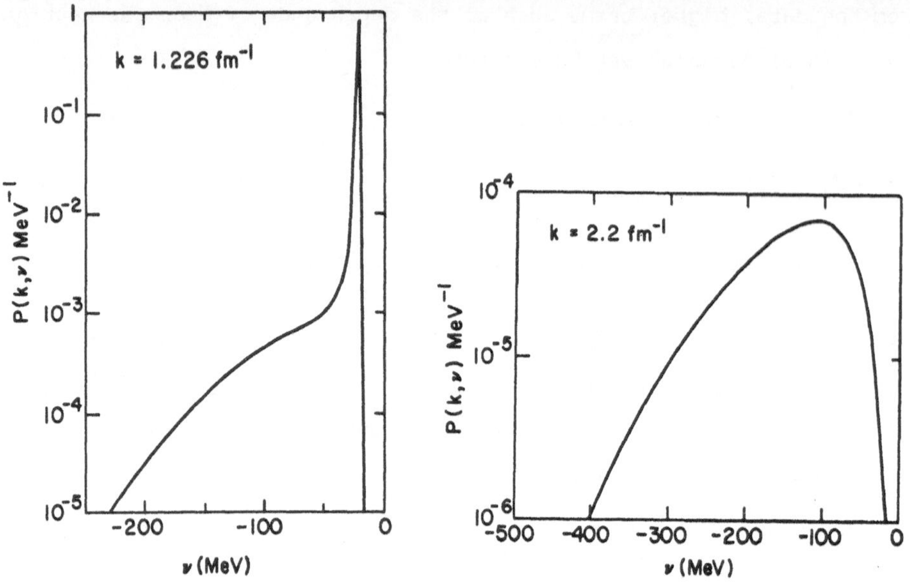

Fig. 24. $P(k,\nu)$ at $k = 1.226 \mathrm{fm}^{-1}$ ($<k_F$) and $k = 2.2\ \mathrm{fm}^{-1}$ ($> k_F$) in nuclear matter at $k_F = 1.33\ \mathrm{fm}^{-1}$.

It can be calculated directly from the density matrix[107] and it is shown in Fig. 25. Approximately 20% of the particles have momenta $> k_F$ in nuclear matter. The depletion of $n(k < k_F)$ is the familiar convergence parameter κ of Brueckner-Bethe theory of nuclear matter.[28]

In relativistic theories the inclusive electron scattering cross section on a nucleon is given by[108,109]:

$$\frac{d\sigma_N}{d\Omega d\omega} = \frac{E'}{E} \left(\frac{e^2}{2\pi q^2}\right)^2 L^{\mu\nu} W_{\mu\nu}, \tag{7.25}$$

$$L^{\mu\nu} = \frac{1}{2} \left(l^\mu l'^\nu + l^\nu l'^\mu - g^{\mu\nu} l \cdot l'\right), \tag{7.26}$$

$$W_{\mu\nu} = W_1 \left(\frac{q_\mu q_\nu}{q^2} - g_{\mu\nu}\right) + \frac{W_2}{m^2} \left\{p_\mu - \frac{p\cdot q}{q^2} q_\mu\right\} \left\{p_\nu - \frac{p\cdot q}{q^2} q_\nu\right\}, \tag{7.27}$$

where p, l and l' are the four momenta of the initial nucleon, electron and the scattered electron (Fig. 26), q is the four momentum transfer, and $L^{\mu\nu}$ and $W_{\mu\nu}$ are the lepton and hadron tensor. The scattering from nuclear matter is given in the impulse approximation by:

$$\frac{d\sigma}{d\Omega d\omega} = \frac{E'}{E} \left(\frac{e^2}{2\pi q^2}\right)^2 \int \frac{d^4p}{(2\pi)^4} P(p)\ L^{\mu\nu} W_{\mu\nu}(p,q), \tag{7.28}$$

where P(p) is the probability of finding a nucleon with four momentum p in nuclear matter. It is obtained from the spectral function:

$$P(p) = P(\vec{p}, p_o) = P(\vec{k} = \vec{p},\ \nu = p_o - m). \tag{7.29}$$

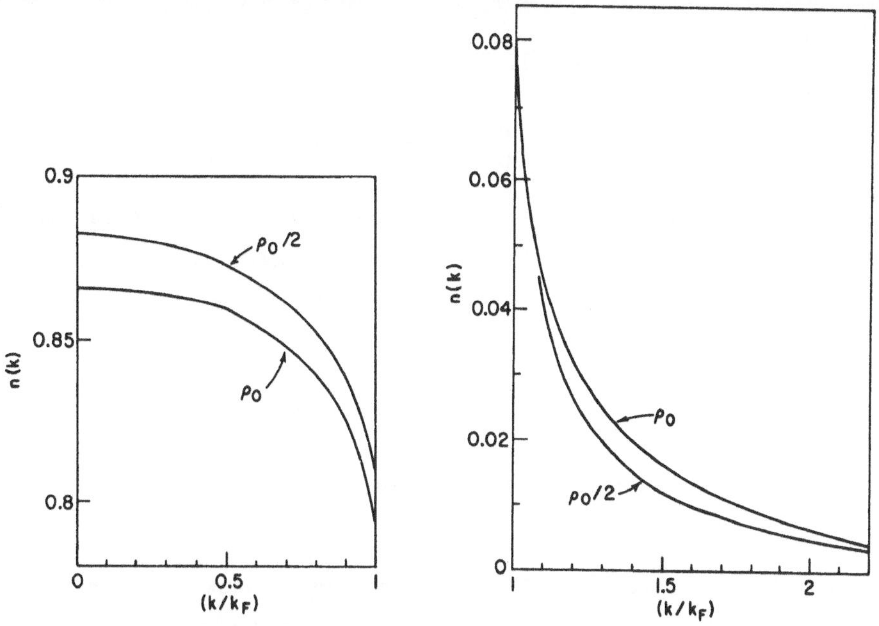

Fig.25. The momentum distribution of nucleons in nuclear matter at equilibrium density $\rho_o = 0.16\ \mathrm{fm}^{-3}$ and at $\rho_o/2$.

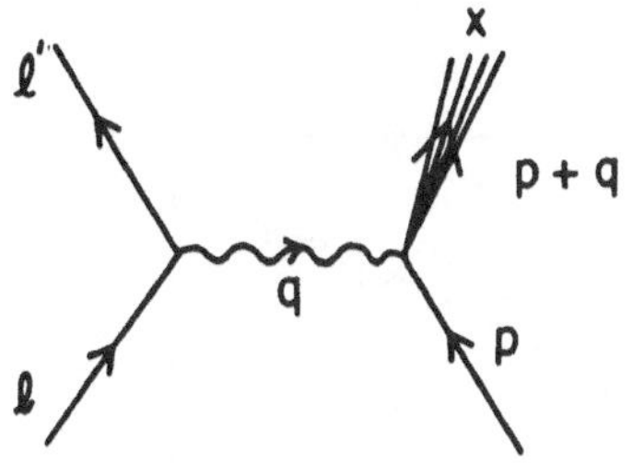

Fig.26. Electron-nucleon scattering.

It is necessary to know the structure functions W_1 and W_2 to evaluate the integral (7.28). These are functions of Lorentz scalars formed with p and q. In studies of electron-nucleon scattering[109] the initial nucleon is free (or it is the almost free neutron in the deuteron), and hence $p^2 = m^2$. Thus q^2 and $p\cdot q$ are the only two scalars available, and W_1 and W_2 have been tabulated as functions of q^2 and $p\cdot q$.[109] In the impulse approximation (IA) the initial nucleon does not have $p^2 = m^2$; however when the struck hadron X is a nucleon it has

$$s = (p+q)^2 = m^2 \tag{7.30}$$

because the interactions between the final hadronic state X with the state $|I\rangle$ of the A-1 nucleons are neglected. Thus the contribution of the X = N states to W_1 and W_2 can be taken as:

$$W_1^{X=N} = -\frac{q^2}{2m} G_M^2(q^2)\ \delta(s-m^2), \tag{7.31}$$

$$W_2^{X=N} = \frac{2m}{1-q^2/4m^2} [G_E^2(q^2) - \frac{q^2}{4m^2} G_M^2(q^2)]\ \delta(s-m^2). \tag{7.32}$$

In this region of q the Δ and other baryon resonances dominate the $X \neq N$ response of free nucleons.[109] These resonances also have well-defined values of s such as m_Δ^2 etc. Hence in IA calculations we can consider W_1 and W_2 as function of q^2 and s:

$$W_{1,2}(q^2,s) = W_{1,2}^{free}(q^2, p\cdot q = \frac{1}{2}(s-m^2-q^2)), \tag{7.33}$$

where $W_{1,2}^{free}$ are taken from experiments on protons and deuterons.[109]

The Σ_V (E = 3.6 GeV, θ = 25°, ω) calculated[105] with the IA is compared with the data[103] in Fig. 27. The figure shows the contributions of $X \neq N$ states as well as the total. As expected the X = N states dominate at low energies, and the two contributions become equal at $\omega \sim$ 1.1 GeV. The large cross section at $\omega > 0.7$ GeV is well reproduced by the theory, but the small cross-section at small ω is underestimated by a factor of ~ 3.

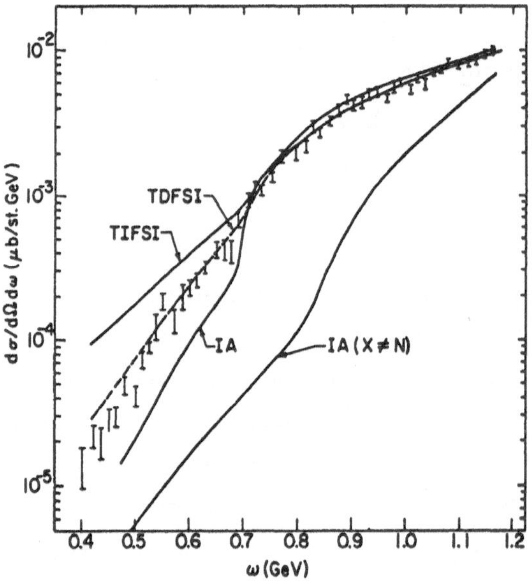

Fig. 27. The experimental $\Sigma_V(\omega)$ at E = 3.6 GeV and θ = 25° is compared with the predictions of impulse approximation (IA), IA and time independent (TIFSI) and time dependent (TDFSI) final state interactions.

Final state interactions (FSI) can have a relatively large effect on the $\sigma(q,\omega)$ at small ω where the $\sigma_{IA}(q,\omega)$ is small. The $\sigma(q,\omega)$ is proportional to:

$$\sigma(q,\omega) \propto \sum_f |\langle f|\gamma_q|0\rangle|^2\, \delta(E_f-E_o-\omega), \tag{7.34}$$

where $|f\rangle$ are the final states of the nucleus and γ_q represents the interaction of the photon. It is convenient to express Eq. (7.31) as a Compton scattering amplitude:

$$\sigma(q,\omega) \propto 2\mathrm{Re} \int_0^\infty e^{i\omega t}\, dt\, \langle 0|\gamma_q^{\dagger} e^{-iHt}\gamma_q|0\rangle, \tag{7.35}$$

$$H = H_o + H_I, \tag{7.36}$$

where H_o is the sum of the Hamiltonians of the (A-1) nucleons (whose eigenstates are $|I\rangle$) and the struck nucleon (whose eigenstates are $|X\rangle$), and H_I represents the interaction between $|I\rangle$ and $|X\rangle$. The IA is obtained by neglecting H_I so that

$$\sigma_{IA}(q,\omega') \propto 2\mathrm{Re} \int_0^\infty e^{i\omega t}\, dt\, \langle 0|\gamma_q^{\dagger} e^{-iH_o t}\gamma_q|0\rangle \tag{7.37}$$

$$\therefore \langle 0|\gamma_q^{\dagger} e^{-iH_o t}\gamma_q|0\rangle \propto \frac{1}{2\pi} \int_{-\infty}^{\infty} d\omega'\, e^{-i\omega' t}\, \sigma_{IA}(q,\omega'), \tag{7.38}$$

where ω' is the energy of the intermediate state $|I,X\rangle$.

At small ω the response is dominated by X = N states having a high momentum nucleon travelling in nuclear matter. The shift V_R in the real part of its energy is small and positive. Dirac optical model fits to the scattering of 0.5 to 1 GeV protons[72] suggest that $V_R \sim 25$ MeV and has a small effect on the response. The main effect is due to the damping of its motion, and on including it we may approximate:

$$\langle 0|\gamma_q^{\dagger} e^{-iHt}\gamma_q|0\rangle \propto \frac{1}{2\pi} \int_{-\infty}^{\infty} d\omega'\, e^{-i(\omega'+V_R)t}\, e^{-t/2\lambda}\, \sigma_{IA}(q,\omega'). \tag{7.39}$$

Substituting the above in Eq. (7.31) one obtains:

$$\sigma(q,\omega) = \int_0^\infty F(\omega-\omega')\, \sigma_{IA}(q,\omega'-V_R)\, d\omega', \tag{7.40}$$

$$F(\omega-\omega') = \frac{1}{\pi}\, \mathrm{Re} \int_0^\infty e^{i(\omega-\omega')t}\, e^{-t/2\lambda}\, dt \tag{7.41}$$

so that the effect of FSI is approximated by folding the $\sigma_{IA}(q,\omega')$ with the function $F(\omega-\omega')$.

The simplest approximation is to take the mean free path λ as

$$\lambda^{-1} = \rho\sigma v, \tag{7.42}$$

where v is the velocity of the struck nucleon, σ is the $NN \rightarrow NN$ cross section and ρ the density of nuclear matter. For the case under consideration ($E=3.6$ GeV, $\theta = 25°$) $v \sim 0.87c$, $\sigma \sim 30$ mb and $\rho \sim 0.16$ fm^{-3} giving $\lambda = 2.4$ fm. The folding function $F_L(\omega-\omega')$ obtained in this approximation is a Lorentzian:

$$F_L(\omega-\omega') = \frac{1}{2\pi}\frac{1}{\lambda}\frac{1}{(1/4\lambda^2+\omega-\omega')^2}\,; \tag{7.43}$$

which has unrealistically long tails.

A better approximation, used in the theory of deep inelastic neutron scattering by helium liquids[110], is obtained by recognizing that the density of liquid at a distance r from a particle in the liquid is not the equilibrium density ρ, but is given by $g(r)\rho$ where $g(r)$ is the pair distribution function.[82] This gives

$$\lambda^{-1}(t) = g(vt)\sigma v, \tag{7.44}$$

and the $F_g(\omega-\omega')$ obtained from it does not have long tails. The Σ obtained in this approximation is shown by the curve TIFSI (time independent FSI) in Fig. 27. It is in excellent agreement with data $\omega > 0.7$ GeV, but it is ~ 3 times the data at $\omega \sim 0.5$ GeV. The effect of FSI calculated in this approximation seems to be too large.

It is possible that this approximation is too crude, and a much better agreement with the data can be obtained within the conventional approach by improving upon it. Nevertheless, in a highly speculative fashion, one could consider other effects that may reduce the $\Sigma(q,\omega)$ at low ω. The effect that is rather simple to consider in the present formalism is that the effective $NN \rightarrow NN$ cross section may depend upon the time t due to quark effects. A quark in the struck nucleon absorbs the photon at $t = 0$. Thus at small t the system has a high momentum quark moving essentially inside the struck nucleon. It is only after time T of order ~ 1/2 fm, in the rest frame of the struck hadron, that the description in terms of a hadron moving in matter may be applicable. Thus it may be more appropriate to consider σ as a function of t,

$$\sigma(t) = \sigma(1-e^{-t/\beta T}), \tag{7.45}$$

where $\beta = (1-v^2/c^2)^{-1/2}$ is the time dilation factor. In this approximation one obtains the curve TDFSI (time-dependent FSI) that is in fair agreement with the data (Fig. 27).

The argument used above is similar to that used in the phenomenon[111] called "color transparency". There it has been argued that elastic electron-nucleon scattering at large values of $\vec{q}$ occurs when the nucleon is in a point-like configuration. It has very weak interactions with the other nucleons immediately after absorbing the photon. It expands with a time scale of ~ 1/2 fm and recovers its normal properties. In this approach at $t/\beta \lesssim 1/2$ fm the struck hadron has to be considered as a superposition of many hadronic states necessary to produce a point-like configuration. In any case at $t/\beta \lesssim 1/2$ fm the struck hadron cannot be well approximated by a moving nucleon, and thus it may be incorrect to use NN → NN cross sections to estimate its interaction in matter.

VII.5 EXCLUSIVE e,e'p REACTIONS

The quasi-particle or overlap orbital $\chi_\alpha(x)$ is defined as:

$$\chi_\alpha(x) = \langle (A-1)_\alpha \, | a(x) | A \rangle, \quad (7.46)$$

where $|A\rangle$ is the ground state of a nucleus with A nucleons and $|(A-1)_\alpha\rangle$ is a state of the nucleus with (A-1) nucleons. The χ_α is labeled with the quantum numbers α of the state of the (A-1) nucleus, and the renormalization, or the spectroscopic factor is given by:

$$Z_\alpha = \langle \chi_\alpha(x) | \chi_\alpha(x) \rangle \quad (7.47)$$

In mean field theories $\chi_\alpha(x)$ is close to the mean field single-particle wave function $\phi_\alpha(x)$ and Z_α is close to unity. The correlations[112] make $\chi_\alpha(x) \neq \phi_\alpha(x)$ and also reduce Z_α. In principle the exclusive e,e'p reaction is sensitive to both the shape of $\chi_\alpha(x)$ and the Z_α, and thus can provide interesting information on nuclear wave functions.[113]

Theoretically, starting from realistic nuclear forces, the $\chi_\alpha(x)$ and Z_α have been calculated for ^{3}He and ^{4}He (ref. 29) and for nuclear matter.[106] In nuclear matter the quasi-hole states α can be labeled with their momentum k. The spectral function of nucleons having $k<k_F$ has a peak due to the contribution of h=k states (fig. 24). As k approaches k_F from below this peak gets narrow and distinct. It is then associated with the Landau's quasi-hole state, and its integral equals Z(k). Defining

$$k_{F\pm} = k_F \pm \varepsilon \quad \lim \varepsilon \to 0^+, \quad (7.48)$$

Migdal[114] has shown that

$$Z \equiv Z(k_{F-}) = n(k_{F-}) - n(k_{F+}), \quad (7.49)$$

for normal Fermi liquids.

The Z is also related to the enhancement of the effective mass at k_F. Mahaux and Sartor[115] have estimates Z_α in ^{208}Pb with a dispersion theory approach starting from empirical optical potentials. They find that near the Fermi energy Z~0.6±0.1. The observed quenching[116] of the single particle contributions to various observables in the lead region also suggest that Z~0.6 in ^{208}Pb.

In absence of FSI the e,e'p reaction would provide a direct measure of $\chi_\alpha(x)$ and Z_α. Unfortunately the FSI are not negligible even in a small nucleus like ^{4}He (ref. 117). Much of the uncertainties in the Z's deduced from e,e'p reactions comes from the treatment of the FSI.[118]

The present experimental and theoretical situation is reviewed by DeWitt Huberts.[113] The experimental (theoretical) values of Z at E_F decrease from 0.8 to 0.5 (0.8 to 0.6) from the helium to lead nuclei. The energy dependence of Z in ^{208}Pb has also been studied. The experimental (theoretical) Z(E) increases from ~0.45 (0.55) to 0.8 as E decreases from E_F to -30 MeV. Thus there is, at present a decent agreement between experiment and theory. Nevertheless it should be pointed out that fully consistent calculations of Z at E_F have only been carried out for ^{3}He, ^{4}He and nuclear matter. The calculated value of $Z(E_F)$ in nuclear matter is ~0.7, and it is estimated[119] that it will decrease to ~0.55 in ^{208}Pb by surface effects.

ACKNOWLEDGEMENTS

I would very much like to thank O. Benhar, J. Carlson, A. Fabrocini, S. Fantoni, S. Pieper, D. Riska, R. Schiavilla and R. Wiringa for illuminating discussions and substantial contributions to the theory and results presented here. This work is supported by the US National Science Foundation via grant PHY-84-15064.

REFERENCES

1. V. R. Pandharipande and D. G. Ravenhall, In Proc. of Les Houches Winter School 1989, Ed. M. Soyeur, Plenum press.
2. R. B. Wiringa, V. Fiks and A. Fabrocini, Phys. Rev. C38, 1010 (1988).
3. R. A. Aziz, et al., J. Chem. Phys. 70, 4330 (1979). M. H. Kalos, et al., Phys. Rev. B24, 115 (1981)
4. I. E. Lagaris and V. R. Pandharipande, Nucl. Phys. A359, 331 (1981).
5. R. B. Wiringa, R. A. Smith and T. L. Ainsworth, Phys. Rev. C19, 1207 (1984).
6. M. Lecombe, et al., Phys. Rev. C21, 861 (1980).
7. B. D. Day and R. B. Wiringa, Phys. Rev. C32, 1057 (1985).
8. J. R. Bergervoet, et al., Phys. Rev. Lett. 59, 2255 (1987).
9. A. W. Thomas, In this volume.

10. W. N. Cottingham, et al., Phys. Rev. D8, 800 (1973).
11. R. Machleidt, K. Holinde and Ch. Elster, Phys. Repts. 149, 1 (1987).
12. I. E. Lagaris and V. R. Pandharipande, Nucl. Phys. A359, 349 (1981).
13. B. D. Serot and J. D. Walecka, Adv. in Nucl. Phys. 16, 1 (1986).
14. D. O. Riska, Phys. Scrip. 31, 107 (1985) and 31, 471 (1985).
15. F. Bergsma, et al., Phys. Lett. 123B, 269 (1983).
16. J. D. Bjorken and S. D. Drell, Relativistic Quantum Mechanics, McGraw-Hill (1964).
17. G. Höhler, et al., Nucl. Phys. B114, 505 (1976).
18. F. Iachello, A. D. Jackson and A. Landè, Phys. Lett. B43, 191 (1973).
19. M. Gari and W. Krümpelmann, Phys. Lett. B173, 10 (1986).
20. J. F. Mathiot, Phys. Repts. 173, 63 (1989).
21. R. Schiavilla, V. R. Pandharipande, and D. O. Riska, Phys. Rev. C, in press (1989).
22. R. Schiavilla, V. R. Pandharipande and D. O. Riska, Phys. Rev. C, in press (1989).
23. K. Ohta, Nucl. Phys. A495, 564 (1989).
24. C. R. Chen, et al., Phys. Rev. C33, 1740 (1986). T. Sasakawa and S. Ishikawa; Few-Body Sys. 1, 3 (1986).
25. H. Witala, T. Cornelius and W. G. Glöckle, Few-Body Syst. 3, 123 (1988).
26. J. Carlson, Phys. Rev. C36, 2026 (1987) and C38, 1879 (1988).
27. E. Feenberg, Theory of Quantum Liquids, Academic Press (1969).
28. C. Mahaux, In Proc. of Les Houches Winter School 1989, Ed. M. Soyeur, Plenum Press.
29. R. Schiavilla, V. R. Pandharipande and R. B. Wiringa, Nucl. Phys. A449, 219 (1986).
30. R. B. Wiringa, private comm. (1989).
31. J. Lomitz-Adler and V. R. Pandharipande, Nucl. Phys. A342, 404 (1980).
32. M. H. Kalos and P. A. Whitlock, Monte Carlo Methods, Vol. 1 Basics, J. Wiley & Sons (1986).
33. J. Carlson, In Electron-Nucleus Scattering, eds. A. Fabrocini, et al., World Scientific 1989.
34. R. V. Reid, Ann. of Phys. 50, 411 (1968).
35. V. R. Pandharipande, In Proc. of NATO Summer School on the Nuclear Equation of State, Editor, W. Greiner, Plenum press (1989).
36. V. R. Pandharipande and R. B. Wiringa, Rev. Mod. Phys. 51, 831 (1979).
37. S. Fantoni and S. Rosati, Nuovo Cim. A25, 593 (1975).
38. H. Kümmel, K. H. Lührmann and J. G. Zabolitzky, Phys. Rept. 36C, 1 (1978).
39. V. R. Pandharipande, S. C. Pieper and R. B. Wiringa, Phys. Rev. B34, 4571 (1986).
40. S. C. Pieper, R. B. Wiringa, and V. R. Pandharipande, to be published.
41. B. A. Friedman and V. R. Pandharipande, Phys. Lett. 100B, 205 (1981).
42. S. Fantoni and V. R. Pandharipande, Phys. Rev. C37, 1697 (1988).
43. S. Fantoni, B. L Friman and V. R. Pandharipande, Nucl. Phys. A399, 51 (1983).
44. J. W. Clark, Prog. Part. Nucl. Phys. 2, 89 (1979).
45. C. R. Chen, et al., Phys. Rev. C31, 2266 (1985), J. L. Friar, B. F. Gibson and G. L. Payne, Phys. Rev. C37, 2869 (1988).
46. T. Sasakawa, Nucl. Phys. A463, 327C (1987).
47. S. Ishikawa and T. Sasakawa, Phys. Rev. C36, 2037 (1987).
48. R. A. Brandenberg, et al., Phys. Rev. C38, 1397 (1988).

49. B. F. Gibson and B. H. J. McKellar, Few Body Syst. 3, 143 (1988).
50. I. Fujita and H. Miyazawa, Prog. Theo, Phys. 17, 360 (1957).
51. B. L. Berman and B. F. Gibson, Editors vol. 260 (1986), Lect. Notes in Phys.
52. V. R. Pandharipande, Lect. Notes in Phys. 260, 59 (1986).
53. P. U. Sauer, Lect. Notes in Physics 260, 107 (1986).
54. I. E. Lagaris and V. R. Pandharipande, Nucl. Phys. A359, 349 (1981).
55. B. A. Friedman and V. R. Pandharipande, Nucl. Phys. A361, 502 (1981).
56. B. D. Keister and R. B. Wiringa, Phys. Lett. B173, 5 (1986).
57. T. L. Ainsworth, et al., Nucle. Phys. A464, 740 (1987).
58. C. J. Horowitz and B. D. Serot, Nucl. Phys. A464, 613 (1987).
59. R. Malfliet, Nucl. Phys. A488, 721c (1988).
60. J. P. Blaizot, Phys. Rept. 64, 171 (1980).
61. M. M. Sharma et al., Phys. Rev. 38C, 2562 (1988).
62. G. E. Brown, Nucl. Phys. A488, 689c (1988).
63. A. Bohr and B. Mottelson, Nuclear Structure, vol. I. W. A. Benjamin (1969).
64. AIP Confernece Prodeedings No. 97, Ed. H. O. Meyer, "The Interaction between Medium Energy Nucleons in Nuclei," 1982.
65. J. W. Negele and K. Yazaki, Phys. Rev. Lett. 47, 71 (1981).
66. S. Fantoni, B. L. Friman and V. R. Pandharipande, Phys. Lett. B104, 89 (1981).
67. J. P. Jeukene, A. Lejeune and C. Mahaux, Phys. Rep. 25C, 83 (1976).
68. S. Fantoni, B. L. Friman and V. R. Pandharipande, Nucl. Phys. A386, 1 (1982).
69. G. F. Bertsch and S. Das Gupta, Phys. Rep. 160, 189 (1988).
70. R. B. Wiringa, Phys. Rev. C38, 2967 (1988).
71. C. Gale, et al., preprint (1989).
72. E. D. Cooper, et al., Phys. Lett. B206, 588 (1988).
73. C. Mahaux, et al., Phys. Rep. 120, 1 (1985).
74. S. Fantoni, V. R. Pandharipande and K. E. Schmidt, Phys. Rev. Lett. 48, 878 (1982).
75. P. L. Chung, et al., Phys. Rev. C37, 2000 (1988).
76. W. Strueve, et al., Nucl. Phys. A465, 651 (1987).
77. S. Auffret, et al., Phys. Rev. Lett. 55, 1362 (1985).
78. E. M. Nyman and D. O. Riska, Phys. Rev. Lett. 57, 3007 (1986).
79. T. deForest and J. D. Walecka, Adv. Phys. 15, 1 (1966).
80. C. Ciofi degli Atti, Prog. in Part. and Nucl. Phys. 3, 163 (1980).
81. K. Batzner, et al., Phys. Lett. 39B, 575 (1972).
82. R. Schiavilla, et al., Nucl. Phys. A473, 267 (1987).
83. R. Schiavilla, A. Fabrocini and V. R. Pandharipande, Nucl. Phys. A473, 290 (1987).
84. R. Schiavilla, V. R. Pandharipande, and A. Fabrocini, Phys. Rev. C40, 1484 (1989).
85. R. Schiavilla, Nucl. Phys. A499, 301 (1989).
86. K. Dow, et al., Phys. Rev. Lett. 61, 1706 (1988); K. Dow, MIT thesis (1987).
87. S. Dytman, private comm. (1988).
88. C. Marchand, et al., Phys. Lett. 153B, 29 (1985). CEN Scalay report CEA-N-2439 and private comm.
89. H. Arenhövel and A. Bernstein, private comm. (1989).
90. D. Beck, private comm. (1989).
91. P. Barreau, et al., Nucl. Phys. A402, 515 (1983).
92. Z. E. Meziani, et al., Phys. Rev. Lett. 52, 2130 (1984).
93. M. Traini, Phys. Lett. 213, 1 (1988); in "Electron Nucleus Scattering" Eds. A. Fabrocini, et al., World Scientific (1988).
94. J. Noble, Phys. Rev. Lett. 42, 412 (1981).
95. C. M. Shakin, Nucl. Phys. A446, 323c (1985).

96. I. Sick, In "Electron Scattering in Nuclear and Particle Science" API Conf. Proc. 161, 58 (1987).
97. J. A. Tjon, In "Electron Nucleus Scattering" Eds. A. Fabrocini, et al., World Scientific (1988).
98. R. Schiavilla and V. R. Pandharipande, Phys. Rev. C36, 2221 (1987).
99. R. Schiavilla, Phys. Lett. 218, 1 (1989).
100. S. Fantoni and V. R. Pandharipande, Nucl. Phys. A473, 234 (1987).
101. A. Fabrocini and S. Fantoni, preprint (1989).
102. C. C. Blatchley, et al., Phys. Rev. C34, 1243 (1986).
103. D. B. Day, et al., Phys. Rev. C40, 1011 (1989).
104. M. Butler and S. Koonin, Phys. Lett. B205, 123 (1988).
105. O. Benhar, et al., work in progress (1989).
106. O. Benhar, A. Fabrocini and S. Fantoni, in "Electron Nucleus Scattering" Eds. A. Fabrocini, et al.,World Scientific (1988).
107. S. Fantoni and V. R. Pandharipande, Nucl. Phys. A427, 473 (1984).
108. C. Itzykson and J-B. Zuber, Quantum Field Theory, McGraw-Hill (1980).
109. J. I. Friedman and H. W. Kendall, Ann. Rev. Nucl. Sci. 22, 203 (1972).
110. L. J. Rodriguez, H. A. Gersch and H. A. Mook, Phys. Rev. A9, 2085 (1974).
111. G. R. Farrar, In "Topological Conf. on Nucl. Chromodynamics" Eds. J. Qiu and D. Sivers, World Scientific (1988).
112. D. Lewart, V. R. Pandharipande and S. C. Pieper, Phys. Rev. B37, 4950 (1988).
113. P. K. A. DeWitt Huberts, Nucl. Phys. (in press), In Proc. of Symp. on the Nuclear Shell Model, Argonne (1989).
114. A. B. Migdal, Sov. Phys. JETP 5, 333 (1957).
115. C. Mahaux, and R. Sartor; Nucl. Phys. (ßin press), In Proc. of Symp. on the Nuclear Shell Model, Argonne (1989).
116. V. R. Pandharipande, C. N. Papanicolas and J. Wambach, Phys. Rev. Lett. 53, 1133 (1984).
117. A. Magnon, et al., Phys. Lett. B222, 352 (1989).
118. P. K. A. DeWitt Huberts, In "Electron Nucleus Scattering" Eds. A. Fabrocini, et al., World Scientific (1988).
119. O. Benhar, A. Fabrocini and S. Fantoni, preprint INFN ISS (1989)/2.

LIMITS ON THE COSMIC BARYON DENSITY

Hubert Reeves

Service d'Astrophysique
C.E.N. Saclay, 91191, Gif sur Yvette
Institut d'Astrophysique de Paris
France

1 - The Q-H phase transition in cosmology

The astrophysical importance of the Q-H phase transition comes from the fact that, below the critical temperature, it may induce in the cosmic fluid baryon density inhomogeneities, together with inhomogeneities in the ratio of neutrons to protons. These inhomogeneities could later modify the Big Bang nucleosynthesis yields of the four cosmological nuclides: D, ^{3}He, ^{4}He, ^{7}Li, over the values obtained in the case of the Standard Big-Bang (homogeneous density).

The course of events depends strongly on the values of the parameters describing the transition. Most important is the order of the transition. These inhomogeneities are only created if the transition is first order. Only then, indeed, do we have the possibility of overcooling and nucleation at temperature below critical, inducing the baryonic number inhomogeneities potentially responsible for the modification of the yields over the standard BBN values (Beaudet and Reeves 1984) (Yang et al 1984).

There was , up to recently a consensus among the experts that the transition is indeed first order. However several recent calculations to be discussed here indicate that the question is not yet settled.

It will be a long time before we have final results on BBN yields including all the effects of the Q-H transition. First the properties of the transition have to be calculated to a good accuracy by QCD calculations on lattice . Then the evolution of the bubbles will have to be followed dynamically all the way down to BBN, including proton and neutron diffusion. For the time being, we have only partial treatments of each of these phases, with uncertainties attached.

2- A simple model of the Q-H phase transition

A first estimation of the properties of the Q-H phase transition can be obtained with the help of an simplified model of the two phases. We assume no net baryon number (equal number of

Hadrons and Hadronic Matter
Edited by D. Vautherin *et al.*
Plenum Press, New York, 1990

particle and antiparticles) . In other words we set the baryon chemical potential m equal to zero. Ideal gas properties are assumed for both phases.

In the low temperature case we have a gas of massless, non-interacting pions (three varieties ; π^0, π^+, π^- thus the factor 3 in the following equations). The energy density is then given by

$$\varepsilon = 3\,(\pi^2/30)\,T^4$$

The relativistic pressure is one third of the energy density:

$$P = (\pi^2/30)\,T^4$$

At high temperature we have a relativistic gas of zero mass quarks and gluons. To obtain free quarks, one adds to the kinetic energy density the B term introduced previously. The Lorentz invariance of the stress-energy tensor then forces us to add a negative term (-B) to the pressure. (This can also be obtained simply from the relation dE = - PdV and E = BV with B = cste) This term will play the role of an effective cosmological constant in our astrophysical discussion. (The term B is often assigned to the confined phase instead of the unconfined phase as here. In this case the (then positive) pressure B has the physical meaning of a pressure from the vacuum to keep the quarks inside their confinement volume.)

$$\varepsilon = 37\,(\pi^2/30)\,T^4 + B$$

$$P = 37/3\,(\pi^2/30)\,T^4 - B$$

The multiplicity factor 37 has the following interpretation. For the gluons: 16 [2 for the spin times 8 for the number of color charges]. For the quarks: 21 [3 for the colors, times 2 for the quarks and antiquarks, times 2 for the savors (only the u and the d are considered here), times 2 for the spin, times a factor 7/8 from the nondimensional integral appropriate to the fermions.]

In the figure 1a the pressure of the two phases is plotted as a function of the temperature. The phase with the higher pressure is the stable one. The transition takes place at the critical temperature T_c at which the two pressures are equal.

$$B = (37-3)\,(\pi^2/30)\,/3\;\;T_c^4\;;\quad T_c = 0.72\,B^{1/4}$$

In the figure 1b the energy densities of the two phases are plotted in the same units. At T_c there is a density energy jump of

$$\Delta\varepsilon = 37\,(\pi^2/30)\,T_c^4 + B - 3\,(\pi^2/30)\,T_c^4 = 4B$$

The fig 1c gives the equation of state $P = P(\varepsilon)$. The entropy

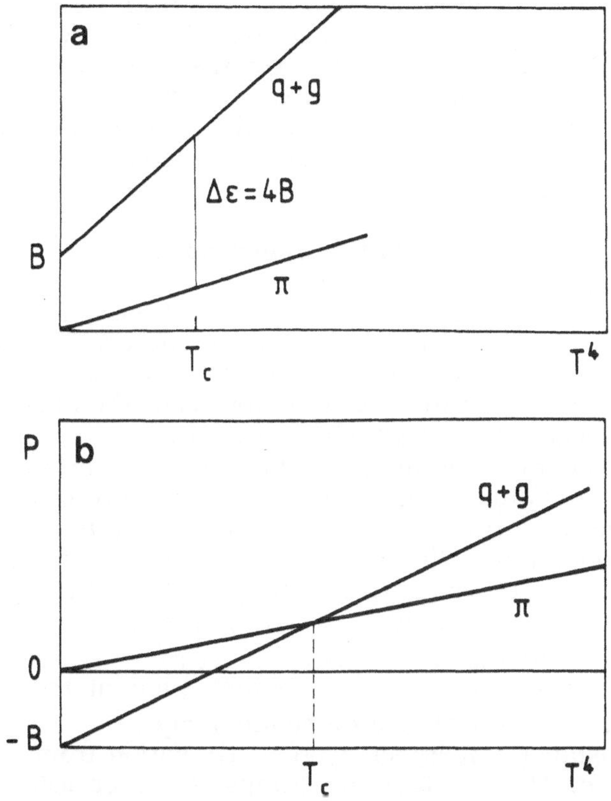

Figure 1. Illustration of the ideal gas model of the transition
1a . Pressure in the two phases vs temperature (T^4) The phase with the higher pressure is the stable one. The transition takes place at the critical temperature T_c at which the two pressures are equal.
1b . Energy densities of the two phases in the same units.

density is given by: $s = 1/V\ dP/dT\ |_V = 4\ (\pi^2/\ 30)\ /3\ T^3$ for each boson and the same number x 7/8 for each fermion,

Thus $s(q \text{ and } g) = 37\ (4\ /3)\ (\pi^2/\ 30)\ T$; $s(\text{pions}) = 4(\pi^2/\ 30)\ T^3$

In fig 1d, s is plotted as a function of T^3. The jump of s at the critical temperature is the sign of a first order transition (in this model...). One finds $\Delta e = T_c\ \Delta\ s$ as expected.

This simple model serves to illustrate the physics. In the high temperature phase, the increase in the number of degree of freedom (37 compared to 3) is counterbalanced by the vacuum density energy needed to free the quarks. The transition takes place at the critical temperature where the two effects compensate each other. The difference between the number of degree of freedom gives a measure of the entropy density to be released during the transition.

3 - Order parameters of the phase transition

An interesting analogy can be drawn with electric conductivity (Satz 85). A dilute gas of atoms is an insulator. The electrons, bound to their atomic nucleus, are not free to jump from one atom to another. However as the gas is compressed, the presence of the other atoms becomes increasingly important.

When the average distance between the atoms becomes comparable to the atomic radius, the orbital electrons starts to feel the presence of other electrons. Their atomic electrostatic long-range potential , $V(r) = 1/r$, is effectively screened by the presence of these other charges. This is described in terms of the Debye screening radius ($r_d = n^{-1/3}$) where n is the number density, by introducing a modified short-range potential $V(r) = (\exp -r/r_d)/r$. The net result is the ionization of the outermost electrons and the onset of electric conductivity .

The substance undergoes a phase transition from insulation to conduction. The conductivity jumps, more or less suddenly, from zero to a finite value. Thus the conductivity can be considered as an order-parameter of the transition, in the usual sense of statistical physics.

In a similar fashion, the long-range potential between two given quarks in a low density nucleonic gas is modified by the presence of intervening quarks and gluons, when the gas is compressed to nuclear densities. As we try to separate two quarks, the potential V between them diverges rapidly at low density but remains finite at high density i.e. when the mean interquark distance becomes smaller than one fermi.

To describe the transition, we define an order parameter $L = \exp(-V(r)/T)$ where T is the temperature. This parameter (the Polyakoff loop) jumps from the zero value at low density to a finite value at high density. This jump, at T_c, is the signature of the confinement phase transition.

As mentionned before, another phase transition occurs during this time. In the standard QCD theory, the quarks are, by definition,

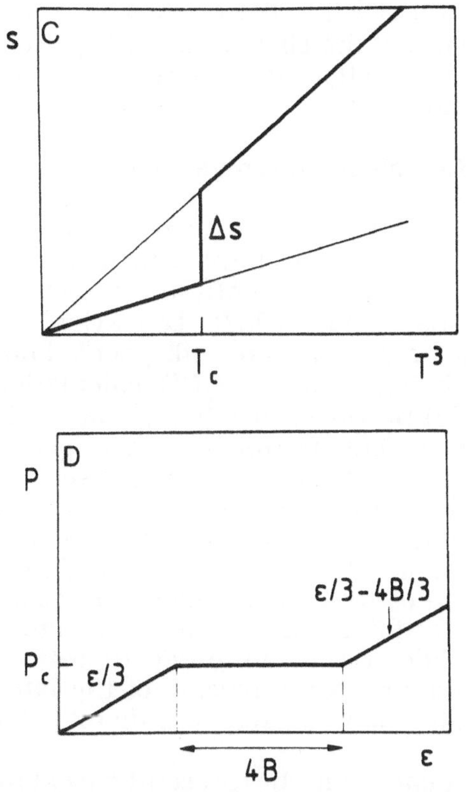

Figure 1c . The equation of state $P = P(\varepsilon)$.
1d . The entropy density s as a function of T^3

massless. The lagrangian is chiral-symetric (invariant to the transformation ($\Psi \to \gamma_5 \Psi$). This implies that it does not contain a term in ($\Psi\Psi^+$), which would not be invariant under this transformation. This is equivalent to the statement that the quark masses are zero.

In the glue, because of the very high energy density, the quarks effectively behave as massless particles. The left and right parity components of their fields do not mix. But this is not the case in our low-temperature world, where the quarks do have a mass. In the frame of the theory, the mass can only be acquired by a spontaneous breaking of chiral symetry.

The extent of the symmetry breaking can be evaluated through the value of the term ($\Psi\Psi^+$). The expectation value of such a term is, in fact, a measure of the effective mass (square) of the quarks in the actual physical situation. It it used as the order parameter of the chiral transition.

4 - Reports on QCD calculations on a lattice

The physics of the quark-hadron phase transition (or transitions, since there is a chiral transition and a confinement transition, Witten 1984) is presently the object of intense studies (Iso et al. 1986; Satz 1985, 1987; Leutwyler 1988) Many of the parameters of the transitions are still poorly known despite the vigourous effort being made in QCD calculations on networks (Irbäck et al (1988),Brown et al (1988), Bacilieri et al (1988)). A summary of the present status has been presented by Ukawa (1988) where an extended bibliography is given.

A complete numerical description of the influence of the Big-Bang Q-H phase transition on the yields of the cosmological nuclides would require dynamical QCD calculations during cosmic expansion. The lattice techniques allow approximate calculations of QCD processes at finite temperature. The standard technique of introducing euclidean time in order to estimate the thermal effects (the temperature is the inverse of the lattice extent in the euclidean time direction) makes it difficult to treat dynamical nuclear processes.

As a consequence, all the presently available results,to be described in this chapter, are by definition static. We have no reliable mean to estimate the alteration required to follow the non-equilibrium conditions created during the expansion. The problem is not with the timescale of the expansion itself (which is slow enough not to pose any real problem) but with the timescale of the hydrodynamical processes accompanying the transition, such as the nucleation and growth of the hadronized phase.

In order to appreciate the uncertainties attached to the parameters of the quark-hadron transition I shall describe briefly the method of QCD calculations on lattice. Real space-time is replaced by a finite number of points located on the corners of a grid with spacing a . A typical calculation will be described , for instance , by the numbers: $L^3 \times L_t = 24^3 \times 8$, meaning that the lattice has L = 24 grid points in each of the 3-space cartesian coordinates, and L_t = 8 points in the euclidean time coordinate $t_{euc} = i\, t_{real}$. The identification of euclidean time with the finite temperature of the system is made by the relation : $aL_t = 1/T$. There are periodicity conditions imposed on the lattice, of the form $f(x, t) = \pm f(x, t + aL_t)$, (thus transforming the lattice into a cylinder curved in the time direction) .

The lattice is used for the calculations of the partition function associated to any particular configuration of the system. In standard thermodynamics, the partition function Z is a sum of exponential terms over all possible states of a system. It given by (H is the hamiltonian of the system):

$$Z = \mathrm{Trace}\ (\ \exp\ (\ -\ H\ /kT))$$

It can be shown that the evaluation of the trace is equivalent to the computation of a product over terms computed on the grid-points. In actual life, the number of grid points is, of necessity, severely limited. Important uncertainties are introduced by the finite-size of the spacing. Their relevance in the study of the Q-H transition will be described here.

In the first generation of computations, called " pure-gauge" the dynamical influence of the quarks were neglected. In the last few years this situation has been progressively improved by the introduction of a parametrized number n_f of quarks with n_f increasing progressively from zero (pure gauge).

In order to control the computational difficulties, the mass of the quark is left as a parameter. A value of m_q equal to infinity gives back the pure gauge situation. Computations are made at various values of the dimensionless ratio m_q/T where T is the critical temperature (approximately 200 MeV) . Realistic values of m_q/T should be a few per cent for the u and d quarks. Results obtained at larger values (where the computations are less time-consuming) are used for extrapolation purposes.

5 - The order of the transition and the critical temperature

The critical temperature of the transition can in principle be determined by a computation of the entropy density $(\varepsilon + P) / T$ of nuclear matter as a function of temperature , a transition being signaled by a jump in the ratio $((\varepsilon + P) / T^4)$. The physics can be understood from the simple model presented earlier. In practice, the search for the critical temperature is made by varying the coupling constant of the lattice : g^2(lattice). The equations of the renormalization group give the approximate relation:

$$g^2(\text{ lattice}) \approx (2\, b_o \ln a^{-1})^{-1}$$

where b_o is a numerical constant attached to the QCD model, n_f being the number of flavors : $b_o = (\ 48\pi^2)^{-1}(\ 33 - 2n_f)$.

The next problem is the calibration, in physical units, of lattice energy unit a^{-1}, corresponding to the value g^2(lattice) at

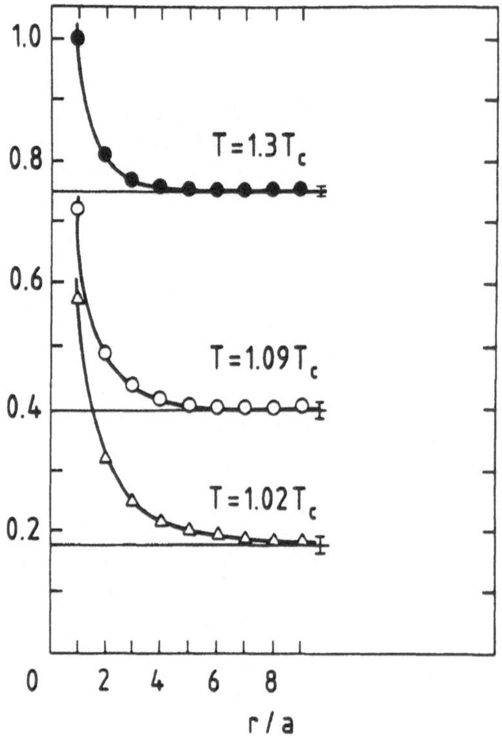

Figure 2. Fit of the computed correlation functions with exponentials of the type (exp - r/ r_0) with r_0 in the lattice units a . At large r/a the value decreases toward zero as one gets close to the critical temperature, signaling the gradual approach to confinement. The correlation length parameter r_0 increases gradually as one approaches the critical temperature. It is already quite large (several lattice units) at a temperature of more ten per cent of the critical temperature .

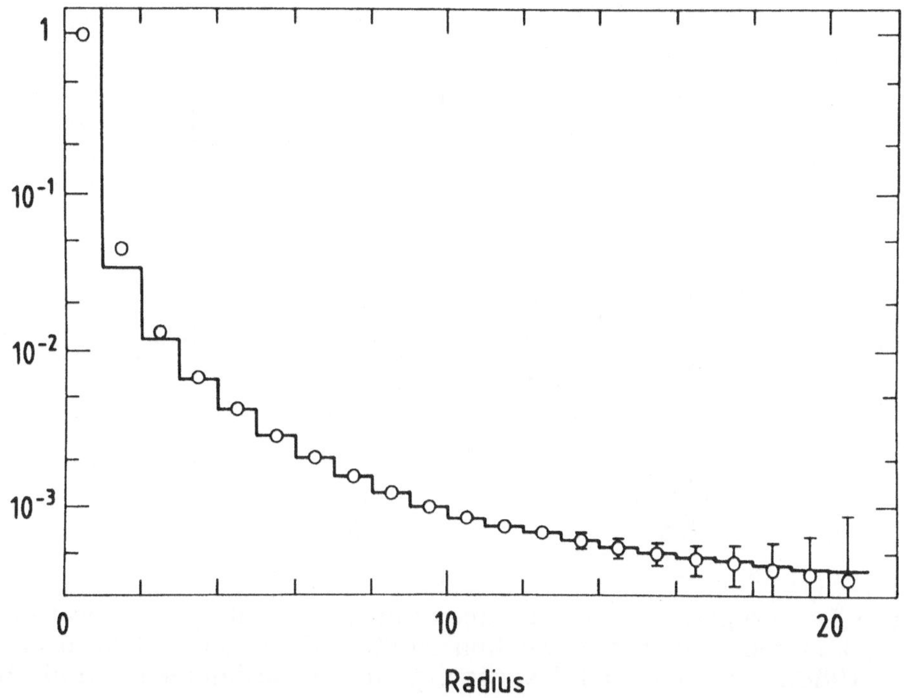

Figure 3. Correlation lengths computed by Bacilieri et al 1988 around the critical temperature.

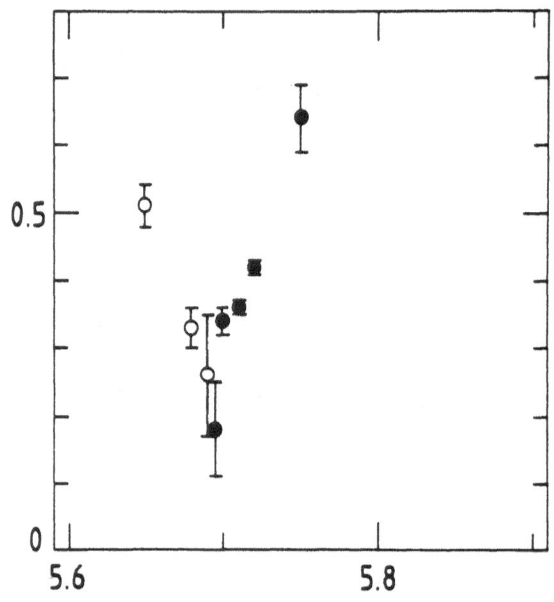

Figure 4 . Properties of the matter around the critical temperature , extracted from the correlation length calculations (Brown et al 1988). The open circles correspond to confinement and the quantity plotted is the product of the string tension σ and L_t. The solid circles correspond to deconfinement and show the Debye value of the screening mass m.

the jump in the function $((\varepsilon + P) / T^4)$. This is done by computing the mass of an hadron, for instance the r meson, in lattice units a^{-1}, at T = 0 and at the same g^2. The accuracy of the evaluation of the critical temperature, obtained from $a^{-1} = T_c \times L_t$, suffers from the many uncertainties attached to this indirect procedure.

Calculations have been recently reported by groups at Columbia University in New York (Brown et al 1988), at Rome (Bacileri et al 1988) and at Bielefeld (Irbäck et al 1988) in Germany. The jump in the $((\varepsilon + P) / T^4)$ is indeed clearly evident with the expected order of magnitude.

One of the major difficulties attached to the lattice technique is the fact that the finite size of the spacing smooths up the jump in the $((\varepsilon + P) / T^4)$, thus making it difficult to determine the order of the transition (a first order transition implies a step-function while a second order transition should be more gradual). Brown et al (1988) conclude to a "weak" first order transition, the exact meaning of these words being left rather unclear.

Another related test is the search by computations for so-called metastable states (the coexistence of two states at the critical temperature), a sign for a first order phase transition. The result is also a function of the assumed quark mass (more exactly of the ratio of the mass over the critical temperature) The "signal" for the existence of a metastable state is , computationally speaking, a function of the assumed number of flavors n_f of quarks. Irback et al (88) , report that the signal is found but that it weakens considerably as n_f is increased from 0.5 to 1 to 2 "thus putting doubt on transition being of first order".

More information on this subject can be obtained from another quantity which could be considered as an "order parameter" of the confinement transition. It is related to the amount of energy required to remove a quark from a plasma of gluons . In the confined state this amount is infinite , but not in the deconfined state.

Formally one computes the difference in the free energy F between of a field of gluons with one quark (F_1) and without quarks (F_0) The appropriate order parameter of the transition is then the so-called Polyakoff loop.

$$P(x) = \exp(-(F_1 - F_0) / kT)$$

As a first approximation , one computes the free energy of a *static* quark (neglecting the interaction of this quark with the gluons in which it is embedded). The calculation implies an integration of the time component of the gluon field at a point x in the plasma. This integration is replaced by a product of terms taken along the cyclic (thus the word "loop") euclidean time direction of the lattice. Monte Carlo simulations are used and averaged over.

In the confined state $(F_1 - F_0)$ is infinite and $< P(x) >$ is zero; in the unconfined state $(F_1 - F_0)$ in finite and $< P(x) >$ is unity (with

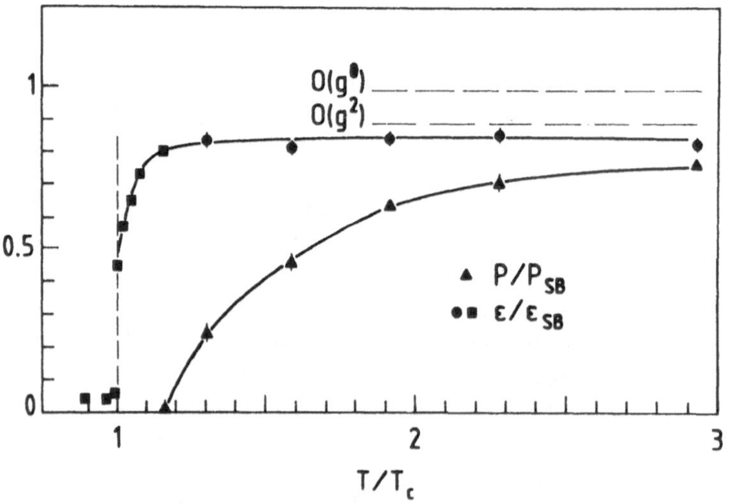

Figure 5. The gluon energy density and the pressure in the pure gauge theory (computations by Karsh) . The pressure does not reach rapidly the $P = \varepsilon / 3$ behaviour . The slow rise of P with T / Tc is another indication of the importance of non-weak-couplings effects at quite a large range of temperature above criticality.

proper normalization, as required for a bona fide order parameter). The symbols < > means the average over many numerical runs. From this, one introduces a most useful parameter : C(x) = <P*(x)P(0) > ; the correlation function between two points of the fluid . C(x) is related to the free energy required to remove a quark at x given that there is another quark at x = 0. It gives important information on the behaviour of the fluid below and above the critical temperature.

Many authors have fitted the results of their computated correlation functions with exponentials (figures 2,3,4). In the deconfined state the numerical value of C(•) decreases gradually as one gets near the critical temperature. This effect is visible in fig 2. At the same time, the correlation length parameter r_0 increases gradually. It is already quite large (several lattice units) at a temperature of more ten per cent of the critical temperature. In the deconfined phase it can be interpreted as a Debye mass $(m)^{-1}$ of the gluons .

Below the critical temperature, the correlation function can be written as $\{\exp - (\sigma/T)\ (r/a)\}$ giving a measure of the string tension s between the quarks, fundamentally responsible for the confinement (phenomenologically represented by an attractive force linearly proportionnal to the separation distance) .This string tension increases gradually as the temperature is lowered below the critical temperature(figure 4) .

The order of the transition can, in principle, be determined by the finite (first order) or infinite size (second order) of the correlation length at the critical temperature. Here again the finite mesh-size of the lattice creates problem. Bacilieri et al (88) use the fact that their computed correlation increases with the size of the lattice to argue that the transition is probably second order. According to these authors "this result excludes the presence of a strong first order transition". The results of a similar analysis is used by Brown et al (88) to conclude, again , to a weak first order phase transition.

Despite all these difficulties , one feature emerges from these computations , which is most likely relevant for cosmology. *The transition is not very sharply concentrated at the critical temperature* . At a given distance x, the correlation function starts to increase when the temperature is still appreciably larger than T_c (and not only very close to T_c as expected from simplified models) Thus the traditional ideal quark-gluon gas may not be a valid approximation just above T_c. Also the string tension does not take its full value until the temperature is appreciably lower than T_c, as seen in fig 3 and 4

Another interesting feature of the calculation is the fact that the pressure does not reach the $P = \varepsilon /3$ behaviour (figure 5) until $T = 1.5$ to $2\ T_c$. This also suggests that the high temperature phase is not simply a plasma of weakly interacting quarks and gluons at least close to the critical temperature. More important for cosmology, the entropy density is reduced close to the critical temperature, over the value estimated from an ideal gas.

These features of the near-criticality region may have important effects on the course of events taking place as the universe cools . They may invalidate to some extent the simplified thermodynamics and the standard bubble-forming mechanism to be described shortly. Indeed the entropy released after nucleation may be appreciably less than previously estimated .

I recall one important approximation of the calculations reported here: the neglect of the quark interaction with the gluon sea in the computation of the free energy pertinent to the Poliakoff loop (the so-called static quark approximation) . It appears that this effect can be taken into account by a small change in the value of $1/g^2$, i.e by a small correction to the corresponding value of T_c.

To conclude: the QCD computations on a lattice investigate the transition by different techniques : the jump in the energy density, the search for metastability and the behaviour of the quark correlation function . These computations do not appear to give an unambiguous answer to the question of the order of the transition, the main difficulty being the correct interpretation of the finite mesh-size effects. Furthermore the numerical value of the critical temperature of the chiral transition is made uncertain by the calibration of the spacing unit. The results are not the same when the calibration particle is chosen to be the nucleon (939 MeV) or the ρ (770 MeV). The introduction of the dynamical quarks appears to reduce the estimate of the critical temperature below the values obtained previously for pure gauge. However according to Akura this may be an artefact of the computation. According to Satz (1987) the best estimate is $T_c = 200 +- 50$ MeV. Similar values are quoted by Akura (1988) Brown et al (1988) evaluate a latent heat jump of $\Delta e / T^4 = 2{,}48 +- 0.24$.

Another approach to the value of the critical temperature has been presented by the Bern group (Gasser and Leutwyler, 1987, 1988, Leutwyler 1988, Gerber and Leutwyler 1988) . The properties of hadronic matter are studied by perturbative methods at low temperature. From low temperature series expansion, an estimate of the critical temperature of the chiral phase transition is obtained. A value of 170 MeV is quoted . It is currently believed that, at low chemical potential, the chiral and the deconfinement transition take place at the same temperature.

6- Cosmic scenario

The effect of the Q-H phase transition on the expansion and cooling rate of the universe can be approximated by the following simplifed formalism . The bag model, is used to fix the equation of state of the matter before and after the transition. We have to add up the cosmic supplement of photons , electrons and neutrinos. In the q phase we have $\varepsilon = g_q (\pi^2/30) T^4 + B$;

$P = g_q (\pi^2/30) T^4 /3 - B$ where $g_q = 51.25$. In the h phase :

$\varepsilon = g_h (\pi^2/30) T^4$, $P = g_h(\pi^2/30) T^4/3$ where $g_h = 17.25$

The ratio $r = g_q/g_h = 2.97$ will be of importance.

For an homogeneous and isotropic universe, the ten Einstein's equations are transformed into

$$(dR/dt)^2 = (8\pi G/3) R^2 \varepsilon$$

$$-dR/R = d\varepsilon / 3(\varepsilon + P)$$

Solving for dR/R one obtains the following fundamental time - energy relation:

$$-d\varepsilon / \{3 \varepsilon^{1/2} (\varepsilon + P)\} = (8\pi G/3)^{1/2} dt$$

For $T > T_c$, the bag energy term B accelerates the expansion with the timescale $t_{qh} = (8\pi GB/3)^{-1/2}$ as we get near to the critical temperature . Numerically with the help of the bag model we find : $t_{qh} = (8\pi GB/3)^{-1/2} = 144\,(100\text{ MeV} / T_c)^2\ \mu\text{sec}$
The energy density in the quark-gluon plasma ε_q varies as :

$$\varepsilon_q = B \coth^2 (2\ t/t_{qh})$$

At $T = T_c$ the energy density is expressed as the sum of the two densities ε_q and ε_h (for the quark-gluon and the hadron phase respectively), the fractional volume f_q being occupied by the quark phase.
The dynamics takes the following form :

$$\{(dR/dt)/R\}^2$$
$$= t_{qh}^{-2}(\varepsilon_q f_q + \varepsilon_h(1-f_q))/B = t_{qh}^{-2} (4 f_q + 6r-3/r-1)$$

In the model considered here the transition takes place at constant total entropy (no irreversible processes) so that we have :

$$d/dt\ (R^3(f_q s_q + (1-f_q) s_h) = 0 \text{ or}$$
$$\{(dR/dt)/R\} = -df_q/dt(r-1) / (3f_q (r-1) + 3)$$

The time change of f_q can be integrated (t_i is the beginning of the transition) .

$$f_q(t) = 1/4(r-1) \{\tan^2 (\arctan (4r-1)^{1/2}$$
$$+ 3(t_i-t)/2\ t_{qh}(r-1)^{1/2}) - 3\}$$

As f_q varies from 1 to 0 the scale factor R changes accordingly .The ratio of R ($f_q = 1$) / R($f_q = 0$) is approximately given by $(r)^{1/3}$ or 1.4 . In other words the Q-H phase transition gives rise to a small inflationary period where the scale factor increases by 40% at constant temperature. There must be an entropy flux from the q-g phase to the hadron phase to conserve the comoving entropy density as f_q decreases.

7 - Overcooling and the mean distance d between the baryonic inhomogeneities

In the case of a first-order phase transition , the new phase is not reached as soon as the system is cooled through the critical temperature. Overcooling occurs in which fluctuations create small volumes of the new phase. The nucleation rate is a very important parameter for BBN. It is related, amongst other things to Σ the surface energy density of the hadronized bubbles. The minimum size of the bubbles is determined by the (unknown) value of Σ; smaller bubbles will "evaporate" back to the glue. If Σ is small , little overcooling is required before the onset of nucleation , the bubbles will then be numerous and the mean distance d between the bubbles will be small. Latent heat is released from the "surviving" bubbles and propagated around, in the form of shock waves moving out at velocity v_s (Kurki- Suonio 1986) . This goes on until the fluid is reheated to the critical temperature.

Fuller et al (1988) have followed the discussion of Kurki-Suonio (1986) to evaluate the interbubble comoving scale d as a function of the QCD parameters T_c and Σ. In terms of the bag model parameter B, Σ can have any value from 0 to $(B^4)^{3/4}$. In units of T_c = 200 MeV and Σ = (200 MeV$)^3$ we obtain :

$$d \approx 50 \text{ m} \left(\Sigma / (200 \text{ Mev})^3 \right)^{3/2} / (T_c / 200 \text{ MeV})^{13/2}$$

This points out the need for better QCD calculations. With the present uncertainty, d could be anywhere from 0 to the value of the horizon scale at the transition : d = 10 km.

8- Baryonic density inhomogeneities

The previous treatment ignored the presence of both matter and antimatter at the Q-H phase transition. At BBN however antimatter has completely disappeared. Thus the important parameter is the contrast R in the baryonic number density (n_b) between the two phases : hadronized bubbles and the quark sea out of which they have formed. In thermodynamic equilibrium, these baryonic number densities (n_b) are related to the chemical potential μ_b in the standard way. The early universe evolves with a value of (μ_b/ T) of approximately 10^{-8} so that only the first term in the expansion of appropriate thermodynamical quantities need to be retained.

Always at the level at the bag model these thermodynamical quantities are obtained by adding to the standard thermodynamic potential

$$W(T, V, \mu) = -T \ln Z$$

a vacuum energy term for the quark-gluon plasma phase:
$(-T \ln Z_{vac}) = B\,V$.

The Z term is the grand partition function for particles and antiparticles with mass m, chemical potential μ, and degeneracy factor g in the large volume limit of a free gas. The parameter $\eta = 1$ for fermions and $\eta = -1$ for bosons..

$$\ln Z(T, V, \mu,) = gV / 6\pi^2 T \int_0^\infty dk\ k^4 / (k^2 + m^2)^{1/2}$$
$$[\{\exp[(k^2 + m^2)^{1/2} - \mu] / T + \eta\}^{-1} +$$
$$\{\exp[(k^2 + m^2)^{1/2} + \mu] / T + \eta\}^{-1}]$$

We are interested in the value of the thermodynamic functions at the critical temperature T_c .The particle-number density n is then given by the relation:

$$n = -1/V (\partial \Omega / \partial \mu)_{V,T}$$

which for a quark-gluon plasma of 3 colors and 2 flavors (i.e. neglecting the contribution of the s quark) , gives, in first order of (μ_b / T_c) :

$$n^q = 2/3\ T^3 (\mu_b / T_c)$$

The corresponding baryon number $n_b{}^q$ in the plasma is 1/3 of this value (3 quarks for 1 baryon).

In the hadronized bubble forming in the quark-gluon plasma, the baryon number $n_b{}^h$ is given, for each type of baryons of mass m, (in the present approximation where the hadronic interactions are neglected) :

$$n_b{}^h = (8 / \pi^3)^{1/2}\ T^3\ (\mu_b / T_c)\ (m/T_c)^{3/2} \exp(-m/T_c)$$

The ratio R of the baryon numbers densities in the high and low density regions is obtained in this simple case, by taking the ratio $n_b{}^q / n_b{}^h$ and summing over all the hadron species. We note that, in this approximation the value of R is independant of (μ_b / T_c) but that it is a function of T_c , reaching very high values ($\sim 10^2$) if T_c is close to 100 MeV (illustrating the importance of obtaining accurate value of this parameter from QCD calculations). The value and behaviour of R (T_c) reflects the fact that the masses of the quarks (a few MeV) are much smaller than the critical temperature which is itself much smaller than the masses of the baryons. As a consequence more degrees of freedom are available for the quarks than for the hadrons at the critical temperature. Much of the physics is contained in this simple statement.

Calculations of R have been made by several groups ((Sale and Matthews 1986), (Applegate and Hogan 1985), (Applegate, Hogan and Sherrer 1987, 1988), (Alcock *et al.* 1987),(Alcock et al 1988), (Fuller *et al.* 1988), neglecting the interactions between hadrons , and by Kapusta and Olive (1988) with a simplified treatment of these interactions. The results are in agreement for $T_c < 150$ MeV.

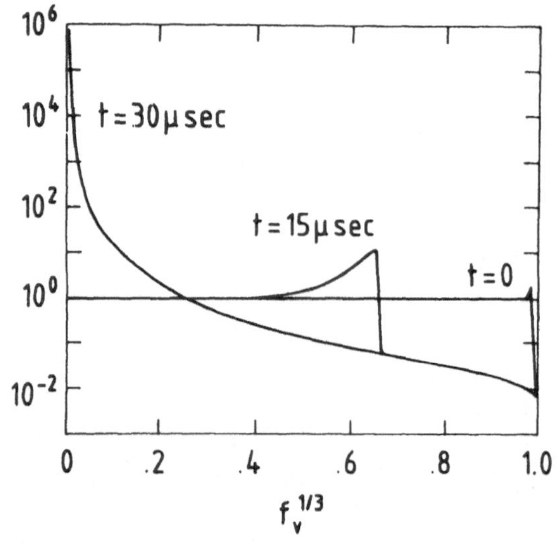

Figure 6. Evolution of the profile of the baryon number density (normalized to the pre-transition value) during the transition, as a function of $(f_V)^{1/3}$, a measure of the comoving fluctuation radius. The graph illustrates the migration of the bubble wall (the jump) at three different times, as the transition is progressing toward completion. From Fuller et al (1988).

At higher T_c, the value of R goes toward 1 in the model without interactions, while it remains at values around 7 to 10 when the interactions are considered. This is to be qualitatively expected , although more detailed treatments of the nuclear forces would be needed before the profile of R (T_c) can be confidently calculated.

9- Baryon number transport and the density profile

As the bubbles expand, entropy and baryon number must be transported across the phase boundary.

One important poorly known parameter is the m.f.p of a quark in quark matter at temperatures close to the critical temperature. (As discussed before there are good computational reasons to doubt the validity of a model of non-interacting particles.) If the cross-section is large, then as the bubble wall moves across space, we expect the baryon ratio R computed earlier with the assumption of chemical potential equilibrium to apply only at the wall itself. As a function of time, the baryon number is evacuated from the hadronized phase and piles up on the wall of the quark bubbles, the amount of piling-up being inversely proportional to the quark m.f.p. in the quark phase. At the end of the transition we may expect each bubble to be characterized by a dense center with baryon number density gradually decreasing outward. (Kajantie and Kurki - Suonio 1986, Kurki-Suonio 1988, Matthews et al 1988) . The figure 6 from Fuller et al (1988) may serve as an illustration.

Many uncertainties are attached to the calculation of these profiles. An unknown fraction of the entropy is transported by neutrinos. (Miller and Pantano 1988). The transport across the border and the quark m.f.p. are also very poorly known parameters of the calculation. In the nucleosynthesis calculations to be described later, the density profile is approximated by parameters such as f_V , the fractional volume at a given baryon number density contrast R. Models have been made with two zones and also with several intermediate zones.

Kurki- Suonio (1988) has recently argued that large values of the contrast R can be created only if d , the interbubble distance is small ($d < 1$ lh) This result is based on an ideal gas equation of state for both phases.

The recent calculations reported here and illustrated in figures 2,3 and 4 imply that the correlation functions are large on both sides of the critical temperature. This cast doubt on the applicability of an ideal gas model.

10 - From the Q-H phase transition to BBN

As the universe cools from T_c at approximately 20 μsec, to one MeV at one second, the neutron to proton ratio (n / p), governed by weak processses is given by the Boltzmann formula of mass-action. Below one MeV the weak processes are no more in thermal equilibrium. The neutrons diffuse from high density phases into low density phases, changing both their density and their (n / p) ratio. The extent of neutron diffusion is a function of both the fractional volume in each phases and of the mean distance d between the high density blobs.

A convenient unit is the present value of d in light-hours (h). One lh today corresponds to 2.5×10^5 cm at $T_9 = 1$ and approximately one meter at the Q-H phase transition, when the horizon scale was approximately ten km. At large values of d ($d \sim 10^4$ lh) the neutron diffusion could not diffuse before BBN. Computations made with the assumption of large values of d would give the same results as computations based on a density-inhomogeneous standard model. At the lower end of the scale, d < 0,1 lh or so, proton diffusion becomes important Computations should recover the results of the standard homogeneous density BBN.

11 - Calculations of BBN yields

Several generations of models have already been published , based on increasingly realistic models. The first generation (Applegate and Hogan 1985) (Malaney and Fowler 1988) (Reeves et al 1988) were two-phase models characterized by a contrast density R and a fractional volume of the high density phase f_v, with the further assumptions of 1) complete neutron homogeneization between the two phases before the onset of BBN and 2) no further neutron diffusion during BBN. Some of these models appeared to be able to reconcile the D, ^{3}He and ^{4}He calculated at critical baryon density $\Omega_b = 1$ with the observations, with however important overproduction of ^{7}Li.

A second generation of models take into account the effect of the interbubble comoving distance d on the neutron and proton diffusion before and during BBN. Several sets of calculations have been published for selected sets of points in the parameter space composed of ρ_b, R, f_q and d (Terasawa and Sato 1988 (two-zones), Mathews et al 1988, Fuller et al 1988) (eight zones) Kurki-Suonio et al 1988, Kurki-Suonio and Matzner 1988 (64 zones). The many-zone calculations of the last papers are potentially important in resolving eventual structure generated by diffusion during BBN.

For the discussion to be presented here we have also used new results extending the work of Terasawa and Sato (Reeves et al 1989). These are two-zone calculations covering the range: $1 < R < 10^4$; $0.5 \times 10^{-31} < \rho_b < 50 \times 10^{-31}$; $0 < f_v < 1$; d = 1, 3 and 10 lh. The agreement between the various calculations is good enough for our following discussion to be of relevance.

12 - Results and Conclusions

It may be a long time before we get definite results on the effect of the quark-hadron phase transitions on the formation rate of the cosmological nuclides. Nevertheless the computations described before hopefully give us the general trends. Our present ignorance of the exact values of many relevant parameters of the Q-H transition can be assimilated to corresponding uncertainties on the final results. These uncertainties are likely to decrease as more detailed studies of the transition become available.

Many authors (Boesgaard and Steigman 1985, Cayrel 1988, Pagel 1987, Kawano and Schramm 1988) have recently discussed the question of the relevant abundances of the light nuclides D, ^3He, ^4He ^{7} Li to be used for comparison with BBN calculations. The differences between these authors are mostly with the allowed fork of uncertainties. Here I will tend to use rather large forks.

The case of deuterium is still plagued with the problem of astration since we have no data prior to the birth of the solar system. Several models of galactic evolution have been used to set constraints on the primordial abundances (Rocca-Volmerange and Schaeffer1988) (Vangioni-Flamm and Audouze 1988). A deuterium fork of 10^{-5} (minimum astration) < D/H < 10^{-4} (maximum astration) is selected.

The uncertainties on the cosmological abundance of ^3He are even larger. So large that this isotope, taken alone or summed with D does not appear to yield interesting information for our quest.

For Y, the mass fraction of 4 He, I have used the rather large fork : 0.23 <Y< 0.26 . For ^{7} Li a fork of $10^{-10} < ^7Li / H < 10^{-9.5}$ is my best choice. In view of the possibility of a rotational destruction of ^7Li in the enveloppe of Pop II stars (Vauclair (1988) I have also considered an upper limit of 10^{-9}. However the recent upper limit on the abundance of lithium in the line of sight towards the SN 1987 (Magain. 1989) (Sahu et al 1989) gives support to the value of $10^{-9.5}$ as a more appropriate upper limit (taking into account uncertainties in the possible depletion of lithium and in the evaluation of the relative population of its different atomic states).

It will appear later that the upper limits on the allowed range of both ^{4} He and ^{7} Li are critical for the evaluation of the upper limit on the baryonic density. For instance if, as many authors have already claimed, one can already exclude the range Y > 0.25 and /or the $^7Li / H > 10^{-9.5}$ the baryonic range will be considerably reduced.

First I discuss the acceptable range of baryonic density (always given here in units of 10^{-31} g cm^{-3} , the corresponding value of η is $\eta = 1{,}5 \times 10^{-10} \rho_b$ in these same units, for T = 2.7K). Agreement between calculations and observations is found for *some values* of the parameters *in the whole range* $2 < \rho_b < 50$, in the range 1 lh < d < 10 lh

Deuterium and helium-4 are the limiting factors in the lower part of the range. At $\rho_b < 2$, ^4He is too small and D is too large unless the fractional D destruction during galactic life is more than a factor of ten which is probably the upper limit tolerable by galactic models.

In the range $2 < \rho_b < 5$ the observational abundances can be met for quite a large range of values of the parameters. As we move to higher baryonic densities the allowed range of parameters is progressively restricted. Increasingly larger values of R are needed together with rather large values of the fractional volume (applicable to the two-zone calculations) $f_V \approx 0.3$ to 0.6 . For instance at $\rho_b = 7$ one needs $R > 5$; and $\rho_b = 10$ one needs $R > 9$.

The range $\rho_b > 20$ requires rather marginal conditions: very large values of $R > 100$, a value of f_V close to 0.3 together with the extreme observational limits of $Y \approx 0.26$ and $^7Li / H = 10^{-9.0}$. This range is excluded if the Pop II lithium value is the correct one .

Even fragmentary knowledge of the likely values of these parameters will help in restricting this range . The best estimate of the critical temperature of the phase transition (180 MeV < T_c < 220 MeV) leads to values of R between 5 and 10 at chemical potential equilibrium. The effective R, taking into account hydrodynamical effects should be somewhat but not much larger (Fuller et al 1988), thus favoring the lower part of the density range.

The same conclusion applies if f_V turns out to be small (f_V < 0.1), as suggested by the computations of Fuller et al (1988).

The previous remarks suggest that although there is some possible agreement between the calculations and the observations all through the range $2 < \rho_b < 50$, we may reasonably already exclude the upper range and select a reasonable fork of $2 < \rho_b < 10$ (x 10^{-31} g cm^{-3}) or $3 < \eta < 15$ (x 10^{-10} This tentative conclusion will be the base of the following cosmological discussion.

An recent analysis by Kurki-Suonio et al (1988) has yielded quite similar conclusions. Their slightly different (narrower) choice of the range of η or r_b compatible with BBN is related to their different estimations of the reliability of the various parameters of the problem. Their paper mention another potentially interesting constraint on the parameters of the quark-hadron transition : if R is less than 100 or so then d must be less than 150 lh .

On the nuclear physics sector , the following parameters are in need of improvements: a) the value of the critical temperature, b) the order of the transition and the equation of state on both sides of the critical temperature, c) the value of the density contrast R at chemical potential equilibrium, d) the surface energy of the bubbles, e) the complete profile of the baryonic density inhomogeneities prior to BBN, including the determinations of the two computational parameters d and f_V discussed previously.

On the astrophysical sector, we need improved galactic evolution models, incorporating all the present data on stellar populations and chemical abundances, in order to estimate the possible amount of astration and fractional destruction of primordial deuterium, in order to obtain a firmer lower limit to the

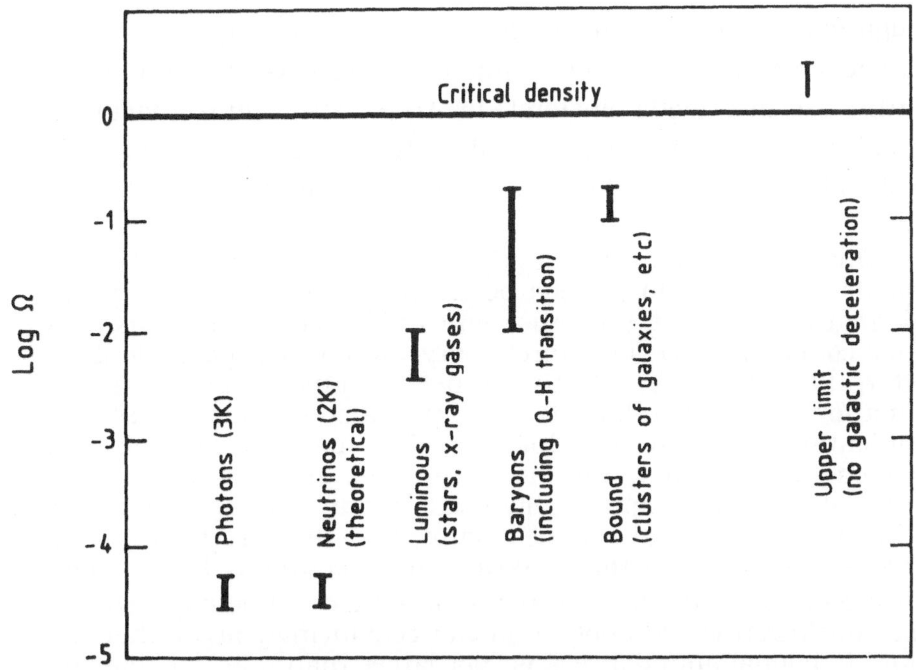

Figure 7. Cosmic densitology The units are the density with respect to the critical density $(10^{-29}$ g cm^{-3} if H is 75 km sec^{-1} kpc^{-1}. plotted in this diagram are the estimations with their uncertainties of various densities .

range of baryonic density. Is a value as low as $\rho_b = 10^{-31}$ g cm^{-3} really excluded ? Better knowledge of the possible effect of stellar rotation on the surface abundance of lithium is needed in order to obtain a firmer upper limit on the primordial abundance of lithium and hence on the upper range of the baryonic density. Better limits on the primordial helium abundance would help at both ends of the baryonic range.

13 - Cosmic densitology

Given all the uncertainties, including those on the Hubble parameter, the range of Ω_b selected here goes at most from 0.2 to 0.01. Although this is appreciably larger than in the case of a homogeneous density universe this does not appear to be large enough to allow the baryons to close the universe ($\Omega_b < 1$).

The cosmic density of luminous matter (stars and X-ray emitting gas in galactic clusters) is $\Omega_L = 0.01$ whitin a factor of two, while the density of clustered matter needed to account for the stability of clusters of galaxy or large scale motions is $\Omega_G = 0.1$ to 0.2.

Thus, within the uncertainties, at one end of the scale the baryonic matter could (barely) be entirely luminous (no baryonic dark matter) while at the other end of the scale, the clustered matter could (barely) be entirely baryonic (no non-baryonic dark matter) (figure 7). As discussed before, any progress in the estimations of the relevant QCD parameters will results in modification of these conclusions. For instance, if one can convincingly argue that the mean present distance d between overdensity regions is appreciably smaller than one light-hour , *or* if the mean fractional volume occupied by these regions is very small ($f_v < 0.1$) , *or* if the effective density contrast R is smaller than five or so, then non-baryonic dark matter will be required.The same conclusion would apply if on can convincingly argue that the helium fractional abundance Y is not larger than 0.25 or that the lithium primordial abundance is not larger than 3×10^{-10}. At the other end of the scale, if large astration of D can be ruled out ($D/H < 10^{-4}$), baryonic dark matter will also be required.

I want to thank André Morel, Jean Paul Blaizot and Madeleine Soyeur from Saclay for their precious help on the physics of the quark-hadron phase transition.

REFERENCES

Alcock, C.R., Fuller, G.M., and Mathews, G.J. 1987, Ap. J., **320**, 439.

Alcock , C.R., Fuller , G., Mathews G.J., and Meyer, B. in Proceedings 7th International Conference on Ultra-Relativitic Nucleus-Nucleus Collisions (Quark Matter' 88) Lenox MA Sept 26-30 1988.

Applegate, J.H., and Hogan, C. 1985, Phys. Rev., **D30**, 3037.

Applegate, J.H., Hogan, C.,and Sherrer R.J. 1987, Phys Rev., **D35**, 1151.

Applegate, J.H., Hogan, C.,and Sherrer R.J. 1988, Ap.J. **329** , 572,

Bacilieri, P., Remiddi, G., Todesco, G.M., Bernaschi, M., Cabasino, S. Cabbibo, N., Fernandez, L.A. Marinari, E., Paolucci, P., Parisi, G., Salina, G., Tarancon , A., Coppola, F., Lombardo, M.P., Simeone, E. Trippiccione, R., Fiorentini, G., Lai, A., Marchessini, P.A., Marzano, F., Rapuano, F., Tross, W. preprint, ROM2F 88 020

Beaudet, G., and Reeves H. 1984, *Astr. Astrophys.*, **134**, 240.

Blaizot, J.P., Acta Physica Polonica, **B18**, 661, (1988)

Boesgaard, A.M. and Steigman G. 1985, *Ann. Rev. Astr. Ap.*, **23**, 319.

Brown, F.R., Christ, N.H., Yuefan Deng, Y., Gao, M., Woch, T.J., preprint Columbia University.

Cayrel, R. 1988, Proceedings of the Alpbach Summer school. Austria

Fuller, G.M., Mathews, G.J., and Alcock, C.R. 1988, *Phys. Rev.*, **D37**, 1380.

Fuller, G.M., Mathews, G.J., and Alcock, C.R. 1988, preprint UCRL 98942 Proceedings 8th Moriond Meeting "Dark Matter" in les Arcs , France

Gasser , J and Leutwyler, H., Light quarks at low temperature. Phys Lett B 184, 83 , 1988.

Gasser , J and Leutwyler, H., Thermodynamics of chiral symmetry . Phys Lett B 188, 477 , 1987.

Gerber, P., and Leutwyler, H., preprint, BUTP 88/30

Irbäck, A., Karsh, F., Peterson , B., and Wyld , H.W.. preprint, CERN TH 5130/88

Iso , K., Kodama , H., and Sato, K., 1986 Phys. lett. 169B 337.

Kajantie , K., and Kurki-Suonio , H. 1988, Phys. Rev . **D 34** 1719

Kawano, L., and Schramm, D. 1988, *Ap. J.*, **327**, 750.

Kapusta ,J.I. and Olive K.A. University of Minnesota preprint UMN- TH-647 (1988)

Kurki-Suonio, H., Matzner, R.A.., Centrella, J.M., Rothman, T., and Wilson, J.R. 1988, *Physical Review D*. **38** 1091 (1988)

Kurki-Suonio, H. *Physical Review D*. **37** , 2104, (1988)

Kurki-Suonio, H., Matzner, R.A. preprint (1988)

Leutwyler, H., QCD: low temperature expansion and finite -size effects. Proceedings of the Seillac Conference April 88.

Malaney, R.A.,and Fowler, W.A., in Origin and Distribution of the Elements, ed. Mathews, G., J., World Scientific, Singapore 1988

Magain, P. 1989, Astr,Ap. **146** 95

Matthews, G.C., Fuller, G.M., Alcock, C.R., Kajino, T. preprint UCRL 98943 (1988)

Miller , J.C., and Pantano, O, preprint SISSA Trieste.

Pagel , B.E.J., 1987 in "A unified view of the macro- and micro-cosmos" First International School on Astroparticle Physics, Erice (Sicily, Italy) Ed: A. De Rujula, N.V. Nanopoulos, P.A. Shaver , World Scientific Singapore.

Reeves, H. 1987, Varenna School "Confrontations between Observations and Theories in Cosmology", eds J.A.Audouze and F Melchiorri, to be published.

Reeves, H., Delbourgo-Salvador, P, Audouze, J., and Salatti, P. 1988, *European Journal of Physics* ,**9**, 179.

Reeves, H., Richer, J., Sato, K., and Terasawa, N, to appear in ApJ. 1989

Rocca-Volmerange, B., and Schaeffer , R., preprint.
Ryan,S.G., Bessel,M.S., Sutherland,R.S., Norris, J.S. submitted to ApJ .1989.
Sahu, K.C., Sahu, M., and Pottasch,S.R., Astr. Astrophys 207 L1
Sale, K.E. and Mathews, G.J. 1986, _Ap.J._, **309**, L1.
Satz ,H., Ann. Rev. Nucl. Sci. **35** 1985
Satz . H ., Proceedings of the Strasbourg Symposium on the quark-hadron phase transition. . July 1987
Terasawa, N., and Sato, K. 1988, preprint UTAP 79
Terasawa, N., and Sato, K. 1989 Prog. Theor . Phys. Lett 81, 254
Thomas, A.W. 1984 in Advances in Nuclear Physics **13** (Negele, J.W. and Vogt, E. eds) pp 1-137
Ukawa, A, " Status of lattice QCD at efinite temperature" CERN -TH 5266 - 1988
Vauclair, S., 1988, Ap . J.335, 971.
Vangioni-Flamm ,E, and Audouze, J., Astron. Astrophys. 193, 81, **1988**
Witten E., 1984 Phys. Rev. **D30** 272
Yang, J., Turner, M.S., Steigman, G., Schramm, D.N., and Olive, K. 1984, _Ap.J._, **281**, 493.

QCD AND HADRONS ON A LATTICE *

John W. Negele

Center for Theoretical Physics
Laboratory for Nuclear Science and Department of Physics
Massachusetts Institute of Technology
Cambridge, Massachusetts 02139 U.S.A.

ABSTRACT

These lectures provide an elementary introduction to the study of hadronic physics using QCD on a lattice. Path integrals are reviewed to show how field theory on a lattice is reduced to quadrature. The Wilson lattice action for gluon fields is motivated and explained, and evidence is presented showing that lattice calculations in the pure gauge sector approach the continuum theory. Difficulties arising when Fermion fields are placed on a lattice are discussed, and the principal methods of circumventing these problems are described. A brief guide to the current literature is given for pertinent aspects of the Monte Carlo method and relevant results of calculations of the properties and phases of hadronic matter and the structure of hadrons.

I. INTRODUCTION

The motivation for lattice gauge theory is to solve, rather than model, QCD. Thus, at this school devoted to the study of hadrons and hadronic matter, it is useful to complement the descriptions of the various models which are currently applied to hadrons with a discussion of how QCD may be solved on a lattice and what has been and may be learned from lattice calculations.

These pedagogical lectures are deliberately elementary and are directed at non-specialists who wish to start at the beginning and understand the basic ideas with a minimum of technical details. Since there appears to be a gap in the literature at this level, I will start at a level even more elementary than the excellent introduction by Creutz[1] and will attempt to fill in the essential background needed by a newcomer. I thus hope to explain the basic concepts so that the reader will appreciate the essential ideas, understand both the potential and limitations of lattice calculations, and be prepared to read the more advanced literature where many of the details I have had to omit may be found. The numerical results presented in the oral lectures and cited in the last section of this manuscript were selected to make specific pedagogical points or to emphasize aspects particularly relevant to this school, and are not intended to be a systematic or exhaustive survey of current calculations.

* This work is supported in part by funds provided by the U. S. Department of Energy (D.O.E.) under contract #DE-AC02-76ER03069.

As will be clear from the status of current calculations, the motivation for learning about lattice QCD is not so much from results yet obtained, as from the challenges and opportunities it provides for the future. At virtually every phase of our quest to understand the structure and interactions of hadrons, nuclei, and hadronic matter, we eventually reach an impasse from the inability to solve QCD. Although an effective hadronic theory of the form described in Pandharipande's lectures is extremely successful in describing a wide range of low-energy nuclear phenomena, we have no way at present of calculating the potentials in this theory from QCD, understanding their domain of validity, or knowing their applicability to matter at higher density. Similarly, as described in Thomas's lecture, although the operator product expansion provides a powerful tool for identifying the leading contributions to deep inelastic scattering at high Q^2 and understanding the scale dependence of moments of structure functions, in the end one is still left with the problem of calculating the fundamental hadron structure functions non-perturbatively from QCD. The search for a quark-gluon plasma discussed in McLerran's lectures involves many questions which hinge upon quantitative solution of QCD, ranging from knowledge of the thermodynamic properties of the plasma itself to an understanding of the process of hadronization so that an unambiguous signature of plasma formation can be identified. Finally, in the domain of astrophysics discussed in Alcock's lectures, the regimes of hadronic matter of interest are minimally constrained by terrestrial observables, and a fundamental understanding of high density matter in the interior of neutron stars, the stability of strange matter, the equation of state governing supernova collapse, and the behavior of hadronic matter in the early universe require quantitative QCD calculations.

At present, it is not clear how many of these problems will be directly amenable to lattice calculations. However, these lectures, as well as the active current research in this field, are motivated by the belief that lattice QCD can ultimately provide substantial insight into the structure of the vacuum and of hadrons. I believe it will provide a useful tool to discriminate between and understand more deeply the plethora of bag, non-relativistic quark, flux tube, chiral, and Skyrme models currently applied to hadronic physics. Furthermore, it is also possible that lattice calculations of correlation functions and other features of hadronic structure which are inaccessible experimentally, will guide and stimulate new approximations and hadronic models.

There are four key ideas underlying this approach. The first is the use of path integrals. One of the great contributions of Feynman to theoretical physics was the formulation of quantum mechanics in terms of path integrals, which provides both a physical picture of quantum evolution in terms of sums of time histories and a powerful computational framework. For the present application, we will make use of the fact that the path integral eliminates the non-commuting operators of quantum mechanics or field theory by introducing an integral over an additional continuous variable, and thus effectively reduces the problem of quadrature.

The second major idea is the introduction of Euclidean time. The basic idea is to write $|\psi\rangle = e^{-\beta H}|\phi\rangle$, where $e^{-\beta H}$ acts as a filter to project the ground state $|\psi\rangle$ out of an arbitrary state $|\phi\rangle$ having the desired set of quantum numbers, so the continuous variable in the path integral is imaginary or Euclidean time. The resulting theory has important connections with statistical mechanics. In the case in which one sums over a complete set of states and calculates the trace $\mathrm{Tr}\, e^{-\beta H}$, one is solving field theory at finite temperature and β corresponds to the physical inverse temperature. The corresponding path integral has the structure of classical statistical mechanics in $d+1$ dimensions. Many familiar ideas from statistical physics concerning critical behavior, order parameters, and Landau's theory of phase transition turn out to be useful.

The third principal idea is lattice regularization, which replaces continuum field theory by a finite quantum many-body problem on a lattice. For any finite lattice spacing a, the maximum momentum which can arise on the lattice is $p_{\max} \sim \frac{\pi}{a}$, so that the lattice effectively imposes a momentum cutoff of order $p_{\max}$ which goes to infinity as the lattice spacing goes to zero. The major practical issue will be to provide convincing evidence that the lattice spacing is small enough to provide a good approximation to the underlying continuum theory.

The last key idea is the use of stochastic, or Monte Carlo, methods to evaluate the lattice path integrals. One should note at the outset that the common misnomer of Monte Carlo "simulations" is quite misleading. In fact, we are not simulating anything. Rather, we are solving an equation in the same sense as one always uses numerical analysis to solve equations. That is, one first selects a desired level of precision, and then using appropriate theorems, determines an algorithm and a number of independent samples which yields that precision. It is also useful to note that although Monte Carlo techniques have only recently been applied to field theory, they have been exploited for decades in other fields such as condensed matter physics, quantum chemistry, and nuclear physics. Since the seminal paper of Metropolis *et al.*[2] 36 years ago, there is now extensive experience upon which to draw.

For readers who wish to go beyond the scope of the present lecture, I recommend several basic references. Much of the background material is discussed in more detail in a text co-authored with Orland.[3]. In particular, the reader is referred to Chapter 1 for treatment of coherent states and Grassmann variables, Chapter 2 for discussion of path integrals, and Chapter 8 for a detailed explanation of stochastic methods. A terse introduction to lattice gauge theory is provided by Creutz[1] and more details may be found in the reprint volume edited by Rebbi[4] which includes all the key articles through 1983. Finally, an extensive summary of many recent results is provided in the reviews by Fukugita.[6]

II. PATH INTEGRALS

Feynman Path Integral

The basic idea of the path integral is illustrated by considering the Feynman path integral for a single degree of freedom. The evolution operator e^{-iHt} is broken up into a large number of "time slices" separated by time interval ϵ, and a complete set of states is inserted between each interval

$$e^{-iHt} = e^{-iH\epsilon} \int dx_n |x_n\rangle\langle x_n| \, e^{-iH\epsilon} \int dx_{k-1} |x_{n-1}\rangle\langle x_{n-1}| e^{-iH\epsilon} \ldots \qquad (2.1)$$

Then, the non-commutivity of the kinetic and potential energy operators is treated by the following approximation which becomes exact in the limit $\epsilon \rightarrow 0$

$$\begin{aligned} \langle x_{k+1}| e^{-i\epsilon\left(\frac{\hat{p}^2}{2m}+V(\hat{x})\right)} |x_k\rangle &\sim \langle x_{k+1}| e^{-i\epsilon\frac{\hat{p}^2}{2m}} \int dp |p\rangle\langle p| e^{-i\epsilon V(\hat{x})} |x_k\rangle \\ &= \int dp \, e^{ip(x_{k+1}-x_k) - i\epsilon\frac{p^2}{2m} - i\epsilon V(x_k)} \qquad (2.2) \\ &= \sqrt{\frac{2m\pi}{\epsilon}} \, e^{i\epsilon \sum_k \left[\frac{m}{2}\left(\frac{x_{k+1}-x_k}{\epsilon}\right)^2 - V(x_k)\right]} + \mathcal{O}(\epsilon^2) \end{aligned}$$

Hence the evolution operator may be expressed as the sum over all paths of the exponential of the classical action

$$\begin{aligned}\langle x_f | e^{-iHt} | x_i \rangle &= \int \mathcal{D}(x_1 \dots x_n)\, e^{i\epsilon \sum_k \left[\frac{m}{2}\left(\frac{x_{k+1}-x_k}{\epsilon}\right)^2 - V(x_k)\right]} \\ &\to \int_{x(0)=x_i}^{x(t)=x_f} \mathcal{D}(x_1 \dots x_n)\, e^{i\epsilon S_{\text{classical}}(x(t))}\end{aligned} \tag{2.3}$$

The quantum mechanics of the non-commuting operators $\hat{x}$ and $\hat{p}$ has thus been represented by an ordinary integral over an additional time variable. This result may be generalized to many degrees of freedom as follows

$$\left\langle x_1^f \dots x_N^f \left| e^{-iHt} \right| x_1^i \dots x_N^i \right\rangle = \int_{x_1^i \dots x_N^i}^{x_1^f \dots x_N^f} e^{i\epsilon \sum_k \left[\sum_i \frac{m}{2}\left(\frac{x_i^{k+1}-x_i^k}{\epsilon}\right)^2 - \frac{1}{2}\sum_{ij} v(x_i^k - x_j^k)\right]} \tag{2.4}$$

where a complete set of states $|x_1 \dots x_N\rangle \langle x_1 \dots x_N|$, is inserted at each time slice.

One important property of the path integral is that a time-ordered product is represented as follows:

$$\begin{aligned}T\, \mathcal{O}(t_1)\mathcal{O}(t_2)\, e^{-i\int_0^T dt H(t)} &= e^{-iH(T-t_2)}\mathcal{O}(t_2)\, e^{-iH(t_2-t_1)}\mathcal{O}(t_1)\, e^{-iH(t_1-0)} \\ &\to \int \mathcal{D}(x_1 \dots x_n)\, e^{iS(x_1 \dots x_n)}\mathcal{O}(x_{k_2})\mathcal{O}(x_{k_1})\end{aligned} \tag{2.5}$$

Hence, any path integral composed of e^{iS} and a sequence of operators automatically corresponds to a time-ordered product.

The classical limit is obtained by including the factors of $\hbar$ which have been suppressed thus far and applying the stationary phase approximation

$$\int \mathcal{D}(x)\, e^{\frac{i}{\hbar}S(x)} \underset{\text{SPA}}{\longrightarrow} e^{\frac{i}{\hbar}S(x_{\text{cl}})} \left(\frac{1}{\sqrt{\det\left(m\frac{d^2}{dt^2} + V''(x_0(t))\right)}} + \mathcal{O}(\hbar) \right) \tag{2.6}$$

in which case the path integral represents the sum of all quadratic fluctuations around the classical path.

It is important to note that there is nothing sacred about the physical time, and any continuous variable may be "sliced" to treat the non-commutivity of $\hat{x}$ and $\hat{p}$. A common case is the Euclidean path integral, in which real time is replaced by imaginary time or temperature, with the result

$$e^{-\beta H} = \prod e^{-\epsilon H} \Longrightarrow \int \mathcal{D}(x)\, e^{-\sum_k \epsilon \left[\frac{m}{2}\left(\frac{x_{k+1}-x_k}{\epsilon}\right)^2 + V(x_k)\right]} \tag{2.7}$$

In this analytic continuation in which $it \to \tau$, the Lagrangian is effectively replaced by the Hamiltonian in the exponent

$$\int dt \left[\frac{m}{2}\dot{x}^2 - V\right] \underset{it\to\tau}{\longrightarrow} \int d\tau \left[\frac{m}{2}\dot{x}^2 + V\right] \tag{2.8}$$

Salient properties of this Euclidean path integral are the fact that it is purely real, it has a well-defined measure, the Wiener measure, and it has the structure of the partition function of statistical mechanics with one extra dimension.

The boundary conditions on the path integral are specified by the specific matrix element or elements under consideration. For example, the thermodynamic trace has the form

$$\mathrm{Tr}\, e^{-\beta H} = \int dx \left\langle x \left| e^{-\beta H} \right| x \right\rangle = \int \mathcal{D}(x_0 x_1 \ldots x_n) e^{-S(x_0 \ldots x_n)} \tag{2.9a}$$

where

$$\begin{aligned} S(x_0 \ldots x_n) = \epsilon \Bigg[& \frac{m}{2} \frac{(x_0 - x_n)}{\epsilon} + V(x_n) + \frac{m}{2} \frac{(x_n - x_{n-1})^2}{\epsilon} + V(x_{n-1}) \\ & + \ldots + \frac{m}{2} \frac{(x_1 - x_0)^2}{\epsilon} + V(x_0) \Bigg] \end{aligned} \tag{2.9b}$$

and thus has periodic boundary conditions. For specific matrix elements however, we obtain the alternative form

$$\left\langle \phi_f \left| e^{-\beta H} \right| \phi_i \right\rangle = \int \mathcal{D}(x_0, x_1 \ldots x_n, x_{n+1})\, e^{-S(x_0, \ldots x_{n+1})} \tag{2.10a}$$

where

$$\begin{aligned} S(x_0 \ldots x_{n+1}) = & - \ln \phi_f(x_{n+1}) + \frac{m}{2} \left(\frac{x_{n+1} - x_n}{\epsilon} \right)^2 + V(x_n) + \ldots \\ & + \frac{m}{2} \left(\frac{x_1 - x_0}{\epsilon} \right)^2 + V(x_0) - \ln \phi_i(x_0) \quad . \end{aligned} \tag{2.10b}$$

Scalar Field Theory

Using this knowledge of the Feynman path integral, it is now easy to generalize to scalar field theory on a lattice. Let the continuum coordinate $\vec{r}$ be replaced by discrete lattice coordinates $\vec{n} \equiv (n_1, n_2, n_3)$ where the n_i are integers and lengths will be understood to be in units of the lattice spacing a. Then one simply views the lattice field theory as a quantum many-body problem where the canonical coordinate and momentum operators $\hat{x}_i$ and $\hat{p}_i$ are replaced by $\hat{\phi}(\vec{n})$ and $\hat{\pi}(\vec{n})$ and the position eigenstates $\hat{x}_i|x\rangle = x_i|x\rangle$ are replaced by eigenstates $\hat{\phi}(\vec{n})|\phi\rangle = \phi(\vec{n})|\phi\rangle$. Then, on the spatial mesh the Hamiltonian density becomes

$$\begin{aligned} & \int d^3 r \left\{ \frac{1}{2} \pi^2(r) + \frac{1}{2} \left| \nabla \phi(r) \right|^2 + V(\phi) \right\} \\ & \Longrightarrow \sum_{\vec{n}} \left\{ \frac{1}{2} \pi^2(\vec{n}) + \frac{1}{2} \sum_{i=1}^{3} \left| \phi(\vec{n} + \mu_i) - \phi(\vec{n}) \right|^2 + V\left(\phi(\vec{n})\right) \right\} \end{aligned} \tag{2.11}$$

where μ_i denotes a displacement by one lattice site in the ith direction, $\sum_{\vec{n}} \left\{ \frac{1}{2} \pi^2(\vec{n}) \right\}$ corresponds to the kinetic energy $\sum_j \frac{1}{2m} \hat{p}_j^2$, and the remaining terms, which we will denote as $F[\phi(\vec{n})]$ to avoid confusion with $V[\phi(n)]$ above, correspond to a sum of one- and two-body potentials $\sum_{ij} v(\hat{x}_i, \hat{x}_j)$. Introducing time slices as before yields

$$e^{-\beta \sum_{\vec{n}} \left\{ \frac{1}{2} \pi^2(\vec{n}) + F[\phi(\vec{n})] \right\}} = \int \mathcal{D}(\phi_k(\vec{n}))\, e^{-\epsilon \sum_{k,\vec{n}} \frac{1}{2} (\phi_{k+1}(\vec{n}) - \phi_k(\vec{n}))^2 + F[\phi_k(\vec{n})]} \tag{2.12}$$

The result is a path integral defined on a four-dimensional lattice, for which we may introduce the obvious notation $n = (n_0, n_1, n_2, n_3)$ where n_0 denotes the time label and n_i denotes the spatial label. One observes that time slicing replaces $\hat{\pi}(\vec{n})$ by $|\phi_{k+1}(\vec{n}) - \phi_k(\vec{n})|^2 \equiv |\phi(n+\mu_0) - \phi(n)|^2$ which has the same structure as the discrete spatial derivative $|\nabla\phi|^2 = \sum_{i=1}^{3} |\phi(n+\mu_i) - \phi(n)|$. Hence, a general time-ordered product acquires the simple form

$$T\mathcal{O}(\phi)\, e^{-\beta \int d^3r \{\frac{1}{2}\pi^2 + \frac{1}{2}(\nabla\phi)^2 + V(\phi)\}} \to \int \mathcal{D}(\phi(n))\, \mathcal{O}(\phi)\, e^{-S_{\mathrm{Eucl.}}(\phi)} \tag{2.13a}$$

where the Euclidean action is

$$S_{\mathrm{Eucl.}}(\phi) \equiv \sum_n \left\{ \frac{1}{2} \sum_{i=0}^{3} (\phi(n+\mu_i) - \phi(n))^2 + V(\phi(n)) \right\} \tag{2.13b}$$

This result merits several comments. Note that $S_{\mathrm{Eucl.}}(\phi)$ is completely symmetric in space and time, even though the first differences in space variables arose from a finite-difference approximation to the spatial derivatives whereas the time differences arose from the path integral time slicing. Of course, we are always free to pick different mesh spacings, a_x and a_t, in the space and time directions, respectively. Although in this derivation, we have gone from H to S using a discrete transfer matrix for evolution from one time slice to the next, it will often be useful to go backwards in the other direction to think of the lattice action as describing evolution of specific states under the Hamiltonians H from one time slice to another in order to interpret lattice observables. Depending on the problem, we may be led to apply different boundary conditions in x and t. In the case in which all boundary conditions are periodic, the physical problem corresponds to finite temperature field theory in a periodic three-dimensional box and the shortest side of the four-dimensional box will effectively act as the temperature.

Coherent States

We now need to generalize this scalar field result for general second quantized Fermion or Boson fields. Recall that the Feynman path integral, and hence the scalar field path integral, used two basic ingredients: eigenstates of $\hat{x}$, $\hat{x}|x\rangle = x|x\rangle$, and the resolution of unity $1 = \int dx |x\rangle\langle x|$. The analogs of these relations for creation and annihilation operators are provided by Boson coherent states.

The basic idea is seen most simply for a single creation operator $\hat{a}^\dagger$, which for example may correspond to a single harmonic oscillator, for which

$$\begin{gathered} [\hat{a}, \hat{a}^\dagger] = 1 \\ \left\{ \begin{matrix} \hat{a}^\dagger \\ \hat{a} \end{matrix} \right\} |n\rangle = \left\{ \begin{matrix} \sqrt{n+1} \\ \sqrt{n} \end{matrix} \right\} |n \pm 1\rangle \\ |n\rangle = \frac{1}{\sqrt{n!}} \left(\hat{a}^\dagger\right)^n |0\rangle \ . \end{gathered} \tag{2.14}$$

The coherent state $|Z\rangle$ is defined

$$|Z\rangle \equiv e^{Z\hat{a}^\dagger} |0\rangle = \sum_n \frac{Z^n}{n!} \left(\hat{a}^\dagger\right)^n |0\rangle = \sum_n \frac{Z^n}{\sqrt{n!}} |n\rangle \tag{2.15}$$

and has the following properties

$$\hat{a}|Z\rangle = \sum_n \frac{Z^n}{\sqrt{n!}}\hat{a}|n\rangle = Z\sum_n \frac{Z^{n-1}}{\sqrt{(n-1)!}}|n-1\rangle = Z|Z\rangle \tag{2.16a}$$

$$\langle Z|Z'\rangle = \sum_{mn}\langle m|\frac{Z^{*m}}{\sqrt{m!}}\frac{Z'^n}{\sqrt{n!}}|n\rangle = e^{Z^* Z'} \tag{2.16b}$$

$$\langle Z| :A(\hat{a}^\dagger,\hat{a}): |Z'\rangle = e^{Z^* Z'} A(Z^*, Z') \tag{2.16c}$$

$$\int \frac{dZ\, dZ^*}{2\pi i} e^{-Z^* Z}|Z\rangle\langle Z| = 1 \tag{2.16d}$$

The last relation is most easily demonstrated by writing the complex variable in polar form $Z = \rho e^{i\phi}$ and performing the ϕ integral first. Analogous results are straightforwardly obtained[3] for a complete set of creation operators $\hat{a}^\dagger_\alpha$

$$\begin{aligned} |Z\rangle &= e^{\sum_\alpha Z_\alpha \hat{a}^\dagger_\alpha}|0\rangle \\ \hat{a}_\alpha|Z\rangle &= Z_\alpha|Z\rangle \\ \langle Z| : A(\vec{a}^\dagger,\vec{a}) : |Z'\rangle &= e^{\sum_\alpha Z_\alpha Z'_\alpha} A(\vec{Z}^*,\vec{Z}) \\ \int\prod_\alpha \frac{dZ^*_\alpha dZ_\alpha}{2\pi i} e^{-\sum_\alpha Z^*_\alpha Z_\alpha}|Z\rangle\langle Z| &\equiv \int d\mu(Z)|Z\rangle\langle Z| = 1 \end{aligned} \tag{2.17}$$

Proceeding as before, we obtain a path integral by time slicing

$$\langle Z_f|e^{-\beta H}|Z_i\rangle = \langle Z_f|e^{-\epsilon H}\int d\mu(Z_n)|Z_n\rangle\langle Z_n|e^{-\epsilon H}\int d\mu(Z_{n-1})\ldots \tag{2.18}$$

and the matrix element of the infinitesimal evolution operator is

$$\begin{aligned} d\mu(Z_k)\langle Z_k|e^{-\epsilon H}|Z_{k-1}\rangle &= \prod_\alpha \frac{dZ^*_{k\alpha}dZ_{k\alpha}}{2\pi i} e^{-\sum_\alpha Z^*_{k\alpha}Z_{k\alpha}} \\ &\quad\times \langle Z_k| : e^{-\epsilon H(a^\dagger a)} : +\mathcal{O}(\epsilon^2)|Z_{k-1}\rangle \\ &= \prod_\alpha \frac{dZ^*_{k\alpha}dZ_{k\alpha}}{2\pi i} e^{-\sum_\alpha Z^*_{k\alpha}(Z_{k\alpha}-Z_{(k-1)\alpha})-\epsilon H(Z^*_{k\alpha},Z_{(k-1)\alpha})} \end{aligned} \tag{2.19}$$

with the result

$$\langle Z_f|e^{-\beta H}|Z_i\rangle = \int \mathcal{D}(Z^*_{k\alpha}, Z_{k\alpha})\, e^{-S(Z^*_{k\alpha},Z_{k\alpha})} \tag{2.20a}$$

where

$$S(Z^*,Z) = \sum_k \epsilon\left\{\sum_\alpha Z^*_{k\alpha}\left(Z_{k\alpha} - Z_{(k-1)\alpha}\right) + H\left(Z^*_{k,\alpha}, Z_{k-1,\alpha}\right)\right\} \tag{2.20b}$$

For Fermions with creation and annihilation operators $c^\dagger_\alpha$ and c_α one must take an additional step and introduce anticommuting Grassmann variables ξ, so that if $\hat{c}_\alpha|\xi\rangle = \xi_\alpha|\xi\rangle$ and $\hat{c}_\beta|\xi\rangle = \xi_\beta|\xi\rangle$, then we can have $\hat{c}_\alpha\hat{c}_\beta|\xi\rangle = \xi_\alpha\xi_\beta|\xi\rangle = -\xi_\beta\xi_\alpha|\xi\rangle = -\hat{c}_\beta\hat{c}_\alpha|\xi\rangle$. For our present purposes, one may regard this construction as a set of purely formal definitions. Since $\xi^2_\alpha = 0$, the only allowable functions are monomials, functions are defined by the non-vanishing terms of their Taylor series, and the definite integral is defined by the properties $\int d\xi_\alpha = 1$ and $\int d\xi_\alpha \xi_\alpha = 1$. Fermion coherent states are then defined by

$$|\xi\rangle = e^{-\sum \xi_\alpha c_\alpha^\dagger}|0\rangle \tag{2.21}$$

and satisfy relations analogous to (2.17) and yield a path integral of the form (2.18). Although there are a few technical details which may be found in Ref. [3], the essential point is that Grassmann coherent states and path integrals have essentially the same form as for Bosons, except for a few crucial minus signs which do all the correct bookkeeping for the difference between Bosons and Fermions.

Gaussian Integrals

Recall the general formula for the Gaussian integral over complex variables

$$\int \prod_i \frac{dx_i^* \, dx_i}{2\pi i} e^{-x_i^* H_{ij} x_j + J_i^* x_i + J_i x_i^*} = [\det H]^{-1} \, e^{J_i^* H_{ij}^{-1} J_j} \tag{2.22}$$

which may be proved by changing to a basis in which H is diagonal and using $\int dx \, e^{-ax^2} = \sqrt{\pi/a}$. An analogous result is obtained for Grassmann variables by noting that

$$\int d\xi^* d\xi \, e^{-\xi^* a \xi} = \int d\xi^* d\xi (1 - \xi^* a \xi) = a \tag{2.23}$$

Hence

$$\int \prod_i d\xi_i^* d\xi_i \, e^{-\xi_i^* H_{ij} \xi_j + \eta_i^* \xi_i + \eta_i \xi_i^*} = [\det H] \, e^{\eta_i^* H_{ij}^{-1} \eta_j} \tag{2.24}$$

and we see that the only difference between complex variables and Grassmann variables is that $\det H$ appears to the power -1 and 1, respectively.

With this result, we are prepared to integrate out the Grassmann variables from the path integral. Suppose the action has the form

$$S(\xi^*, \xi, \phi) = \xi_i^* M(\phi)_{ij} \xi_j + S_B(\phi) \tag{2.25}$$

where, for example, ξ^*, ξ might represent the Fermions $\bar{\psi}$, ψ in $\bar{\psi}(\not{p} - \not{A} + m)\psi + F_{\mu\nu}(A)^2$ and ψ represents the real Bose field A. Then

$$\int d\xi^* d\xi \, d\phi \, e^{\xi^* M(\phi) \xi + S_B(\phi)} = \int d\phi \, e^{\ln \det M(\phi) + S_B(\phi)} \tag{2.26}$$

and we are left with an integral over the real field ϕ of an effective action

$$S_{\text{eff}}(\phi) = \ln \det M(\phi) + S_B(\phi) \tag{2.27}$$

In the same way, we can perform the Gaussian integrals for propagators. Consider first the propagator (or contraction in the language of Wick's theorem) corresponding to the thermodynamic average of the time-ordered product of field annihilation and creation operators at space-time points $i = (x_i t_i)$ and $j = (x_j t_j)$, respectively:

$$\begin{aligned} \langle T\psi_i \bar{\psi}_j \rangle &= \operatorname{Tr} T\psi_i \bar{\psi}_j \, e^{-\bar{\psi} M(\hat{\phi}) \psi + S_B(\hat{\phi})} \\ &= \int d\xi^* d\xi \, d\phi \, \xi_i \xi_j^* \, e^{-\xi^* M(\phi) \xi + S_B(\phi)} \\ &= \int d\phi \, M^{-1}(\phi)_{ij} e^{S_{\text{eff}}(\phi)} \ . \end{aligned} \tag{2.28}$$

The last line is obtained by differentiating Eq. (2.24) with respect to η_i and η_j which brings down the Grassmann variables ξ_i^* and ξ_j^* on the left and the inverse matrix on the right. The general integral with n pairs of creation and annihilation operators follows similarly from taking n pairs of derivatives and yields the general form of Wick's theorem:

$$\begin{aligned}
&\int \mathcal{D}(\xi^*\xi)\xi_{i_1}\ldots\xi_{i_n}\xi_{j_n}^*\ldots\xi_{j_1}^*\, e^{-\xi^* M\xi} \\
&\quad = \frac{\delta^{2n}}{\delta\eta_{i_1}^*\ldots\delta\eta_{i_n}^*\delta\eta_{j_n}\ldots\delta\eta_{j_1}} \int \mathcal{D}(\xi^*\xi)\, e^{-\xi_i^* M_{ij}\xi_j+\eta_i^*\xi_i+\eta_i\xi_i^*}\Bigg|^{\eta=\eta^*=0} \\
&\quad = \frac{\delta^{2n}}{\delta\eta_{i_1}^*\ldots\delta\eta_{i_n}^*\delta\eta_{j_n}\ldots\delta\eta_{j_1}} \det H\, e^{\eta_i^* M_{ij}\eta_j}\Bigg|^{\eta=\eta^*=0} \\
&\quad = \sum_P (-1)^P M^{-1}_{i_{P_n}j_n}\ldots M^{-1}_{i_{P_1}j_1}\, e^{\ln\det H}
\end{aligned} \tag{2.29}$$

where P denotes a permutation of the n indices. Hence, the Fermions may be integrated out of any physical observable when the action has the form (2.25), leaving the sum of all possible contractions weighted by the effective action (2.27), and we are left with an effective theory containing only Bosonic degrees of freedom.

Boundary Conditions

Thus far, we have not been specific about the boundary conditions arising in the action (2.20). For an arbitrary matrix element in a basis of Boson or Fermion coherent states, we have

$$\langle Z_F|e^{-\beta H}|Z_I\rangle = \int \mathcal{D}(Z^*Z)e^{-S(Z^*Z)} \tag{2.30a}$$

where

$$\begin{aligned}
S(Z^*Z) = &-Z_F^*Z_n + H_{F,n} + Z_n^*(Z_n - Z_{n-1}) + H_{n,n-1} \\
&+\ldots + Z_1^*(Z_1 - Z_I) + H_{z,I} \;.
\end{aligned} \tag{2.30b}$$

For the thermodynamic trace, the same argument which led to the completeness relation in (2.17) yields

$$\mathrm{Tr}\, e^{-\beta H} = \int dZ_0^* dZ_0\, e^{-Z_0^*Z_0}\langle \pm Z_0|e^{-\beta H}|Z_0\rangle \tag{2.31a}$$

where the upper and lower signs refer to Bosons and Fermions, respectively, and the extra minus sign for Fermions arises from the anticommutation of the Grassmann variable $\langle\psi|\xi\rangle\langle\xi|\chi\rangle = \langle -\xi|\chi\rangle\langle\psi|\xi\rangle$. Hence, for the trace, the boundary condition in the action (2.30b) is periodic or antiperiodic:

$$\begin{aligned}
S(Z^*Z) = &\pm Z_0^*(\pm Z_0 - Z_n) + H_{0,n} + Z_n^*(Z_n - Z_{n-1}) \\
&+ H_{n,n-1} + \ldots + Z_1^*(Z_1 - Z_0)H_{1,0} \;.
\end{aligned} \tag{2.31b}$$

Every time slice is physically equivalent to every other, and we will regard the lattice as being on a cylinder or torus.

It will also be useful to consider matrix elements between zero-particle states, which in the case of normal ordered field operators, corresponds to no quarks or antiquarks relative to the vacuum. Since $|Z\rangle \equiv e^{Z_a^\dagger}|0\rangle$, the zero-particle state is just the state with $Z = 0$. Hence, the matrix element $\langle 0|e^{-\beta H}|0\rangle$ is given by the action

$$S(Z^*Z) = Z_n^*(Z_n - Z_{n-1}) + H_{n,n-1} + \ldots + Z_2^*(Z_2 - Z_1) + H_{2,1} + Z_1^*Z_1 \;. \tag{2.32}$$

In contrast to the periodic or antiperiodic action (2.31b), the action (2.32) has a "hard wall" boundary condition which prevents particle propagation outside the interval $(0,\beta)$.

III. LATTICE QCD FOR GLUONS

It is useful to begin the study of lattice gauge theory with the simplest possible gauge theory, and gradually increase the generality and complexity one step at a time. Hence, in this chapter we will completely ignore Fermions, and concentrate only on the pure gluon sector. This will correspond to the physical limit in which the quark mass goes to infinity and quarks cease to be dynamical degrees of freedom. Furthermore, we will begin with the simplest possible gauge group, $U(1)$ corresponding to QED, and only after motivating and displaying the Wilson action for this case will we move to the non-Abelian $SU(N)$ gauge theory.

U(1) Gauge Theory and the Wilson Action

To motivate the way gauge theory will be formulated on a discrete space-time lattice, it is useful to recall the essential ideas underlying continuum gauge theory, and how the entire theory may be viewed as arising from the principle of gauge invariance. Therefore, let us consider a Lagrangian for a complex scalar field

$$L = \partial_\mu \phi^* \partial_\mu \phi - V(\phi^* \phi) \tag{3.1}$$

and look for the simplest extension of the theory which is consistent with local gauge invariance. Note that since our final goal will be to calculate Euclidean path integrals, we will always write the Lagrangian and action in Euclidean form, with the result that $g_{\mu\nu} = \delta_{\mu\nu}$ and upper and lower Dirac indices are equivalent. Whereas L is manifestly invariant under the global gauge transformation $\tilde{\phi}(x) = -e^{-i\alpha}\phi(x)$, the derivatives in L yield new terms in the case of a local transformation $\tilde{\phi}(x) = e^{-i\alpha(x)}\phi(x)$

$$\partial_\mu \tilde{\phi}^* \partial_\mu \tilde{\phi} = \left[(\partial_\mu - i(\partial_\mu \alpha))\,\phi\right]^* (\partial_\mu - i(\partial_\mu \alpha))\,\phi \tag{3.2}$$

If we adopt the principle of local gauge invariance, that the theory should be independent of the arbitrary phase choice $\alpha(x)$ that various observers might choose at different points in space, then we may repair the theory by adding a "compensating" field $A_\mu(x)$ such that

$$\tilde{A}_\mu(x) = A_\mu(x) + \frac{1}{g}\partial_\mu \alpha(x) \tag{3.3}$$

If we now replace the derivative ∂_μ in L by the covariant derivative

$$D_\mu \phi(x) \equiv (\partial_\mu + igA_\mu(x))\,\phi(x) \tag{3.4}$$

we observe that the transformation of $A_\mu(x)$ exactly compensates for the undesired derivative of α and yields an invariant Lagrangian.

Whereas the coupling of the new field A to ϕ was determined from gauge invariance, the only guiding principle for determining the action for A itself is simplicity and economy. Thus, we seek the simplest action for A_μ involving the least number of derivatives which is consistent with gauge invariance and Lorentz invariance. Noting that

$$\partial_\mu\left(A_\nu + \frac{1}{g}\partial_\nu \alpha\right) - \partial_\nu(A_\mu + \frac{1}{g}\partial_\mu \alpha) = \partial_\mu A_\nu - \partial_\mu A_\mu \equiv F_{\mu\nu} \tag{3.5}$$

we observe that $F_{\mu\nu}$ is gauge invariant so that ${F_{\mu\nu}}^2$ is both gauge and Lorentz invariant and we are thus led automatically to Maxwell's equations and the complete Lagrangian

$$L = -\frac{1}{4e^2}\left(\partial_\mu A_\nu - \partial_\nu A_\mu\right)^2 + \left(D_\mu \phi\right)^* \left(D_\mu \phi\right) - V\left(\phi^* \phi\right) \quad . \tag{3.6}$$

For subsequent treatment on a lattice, it is useful to note that the appropriate operator to compare fields at two different points x and y is the link variable

$$U(y,x) = e^{i\int_x^y dx_\mu g A_\mu(x)} \tag{3.7}$$

which simply removes the arbitrary phases between the two points and yields a gauge invariant result.

We now consider how to approximate this continuum theory on a space-time lattice. Often in numerical analysis, one may allow discrete approximations to break fundamental underlying symmetries. For example, when one solves the time-independent Schrödinger equation on a spatial mesh, one violates translational invariance. There may be small spurious pinning forces which reflect the fact that the energy is slightly lower when the solution is centered on a mesh site or centered between mesh sites, but there are no major qualitative errors and the quantitative errors may be strictly controlled. When one solves the time-dependent Schrödinger equation or time-dependent Hartree–Fock equation, however, one finds that it is important to enforce certain properties such as energy conservation and unitarity when discretizing the problem in time. In the case of lattice gauge theory, since by the previous argument gauge invariance plays such a crucial role in defining the theory, it is desirable to enforce it exactly in the lattice action. In contrast, as in the case of the Schrödinger equation, we will settle for an action which breaks Lorentz invariance, and simply insist on making the lattice spacing small enough that the errors are acceptably small.

Following Wilson, we define the action in terms of directed link variables assigned to each of the links between sites of the space time lattice. For $U(1)$, we define the link variable from site n in the μ direction to site $n+\mu$ as a discrete approximation to the integral $e^{ig\int_n^{n+\mu} dx\, A_\mu}$ which we denote

$$U_\mu(n) = e^{i\theta_\mu(n)} \quad . \tag{3.8}$$

Thus $\theta_\mu(n)$ is a discrete approximation to $g\int_n^{n+m} dx\, A_\mu$ along the direction of the link, and when the direction is reversed, $U_n(n) \to U_n(n)^\dagger$ and $\theta_\mu(n) \to -\theta_\mu(n)$. The link variable is then a group element of $U(1)$ and the compact variable $\theta_\mu(n)$ will be associated with $agA_\mu(x)$ in the continuum limit. With these link variables, the integral over the field variables in the path integral is replaced by the invariant group measure for $U(1)$, which is $\frac{1}{2\pi}\int_{-\pi}^{\pi} d\theta$.

The fundamental building blocks of the lattice action are products of directed link variables taken counter-clockwise around each individual plaquette of the lattice. By construction, this product is gauge invariant, ensuring gauge invariance of the resulting action. A typical plaquette is sketched in Fig. 1, where n is an arbitrary site and μ and ν denote displacements by one site in the horizontal and vertical directions. By convention, the compact variables θ_μ and θ_ν are associated with links directed in the positive μ and ν directions so that $-\theta_\mu$ and $-\theta_\nu$ must be associated with links in the negative μ and ν directions. The product of the four group elements around the plaquette may thus be written

$$\begin{aligned} U_\square &= \prod_\square U\,U\,U\,U \\ &= e^{i\theta_\mu(n)}\, e^{i\theta_\nu(n+\mu)}\, e^{-i\theta_\mu(n+\nu)}\, e^{-i\theta_\nu(n)} \\ &\equiv e^{i\Sigma_{\mu\nu}} \end{aligned} \tag{3.9a}$$

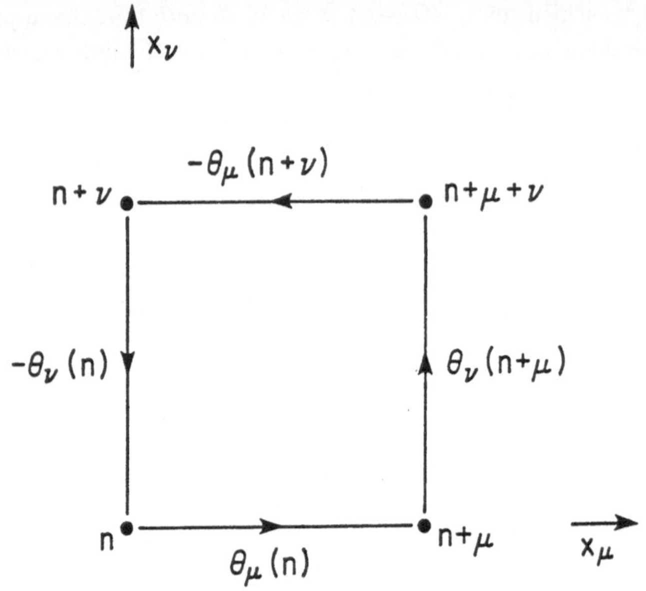

Fig. 1. An elementary plaquette of link variables.

where

$$\begin{aligned}\Sigma_{\mu\nu} &= \theta_\nu(n+\mu) - \theta_\nu(n) - (\theta_\mu(n+\nu) - \theta_\mu(n)) \\ &\equiv \Delta_\mu\theta_\nu - \Delta_\nu\theta_\mu \quad .\end{aligned} \tag{3.9b}$$

Here Δ_μ denotes the discrete lattice difference operator which becomes a derivative in the continuum limit, so that $\Sigma_{\mu\nu}$ is a discrete approximation to the curl on the lattice and is proportional to $\partial_\mu A_\nu - \partial_\nu A_\mu = F_{\mu\nu}$ in the continuum. Since each plaquette generates an approximation to $F_{\mu\nu}$, an action which corresponds to QED in the continuum limit may be constructed by choosing a function of $U_\square$ which yields $F^2_{\mu\nu}$ plus terms which are negligible in the continuum limit. Using conventional notation, the action is written

$$\begin{aligned}S &= \beta \sum_\square (1 - \mathrm{Re}\, U_\square) \\ &= \beta \sum_\square (1 - \cos \Sigma_{\mu\nu}) \quad .\end{aligned} \tag{3.10a}$$

Continuum QED is recovered by defining the new variables $\beta = \frac{1}{g^2}$ and $\theta_\mu(n) = agA_\mu(n)$, and expanding $\theta_\nu(n+\mu) = \theta_\nu(n) + a\partial_\mu\theta_\nu(n) + \mathcal{O}(a^2)$, with the result that the leading contribution as $a \to 0$ is

$$\begin{aligned}S &\sim \frac{1}{g^2}\sum_\square (1 - \cos[a(\partial_\mu\theta_\nu - \partial_\nu\theta_\mu)]) \\ &\sim \frac{1}{g^2}\sum_\square (1 - \cos(a^2 g F_{\mu\nu})) \\ &\sim \frac{1}{g^2}\sum_{n\{\mu\nu\}} \left(\frac{a^4 g^2}{2} F^2_{\mu\nu}(n) + \ldots\right) \\ &\sim \frac{1}{2}\sum_n a^4 \sum_{\{\mu\nu\}} F^2_{\mu\nu}(n) \\ &\to \frac{1}{4}\int d^4x F_{\mu\nu}(x) F_{\mu\nu}(x) \quad .\end{aligned} \tag{3.10b}$$

In the third line $\{\mu\nu\}$ denotes the sum over all pairs of μ and ν arising from the sum over plaquettes and the extra factor of $1/2$ in the last line accounts for the fact that each pair occurs twice in the double sum over repeated indices $F_{\mu\nu}F_{\mu\nu}$. The terms higher order in the lattice cutoff a vanish in the classical continuum limit and may give rise to finite renormalization of the coupling constant in quantum field theory. The lattice gauge theory defined by (3.10) is in a form which may be solved directly using the Metropolis or heat bath methods described in Section V for updating the global action.

An alternative form of lattice gauge theory which is useful in the pure gauge sector is the Hamiltonian form. Consider the generalization of the action (3.10a) to the case of unequal lattice spacings a_s in the space direction and a_t in the time direction. By repeating the steps in Eq. (3.10), it is clear that in order to retain the continuum limit with unequal spacings, the action must be

$$S = \frac{a_s}{a_t}\beta\sum_{\Box_t}(1-\cos\Sigma_{\mu\nu}) + \frac{a_t}{a_s}\beta\sum_{\Box_s}(1-\cos\Sigma_{\mu\nu}) \tag{3.11}$$

where $\Box_t$ denotes a space-time plaquette and $\Box_s$ denotes a space-space plaquette and we must have $\theta_0(n) = a_t g A_0(n)$ and $\theta_i(n) = a_s g A_i(n)$.

Choosing the temporal gauge in which U is set to unity on time links, so that $\theta_0 = 0$, we then obtain for the space-time plaquettes,

$$\frac{a_s\beta}{a_t}\,\mathrm{Re}\,U_{\Box} = \frac{a_s\beta}{a_t}\cos\left(1+\theta_i(n+\mu_0)-1-\theta_i(n)\right) \sim \frac{a_s\beta}{2a_t}\left(\theta_i(n+\mu_0)-\theta_i(n)\right)^2 \ . \tag{3.12}$$

Recalling that in the ordinary path integral, (2.2), evolution for infinitesimal time ϵ with the kinetic energy operator $\hat{p}^2 = -\left(\frac{\partial}{\partial x}\right)^2$ gives rise to the contribution to the action $\frac{(x_{k+1}-x_k)^2}{\epsilon}$, we observe that the Hamiltonian which produces the action (3.12) under evolution for infinitesimal time a_t is

$$H = -\frac{1}{2a_s\beta}\sum_{i,n}\frac{\partial^2}{\partial\theta_i(n)^2} - \frac{\beta}{a_s}\sum_{\{i,j\},\,n}(1-\cos\Sigma_{ij}) \tag{3.13}$$

where i and j run over the spatial directions and n runs over the spatial lattice sites. The continuum limit is verified by using $\beta = \frac{1}{g^2}$ and $\theta(n) = ga_sA_i(n)$ and noting that the properly normalized commutation relation on the lattice

$$[A_i(n), E_j(m)] = \frac{\delta_{ij}\delta_{nm}}{a_s^3} \tag{3.14a}$$

requires

$$E_i = -\frac{i}{a_s^3}\frac{\partial}{\partial A_i} \tag{3.14b}$$

with the result

$$\begin{aligned} H &= a_s^3\sum_{i,n}\frac{1}{2}E_i^2(n) - \frac{1}{g^2 a_s}\sum_{\{i,j,\},\,n}\left(1-\cos a_s^2 g\left(\partial_i A_j - \partial_j A_i\right)\right) \\ &\approx a_s^3\sum_{i,n}\frac{1}{2}E_i^2(n) + a_s^3\sum_{k,n}\frac{1}{2}B_k^2(n) \\ &\to \int dx\,\frac{1}{2}\left(E^2+B^2\right) \end{aligned} \tag{3.15}$$

The lattice Hamiltonian (5.14) may be viewed as a many-body Schrödinger equation with coordinates θ_i, and a four-body potential. The initial value Monte Carlo method[3] therefore provides a useful alternative to the usual global sampling of the Lagrangian action, and has been exploited for $U(1)$ and $SU(N)$ gauge theories.[6]

In discussing the relation between the Hamiltonian and Lagrangian forms of lattice gauge theory, it is useful to examine the role of Gauss' law and how the presence of external charges is manifested in the theory. The basic ideas are most easily sketched in the continuum theory. Since the Hamiltonian does not constrain the charge state of the system, we must project the states appearing in the path integral onto the space satisfying $\vec{\nabla}\cdot\vec{E} = \rho$ with a specific background charge ρ, which may be accomplished by writing a δ-function in the form $\int \mathcal{D}\chi\, e^{i\int dx\,dt\,\chi(\vec{\nabla}\cdot\vec{E}-\rho)}$. Remaining in temporal gauge $A_0 = 0$ and using the form of the path integral (2.2) in which both the coordinate $x \to A$ and momentum $p \to E$ appear, the path integral for the partition function projected onto the space with external source ρ may be written

$$\begin{aligned} Z &= \int \mathcal{D}\chi\,\mathcal{D}\vec{A}\,\mathcal{D}\vec{E}\, e^{\int dx\,dt\left[i\vec{E}\cdot\dot{\vec{A}}-\frac{1}{2}(E^2+B^2)+i\chi(\vec{\nabla}\cdot\vec{E}-\rho)\right]} \\ &= \int \mathcal{D}\chi\,\mathcal{D}\vec{A}\, e^{-\int dx\,dt\left\{\frac{1}{2}\left[\left(\dot{\vec{A}}-\vec{\nabla}\chi\right)^2+B^2-i\chi\rho\right]\right\}} . \end{aligned} \tag{3.16a}$$

Equation (3.16a) is an important result. Having started in temporal gauge $A_0 = 0$, we see that enforcing Gauss' law gives rise to a projection integral over an additional field χ which enters into the final action just like the original A_0 field. Indeed, renaming $\chi = A_0$ so that $\dot{A}_i - \partial_i A_0 = F_{0i}$ and writing the source as a set of point charges $\rho(x) = \sum_n q_n \delta(x - x_n)$, we obtain

$$z = \int \mathcal{D}A_\mu\, e^{-\int dx\,dt\frac{1}{4}F_{\mu\nu}F_{\mu\nu}} \prod_n e^{-iq_n\int dt A_0(x_n,t)} . \tag{3.16b}$$

Thus, the Hamiltonian path integral with projection is precisely the Lagrangian path integral with a line of $\pm A_0$ fields at the positions of the fixed external $\pm$ charges. In the case of no external charges, we may think of the Lagrangian path integral including the A_0 integral as the usual filter $e^{-\beta H}$ selecting out the ground state. In the presence of charges, the path integral augmented by lines of A_0 at the positions of the charges filters out the ground state in the presence of these sources.

SU(N) Gauge Theory

The generalization to non-Abelian gauge theory is straightforward. The link variables become group elements of $SU(N)$

$$U_\mu(n) = e^{i\,ag\frac{1}{2}\lambda^c A_\mu^c(n)} \equiv e^{i\,ag\tilde{A}_\mu(n)} \tag{3.17}$$

where the λ^c are Pauli matrices or Gell-Mann matrices for $SU(2)$ or $SU(3)$ and c is a color label which runs over the $N^2 - 1$ generators λ^c. The integration in the path integral is defined by the invariant group measure which we will denote by $\mathcal{D}(U)$.

Using the same labeling conventions as in Fig. 1 with θ_μ replaced by $ag\tilde{A}_\mu$, the product of $SU(N)$ group elements around an elementary plaquette is

$$\begin{aligned} U_\square &= \prod_\square U\,U\,U\,U \\ &= e^{iag\tilde{A}_\mu(n)} e^{iag\tilde{A}_\nu(n+\mu)} e^{-iag\tilde{A}_\mu(n+\nu)} e^{-i\tilde{A}_\nu(n)} \end{aligned} \tag{3.18}$$

The continuum contribution is obtained by expanding $\tilde{A}(n+\mu)$ as before and applying the Baker–Hausdorff identity $e^{\hat{X}}e^{\hat{Y}} = e^{\hat{X}\hat{Y}+\frac{1}{2}[\hat{X},Y]+\ldots]}$ to each quantity below in curly brackets with the leading order result

$$\begin{aligned} U_{\Box} &\sim \left\{e^{iag\tilde{A}_\mu}e^{iag(\tilde{A}_\nu+a\partial_\mu\tilde{A}_\nu)}\right\}\left\{e^{-iag(\tilde{A}_\mu+a\partial_\nu\tilde{A}_\mu)}e^{-iag\tilde{A}_\nu}\right\} \\ &\sim \left\{e^{iag(\tilde{A}_\mu+\tilde{A}_\nu+a\partial_\mu\tilde{A}_\nu+\frac{1}{2}iag[\tilde{A}_\mu,\tilde{A}_\nu])}e^{iag(-\tilde{A}_\mu-\tilde{A}_\nu-\partial_\nu\tilde{A}_\mu+\frac{1}{2}iag[\tilde{A}_\mu,\tilde{A}_\nu])}\right\} \\ &\sim e^{ia^2g(\partial_\mu\tilde{A}_\nu-\partial_\nu\tilde{A}_\mu+ig[\tilde{A}_\mu,\tilde{A}_\nu])} \\ &\equiv e^{ia^2g\tilde{F}_{\mu\nu}} \quad . \end{aligned} \tag{3.19}$$

Since each plaquette generates the correct non-Abelian $F_{\mu\nu}$, we may again define a discrete lattice action by choosing a function of $U_{\Box}$ which yields $F^2_{\mu\nu}$ plus terms which become negligible in the continuum limit. For $SU(N)$, we thus define

$$\beta \equiv \frac{2N}{g^2} \tag{3.20}$$

and

$$S(U) = \beta\sum_{\Box}\left(1-\frac{1}{N}\,\mathrm{Re}\;\mathrm{Tr}\,U_{\Box}\right) \quad . \tag{3.21}$$

Substitution of $U_{\Box}$ from (3.18) in the action (3.20) yields the desired continuum action in leading order

$$\begin{aligned} S(U) &= \beta\sum_{\Box}\left(1-\frac{1}{N}\,\mathrm{Re}\;\mathrm{Tr}\left(1+ia^2g\tilde{F}_{\mu\nu}-\frac{1}{2}a^4g^2\tilde{F}^2_{\mu\nu}\ldots\right)\right) \\ &\sim \frac{1}{2}\beta a^4g^2\sum_{n,\{\mu\nu\}}\frac{1}{N}\,\mathrm{Tr}\left(\frac{1}{2}\lambda^c F^c_{\mu\nu}(n)\frac{1}{2}\lambda^b F^b_{\mu\nu}(n)\right) \\ &\sim \beta\frac{g^2}{2N}\sum_n a^4\sum_{\{\mu,\nu\}}\frac{1}{2}F^c_{\mu\nu}(n)F^c_{\mu\nu}(n) \\ &\to \frac{1}{4}\int d^4x\, F^c_{\mu\nu}(x)F^c_{\mu\nu}(x) \quad . \end{aligned} \tag{3.22}$$

Summation over repeated indices is implied everywhere except where $\sum_{\{\mu\nu\}}$ denotes the sum over distinct pairs μ and ν as in Eq. (3.11). The field strength $\tilde{F}_{\mu\nu}$ has been expanded using the generators λ^c and the fact that the generators are traceless has been used to eliminate the linear term in $\tilde{F}$ in the second line and the property $\mathrm{tr}\,\lambda^b\lambda^c = 2\delta_{bc}$ has been used in the third line.

Having seen how the Wilson action (3.20) reproduces the continuum action plus terms vanishing in the continuum limit, it is clear that there is a great deal of freedom to construct other expressions with the same continuum limit. Consider, for example, a product of link variables $U_{\Box}(jk)$ around a larger rectangle of length ja in one direction and ka in the other. Repeating the steps leading to (3.22) and retaining the leading correction leads to an expression of the general form

$$\begin{aligned} \frac{1}{a^4}\left[1-\frac{1}{N}\,\mathrm{Tr}\,U_{\Box}(jk)\right] &= c_{jk}F^2_{\mu\nu} \\ &\quad + a^2\sum_m d^m_{jk}I^m\left(D_\mu,D_\nu,F_{\alpha\beta},F_{\gamma\delta}\right) \\ &\quad + \mathcal{O}(a^4) \end{aligned} \tag{3.23}$$

where I^m denotes the mth invariant which can be constructed by contracting the indices in two D's and two F's and the c's and d's are calculable coefficients depending on the lengths j and k. This freedom can be exploited to construct higher order actions by taking linear combinations of the action for various $U_{\square}(jk)$ such that the leading term in $F_{\mu\nu}^2$ and the a^2 terms identically cancel.[7] This procedure is analogous to deriving the five-point formula for a second derivative by taking the linear combinations

$$\begin{aligned}
\frac{4}{3}\left[\frac{1}{a^2}(f_1 - 2f_0 + f_{-1}) = \frac{d^2 f}{dx^2} + \frac{1}{12}a^2\frac{d^4 f}{dx^4} + \mathcal{O}(a^4)\right] \\
-\frac{1}{3}\left[\frac{1}{(2a)^2}(f_2 - 2f_0 + f_{-2}) = \frac{d^2 f}{ax^2} + \frac{1}{12}(2a)^2\frac{d^4 f}{ax^4} + \mathcal{O}(a^4)\right] \quad .
\end{aligned} \tag{3.24a}$$

with the result

$$\frac{1}{12a^2}(-f_2 + 16f_1 - 30f_0 + 16f_{-1} - f_2) = \frac{d^2 f}{dx^2} + \mathcal{O}(a^4) \quad . \tag{3.24b}$$

Just as numerical calculations are often more efficient using the more complicated five-point formula with fewer mesh points, it is reasonable to consider higher order actions for serious numerical calculations.

Wilson Loops and Lines

Rephrasing the original discussion of gauge fields in terms of lattice variables, if there were a quark field defined on a lattice, then under a local gauge transformation, the field ψ_i at each site would be multiplied by a group element g_i. The link variables were explicitly introduced to compensate such a gauge transformation, so the link variable U_{ij} going from site i to j is multiplied by g_i and g_j^{-1}. Thus, the overall effect of a gauge transformation is the following:

$$\begin{aligned}
U_{ij} &\rightarrow g_i U_{ij} g_j^{-1} \\
\psi_i &\rightarrow g_i \psi_i \\
\bar{\psi}_i &\rightarrow \bar{\psi}_i g_i^{-1} \quad .
\end{aligned} \tag{3.25}$$

In the pure gauge sector, where the only variables are link variables, it is clear from (3.25) that the only gauge invariant objects which can be constructed are products of link variables around closed paths, for which the factors of g and g^{-1} combine at each site. The Wilson loop is therefore defined as the trace of a closed loop of link variables

$$\begin{aligned}
W &\equiv \mathrm{Tr} \prod_{ij \in c} U_{ij} \\
&= \mathrm{Tr}\, U_{ij} U_{jk} \ldots U_{mn} U_{ni}
\end{aligned} \tag{3.26}$$

and specifies the rotation in color space that a quark would accumulate along the loop c from the path-ordered product $P_c\, e^{\int ig\tilde{A}}$.

To understand the physical significance of a space-time Wilson loop, it is useful to note that by (3.25), a quark creation operator at site j and an annihilation operator at site i transform under gauge transformations like a product of link variables connecting sites i and j

$$\begin{aligned}
\psi_i \bar{\psi}_j &\rightarrow g_i \psi_i \bar{\psi}_j g_j^{-1} \\
U_{ik} U_{k\ell} \ldots U_{mj} &\rightarrow g_i U_{ik} U_{k\ell} \ldots U_{mj} g_j^{-1} \quad .
\end{aligned} \tag{3.27}$$

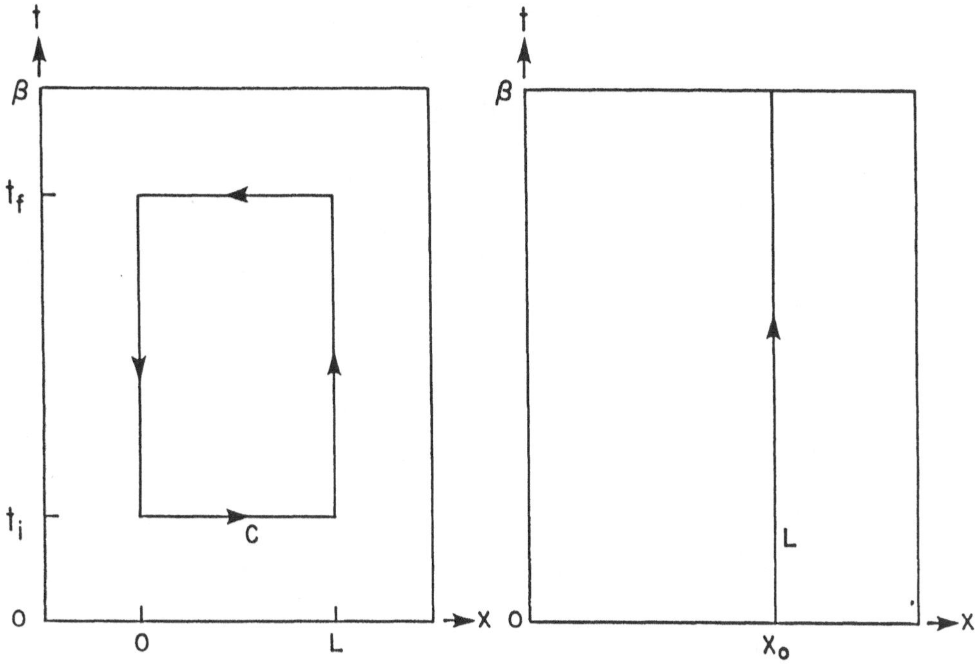

Fig. 2. A space-time Wilson loop defined by the chain of link variables C on a finite lattice (left) and a Wilson or Polyakov line defined by the chain of link variables L on a finite lattice (right).

Thus, as far as the gluon fields are concerned, the ends of a chain of link variables are equivalent to an external quark-antiquark source, and the presence of such a chain of link variables therefore measures the response of the gluon fields to an external quark-antiquark source.

Now, consider the time evolution of the system corresponding to the expectation value of the Wilson loop drawn in Figure 2

$$\langle W \rangle = \frac{\int dU\, e^{-S(U)}\, \mathrm{Tr} \prod_{c} U_{ij}}{\int dU\, e^{-S(U)}} \quad . \tag{3.28}$$

Prior to the time t_i, there are no color sources present, so evolution filters out the gluon ground state in the zero charge sector, $|0\rangle = e^{-t_i H}\, |Q = 0\rangle$. At time t_i, the line of link variables between 0 and L creates an external antiquark source at 0 and a quark source at L. As discussed in connection with Eq. (3.16b), the links in the time direction between t_i and t_f maintain these sources at 0 and L. Hence, for any t between t_i and t_f, the evolution filters out the lowest gluon configuration in the presence of external quark-antiquark sources producing the state $|\psi\rangle = e^{-(t-t_i)H}\psi(0)\bar{\psi}(L)|0\rangle$. Finally, at time t_f, the external sources at 0 and L are removed by a line of links from L to 0, and the system is returned to the zero charge sector. Using Feynman's picturesque language of antiquarks corresponding to quarks propagating backwards in time, one may succinctly characterize the Wilson loop as measuring the response of the gluon fields to an external quark-like source traveling around the perimeter of the space-time loop in the direction of the arrows.

Quantitatively, if $t_f - t_i$ is large enough, the lowest gluon state in the presence of quark and antiquark sources separated by L will dominate, and $\langle W \rangle$ will be proportional to $e^{-(t_f-t_i)V(L)}$ where $V(L)$ is the static quark-antiquark potential. Physically, this potential corresponds to the potential arising in heavy quark spectroscopy, and in

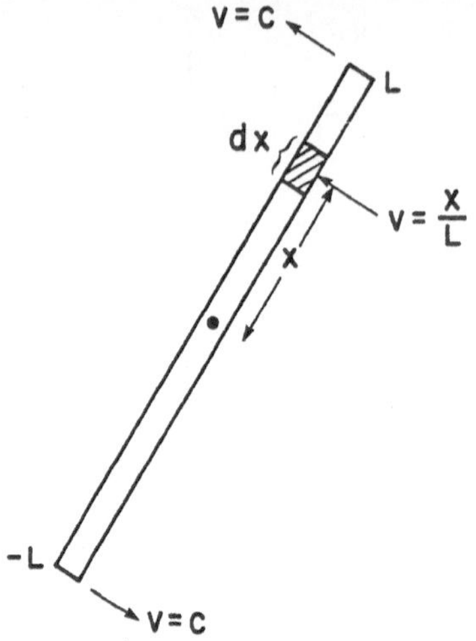

Fig. 3. Sketch of a relativistic flux tube. A massless quark at L and anti-quark at −L moving at the velocity of light are connected by a flux tube with string tension σ.

a later section lattice calculations will be compared with a phenomenological quark-antiquark potential. Furthermore, at large distances, the potential in the pure gluon sector becomes linear (since the flux tube cannot be broken by $q\bar{q}$ pair creation) so the Wilson loop enables direct numerical calculation of the string tension. If the Wilson loop has I links in the time direction and J links in the space direction, then

$$\begin{aligned} W(IJ) = \left\langle \mathrm{Tr} \prod_{C_{IJ}} U \right\rangle & \underset{I\to\infty}{\sim} e^{-aIV(aJ)} \\ & \underset{I,J\to\infty}{\sim} e^{-a^2\sigma IJ} \quad . \end{aligned} \tag{3.29}$$

The exponent is thus proportional to the area for large loops, and this area behavior is a signature of confinement, since it arises directly from the linearly rising potential. Although the preceding physical argument was framed in Hamiltonian form with evolution in the time direction, it is clear that because of the symmetry of the Euclidean action, all space-time dimensions are equivalent and the area law reflects this symmetry.

Since the string tension can be calculated directly from Wilson loops, it is useful to relate it to an experimentally measurable quantity, the slope of Regge trajectories. It is an empirical fact that families of meson states with a given set of internal quantum numbers have mass dependence on the total angular momentum J which is accurately described by the Regge formula

$$M_J^2 = \frac{1}{\alpha'} J \qquad \alpha' = 0.9\,\mathrm{GeV}^{-2} \quad . \tag{3.30}$$

To see how the slope α' is related to the string tension, it is useful to consider a very simple model in which a massless quark and antiquark are connected by a string or flux tube of length $2L$ as sketched in Fig. 3. Since the quarks are massless, they must move at the speed of light, and the velocity of the segment of string a distance x from the origin is $v = \frac{x}{L}$. If σ is defined as the energy per unit length of the flux tube in its rest frame, then the contributions of the element of length dx at point x to the energy and angular momentum are

$$dE = \gamma \sigma \, dx \ , \qquad dJ = \gamma \sigma \, v x \, dx \tag{3.31a}$$

where $v = \frac{x}{L}$ (with $c = 1$) and $\gamma = \left(1 - v^2\right)^{-1/2}$. Hence,

$$M = \int_{-L}^{L} \frac{\sigma \, dx}{\sqrt{1 - \left(\frac{x}{L}\right)^2}} = \pi \sigma L \ , \quad J = \int_{-L}^{L} \frac{\sigma \frac{x^2}{L} dx}{\sqrt{1 - \left(\frac{x}{L}\right)^2}} = \frac{\pi}{2} \sigma L^2 \tag{3.31b}$$

and

$$M^2 = \pi^2 \sigma^2 L^2 = 2\pi\sigma J \tag{3.31c}$$

so that

$$\sqrt{\sigma} = \left[2\pi\alpha'\right]^{-1/2} = 420 \, \text{MeV} \ . \tag{3.32}$$

Clearly this flux tube model is a drastic oversimplification, especially for low angular momentum states for which the finite width of the tube, the structure of the end caps, and the lack of localization of the quarks could all produce large corrections. At large angular momentum, however, the picture is somewhat more convincing. In the context of this model, I will therefore take the point of view that the accurate Regge behavior at low angular momentum is accidental and that σ is primarily determined by high angular momentum states. However, since the ultimate theory is undoubtedly much more complicated, I will not regard it as a serious problem if lattice parameters determined from this rough argument disagree at the 5 – 10% level relative to those determined from other observables such as direct calculation of hadron masses.

Although the area law behavior of large Wilson loops is clear from (3.29), in practical calculations on finite lattices, there are significant corrections, including a term proportional to the perimeter arising from the self-energy of the external sources and a constant arising from gluon exchanges at the corners, so that

$$-\ln W(I, J) \sim C + D(I + J) + a^2 \sigma I J \ . \tag{3.33}$$

To eliminate the constant and perimeter terms, the following ratio of Wilson loops having the same perimeter is calculated to cancel out the C and D terms in (3.33)

$$\chi(I, J) = -\ln \left(\frac{W(I, J) W(I-1, J-1)}{W(I, J-1) W(I-1, J)} \right) \sim a^2 \sigma \tag{3.34}$$

and we will subsequently show results for this quantity.

A Wilson or Polyakov line is another form of gauge invariant closed loop which can be placed on a periodic lattice. In this case, as sketched in Fig. 2, the links are located at a fixed position in space x_0 and run in the time direction from the first time slice to the last, which by periodicity is equivalent to the first and thus renders the product gauge invariant. If the length of the lattice in the time direction is β_t (were the subscript t distinguishes it from the inverse coupling constant $\beta_g \equiv \frac{2N}{g^2}$ to which we will append a subscript g when necessary), the expectation value of the Wilson line yields the partition function for the gluon field in the presence of a single fixed quark at inverse temperature β_t and thus specifies the free energy F_{quark} of a single quark.

$$\begin{aligned} \left\langle \hat{L} \right\rangle &= \frac{\int \mathcal{D}U \, e^{-S(U)} \, \text{Tr} \prod_L U_{ij}}{\int \mathcal{D}U \, e^{-S(U)}} \\ &= e^{-\beta_t F_{\text{quark}}} \ . \end{aligned} \tag{3.35}$$

This quantity will be useful shortly as an order parameter for the deconfinement phase transition. Note that because the periodic lattice is a four-dimensional torus and L winds around the lattice once in the time direction, it is characterized by a winding number and is thus topologically distinct from a Wilson loop which has winding number 0. By the preceding argument, two lines in opposite directions, one at $x = 0$ and one at $x = L$, will produce the free energy of a quark and antiquark separated by distance L, and as $\beta_t \to \infty$ this provides an alternative means of calculating the static quark-antiquark potential

Strong Coupling Expansion

One can obtain a useful physical picture of what happens when one evaluates the expectation value of a Wilson loop

$$\langle W \rangle = Z^{-1} \int \mathcal{D}(U)\, e^{-\beta_g \sum_\Box \left(1 - \frac{1}{2N} \mathrm{tr} \left(U_\Box + U_\Box^\dagger\right)\right)} \,\mathrm{tr} \prod_C U \tag{3.36}$$

by expanding the exponent in powers of the inverse coupling constant $\beta_g \equiv \frac{2N}{g^2}$. Since this expansion requires small β_g and thus large g^2, it is called the strong coupling expansion. Note that although it is formally analogous to the high temperature expansion in statistical mechanics, where β_g would be replaced by the inverse temperature, the actual physical temperature of our system is specified by the length of the lattice in the time direction, β_t, and is distinct from β_g.

The structure of the expansion is revealed by considering the integrals over group elements which arise in the path integral (3.36). A general discussion of integration over $SU(N)$ group elements is given by Creutz,[1] but for our present purposes it is sufficient to use the following two results, where Greek indices denote $SU(N)$ matrix indices, not sites

$$\int dU\, U_{\alpha\beta} = 0 \tag{3.37a}$$

$$\int dU\, U_{\alpha\beta} U^{-1}_{\gamma\delta} = \frac{1}{N} \delta_{\alpha\delta} \delta_{\beta\gamma} \tag{3.37b}$$

which follow directly from the orthogonality relation for irreducible matrix representations of the group and are trivially verified for $U(1)$ for which the invariant measure is $\int dU = \frac{1}{2\pi} \int_{-\pi}^{\pi} d\theta$ and $U = e^{i\theta}$.

Now consider the diagrams which result from drawing the links in the Wilson loop $\prod_C U$ and some set of plaquettes $U_\Box$ and $U_\Box^\dagger$ obtained from expanding the exponential in (3.36). The integral (3.37a) tells us that any diagram which has a single exposed link (that is, a single link between a pair of sites) anywhere on the lattice gives no contribution. Thus, the only non-vanishing terms in the expansion are those in which we manage to mate plaquettes from the exponential with the Wilson loop to eliminate all exposed links. The simplest way to mate two links to obtain a non-vanishing result is to place them between the same sites in opposite directions, which by (3.37b) yields $\frac{1}{N}$. Since each plaquette brings with it a factor of β, the lowest order non-vanishing contribution to $\langle W \rangle$ is obtained by "tiling" the interior of the Wilson loop with plaquettes oriented in the opposite direction as sketched in Fig. 4 for a 3×3 loop. Note that each of the outer links of the original Wilson loop is protected, and the interior links protect each other pairwise. The leading term for an $I \times J$ Wilson loop would thus have $I \times J$ tiles, each contributing a factor $\frac{\beta}{2n}$.

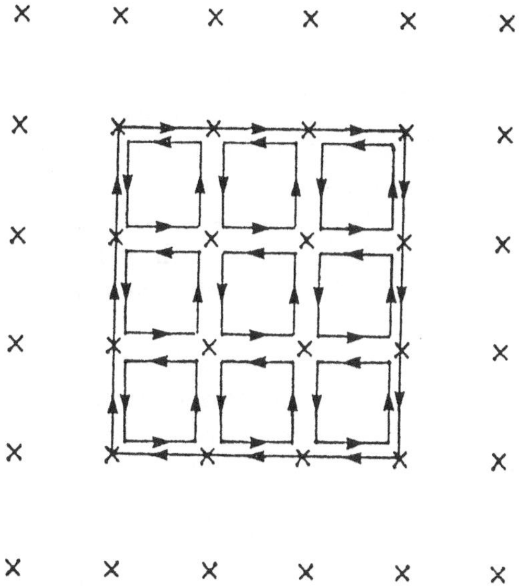

Fig. 4. A 3 × 3 Wilson loop tiled with plaquettes in the strong coupling expansion.

In addition, because of the traces in the plaquettes and Wilson loop in (3.36) and the δ's in Eq. (3.37b) there is a factor of N for each of the $(I+1)(J+1)$ sites and because of the factor $\frac{1}{N}$ in (3.37b) there is a factor $\frac{1}{N}$ for each of the $(2IJ+I+J)$ double bonds. Hence, except for $SU(2)$, where the counting is different because the two orientations of plaquettes are equivalent, the overall contribution goes as

$$W(IJ) \sim \left(\frac{\beta}{2N^2}\right)^{IJ} \tag{3.38a}$$

giving the lowest order contribution to the string tension

$$\sigma \sim -a^{-2} \ln \left(\frac{\beta}{2N^2}\right) \quad . \tag{3.38b}$$

Fancier tilings are also possible if one is willing to use more tiles and thus include more powers of β. For example, one could place five tiles together to make a cubic box with an open bottom and replace one or more tiles with this box. The box could be elongated, or even grown into a tube which connects back somewhere else. Alternatively, one could replace a plaquette oriented in one direction by $(N-1)$ plaquettes oriented in the opposite direction to obtain a non-vanishing $SU(N)$ integral.

The utility of this expansion is twofold. It provides a physical picture of filling in the Wilson loop with a gluon membrane, whose vibrations and contortions represent all the quantum fluctuations of the gluon field. When observed on a particular time slice, the cross section of this surface corresponds to a color flux tube joining the quark-antiquark sources. In addition, in low orders, the individual terms can be calculated explicitly and provide a valuable quantitative check of numerical calculations.

The formal similarity between the Euclidean path integral for pure gauge theory and statistical mechanics in $d+1$ dimensions allows us to apply standard techniques for the symmetry analysis of phase transitions to the confinement transition. Symmetry analysis of gauge theory is treated in detail in the review by Svetitsky,[8] and I will only summarize the main ideas here.

The essential idea is to follow the approach of Landau and identify an order parameter characterizing the transition, construct an effective action in terms of this order parameter, and use symmetry considerations to identify the form of the action.

The Wilson loop satisfies Landau's definition of an order parameter since, by (3.35), $\langle L\rangle = e^{-\beta_t F_{\text{quark}}}$. In the low-temperature confined phase, the free energy of single quark is infinite so that $\langle L\rangle = 0$ whereas in the deconfined high-temperature phase the free energy is finite and thus $\langle L\rangle \neq 0$. Since this behavior appears superficially to be just the opposite of that of a magnet, which has magnetization $\langle M\rangle \neq 0$ at low temperature and $\langle M\rangle = 0$ at high temperature, it is important to recognize once again that the parameter which enters the lattice partition function analogously to the temperature in statistical mechanics is not the physical temperature. For a fixed lattice, the action has the form $e^{-\frac{2n}{g^2}S_\Box}$ so that g^2 plays the role of an effective temperature whereas the physical inverse temperature is $\beta_t = N_t a$ where N_t is the (fixed) number of lattice sites in the time direction. Thus, in order for the physical temperature to increase, a must decrease, which as we shall see subsequently, means that g decreases, so that the effective temperature decreases. Hence, both a magnet and a lattice gauge system have finite order parameters when the relevant effective temperature is low and vanishing order parameters at high effective temperature.

It is useful to define an effective action $S_{\text{eff}}[L]$ such that the original partition function obtained by integrating the lattice action $S(U)$ over the group elements can be written as an integral of $S_{\text{eff}}[L]$ over the order parameter

$$Z = \int \mathcal{D}U\, e^{-S(U)} = \int \mathcal{D}L\, e^{-S_{\text{eff}}[L]} \quad . \tag{3.39a}$$

This effective action is constructed in the standard way by introducing a δ-function requiring that $L(x)$ be equal to the expectation value of a Wilson line at point x:

$$e^{-S_{\text{eff}}[L]} \equiv \int \mathcal{D}U\, e^{-S(U)} \prod_x \delta\left(L(x) - \text{Tr} \prod_{L_x} U_{ij}\right) \tag{3.39b}$$

I will now show that this effective action is symmetric with respect to the center of the gauge group, and that this symmetry has important implications for the confinement phase transition.

Let Z be an element of the center of the group, $Z \in \mathcal{C}$. That is, Z commutes with every element of the group. Suppose every link in the time direction originating on a particular time slice is multiplied by Z. In effect, the lattice is now no longer periodic but rather is periodic to within multiplication by the center element Z. The lattice action $S(U)$ is invariant under this multiplication by Z, since any space-time plaquette containing the affected time links is transformed as

$$\begin{aligned} U_1 U_2 U_3^\dagger U_4^\dagger &\to Z U_1 U_2 U_3^\dagger Z^\dagger U_4^\dagger \\ &= U_1 U_2 U_3^\dagger U_4^\dagger \quad . \end{aligned} \tag{3.40}$$

The last line shows that it is essential that Z commute with all elements of the group. Each Wilson line, however, necessarily contains one factor of Z which may be commuted to this end of the line, so that

$$\prod_{L_x} U_{ij} \to Z \prod_{L_x} U_{ij} \quad . \tag{3.41}$$

A matrix which commutes with every $SU(N)$ matrix must be a multiple of the unit matrix, so we may write $Z = \mathcal{Z}I$. Since the action is invariant under multiplication of an entire time slice by Z whereas the Wilson line is multiplied by Z, the effective action has the symmetry

$$S_{\text{eff}}[L] = S_{\text{eff}}[\mathcal{Z}L] \qquad \mathcal{Z}I \in C \quad . \tag{3.42}$$

Given this symmetry of the effective action with respect to the center, the usual symmetry arguments for analyzing phase transitions apply. The major assumption at this point is that one can integrate out the degrees of freedom at short distance scales and derive a Landau–Ginzburg action of the usual local form

$$S_{\text{eff}} \to \int dx \, \left\{ (\partial_i L)^2 + V(L) \right\} \tag{3.43}$$

where the effective potential has symmetry with respect to the center $V(L) = V(\mathcal{Z}L)$ and the reader is referred to Ref. [8] for details.

We are now ready to examine the implications of the symmetry with respect to the center of the local potential $V(L)$ for the order of the deconfinement transition. First, consider $U(1)$ gauge theory. Since the group is Abelian, the center is the whole group and $V(L)$ in symmetric under $L \to e^{i\theta}L$. This requires that V is a function of $|L|^2$ and has the generic form

$$V_{U(1)}\left(|L|^2\right) = a|L|^2 + b|L|^4 + c|L|^6 + \ldots \quad . \tag{3.44}$$

A phase transition will occur in the region of temperature in which $a(t)$ changes sign, and we note that the value of L at which the minimum occurs will change discontinuously or continuously depending upon whether b is negative or positive. Thus, the order of the phase transition depends on the sign of b, and in particular, is not constrained by the form of (3.44).

Is the non-Abelian case constrained any more strongly by symmetry? It is easy to see that the center of $SU(N)$ is $Z(N)$, that is, the N roots of unity, since $\det(\mathcal{Z}I) = \mathcal{Z}^N = 1$. For $SU(2)$, this means $\mathcal{Z} = \pm 1$ and hence, $V(L) = V(-L)$. As in the case of $U(1)$ this only requires V to be a function of L^2 leading again to the form (3.44) for which the order of phase transition is indeterminate. For $SU(3)$, however, we have $Z(3)$ symmetry so that $V(L) = V\left(e^{i\frac{n2\pi}{3}}L\right)$. In this case, the action may include a cubic invariant $\operatorname{Re} L^3$ in addition to powers of $|L|^2$, so that (for real L) the action has the generic form

$$V_{SU(3)}(L) = aL^2 + bL^3 + cL^4 + \ldots \quad . \tag{3.45}$$

In the presence of a cubic term, the position of the minimum must necessarily jump discontinuously, and we therefore conclude that the phase transition must be first order. This argument, while elegant and compelling, does depend on the formally exact effective action having the essentially local Landau–Ginzburg form (3.43).

There are two essential aspects of the $Z(3)$ symmetry of $SU(3)$ for our purposes. The first is the role of the Wilson line $\langle \hat{L} \rangle$ as an order parameter and measure of spontaneous symmetry breaking. In the confining phase, $\langle \hat{L} \rangle = 0$ and there is no spontaneous symmetry breaking. Thus, a plot of a Monte Carlo calculation of $\langle \hat{L} \rangle$ in the complex phase will produce a graph which is symmetric under rotation by $\frac{2\pi}{3}$, and has points equally distributed along the directions 1, $e^{\frac{2\pi}{3}i}$, and $e^{-\frac{2\pi}{3}i}$. This is analogous to a calculation of the spin in the unbroken symmetry phase of an Ising system, for which the distribution of spins is evenly divided between up and down. In the broken symmetry deconfined phase, however, $\langle \hat{L} \rangle \neq 0$, and calculated values of $\langle L \rangle$ may be clustered around any one of the three axes 1, $e^{\frac{2\pi}{3}i}$ and $e^{-\frac{2\pi}{3}i}$, just as the ordered state of the Ising system will either be concentrated around spin up or down. This behavior is clearly displayed in the Monte Carlo calculations[9] shown in Fig. 5. As the length of the lattice in the time direction β_t, corresponding to the inverse temperature, is increased from 2 to 5 lattice units, one observes a qualitative change from the broken symmetry solution in part (a) clustered around the real axis to the completely symmetric solution in part (d) which is rotationally symmetric around the origin.

The second essential result is that the $SU(3)$ confinement transition is first order. It is difficult to establish definitively the order of a phase transition by numerical calculations on finite lattices, since discontinuous transitions in the thermodynamic limit correspond to continuous transitions on finite lattices. There is, however, very strong evidence that the transition is indeed first order, including the sharpening of the transition as the lattice size is increased, hysteresis observed when a series of calculations with decreasing temperature is compared with a series calculated with increasing temperature, and two phase coexistence in which a sequence of Monte Carlo configurations in one phase persists in apparent equilibrium for a long time followed by an equally persistent sequence of configurations in the other phase.

Particularly suggestive evidence is provided by the microcanonical Monte Carlo results[10] shown in Fig. 6 for $SU(2)$ and $SU(3)$. Recall that for a van der Waals liquid-gas transition, the density is a multi-valued function of the temperature and the Maxwell construction specifies the actual discontinuity in the density occurring in the first-order transition in the thermodynamic limit. However, the S-shaped curve corresponds to metastable states which, although inaccessible in the canonical ensemble, can be explored by microcanonical calculations. For gauge theory on a lattice with a fixed number of sites, the order parameter $\langle L \rangle$ corresponds to the density order parameter and β_g corresponds to the temperature since increasing β_g decreases g which decreases the lattice spacing a and thus increases the temperature. Figure 6 clearly demonstrates that on the same finite lattice, the Wilson line in $SU(3)$ shows the double-valued van der Waals dependence on β_g characteristic of a first-order transition whereas in $SU(2)$ there is no reentrant behavior and the transition appears second order. This result is consistent with and strongly supports the $Z(N)$ symmetry analysis.

As we shall see subsequently, the deconfinement transition is much more difficult to treat in the presence of dynamical Fermions. Physically, it is clear that a Wilson line no longer serves as a rigorous order parameter since the possibility of creating quark-antiquark pairs from the vacuum can lead to screening of the external source and thus a finite rather than infinite free energy. In terms of our symmetry analysis, introduction of gauge invariant quark-gluon coupling terms of the form $\bar{\psi}_i U_{ij} \psi_j$ destroys the $Z(N)$ symmetry of the action since $\bar{\psi}_\tau U \psi_{\tau+1} \rightarrow Z \bar{\psi}_\tau U \psi_{\tau+1}$ at the time slice on which temporal link variables are multiplied by elements of the center.

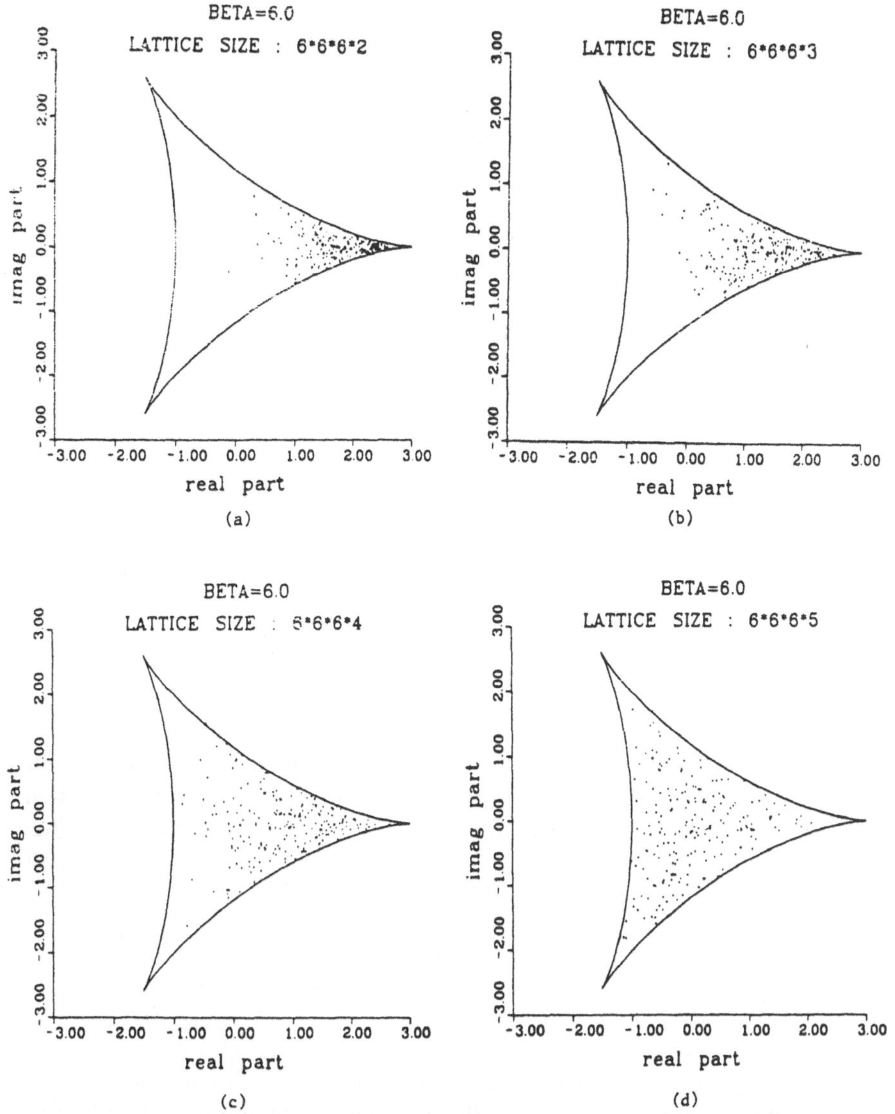

Fig. 5. Monte Carlo calculations from Ref. [4] of the order parameter $\langle L \rangle$ in the complex plane, showing symmetry restoration as the inverse temperature β_t increases from 2 to 5 lattice units.

Continuum Limit and Renormalization

Pure gauge theory on a finite lattice is specified by two parameters: the dimensionless bare coupling constant g and the lattice spacing a corresponding to a momentum cutoff $p_{\max} \sim \frac{\pi}{a}$. As a is changed, the bare g must be changed to keep physical quantities fixed.

In principle, the renormalization procedure on a lattice is very simple and could be carried out as follows. First, pick an initial value of g and calculate some set of dimensionful physical observables $\langle \mathcal{O}_i \rangle$. These observables may be written in the form

$$\langle \mathcal{O}_i \rangle = a^{-d_i} \langle f_i(g) \rangle \tag{3.46}$$

where d_i is the dimension of the operator and f_i is the dimensionless quantity calculated on the lattice using the Wilson action with $\beta = \frac{2N}{g^2}$ and with all lengths expressed in units of the lattice spacing a. For example, we have already seen in

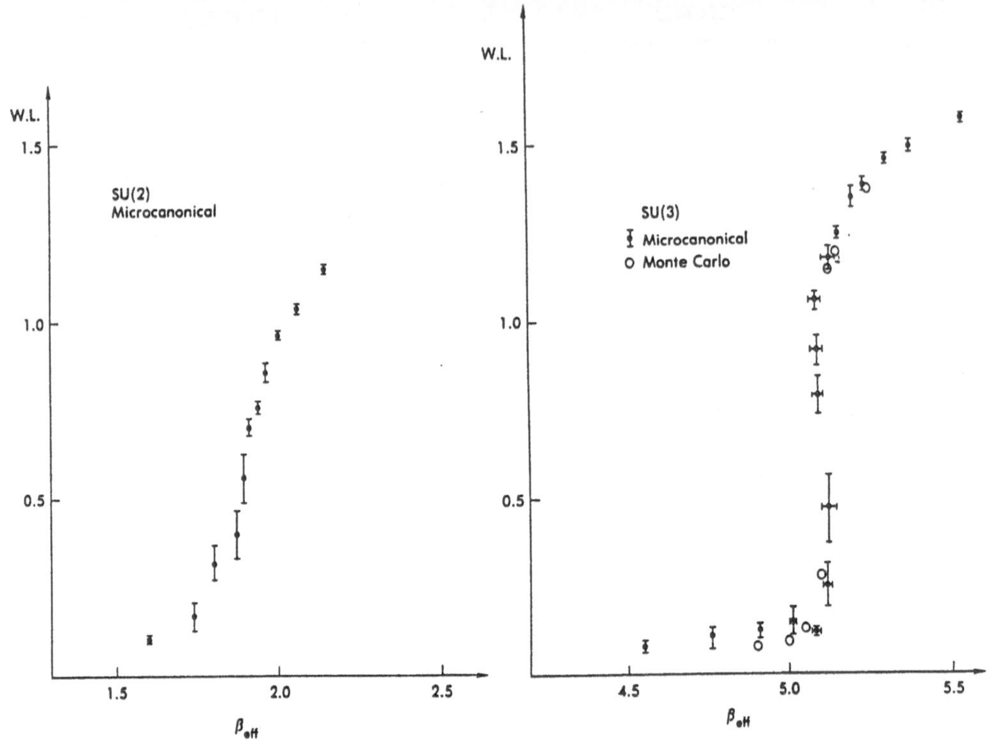

Fig. 6. Microcanonical calculation[10] *of the dependence of the Wilson line order parameter, denoted WL, on* β_{eff}*, which increases with increasing temperature. The reentrant behavior for* $SU(3)$ *indicates a first-order transition whereas the single-valued increase for* $SU(2)$ *indicates a second order transition.*

Eq. (3.34) that the string tension has the form $\sigma = a^{-2}\chi$. Then, use the physical value of one operator, say $\mathcal{O}_1$, to determine the physical value of a corresponding to the selected g. Again, using the string tension example, we could define $a = \sqrt{\chi}/420\,\text{MeV}$. With this value of a determined from $\mathcal{O}_1$, all other observables $\mathcal{O}_2 \ldots \mathcal{O}_N$ are completely specified. One should then repeat this procedure for a sequence of successively smaller and smaller values of g, thereby determining the function $a(g)$ and a sequence of values for the observables $\mathcal{O}_2 \ldots \mathcal{O}_N$. If the theory is correct, then each sequence of observables $\mathcal{O}_i$ $i \neq 1$ should approach a limit as $g \to 0$, and that limit should agree with nature.

In practice, it would be very difficult to carry out a series of calculations as described above to small enough g to make a convincing case. Hence, it is preferable to make use of our knowledge of the relation between the coupling constant and cutoff based on the renormalization group in the perturbative regime, and only carry out explicit lattice calculations down to the point at which the renormalization group behavior is clearly established. The foundation of the argument is the fact that the first two coefficients in the expansion of the renormalization group function $a\frac{dg}{da}$ are independent of the regularization scheme, and thus may be taken from continuum one and two loop calculations:

$$a\frac{dg}{da} = \gamma_0 g^3 + \gamma_1 g^5 + \ldots \quad . \tag{3.47}$$

Integration of this equation yields the desired relation

$$a(g) = \frac{1}{\Lambda_L}\left(\frac{16\pi^2}{11g^2}\right)^{51/121} e^{-\frac{8\pi^2}{11g^2}} \tag{3.48}$$

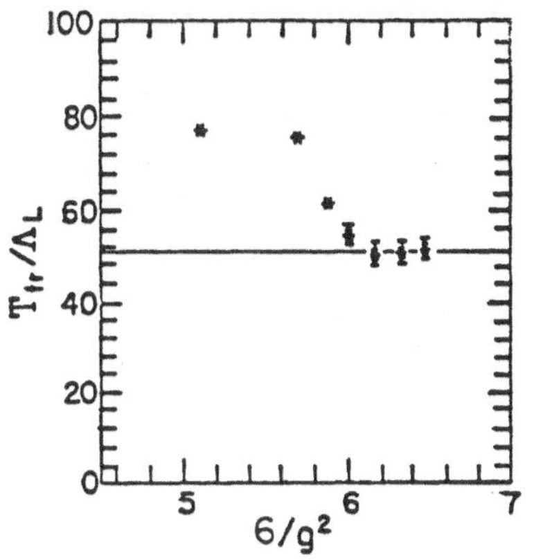

Fig. 7. Ratio of the transition temperature $T_{\rm tr}$ to the lattice scale parameter Λ_L as a function of inverse coupling constant.

where Λ_L is an integration constant and we have used the values of γ_0 and γ_1 for $SU(3)$ with no Fermions.

The constant Λ_L governing the relation between the bare coupling constant and the lattice cutoff can be related by one-loop continuum calculations to the constants $\Lambda_{\rm MOM}$ and $\Lambda_{\overline{MS}}$, which govern the relation between the renormalized coupling constant and continuum cutoff using the momentum space subtraction procedure in Feynman gauge and the minimal subtraction procedure respectively, with the results[11]

$$\Lambda_{\rm MOM} = 83.5\,\Lambda_L \quad , \qquad \Lambda_{\overline{MS}} = 28.9\,\Lambda_L \quad . \tag{3.49}$$

This correspondence is important for two reasons. First, the large coefficients in (3.49) allow us to reconcile our notion that the basic scale $\Lambda_{\rm QCD}$ is of order several hundred MeV with the fact that lattice measurements yield values of $\Lambda_L \sim 4 - 4.6$ MeV, which would otherwise appear astonishingly low. Second, in principle, it will provide a quantitative consistency test if experiments in the perturbative regime of QCD can produce sufficiently accurate values of $\Lambda_{\rm MOM}$ or $\Lambda_{\overline{MS}}$.

There is now substantial numerical evidence that lattice calculations in the pure gauge sector display the correct renormalization group behavior, and thus provide accurate solutions of continuum QCD. Data exist for two independent quantities, $T_{\rm tr}$, the temperature of the deconfinement transition, and the string tension σ. Results for the transition temperature are shown in Fig. 7 taken from Ref. [12] based on data from Ref. [13]. The transition temperature on a lattice with N_t time slices is given by $T_{\rm tr} = \frac{1}{N_t a(g_{\rm tr})}$ where $g_{\rm tr}$ is the value of the coupling for which the transition occurs. If $a(g_{\rm tr})$ is calculated using the perturbative expression (3.48), then once g is small enough that the lattice theory coincides with the continuum theory,the quantity $T_{\rm tr}/\Lambda_L$ should approach a constant. As seen in Fig. 7, $T_{\rm tr}/\Lambda_L$ indeed appears constant above $\beta = \frac{6}{g^2} = 6$.

A second test of continuum behavior is provided by the string tension. One could display the approach to the continuum limit by plotting $\frac{\sigma}{\Lambda_L^2}$ as a function of $\beta = \frac{6}{g^2}$ and observing that, as in Fig. 7, this ratio becomes constant beyond $\beta = 6$. An alternative plot, which will also be useful for subsequent purposes is shown in Fig. 8.

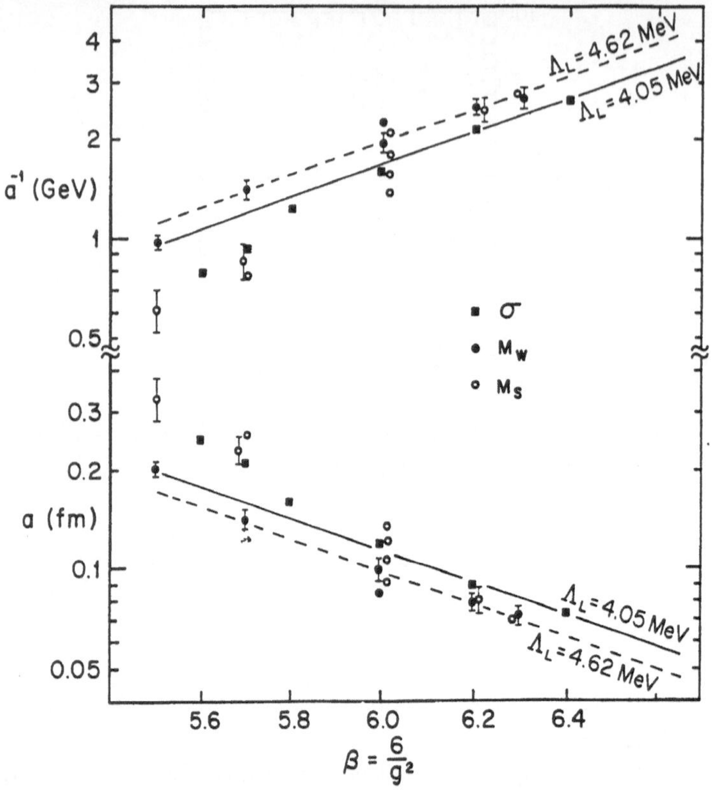

Fig. 8. Physical lattice spacing a(g), determined from the string tension (squares) and hadron masses in the quenched approximation (circles). The solid and dashed lines show the dependence expected in the continuum limit from the renormalization group expression (3.48) for two values of Λ_L. The difference between masses calculated with Wilson Fermions (solid circles) and staggered Fermions (open circles) is discussed in Chapter 4. Data are from references cited in Ref. [5].

For the present, observe only the solid squares, which denote values of a calculated from the surface tension as follows

$$a(g^2) = \left[\frac{\left[\sigma a^2\right]_{g^2}}{\sigma_{\text{expt}}}\right]^{1/2} = \frac{\sqrt{\chi}}{420\,\text{MeV}} \quad . \tag{3.50}$$

For comparison, the renormalization group expression (3.48) is also plotted for the value $\Lambda = 4.05$ MeV which provides the best fit to the last two points. Again, one sees continuum behavior above $\beta = 6$. This graph also allows us to read off directly the values of a and a^{-1} at various values of β and compare values of a

determined from the string tension with those determined from hadron masses to be discussed in a later section. One may also note that over the range of coupling constants relevant to lattice calculations, the effect of the premultiplying factor $g^{-0.84}$ in Eq. (3.48) is indiscernible and $\ln a$ is essentially linear in β. Although space does not permit additional figures pertinent to the continuum limit, it is also instructive to look directly at the renormalization group function $a\frac{dg}{da}$, and a useful graph combining string tension and transition temperature calculations is shown in Ref. [14].

In summary, I believe it is reasonable to regard lattice gauge theory in the pure gauge sector to be quite satisfactory. There are no glaring conceptual or computational problems, and all the numerical evidence to date suggests that one obtains an excellent approximation to the continuum theory for β_g above 6. In contrast, we will now see that full QCD including Fermions is more problematic at both the conceptual and computational levels.

4. LATTICE QCD WITH QUARKS

Significant new problems arise when one attempts to apply the ideas which are so successful for the stochastic evolution of path integrals for Bosons to many-Fermion systems. The fundamental underlying problem is the minus signs arising from antisymmetry. In the absence of projection onto the antisymmetric subspace, $e^{-\beta H}$ filters out the lowest state of any symmetry, and favors the lowest symmetric state of energy E_0^S relative to the lowest antisymmetric state with energy E_0^A by the factor $e^{-\beta(E_0^A - E_0^S)}$. If one attempts to project stochastically, for example by antisymmetrizing path integral Monte Carlo evolution at each step, the projection error is of order $\frac{1}{\sqrt{N}}$ and can never overcome the exponential factor in cases in which E_0^A represents a half filled band, Fermi sea, or Dirac sea and E_0^S corresponds to a Bose condensate in the lowest state. Thus, aside from special cases such as one spatial dimension, the only known alternative is write a path integral with Grassmann variables, introduce integrals over auxiliary fields if necessary to reduce it to quadratic form, and integrate out the Fermion variables as in Eq. (2.26) to obtain a Bosonic action containing a Fermionic determinant. Since the Grassmann integral has been done analytically, the projection onto the antisymmetric space is exact, eliminating one part of the sign problem. The resulting determinant may be positive or negative, so there still remains the danger of catastrophic sign cancellations in the stochastic evaluation of the remaining Bosonic integrals. If, however, as in our present case, there is an even number of Fermion species with the same action, the determinant appears with an even power and this final sign problem is also eliminated.

This major detour to beat the Fermion sign problem comes at a high price. We started with Fermions, which in occupation number representation are represented by a bunch of 1's and 0's, and we seek to deal with them on a digital computer which can only work in terms of 1's and 0's. Yet we must resort to calculating determinants of huge, non-local, nearly-singular matrices involving exceedingly large numbers of floating-point variables and operations. In addition, by virtue of putting the Fermions on a lattice, we encounter an additional unexpected difficulty associated with Fermion doubling, which we will discuss next. It should thus be clear at the outset, that the treatment of Fermions provides fertile ground for new ideas.

To appreciate the essential issues, consider the simplest Hermitian finite difference expression with the desired continuum limit:

$$\begin{aligned}
S_F^{\text{naive}} &= a^4 \sum_{\vec{n}} \Big[\bar\psi(\vec n) m \psi(\vec n) + \frac{1}{2a} \sum_\mu \Big\{ \bar\psi(\vec n)\gamma_\mu U_\mu(n)\psi(\vec n + a_\mu) \\
&\qquad - \bar\psi(\vec n + a_\mu)\gamma_\mu U_\mu^\dagger(n)\psi(\vec n)\Big\}\Big] \\
&\Longrightarrow \int d^4x \Big[\bar\psi m \psi + \frac{1}{2a}\sum_\mu \Big\{\bar\psi \gamma_\mu (1 + igA_\mu)(1 + a\partial_\mu)\psi \\
&\qquad - \bar\psi\left(1 + a\overleftarrow{\partial}_\mu\right)\gamma_\mu (1 - igaA_\mu)\psi\Big\}\Big] \\
&\Longrightarrow \int d^4x\, \bar\psi\left[m + \gamma_\mu(\partial_\mu + igA_\mu)\right]\psi \ .
\end{aligned} \tag{4.1}$$

Note that throughout we will use Euclidean γ-matrices satisfying $\gamma_\mu\gamma_\nu + \gamma_\nu\gamma_\mu = 2\delta_{\mu\nu}$, for which an explicit representation is given in Ref. [1]. Although this naive action appears to have the desired continuum behavior and symmetries, it has an unexpected problem. To see this problem in its simplest form, consider the Hamiltonian corresponding to the action (4.1) in one space dimension for the special case of free quarks ($A = 0$) and zero mass:

$$\begin{aligned}
H^{\text{naive}} &= a\sum_n \psi^\dagger(n)\alpha\frac{1}{i2a}\left(\psi(n+1) - \psi(n-1)\right) \\
&\Longrightarrow \int dx\, \psi^\dagger \alpha \frac{1}{i}\partial_x \psi \ ,
\end{aligned} \tag{4.2}$$

where $\alpha = \gamma_0\gamma_1$ and for convenience we choose the representation $\alpha = \begin{pmatrix} 1 & 0 \\ 0 & -1 \end{pmatrix}$. Note that H^{naive} is Hermitian by virtue of the symmetric difference $\psi(n+1) - \psi(n-1)$. We now transform the field operators to momentum space by writing the Fourier sum

$$\psi(n) = \frac{1}{\sqrt{Na}} \sum_{k=-\frac{\pi}{a}}^{\frac{\pi}{a}} \psi_k e^{ik\,na} \tag{4.3a}$$

where it is understood that for a lattice with N sites and periodic boundary conditions, the sum over momenta in the first Brillouin zone extends over the N momenta

$$k_p = \frac{p\pi}{Na} \qquad -\frac{N}{2} \le p \le \frac{N}{2} \ . \tag{4.3b}$$

The Hamiltonian is diagonal

$$H^{\text{naive}} = \sum_{k=-\frac{\pi}{a}}^{\frac{\pi}{a}} \psi_k^\dagger \alpha \frac{\sin(ka)}{a}\psi_k \tag{4.4a}$$

and thus has the eigenvalue spectrum

$$E_k = \pm\frac{\sin ka}{a} \underset{k\to 0}{\sim} \pm k\left(1 - \frac{(ka)^2}{3} + \ldots\right) \tag{4.4b}$$

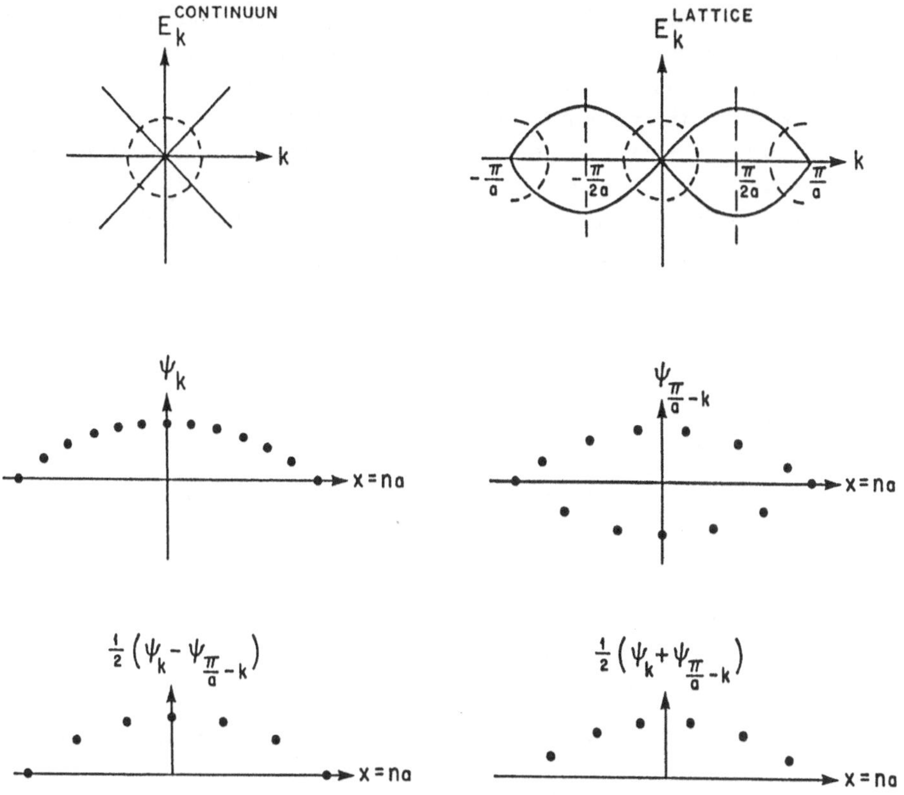

Fig. 9. Fermion doubling in one dimension. The top plots compare the physical continuum spectrum with the spectrum of the lattice Hamiltonian. The middle plots show a half wavelength of the real part of the non-vanishing component of a physical wave function $\Psi_k(x)$ and its degenerate unphysical sawtooth partner $\Psi_{\frac{\pi}{a}-k}(x)$. The bottom plots show the linear combinations $\Psi_k(x) \pm \Psi_{\frac{\pi}{a}-k}$ corresponding to staggered Fermions.

with eigenfunctions

$$\Psi_k^{\pm}(n) = e^{ikna}\chi^{\pm}$$

where $\chi^{\pm}$ is a two component spinor with either an upper or lower component unity and the other component zero. The comparison of the continuum spectrum for a massless Dirac particle $E_k = \pm k$ and the lattice spectrum in the top of Fig. 9 displays the species doubling problem. In the region denoted by the dashed circle centered at the origin, the lattice spectrum (4.4b) yields a good approximation to the linear physical spectrum, and the range of linearity increases as $a \to 0$. However, at the edge of the Brillouin zone, there is a second region in which the spectrum also goes to zero linearly, denoted by the two dashed semicircles. In fact, for every physical mode Ψ_k, there is a precisely degenerate unphysical mode $\Psi_{\frac{\pi}{a}-k}$. Since the partition function blindly counts and weights all modes according to their energies, it is clear that all Fermion loops will be overcounted by a factor of 2 in all physical observables. Note also that since the velocity is $v = \frac{dE}{dk}$, the lattice spectrum necessarily mixes right-moving and left-moving modes.

The origin and structure of the doubled states is simple. The degenerate partner to the state $\Psi_k(n) = e^{ik\,na}\chi$ is the state $\psi_{\frac{\pi}{a}-k}(n) = e^{in\pi}e^{-ik\,na}\chi$, that is, a sawtooth mode in which every other lattice site has an extra factor of -1. The real part of a low k mode Ψ_k and its sawtooth partner $\Psi_{\frac{\pi}{a}-k}$ are sketched for a half wavelength in the middle section of Fig. 9. Note that although there are sufficient points in the half wavelength of Ψ_k to yield an accurate integral with any smooth function, there is no way that the rapidly oscillating wave function $\Psi_{\frac{\pi}{a}-k}$ can represent a mode with momentum near $\frac{\pi}{a}$. Thus, we need some way to eliminate these modes so that they play no role in the continuum limit. The origin of the degeneracy of the physical mode with its sawtooth partner is the symmetric difference approximation to the derivative $\psi' \sim \frac{\psi(n+1)-\psi(n-1)}{2a}$ in the naive Hamiltonian (4.2), which clearly is impervious to the minus signs $e^{-in\pi}$ and thus yields the same magnitude for the derivatives of ψ_k and $\psi_{\frac{\pi}{a}-k}$. The origin of this symmetric difference, in turn, is Hermiticity, since expressions involving only nearest neighbor differences like $\psi^\dagger n(\psi(n+1)-\psi(n))$ are non-Hermitian and yield complex eigenvalues.

We may note in passing that a finite difference approximation to a second derivative strongly breaks the degeneracy between the physical and sawtooth modes. Indeed, a perturbation of the following form

$$\begin{aligned} H' &= -\sum_n \psi^\dagger(n)\gamma_0\left(\psi(n+1)-2\psi(n)+\psi(n-1)\right) \\ &= -a\,a\sum_n \psi^\dagger(n)\gamma_0\left(\frac{\psi(n+1)-2\psi(n)+\psi(n-1)}{a^2}\right) \\ &\underset{a\to 0}{\longrightarrow} -a\int dx\,\bar{\psi}(x)\psi''(x) \underset{a\to 0}{\longrightarrow} 0 \end{aligned} \tag{4.5a}$$

which vanishes in the continuum limit and behaves like a momentum-dependent mass term

$$H' = \sum_k \bar{\psi}_k m(k)\psi_k \tag{4.5b}$$

where

$$\begin{aligned} m(k) &= \frac{2}{a}(1-cos(ka)) \\ &\underset{k\to 0}{\sim} ak^2 \\ &\underset{k\to\frac{\pi}{a}}{\sim} \frac{4}{a}\ . \end{aligned} \tag{4.5c}$$

For the physical modes in the region of the origin, the perturbation has no effect as $a \to 0$, consistent with the fact H' vanishes in the continuous limit. For the spurious sawtooth modes, however, the mass diverges as $\frac{1}{a}$. Thus, one way to remove the unphysical modes would be to include a term of the form (4.5a) in the lattice Hamiltonian to raise their mass so high that they contribute negligibly to the partition function, and this is the idea underlying Wilson's lattice action for Fermions discussed below.

The doubling we have discussed for simplicity in one dimension arises analogously in each of the four Euclidean dimensions of the naive Fermion action, Eq. (4.1), so that we obtain $2^4 = 16$ lattice modes for each physical mode. Again, specializing to the massless case, the momentum space action corresponding to Eq. (4.1) is

$$S_F^{\text{naive}} \equiv \bar{\psi}M\psi = \sum_k \bar{\psi}_k \sum_\mu \gamma_\mu \frac{\sin(k^\mu a)}{a}\psi_k \tag{4.6a}$$

so that the inverse propagator is

$$\langle T\bar{\psi}\psi\rangle^{-1} = M(k) = \sum_\mu \gamma_\mu \frac{\sin(k^\mu a)}{a}\ . \tag{4.6b}$$

This propagator replicates the physical behavior in the region of $k^\mu \sim 0$ fifteen times around points on the edge of the four-dimensional Brillouin zone at which one or more of the components $k^\mu \sim \frac{\pi}{a}$.

We are now prepared to understand both the features giving rise to the doubling problem and the generality of the problem.[15] Whereas the specific function $\sin(k^\mu a)$ in Eqs. (4.4a) and (4.6) is the result of using the lowest-order Hermitian difference formula for the derivatives in the continuum action, the most general form of the chiral symmetric, Hermitian action derived from discrete derivatives on a periodic lattice with the correct continuum limit is

$$S_F = \sum_k \bar{\psi}_k \sum_\mu \gamma_\mu P^\mu(k) \psi_k \tag{4.7}$$

where $P^\mu(k)$ is real for Hermiticity, $P^\mu(k) \underset{k\to 0}{\longrightarrow} 0$ for the correct continuum limit, and $P^\mu(k)$ is periodic under $k^\mu \to k^\mu + \frac{2\pi}{a}$ and continuous for local discrete difference formulae on a lattice. Note that chiral symmetry requires the form $\bar{\psi}\gamma_\mu P^\mu \psi$, so that under a chiral transformation $\psi \to e^{i\alpha\gamma_5}\psi$, the two sign changes from γ_0 in $\bar{\psi}$ and γ_μ leave the action invariant. Since $P^\mu(k)$ is real, continuous and periodic in k^μ with period $\frac{2\pi}{a}$, it must cross the axis at some intermediate point, so that this general discrete action has the doubling observed in Eq. (4.6a) in each of the four Euclidean directions yielding 15 spurious low-mass excitations for each physical excitation. A rigorous version of these arguments is known as the Nielsen–Ninomiya no-go theorem,[16] which proves using homotopy theory that one cannot avoid Fermion doubling in a lattice theory which is simultaneously Hermitian, local and chiral symmetric.

An additional aspect of Fermion doubling is the absence of the axial anomaly. The axial current, which is conserved for gauge theories at the classical level but not conserved at the quantum level in the continuum theory, is conserved for naive lattice Fermions. Again, the culprits are the unphysical lattice duplicates of the physical Fermion excitations, which couple to an external axial current with the opposite chiral charge and effectively cancel the axial anomaly arising from the physical Fermions.

Wilson Fermions

One of the ways out of the no-go theorem is to give up chiral symmetry and, following Wilson, add a second derivative term of the form (4.5) to raise the mass of the unphysical sawtooth modes. In one dimension, combining the naive Hamiltonian (4.2) with a multiple r of the perturbation (4.5a) yields the Wilson Hamiltonian

$$\begin{aligned} H_W &= a\sum_n \psi^\dagger(n)\left[\frac{\alpha}{i}\frac{(\psi(n+1)-\psi(n-1))}{2a} - \frac{ra\gamma_0}{2i}\frac{\psi(n+1)-2\psi(n)+\psi(n-1)}{a^2}\right] \\ &= \sum_k \psi_k^\dagger\left[\alpha\frac{\sin(ka)}{a} - r\frac{\gamma_0}{i}\frac{(\cos(ka)-1)}{a}\right]\psi_k\ . \end{aligned} \tag{4.8a}$$

Using a representation with $\alpha = \sigma_3$ and $-i\gamma_0 = \sigma_1$, we obtain the energy spectrum

$$E^2 = \left(\frac{\sin(ka)}{a}\right)^2 + \left[\frac{r}{a}(\cos(ka)-1)\right]^2 \tag{4.8b}$$

with the limits

$$\begin{aligned} E &\underset{k\to 0}{\longrightarrow} \pm k\left(1 - \frac{1}{6}k^2a^2 + \frac{r^2}{8}k^2a^2\right) \\ E &\underset{k\to \frac{\pi}{a}}{\longrightarrow} \pm\frac{2r}{a}\ . \end{aligned} \tag{4.8c}$$

Thus, for fixed r, the mode for $k \sim 0$ has the correct continuum limit whereas the sawtooth mode for $k \sim \frac{\pi}{a}$ becomes infinitely massive and decouples from the theory.

In four Euclidean dimensions, the corresponding Wilson action is

$$S_W = -a^4 \sum_{\vec{n}} \frac{1}{2a} \sum_{\mu} \left[\bar{\psi}(\vec{n})(r - \gamma_\mu) U_\mu(\vec{n}) \psi(\vec{n} + a_\mu) + \bar{\psi}(\vec{n} + a_\mu)(r + \gamma_\mu) U_\mu^\dagger(\vec{n}) \psi(\vec{n}) \right]$$
$$+ a^4 \sum_{\vec{n}} \left(m + \frac{4r}{a} \right) \bar{\psi}(\vec{n}) \psi(\vec{n}) \tag{4.9}$$

and the propagators for the spurious modes acquire masses which diverge as $\frac{r}{a}$ as in the one-dimensional case.

The Wilson action manifestly breaks chiral symmetry for $m = 0$, since under the transformation $\psi \to e^{i\alpha\gamma_5}\psi$, $\bar{\psi}\psi \to \bar{\psi}\, e^{i2\alpha\gamma_5}\psi$. As long as the contribution of the symmetry breaking term can be made arbitrarily small, its presence does not interfere with the physics of spontaneous symmetry breaking. In the case of a spin system, for example, one defines the spontaneous magnetization as the limit of the thermodynamic trace of the spin in the presence of an external magnetic field in the limit as the external field goes to zero. Heuristically, we expect the contribution of the chiral symmetry breaking term to vanish in the continuum limit even though the masses of the unphysical modes go to infinity since $a \int dx\, \bar{\psi} \Box^2 \psi \underset{a \to 0}{\longrightarrow} 0$, and explicit calculation[15] shows that this is the case. Furthermore, including 15 massive Wilson Fermion partners as well as the physical mode yields the correct axial anomaly in the continuum limit.[15]

The combination $M = \left(m + \frac{4r}{a}\right)$ in Eq. (4.9) enters the action like a mass term. Since this mass term is not protected from renormalization by any symmetry, it must be renormalized by "fine tuning." Thus, by a search involving a series of calculations of the pion mass at fixed coupling constant g and Wilson r, we may determine a critical mass $M_{cr}(g, r)$ such that $m_\pi = 0$ for a chiral symmetric theory and $M(g, r)$ such that $m_\pi = 140$ MeV for the physical theory. To the extent to which it is meaningful to define a quark mass, one may define $M_{quark} \equiv M - M_{cr}$. A more conventional notation is in terms of the hopping parameter

$$\kappa \equiv \frac{1}{2Ma} = \frac{1}{2ma + 8r} \tag{4.10a}$$

and rescaled Fermion fields

$$\Psi \equiv \left(Ma^4\right)^{1/2} \psi \tag{4.10b}$$

for which the action has the form

$$S_W = \sum_{\vec{n}} \Bigg\{ \bar{\Psi}(\vec{n}) \Psi(\vec{n}) - \kappa \sum_{\mu} \Big[\bar{\Psi}(\vec{n})(r - \gamma_\mu) U_\mu(\vec{n}) \Psi(\vec{n} + a_\mu)$$
$$+ \bar{\Psi}(\vec{n} + a_\mu)(r + \gamma_\mu) U_\mu^\dagger(\vec{n}) \Psi(\vec{n}) \Big] \Bigg\} \tag{4.10c}$$

and

$$M_{quark} = \frac{1}{2a} \left(\frac{1}{\kappa} - \frac{1}{\kappa_{cr}} \right) \quad . \tag{4.10d}$$

The fields have been scaled such that the diagonal term is now unity, and the hopping parameter κ specifies the strength of the nearest-neighbor coupling via link variables.

Although the Wilson parameter r is often chosen to be 1, in principle it may be optimized to render the errors from $ar\bar{\psi}\Box^2\psi$ in the physical modes and the contribution of the sawtooth modes comparable.[17]

Just as one can derive the conserved vector current of the continuum theory as the Noether current of the continuum action, one can derive a discrete version of Noether's theorem from the Wilson action (4.9) of the form

$$\Delta^\mu V_\mu(\vec{n}) \equiv V_\mu(\vec{n}) - V_\mu(\vec{n} - a_\mu) = 0 \tag{4.11a}$$

where the conserved vector current on the lattice is

$$\begin{aligned} V_\mu(\vec{n}) = &- \frac{1}{2}\bar{\psi}(\vec{n})(r - \gamma_\mu)U_\mu(\vec{n})\psi(\vec{n} + a_\mu) \\ &+ \frac{1}{2}\bar{\psi}(\vec{n} + \mu)(r + \gamma_\mu)U_\mu^\dagger(\vec{n})\psi(\vec{n}) \quad . \end{aligned} \tag{4.11b}$$

Note that neither the local current $\bar{\psi}(\vec{n})\gamma_\mu\psi(\vec{n})$ nor the point-split current defined by (4.11b) with $r = 0$ is conserved on the lattice. In contrast, since there is no chiral symmetry, there is no unique definition of the lattice axial current, so it is necessary to use perturbation theory to explicitly calculate the difference between any choice for the lattice axial current and the continuum axial current.

In summary, Wilson Fermions provide a framework for completely solving the doubling problem which yields the correct physics in the limit of sufficiently small lattice spacing a. The primary disadvantages are associated with the lack of explicit chiral symmetry for finite a.

Staggered Fermions

For investigations in which it is desirable to maintain explicit chiral symmetry on the lattice, it is sometimes preferable to use an alternative lattice action in which one does not avoid the no-go theorem, but rather thins the Fermion degrees of freedom to reduce the number of Fermion species.[18] The basic idea is to transform the Fermion field operators to a new representation in which the naive Fermion action is diagonal in the Dirac indices, so that the naive Fermions represent N_{D} copies of the new Fermions, where N_{D} is the number of Dirac components. By keeping only one of these N_{D} copies, one effectively thins the degrees of freedom by $1/N_{\text{D}}$.

A convenient choice for the transformation of the Fermion field in four space-time dimensions is

$$\begin{aligned} \psi(\vec{n}) &= \gamma_0^{n_0}\gamma_1^{n_1}\gamma_2^{n_2}\gamma_3^{n_3}\chi(\vec{x}) \\ \bar{\psi}(\vec{n}) &= \bar{\chi}(n)\gamma_3^{n_3}\gamma_2^{n_2}\gamma_1^{n_1}\gamma_0^{n_0} \quad . \end{aligned} \tag{4.12}$$

To see how this transformation renders the naive action (4.1) diagonal, consider a typical term:

$$\begin{aligned} \bar{\psi}(\vec{n})\gamma_2\psi(\vec{n} + a_2) &= \bar{\chi}(\vec{n})\gamma_3^{n_3}\gamma_2^{n_2}\gamma_1^{n_1}\gamma_0^{n_0}\gamma_2\gamma_0^{n_0}\gamma_1^{n_1}\gamma_2^{(n_2+1)}\gamma_3^{n_3}\chi(\vec{n} + a_2) \\ &= (-1)^{n_0+n_1}\bar{\chi}(\vec{n})\chi(\vec{n} + a_2) \end{aligned} \tag{4.13a}$$

where the factor $(-1)^{n_0+n_1}$ arises from anticommuting γ_2 through the product $\gamma_0^{n_0}\gamma_1^{n_1}$ and the remaining γ matrices combine pairwise to unity. By the same argument,

$$\bar{\psi}(\vec{n})\gamma_\mu\psi(\vec{n} + a_\mu) = \eta_\mu(\vec{n})\bar{\chi}(\vec{n})\chi(\vec{n} + a_\mu) \tag{4.13b}$$

where

$$\eta_\mu(\vec{n}) \equiv (-1)^{\sum\limits_{\nu=0}^{\mu-1} n_\nu} \quad . \tag{4.13c}$$

The naive action may thus be written

$$\begin{aligned} S_F = a^4 \sum_{\vec{n}} \Bigg[\bar{\chi}(\vec{n}) m \chi(\vec{n}) + \frac{1}{2a} \sum_\mu \eta_\mu(\vec{n}) \bar{\chi}(\vec{n}) \\ \times \{U_\mu(\vec{n})\chi(\vec{n}+a_\mu) - U_\mu(\vec{n}-a_\mu)\chi(\vec{n}-a_\mu)\} \Bigg] \end{aligned} \tag{4.14}$$

where $\chi(\vec{n})$ now represents any one of the N_D Dirac components and may thus be regarded as a scalar. By (4.12), it is clear that the components of χ are specific linear combinations of the doubled Fermion fields.

It is particularly simple to see how staggered Fermions work in the case of free, massless particles in one dimension described by the Hamiltonian (4.4a). The Dirac equation has two components, and in this case the staggered solutions χ for a given k are just the sum or difference of the physical mode Ψ_k and the sawtooth mode $\Psi_{\frac{\psi}{a}-k}$ and are sketched in the bottom portion of Fig. 9. Note that for each of the staggered solutions, half of the points, either the even or odd points, correspond to the physical Fermion mode and the other half of the points are identically zero. Thus, the thinning of the degrees of freedom corresponds to having essentially doubled the lattice spacing. The maximum momentum which can be sustained on the lattice with spacing $2a$ is $\frac{\pi}{2a}$, so that for both the even site mode and the odd site mode only the portion of the spectrum in the upper right portion of Fig. 9 between $-\frac{\pi}{2a}$ and $\frac{\pi}{2a}$ contributes and there are no spurious low-mass modes. In the special case of one space dimension where there are two Dirac components and two naive Fermion modes, staggered Fermions completely resolve the doubling problem.

If one were to solve Hamiltonian field theory in three space dimensions, reduction of the 2^3 naive Fermions by a factor of 1/4 for the four Dirac components would leave two species of staggered Fermions which one could regard as two degenerate flavors corresponding to up and down quarks. In Lagrangian field theory in 3+1 dimensions, the 16 naive Fermions are only reduced to four flavors of staggered Fermions.

The principle advantage of staggered Fermions is the residual chiral symmetry of the lattice action. The mass is thereby protected from renormalization and it is possible to define a lattice axial current. The primary disadvantage of staggered Fermions is the existence of four flavors. In addition, the lattice resolution is cut in half relative to Wilson Fermions and physical operators are complicated, non-local combinations of χ fields. Hence, for most purposes in studying hadronic physics, it will be desirable to use Wilson Fermions.

Hopping-Parameter Expansion

Just as the strong-coupling expansion provided insight into solutions of lattice QCD in the pure gauge sector, the hopping-parameter expansion provides analogous insight into solutions in the presence of Fermions. Consider the partition function for Wilson Fermions with the rescaled fields and hopping parameter defined in Eq. (4.10), which we write for convenience in the following schematic form:

$$Z = \int \mathcal{D}(\bar{\psi}\psi)\mathcal{D}(U)\, e^{-\bar{\psi}(1+\kappa U)\psi - \beta \sum\limits_{\square} \left(1 - \frac{1}{N} \mathrm{Re\,Tr}\, U_\square\right)} \quad . \tag{4.15}$$

In the strong coupling regime, g^2 is large so that $\beta = \frac{2N}{g^2}$ is small and the U's distributed according to $e^{-\beta S(u)}$ are nearly random. Thus, the average of κU is in some sense small, and we may regard the diagonal term $\bar{\psi}_n \psi_n$ representing static quarks as the dominant term and expand in the hopping term $\bar{\psi}_n \kappa U \psi_{n\pm1}$. Expansion of the exponential $e^{-\sum_n \bar{\psi}_n \kappa U \psi_{n\pm1}}$ yields a sum of terms of the form $\bar{\psi} k U \psi\, \bar{\psi} k U \psi \ldots \bar{\psi} k U \psi$, and the resulting integral over Grassmann variables may be written schematically

$$\begin{aligned} Z(U) &\equiv \int \mathcal{D}(\bar{\psi}\psi)\, e^{-\sum_n \bar{\psi}_n \psi_n}\, e^{-\sum_n \bar{\psi}_n \kappa U \psi_{n\pm1}} \\ &= \int \mathcal{D}(\bar{\psi}\psi)\, e^{-\sum_n \bar{\psi}_n \psi_n} \sum \bar{\psi}\kappa U \psi\, \bar{\psi}\kappa U \psi \ldots \bar{\psi}\kappa U \psi \end{aligned} \tag{4.16a}$$

Using Wick's theorem, this integral is equal to the sum of all contractions, where because the matrix in the exponent is the unit matrix, the contraction $\langle \psi_m \bar{\psi}_n \rangle$ is just δ_{mn}. Thus, each factor $U\kappa$ which connects one site to an adjacent site must be connected to another hopping term $U\kappa$ emanating from the new site, and the net result after integrating out the Fermions is the sum of all possible closed chains of κU in which the U's are oriented head to tail

$$Z(U) = \sum \kappa^k U_{n_1 n_2} U_{n_2 n_3} \ldots U_{n_k n_1} \quad . \tag{4.16b}$$

The full partition function is then the integral over gauge fields of all such loops weighted by the gluon action.

Consider now the hopping parameter expansion for a meson propagator

$$\begin{aligned} &\langle \bar{\psi}(x,t)\Gamma\psi(x,t)\bar{\psi}(0,0)\Gamma\psi(0,0)\rangle \\ &= Z^{-1} \int \mathcal{D}(\bar{\psi}\psi)\mathcal{D}(U) e^{-\sum \bar{\psi}\psi} e^{-\sum \bar{\psi}\kappa U \psi} e^{-\beta \sum_{\Box}\left(1-\frac{1}{N}\,\mathrm{Re\ Tr}\, U_{\Box}\right)} \bar{\psi}\Gamma\psi_{(x,t)}\bar{\psi}\Gamma\psi_{(0,0)} \end{aligned} \tag{4.17}$$

where $\bar{\psi}\Gamma\psi$ represents a combination of Fermion fields of the appropriate flavors and γ matrices to create or annihilate the desired meson state. As before, we expand $e^{-\sum \bar{\psi}\kappa U \psi}$ and apply Wick's theorem to obtain all contractions of $\bar{\psi}$ and ψ. The lowest order (in κ) non-vanishing contribution is obtained by creating two straight chains of U's, one from $(0,0)$ to (x,t) and the other from (x,t) to $0,0$. Higher-order contributions are obtained by elongating these two chains to form any closed path including the points $(0,0)$ and (x,t) and by adding any additional number of separate closed loops of U's. The complete propagator is the sum over all such loops of U's

$$\begin{aligned} \langle \bar{\psi}\Gamma_{(x,t)}\bar{\psi}\Gamma\psi_{(0,0)}\rangle &= Z^{-1}\int \mathcal{D}(U) e^{-\beta\sum_{\Box}\left(1-\frac{1}{2N}\,\mathrm{tr}\,\left(U_{\Box}+U_{\Box}^{\dagger}\right)\right)} \\ &\times \sum_{\text{loops}} \kappa^{N_{\text{links}}} (U\,U \ldots U)_{00,xt}(U\,U \ldots U)\ldots(U\,U\ldots U) \quad . \end{aligned} \tag{4.18}$$

Note that the remnants of the Fermions at this stage of the calculation are just Wilson loops, again underscoring our previous interpretation of Wilson loops as the world lines of quarks. The simple quark model of the meson is described by the sum over all time histories of the loop $(U\,U \ldots U)_{00,xt}$ representing a quark and antiquark propagating from 00 to xt and the excitation of quark-antiquark pairs out of the vacuum is described by the additional quark loops $(U\,U \ldots U)$. The integral over the gauge fields in (4.18) now proceeds precisely as in the case of the Wilson loop discussed in the pure gauge sector. Each of the closed loops in (4.18) must be tiled with plaquettes, with the lowest-order contribution corresponding to the minimum tiling required to eliminate all exposed links and higher-order contributions corresponding to more elaborate surfaces. The general structure which emerges is thus a sum over Fermion loops covered or connected with gluon membranes, with the partition function dictating the optimal compromise between short Fermion paths and minimal membranes favored by small β and small κU and the higher entropy of longer Fermion paths and complicated surfaces. The physical picture of a meson state which emerges when one observes the configuration on a single time slice between $(0,0)$ and (x,t) is a quark-antiquark pair connected by a flux tube. When one works out the explicit factors for $SU(N)$, one can also see the $1/N$ expansion emerge naturally, with the dominant contributions arising from planar diagrams with no additional Fermion loops. Propagators for baryons are similar to those for mesons, with three chains of U's starting at the point $(0,0)$, corresponding to three quarks of the appropriate flavors, and terminating at (x,t). The surface between these chains must again be tiled with plaquettes, and the leading membrane contribution in this case, when cut on a single time slice, corresponds to a Y configuration of flux connecting the three quarks.

Instead of expanding the Fermionic action $\bar{\psi}M(u)\psi = \bar{\psi}(1+\kappa U)\psi$ in powers of κ, one can alternatively write the result of integrating out the Fermions directly in terms of $M(U)$

$$\begin{aligned}
&\langle \psi\psi \ldots \psi\bar{\psi}\bar{\psi} \ldots \bar{\psi}\rangle \\
&= Z^{-1} \int \mathcal{D}(U)\mathcal{D}(\bar{\psi}\psi)\, e^{-\bar{\psi}M(U)-S(U)} \psi\psi \ldots \psi\bar{\psi}\bar{\psi} \ldots \bar{\psi} \\
&= Z^{-1} \int \mathcal{D}(U)\, e^{\ln \mathrm{Det}\, M(U)-S(U)} \sum_{\text{contractions}} M^{-1}(U)M^{-1}(U) \ldots M^{-1}(U) \ .
\end{aligned} \tag{4.19}$$

If one now expands $M^{-1} = (1+\kappa U)^{-1}$ and $\ln \mathrm{Det}\, M(U) = \mathrm{tr}\, \ln(1+\kappa U)$ in κ, one observes that one obtains the previous hopping-parameter expansion with all the quark propagators joining the ψ's and $\bar{\psi}$'s in $\langle \psi\psi \ldots \psi\,\bar{\psi}\bar{\psi} \ldots \bar{\psi}\rangle$ arising from the M^{-1}'s and all the additional closed quark loops arising from expansion of $\ln \mathrm{Det}\, M$. As before, the quark lines thus obtained are tiled with plaquettes of the proper orientation from $S(U)$.

This knowledge of the role of the various terms in Eq. (4.19) allows us to understand the physics of the so-called quenched approximation, which might more properly be called the valence quark approximation. The quenched approximation corresponds to omitting the term $\ln \mathrm{Det}\, M(U)$ in the exponent of (4.19) when performing the integral over gauge fields $\int \mathcal{D}(U)$. From the preceding argument, this approximation omits all time histories in which dynamical quark loops are excited out of the Fermi sea. Hence, only valence quarks connecting the field operators in $\langle \psi\psi \ldots \psi\bar{\psi}\bar{\psi}\bar{\psi}\rangle$ are included, and the integral over gluon fields incorporates the QCD interactions of these valence quarks to all orders. Technically, the motivation for making the quenched approximation is the fact that the stochastic evaluation of the path integral is immensely more difficult when the non-local term $\ln \mathrm{Det}\, M(U)$ is included in the action than when one must only treat the local term $S(U)$.

Observables

Given the methods we have discussed for evaluating quark and gluon fields on a lattice, it is now straightforward to calculate a number of physical observables.

Static sources allow us to calculate the forces between quarks and explore the behavior of color electric and magnetic fields in their presence. We have already discussed the two sources shown in Fig. 2, namely the Wilson or Polyakov line

$$\langle L\rangle = \langle\prod_L U\rangle \tag{4.20a}$$

which serves as an order parameter for infinite quark mass and is used in practice for finite quark mass as well and the Wilson loop

$$\langle W\rangle = \langle\prod_c U\rangle \tag{4.20b}$$

which specifies the static potential between a quark and an antiquark. In the same spirit, one can calculate the static potential between three quarks at positions $\vec{r}_1$, $\vec{r}_2$ and $\vec{r}_3$ as follows. Let $\vec{R}$ denote some convenient lattice site in the vicinity of $\vec{r}_1$, $\vec{r}_2$ and $\vec{r}_3$ and define three staple-shaped space-time paths C_j leading from $\vec{R}$ to $\vec{r}_j$ at time t_i, remaining fixed at the spatial point $\vec{r}_j$ from t_i to t_f, and leading from $\vec{r}_j$ to R at t_f. These staples can be connected to a color singlet at t_i and t_f by combining the color indices with an antisymmetric tensor ϵ_{ijk}, with the result that the three-quark source can be written

$$\langle Q\rangle = \left\langle \epsilon_{ijk}\epsilon_{pqr}\left[\prod_{C_1} U\right]_{ip}\left[\prod_{C_2} U\right]_{jq}\left[\prod_{C_3} U\right]_{kr}\right\rangle \ . \tag{4.20c}$$

The physical ground state in the presence of each of these sources is obtained on intermediate time slices between t_i and t_f which are sufficiently far away from either end. The interaction energy can be measured by comparing sources of different temporal extent $t_f - t_i$ to cancel out end effects and properties of the ground state may be obtained by measuring appropriate observables on intermediate time slices. For example, E^2 and B^2 in the presence of a static source can be measured using the fact that, in the continuum limit, a plaquette in the $\mu\nu$ plane may be expanded

$$\mathrm{Tr}\, U^{\square}_{\mu\nu} = \mathrm{Tr}\, e^{ia^2 gF_{\mu\nu}} \sim 1 - \frac{1}{2}a^4 g^2 F^2_{\mu\nu} \ . \tag{4.21a}$$

Hence, in the presence of a source $\mathcal{O}$, where $\mathcal{O}$ denotes L, W or Q above, the change in E^2 or B^2 relative to the vacuum is given by the space-time or space-space components of

$$\langle F^2_{\mu\nu}(\vec{r})\rangle = \frac{\left\langle \mathcal{O}U^{\square}_{\mu\nu}(\vec{r})\right\rangle}{\langle\mathcal{O}\rangle} - \left\langle U^{\square}_{\mu\nu}\right\rangle \ . \tag{4.21b}$$

The chiral order parameter is calculated by evaluating

$$\begin{aligned}\langle\bar{\psi}\psi\rangle &= Z^{-1}\int \mathcal{D}(U)\mathcal{D}(\bar{\psi}\psi)\, e^{-\bar{\psi}M(U)\psi - S(U)} \\ &= Z^{-1}\int \mathcal{D}(U)\, e^{\ln \mathrm{Det}\, M(U) - S(U)} M^{-1}(U) \ .\end{aligned} \tag{4.22}$$

As discussed in connection with the chiral phase transition, $\langle\bar{\psi}\psi\rangle$ serves as an order parameter for the case of zero quark mass and is also used as an indicator of a phase transition at finite quark mass. So as to avoid spurious chiral symmetry breaking due to the presence of a chiral symmetry breaking term in the lattice action as in the case of Wilson Fermions, it is useful to use staggered Fermions. The price one pays is the necessity of having an integer multiple of four flavors. Thus one cannot directly study the case of three flavors, where the presence of a cubic invariant implies a first-order transition, or the physically relevant case of two flavors of light quarks.

To discuss hadronic observables, it is convenient to discuss idealized calculations in which one filters in the pure gauge sector for a long enough time to filter out the vacuum state $|0\rangle$. (In practice, calculations usually do not completely filter out the ground state in the pure gauge sector, so the expressions here must have $|0\rangle$ replaced by sums over a sequence of states having the quantum numbers of the vacuum. The basic idea, however, is correct and the resulting formulae are more transparent.) One then creates a state $\hat{J}^\dagger|0\rangle$ of the desired quantum numbers by acting on the vacuum with an appropriate source $\hat{J}$, filters the lowest eigenstate with these quantum numbers by evolution in imaginary time, and projects onto specified momentum states as required. A local field operator we may use for a π^+, for example, is given by

$$J_\pi(x) = \bar{d}(x)\gamma_5 u(x) \tag{4.23a}$$

where u and d denote up and down quark fields, respectively, and the γ_5 makes the operator pseudoscalar. As an alternative to this local point source, it may be preferable in some applications susceptible to large stochastic errors to use non-local extended sources. For the proton, a local field with the correct transformation properties is

$$J^P_\mu(x) = \epsilon^{\alpha\beta\gamma} u^\alpha_\mu(x)\left[u^\beta_\nu(x)(C\gamma_5)^{\nu\delta}\, d^\gamma_\delta(x)\right] \tag{4.23b}$$

where $C = \gamma_2\gamma_4$ is the charge conjugation matrix and α, β, γ denote color labels. Several comments concerning this operator may be helpful.[19] Since a quark and antiquark have opposite parity, the combination $uC\gamma_5 d$ transforms as a Lorentz scalar so that J^P_μ transforms like u_μ and thus as a spin 1/2 Dirac spinor. The color variables are explicitly antisymmetrized and the antisymmetry of the Grassmann variables combined with the fact that the local operator carries no orbital angular momentum assures the symmetry of the spin-flavor wave function. The non-relativistic limit also agrees with the non-relativistic quark model, since for the upper components $uC\gamma_5 d = u(-i\sigma^2)d = -u_\uparrow d_\downarrow + u_\downarrow d_\uparrow$ and the symmetrized state $S\{u_\uparrow(u_\uparrow d_\downarrow - u_\downarrow d_\uparrow)\}$ is the $SU(6)$ proton wave function.

Given these hadron sources, it is straightforward to calculate a variety of physical observables. Consider first the two-point function

$$C_{J_\pi J_\pi}(\tau) = \int dx\, \langle 0\left|J_\pi(x,\tau)J^\dagger_\pi(0,0)\right|0\rangle \tag{4.24a}$$

which is calculated by evaluating

$$\begin{aligned} C_{J_\pi J_\pi}(\tau) &= \int dx \int \mathcal{D}(\bar{\psi}\psi)\mathcal{D}(U)e^{-\bar{\psi}M(U)\psi - S(U)}\bar{d}(x,\tau)\gamma_5 u(x,\tau)\bar{u}(0,0)\gamma_5 d(0,0) \\ &= \int dx \int \mathcal{D}(U)e^{\ln \mathrm{Det}\,(U) - S(U)} M^{-1}(U)_{x\tau,00}\gamma_5 M^{-1}_{00,x\tau}\gamma_5 \end{aligned} \tag{4.24b}$$

on a mesh of temporal extent from $-T$ to $T+\tau$ with the hard wall boundary conditions (2.32) at the time boundaries. Selecting the time interval T outside the sources sufficiently large selects the vacuum state $|0\rangle$ discussed above, but in practical calculations it is usually not essential to enforce this filtering.[19] The physical content of $C_{J_\pi J_\pi}$ is clear by transforming $J_\pi(x,\tau)$ with translation operators in x and τ and inserting a complete set of pionic states $|np\rangle$, where n denotes the nth intrinsic excited state of the pion and p denotes its momentum. Thus,

$$C_{J_\pi J_\pi}(\tau) = \int dx \langle 0|e^{\tau\hat{H}-ix\cdot\hat{p}} J_\pi(0,0)e^{-\tau\hat{H}+ix\cdot\hat{p}} \sum_n \int dp \frac{|np\rangle\langle np|}{2E_{np}} J_\pi^\dagger(0,0)|0\rangle \ . \tag{4.24c}$$

The integral over x projects onto zero momentum states $p=0$ and for τ large $\sum\limits_n e^{-\tau E_{n,0}}|n0\rangle\langle n0| \to e^{-\tau M_\pi}|\pi\rangle\langle\pi|$. Hence,

$$C_{J_\pi J_\pi}(\tau) \to \frac{e^{-M_\pi\tau}}{2M_\pi}|\langle\pi|J_\pi|0\rangle|^2 \tag{4.24d}$$

which enables us to read off the pion mass from the large τ decay and provides the factor $|\langle\pi|J_\pi|0\rangle|$ required to normalize other quantities. In practice, since the two-point correlation function coincides with the Bethe–Salpeter wave function discussed subsequently at zero separation, it is not necessary to calculate it as a separate entity. Operationally, as seen in (4.24b), the calculation is straightforward. Having generated a set of gauge field configurations sampling $e^{-S(U)}$ in the quenched case or $e^{\ln \mathrm{Det}\, M(U)-S(U)}$ in the case of full dynamical quarks, the combination of propagators $M^{-1}(U)_{x\tau,00}\gamma_5 M^{-1}(U)_{00,x\tau}\gamma_5$ is simply averaged over these gauge fields. The case of the proton two-point function $C_{J^p J^p}$ is analogous, where one obtains $\frac{e^{-M_p\tau}}{2M_p}\langle p|J^p|0\rangle^2$ and must calculate an average over gauge fields of a propagator for the down quark times the two different contractions of the up quarks.

Given Eq. (4.24), which shows how hadron masses are calculated on the lattice, it is useful to return to Fig. 8 and discuss the data from hadron mass calculations which are displayed in that figure. Recall that the agreement between the renormalization group relation (3.48) and values of the lattice spacing extracted from the string tension in the pure gauge sector as a function of inverse coupling β, denoted by squares, indicated approach to continuum QCD for β above 6. Furthermore, defining the string tension in terms of the Regge slope via Eq. (3.32) yielded the lattice cutoff $\Lambda_L = 4.05$ MeV. For QCD with Fermions at a given value of β there are two unknown parameters, the bare quark mass (or hopping-parameter) and lattice spacing a, which can be fixed by calculation of the physical π and ρ masses. The values of a thus obtained in the quenched approximation are shown in Fig. 8 by the solid circles for Wilson Fermions and the open circles for staggered Fermions. It is noteworthy that the values of a determined from either form of lattice Fermion also display continuum renormalization group behavior above $\beta = 6$, albeit with somewhat larger error bars reflecting the greater difficulty of the lattice measurement of masses relative to Wilson loops. Furthermore, one observes significant deviation from renormalization group behavior for staggered Fermions at roughly half the value of a at which one sees deviation for Wilson Fermions, reflecting the fact that the effective lattice spacing for staggered Fermions is $2a$ compared to a for Wilson Fermions. Finally, the 10% discrepancy between values of Λ_L determined from masses and the string tension is not worrysome, both because of the limitations of the argument relating the string tension to the Regge slope and because of the omission of the contributions of dynamical Fermions in the quenched approximation.

The equal-time gauge-invariant Bethe–Salpeter amplitude is analogous to the two-point function we have just discussed, but the field operators in the annihilation operator are separated by a distance y and connected by link variables. Thus, for the pion, we define

$$C_{\mathrm{BS}}(y) = \int dx \langle 0|\bar{d}(x+y,\tau)\gamma_5 \left(\prod_{x}^{x+y} U\right) u(x,\tau)\ J_\pi^\dagger(0,0)|0\rangle \tag{4.25a}$$

and by the previous argument

$$C_{\mathrm{BS}}(y) \to \langle 0|\bar{d}(y)\gamma_5 \left(\prod_{0}^{y} U\right) u(0)|\pi\rangle \frac{e^{-M_\pi \tau}}{2M_\pi} \langle \pi|J_\pi|0\rangle \tag{4.25b}$$

This gauge-invariant expression specifies the amplitude for finding an up quark and a down antiquark in a π^+ separated by a distance y, and thus informs us about one piece of the Fock space decomposition of the pion. Note that in order to have a gauge invariant quantity, it was necessary to connect the $\bar{d}$ and u by a chain of link variables. In this special case, one could compare the result directly to a calculation of the pion in an axial gauge. In general, one has to make the choice as to whether to gauge fix, and then calculate the conventionally defined wave function, or to take some minimally prejudiced gauge invariant definition. A particularly appealing alternative is to replace the line of U's between u and d by a staple with lines of U's fixed at x and $x+y$ extending far back in the time direction and then connected to each other, so that the resulting amplitude corresponds to $\langle 0(y,0)|\bar{d}(y)u(0)|\pi\rangle$ specifying the overlap between the pion and the ground state of the QCD vacuum in the presence of a static u quark at 0 and a static d antiquark at y. Analogous Bethe–Salpeter amplitudes can be calculated for the proton. In the special case of two quarks at one position and the remaining quark a distance y away, the diquark pair behaves essentially like an antiquark, and this particular component of the proton is essentially identical to the pion.

Expectation values of operators are calculated by using two widely separated sources $J, J^\dagger$ to create a state of the desired quantum numbers placing the operator to be measured in between where the lowest state has been filtered out, and projecting in momentum. Consider, for example, the quark density-density correlation function, where for simplicity we consider the correlation between up and down quarks in the π^+:

$$\begin{aligned} C_{\rho_u\rho_d}(y) &= \int dx \left\langle 0\left|J_\pi(0,\tau)\hat{\rho}_\mu(x+y/2,\tau/2)\,\hat{\rho}_d(x-y/2,\tau/2)\,J_\pi^\dagger(0,0)\right|0\right\rangle \\ &\to \int dp \left\langle \pi p\left|\hat{\rho}_\mu(y/2)\hat{\rho}_d(-y/2)\right|\pi p\right\rangle \frac{e^{-E_\pi(p)\tau}}{2E_\pi(p)} \left|\langle \pi, p|J^\dagger|0\rangle\right|^2 \\ &\propto \left\langle \pi\left|\hat{\rho}_u(y/2)\hat{\rho}_d(-y/2)\right|\pi\right\rangle \quad . \end{aligned} \tag{4.26}$$

Note that the integral over the cm variables x forces the momenta of the two pion states in the matrix element of $\hat{\rho}_u\hat{\rho}_d$ to be equal, and since the operator depends only on the relative coordinate y, the matrix element is independent of p and hence yields the desired density-density correlation function with only one projection integration. As emphasized in Ref. [17], the $\rho_u\rho_d$ correlation function in the pion measures contributions from all the multiquark-antiquark components of the Fock space, and thus, in principle, provides valuable complementary information to that in the Bethe–Salpeter amplitude. In addition to the quark-quark correlation function, one may measure quark-gluon correlation functions by replacing one of the $\hat{\rho}(x)$'s by $\mathrm{Tr}\, U_{\mu\nu}^{\square}(x)$ as in Eq. (4.2) and can thereby explore the degree to which the gluon structure of the hadron resembles that of the bag model. Clearly, the same operators may be evaluated in the proton by summing over appropriate sets of contractions and one can straightforwardly go to higher-order correlation functions with more quarks or gluon operators.

Form factors are analogous to the expectation values of operators just described, with the exception that another projection must be performed to provide the momentum transfer q required by the definition of a form factor. Consider, for example the vector current form factor of the pion:

$$\begin{aligned} C_V(p,q) &= \int dx\, e^{-iqx} \int dy\, e^{-ipy} \left\langle 0 \left| J_\pi(x,\tau) V_\mu(y,\tau/2) J_\pi^\dagger(0,0) \right| 0 \right\rangle \\ &\to \langle \pi, p \,|V_\mu|\, \pi, p+q\rangle \frac{\langle 0|J_\pi|\pi,p\rangle}{2E_\pi(p)} \frac{\langle \pi, p+q|J_\pi^\dagger|0\rangle}{2E_\pi(p+q)} e^{-\frac{\tau}{2}(E_\pi(p)+E_\pi(p+q))} \\ &\propto F(q^2) \ . \end{aligned} \tag{4.27}$$

Thus far, we have ignored the technical complications of renormalization and operator mixing. Ultimately, of course, lattice observables must be related to continuum observables, and if the calculations are performed close enough to the continuum limit, one may hope to calculate the differences between lattice and continuum quantities perturbatively. Operators having the same quantum numbers of exactly conserved symmetries, such as charge, flavor and isospin, can mix under renormalization, potentially complicating the extraction of physical observables from lattice measurements. In practice, one can significantly reduce such complications by calculating operators of the lowest dimension $d_{\min}$ for a given set of quantum numbers, so that the contributions of some other operator of dimension d which mixes with this operator vanishes as $a^{(d_{\min}-d)}$ as $a \to 0$.

V. FURTHER READING

The primary goal of the preceeding sections, as explained previously, has been to present an elementary introduction for non-specialists to the basic concepts of lattice gauge theory relevant to the study of hadronic physics. With this introduction, it should be possible to read more advanced treatments, such as Ref. [1], as well as the research literature including the classic papers collected in Ref. [4].

Limitations of space and time have precluded treatment of two topics which were included in the lectures presented at Cargese: a sketch of the Monte Carlo method and a survey of recent computational results. Whereas both topics will be included in another summer school proceedings,[20] some of the principle references on these topics will be summarized here.

A pedagogical introduction to the stochastic evaluation of path integrals is presented in Chapter 8 of Ref. [3]. The only major topics relevant to lattice QCD which are not included there are the pseudo-heat-bath method for updating gauge fields in quenched calculations, which is briefly described in Ref. [19] and explained in detail in the original article by Cabbibo and Marinari,[21] and the exact hybrid method for including dynamical Fermions.[22] A very readable introduction to Monte Carlo algorithms and their application to QCD is presented in recent lectures by Toussaint.[23]

A comprehensive survey of a number of numerical results for the string tension, finite temperature phase transitions, and hadron masses is presented in the review articles referred to previously by Fukugita.[5] Potential and color electric and magnetic fields in the presence of two and three static quarks are calculated by Flower.[24] A quantitative check of the lattice techniques for hadrons described in the preceeding section is provided in the case of QCD in two space-time dimensions in Ref. [17]. In that work, the Bethe–Salpeter amplitude and density-density correlation for a pion calculated on a lattice are shown to agree in detail with analytic solutions. Preliminary calculations of Bethe–Salpeter amplitudes and density-density correlation functions in four space-time dimensions are presented in Ref. [19] and more complete results are being prepared for publication.[25] A discussion and review of lattice calculations of vector and axial current matrix elements is given in lectures by Woloshyn.[26] The most complete set of calculations of form factors and structure functions of the pion and nucleon and the proton σ-term are described by a series of papers by Martinelli and Sachrajda.[27] The conclusion from all these numerical results is that although computational and technical difficulties abound, lattice QCD has come of age and has much to teach us about hadronic physics.

ACKNOWLEDGEMENTS

It is a pleasure to acknowledge the many fruitful discussions with my collaborators and colleagues Ming Chu, Suzhou Huang, Marcello Lissia, and Janos Polonyi, with and from whom I learned about hadronic physics on a lattice.

REFERENCES

1. M. Creutz, *Quarks, Gluons and Lattices* (Cambridge University Press, Cambridge, UK, 1983).
2. M. Metropolis, W. A. Rosenbluth, M. N. Rosenbluth, A. H. Teller and E. Teller, *J. Chem. Phys.* **21** (1953) 1087.
3. J. W. Negele and H. Orland, *Quantum Many-Particle Systems* (Addison Wesley, Reading, MA, 1988).
4. C. Rebbi, *Lattice Gauge Theories and Monte Carlo Simulations* (World Scientific, Singapore, 1983).
5. M. Fukugita, in *Proceedings of the CCAST Symposium on Lattice Gauge Theory using Parallel Processes*, Beijing (Gordon & Breach, London, 1987); *Proceedings of the International Symposium on Field Theory on the Lattice*, Seillac, 1987, *Nucl. Phys. B* (Proc. Supple. 4) **105** (1988).
6. S. Chin, J. W. Negele and S. E. Koonin, *Annals of Physics* (NY) **157** (1984) 190; S. A. Chin, O. S. van Roosmalen, E. A. Umland and S. E. Koonin, *Phys. Rev.* **D31** (1985) 3201.
7. A. Patel and R. Gupta, *Phys. Lett.* **183B** (1987) 193.
8. B. Svetitsky, *Physics Reports* **132** (1986) 1.
9. R. Gupta and A. Patel, *Nucl. Phys.* **B226** (1983) 152.
10. J. Kogut, J. Polonyi, H. W. Wyld, J. Shigemitsu and D. K. Sinclair, *Nucl. Phys.* **B251** (1985) 311.
11. R. Dashen and D. J. Gross, *Phys. Rev.* **D23** (1981) 2340.
12. B. Svetitsky, *Nucl. Phys. A* **461** (1987) 71c.
13. S. A. Gottlieb *et al.*, *Phys. Rev. Lett.* **55** (1985) 1958.
14. A. Hasenfratz, P. Hasenfratz, U. Heller and F. Karsch, *Phys. Lett.* **143B** (1984) 193.

15. L. H. Karsten and J. Smit, *Nucl. Phys.* **B183** (1981) 103.

16. H. B. Nielsen and M. Minomiya, *Nucl. Phys.* **B185** (1981) 20.

17. S. Huang, J. W. Negele and J. Polonyi, *Nucl. Phys.* **B307** (1988) 669.

18. J. Kogut and L. Susskind, *Phys. Rev.* **D11** (1975) 395.

19. M. Lissia, Ph.D. Dissertation, Massachusetts Institute of Technology (1989).

20. J. W. Negele, in *Proceedings of 1990 National Summer School of Nuclear Physics*, W. Haxton and J. Randrup, eds. (University of California at Santa Cruz), to be published.

21. N. Cabbibo and E. Marinari, *Phys. Lett.* **119B** (1982) 387.

22. K. Bitar, A. D. Kennedy, R. Horsley, S. Meyer and P. Rossi, *Nucl. Phys.* **B313** (1989) 377.

23. D. Toussaint, University of Arizona preprint AZPH–TH/89-6.

24. J. Flower, *Nucl. Phys.* **B289** (1987) 484.

25. M. Chu, M. Lissia and J. W. Negele, to be published.

26. R. M. Woloshyn, in *Proceedings of Lake Louise Winter Institute*, TRIUMF preprint TRI–PP–88–9.

27. G. Martinelli and C. T. Sachrajda, *Phys. Lett.* **B190** (1987) 151 and **B196** (1987) 184; *Nucl. Phys.* **B306** (1988) 865 and **B316** (1989) 355.

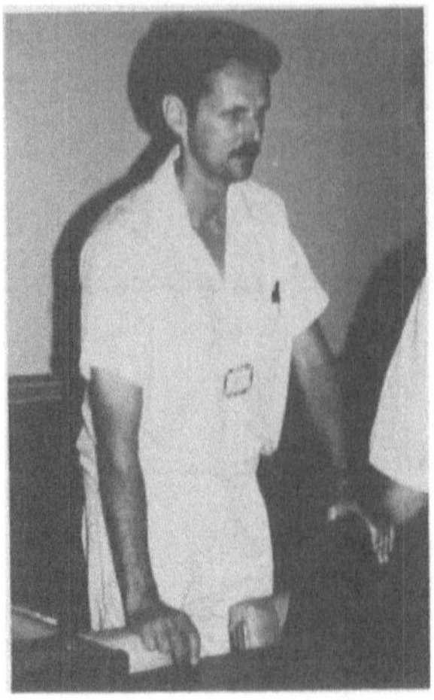

Charles
ALCOCK

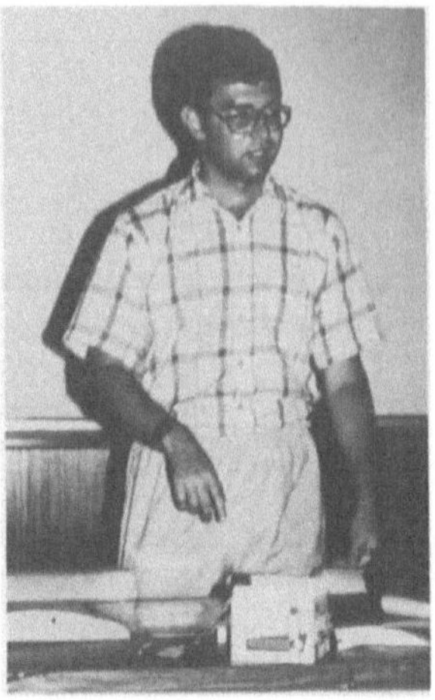

Igor
KLEBANOV

Larry
Mc LERRAN

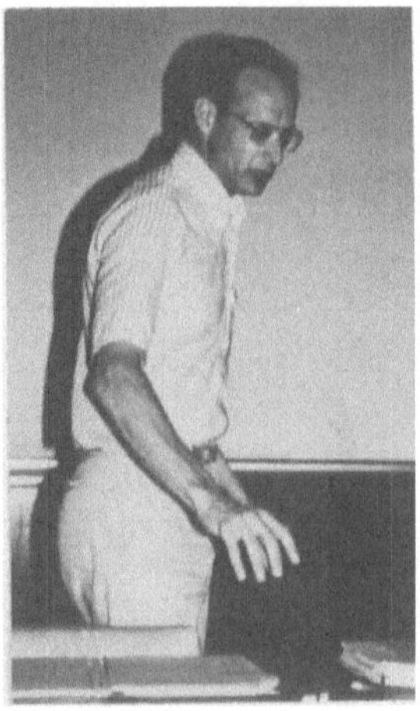

John
NEGELE

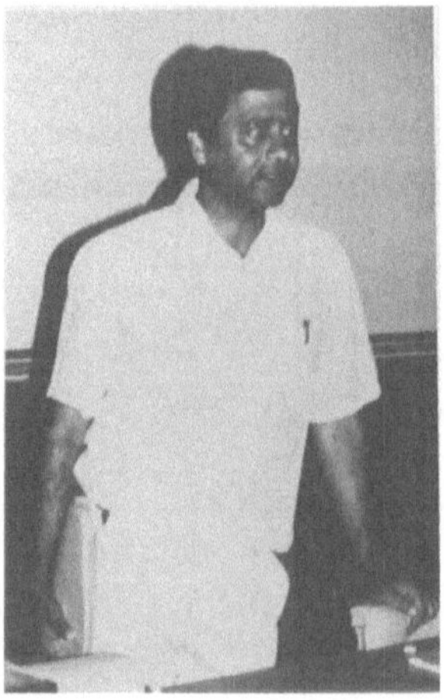

Vijay
PANDHARIPANDE

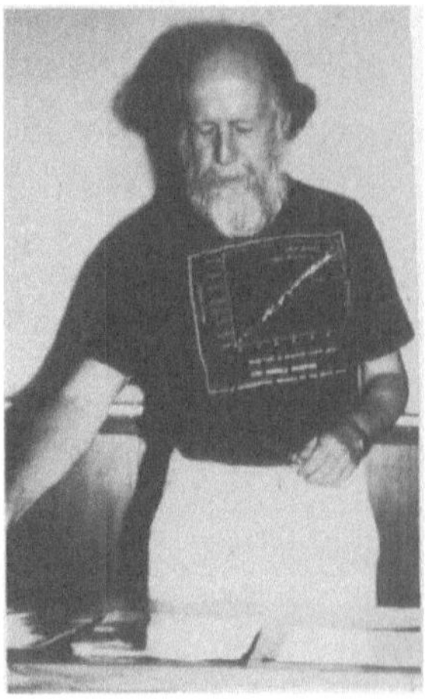

Hubert
REEVES

Anthony
THOMAS

John
NEGELE

Georges
RIPKA

Dominique
VAUTHERIN

INDEX